THE NEUROSCIENCES AND MUSIC

ANNALS OF THE NEW YORK ACADEMY OF SCIENCES
Volume 999

THE NEUROSCIENCES AND MUSIC

Edited by Giuliano Avanzini, Carmine Faienza, Diego Minciacchi, Luisa Lopez, and Maria Majno

The New York Academy of Sciences
New York, New York
2003

Library of Congress Cataloging-in-Publication Data

The neurosciences and music / edited by Giuliano Avanzini ... [et al.].
  p. cm. — (Annals of the New York Academy of Sciences ; v. 999)
Proceedings of a conference on Neurosciences and music : Mutual
interactions and implications on developmental functions, held on
October 25–27 in Venice, Italy.
Includes bibliographical references and indexes.
  ISBN 1-57331-452-8 (cloth : alk. paper) — ISBN 1-57331-453-6 (pbk. :
alk. paper)
  1. Neurosciences. 2. Brain—Localization of functions—Congresses.
3. Music—Physiological aspects—Congresses. 4. Music—Psychological
aspects—Congresses. I. Avanzini, G. II. Series.
  Q11.N5 vol. 999
  [RC321]
  500 s—dc22
  [612.8/

                                                                    2003023564

GYAT/PCP
Printed in the United States of America
ISBN 1-57331-452-8 (cloth)
ISBN 1-57331-453-6 (paper)
ISSN 0077-8923

ANNALS OF THE NEW YORK ACADEMY OF SCIENCES

Volume 999
November 2003

THE NEUROSCIENCES AND MUSIC

Editors and Conference Organizers

GIULIANO AVANZINI, CARMINE FAIENZA, DIEGO MINCIACCHI, LUISA LOPEZ, AND MARIA MAJNO

Scientific Advisors

ISABELLE PERETZ AND ROBERT J. ZATORRE

This volume is the result of a conference entitled **The Neurosciences and Music: Mutual Interactions and Implications on Developmental Functions** held by the Fondazione Pierfranco e Luisa Mariani in cooperation with the International School of Neurological Sciences, San Servolo, Venice, and the Venice International University, San Servolo, and co-sponsored by the New York Academy of Sciences; European Society for the Cognitive Sciences of Music (ESCOM), Belgium; 7th International Conference on Music Perception & Cognition (ICMPC 2002), Australia; and the International Society for Music in Medicine (ISMM); with the participation of the Conservatorio di Musica "Benedetto Marcello," Venice, and held on October 25–27, 2002 in Venice, San Servolo Island, Italy.

CONTENTS

Part I. Cerebral Organization of Music-Related Functions

Part II. Brain Sciences versus Music

General Foreword

GIULIANO AVANZINI,[a] CARMINE FAIENZA,[b] LUISA LOPEZ,[c] MARIA MAJNO,[d]
AND DIEGO MINCIACCHI[e]

[a]*Department of Experimental Research and Diagnostics, National Neurologic Institute
"Carlo Besta," Milan, Italy*

[b]*Department of Pediatrics, University of Parma, Parma, Italy*

[c]*Center for Developmental Disabilities "Eugenio Litta," Rome, Italy*

[d]*Fondazione Pierfranco e Luisa Mariani, Milan, Italy*

[e]*Department of Neurological and Psychiatric Sciences, University of Florence,
Florence, Italy*

As an ideal continuation of "The Biological Foundations of Music," edited by
Robert J. Zatorre and Isabelle Peretz and published in 2001, this volume hosts the
proceedings of the meeting, "The Neurosciences and Music," held in Venice in
October 2002, organized and fully supported by the Fondazione Pierfranco e Luisa
Mariani of Milan, Italy.

The project of an international update on the rapidly evolving field of the
relations between "Music and the Brain"—now more appropriately extended to
encompass the neurological sciences, yet the tribute to Critchley and Henson's
pioneering work of 1977 should still be acknowledged—had already been present in
the Mariani Foundation's programs for a few years. The initial spark had come from
the contact between a neurologist deeply involved in music (Giuliano Avanzini of
the National Neurological Institute Carlo Besta of Milan and current President of the
International League Against Epilepsy) and a musician with multiple interests in
neurology (Maria Majno, program director of the Foundation).

While translating the prospect into a concrete outline, a coincidence was discovered with the simultaneous planning of a conference hosted in New York by the
New York Academy of Sciences in May 2000, upon the proposal of Zatorre and
Peretz who then became the editors of the aforementioned volume 930 in the NYAS
Annals collection. Consequently, the Organizing Committee of the Italian counterpart decided to align its prospects accordingly: while the antecedent had admittedly
and so competently dealt with neurobiology, the next installment would take the
previous results into account and characterize its perspectives in new and mostly
different directions.

Evidently, the intention was also to represent the Fondazione Mariani's aims in
the course plotted for this "neuromusical" follow-up: given the primary dedication
of the Foundation's mission in the field of pediatric neurology (where the activities
range from postgraduate training to funding of research, from highly specialized
seminars to the implementation of services for families and children with special
needs), the developmental perspective was an almost mandatory guideline of the
program.

Ann. N.Y. Acad. Sci. 999: xi–xii (2003). © 2003 New York Academy of Sciences.
doi: 10.1196/annals.1284.070

The most striking peculiarity of the 2002 conference, promoted in cooperation with the International School of Neurological Sciences of San Servolo and the Venice International University, was the combination of three points of view. On the one hand, the basic aspects of music, pitch, timbre, and rhythm together with some features, such as emotions and creativity, were analyzed or even dissected, utilizing traditional scientific approaches. This session was orchestrated by Giuliano Avanzini, a neurologist with a wide range of interests and activities in both clinical and basic research. On the other hand, Diego Minciacchi, neurologist, neuroanatomist, and a composer himself, succeeded in conveying the need for more communication between those who compose or perform music and those who study its components.

Finally, the third day was devoted to the development of musical abilities, also related to acquired and innate competencies, and to cognitive disorders of, childhood. This part of the conference was the result of the collaboration between Carmine Faienza, a child neurologist who has edited another volume on "Music, Speech and the Developing Brain" based on the 1992 international workshop held in Parma; Maria Majno; and Luisa Lopez, a neurophysiologist with a particular interest in pediatric neurology and music. The issues addressed on this third day were thus related to the building of musical elements, from basic sound analysis and infants' perception to the making of a "genius."

In the following pages, a thorough introduction to each single part will be given by those who were involved in developing the program or chairing the round table sessions.

From an overall view, we can only add that the book points out some interesting acquisitions. First, all authors agree that music is made up of much more than what can be measured; yet, the need for a scientific approach to its basic mechanisms has driven research towards a top-bottom approach to single items. From this viewpoint, the designing of new techniques that integrate time and spatial resolution with innovative protocols will surely add a precious contribution to further development in music science.

Second, building a bridge with musicians and new types of music will bring reciprocal advantages to music and science, like a true alphabetization to new styles and modes of producing and broadcasting music.

Third, the issues covered on the third day can be used by those involved in developing educational policies, as correctly pointed out by John Sloboda, to fully explore the children's potentials in building their own musical preferences and skills, and probably using it for nonmusical learning as well.

Fourth, although the final version of this book does not report any contribution on music therapy, a few words need to be said on this issue. Some posters were presented at the meeting that show the interest in this field. Although promising, this field is still pervaded by the lack of evidence-based results. Our belief is that the subsequent meetings addressing neuroscience and music will foster interesting papers by groups employing music therapy as a rehabilitative tool.

In conclusion, the organizing of this conference and the assembling of this book were a great adventure, and our hope is that you will enjoy the outcome as much as we did planning it.

Part I: Cerebral Organization of Music-Related Functions

Introduction

GIULIANO AVANZINI

Department of Experimental Research and Diagnostics,
National Neurologic Institute "C. Besta," Milan, Italy

Over the last decade, we have seen an impressive flourishing of investigations into the cerebral representation of music-related functions. This increased attention is quickly bridging the gap between this field and previously more advanced research areas such as perceptual, language, and motor functions, and our better understanding of the way music is processed within the central nervous system has provided new insights into the neural machinery involved in higher brain functioning.

One important factor contributing to our greater ability to explore the musical competence of the brain has been the improvements made in the morphological resolution and functional investigation capacities of imaging techniques. Positron emission tomography (PET) and functional magnetic resonance imaging (fMRI) studies are complementing neurophysiological investigations that are still unsurpassed for their optimal time resolution and that have benefited from new signal analysis techniques in both electroencephalography (EEG) and magneto-encephalography (MEG).

The integration of these instrumental analyses with neuropsychological tests specifically designed to test musical competencies related to music perception and production has significantly improved our understanding of the brain representation of musical abilities. It is important to remember that "musical" abilities need to be tested by experimental protocols that necessarily address only specific aspects of music: the term "pitch" refers to a perceptual rather than a particular stimulus structure, and we perceive a given pitch whether it is associated with a pure tone or a harmonic stimulus such as a musical instrument or voice. One critical aspect of music appreciation is the analysis of perceptual patterns in which individual percepts are simultaneously combined to form chords and harmonic structures and sequentially to form melodies. Various paradigms based on the presentation to subjects of inappropriate changes ("violations") in musical structure have proved to be particularly fruitful, and a number of protocols have been developed on the basis of violations of authentic cadences, melody contours, temporal organization of sounds, and spatial localization of their source. Specific event-related brain potentials (ERPs) detected by analyzing EEG and MEG signals are associated with different types of violations. These are called early right anterior negativity (ERAN), mismatch negativity (MMN), and P300 depending on the polarity, latency, and topography of the recorded event and the different stimulatory conditions. Their

Ann. N.Y. Acad. Sci. 999: 1–3 (2003). © 2003 New York Academy of Sciences.
doi: 10.1196/annals.1284.069

combination with functional imaging techniques allows a more precise definition of the topographic representation of such events within the brain.

The recent studies described in the following articles investigate the culture and domain specificity of music-related brain activation. The similarities and differences between language and music brain processing in particular have been investigated because, despite the clear-cut difference between language and music in terms of isochrony, certain aspects have cross-domain similarities that are reflected by electrophysiological and functional imaging data.

Transcultural studies of music brain processing are still limited. The available data suggest that the musical quality of a sound object is recognized by the human brain whether or not it belongs to the subject's native culture. Between-culture differences in linguistic rhythm seem to be reflected by differences in musical rhythm.

The processing differences between musicians and nonmusicians, which are qualitative rather than quantitative and relate to the specific type of music ability, have been explored in greater detail. However, although the methods available to neuroscientists interested in music have been considerably refined, certain limitations in terms of time and space resolution need to be taken into account. In principle, neurophysiological recordings provide real-time information and thus fulfill the requirement of capturing the fast events underlying musical processing; however, this property can only be fully explored in subjects with implanted intracranial electrodes for therapeutic purposes. The averaging techniques needed to extract ERPs from surface recordings significantly limit the information contained in the recorded response; however, further developments in autoregressive analysis techniques may help to solve this problem.

PET and fMRI are certainly inferior to neurophysiology in the time domain but superior in terms of topographical definition. However, when interpreting the results of imaging studies of the cortex, anatomical variations in cytoarchitectonic areas with respect to macroscopic boundaries need to be taken into account. The lack of consistency between subjects may partly reflect the fact that consistency can only be defined in relation to macroscopic landmarks rather than the cytoarchitectonic areas that are not accurately delineated by such landmarks.

A classic approach to the analysis of functional representation in the brain is to correlate the location of a brain lesion and the associated functional defects in patients. After the observations of amusias reported at the turn of the century, a significant body of evidence has been gathered that indicates the selective impairment of musical competencies depending on localized or diffuse cortical lesions. The following chapters provide a comprehensive review of the history of such clinical research. A major contribution to the systematic analysis of amusias along melodic and temporal dimensions was made by the development of the Montreal Battery of Evaluation of Amusias (MBEA), and a description of the suitability and validity of the latest diagnostic version is presented.

Studies of the different forms of epilepsy have improved our understanding of the brain activities involved in musical perception and the mechanisms underlying seizure precipitation in terms of the analysis of musicogenic seizures and the musical defects following cerebral resections performed to treat intractable epilepsy. The literature concerning music-induced seizures is reviewed here. The specific musical component responsible for seizure precipitation is still unknown, but an important role has been attributed to the emotional aspect of music. The systematic study of this rare disorder has fostered our knowledge of the nature of the triggering mechanisms responsible for this unique pathological condition.

The last group of papers in this first section are derived from a round table discussion of the elementary components of music by a panel of neuroscientists and musicians. On the one hand, evidence is provided for the separate processing of different subcomponents of the traditional elements upon which musical education is based. On the other hand, the need for interaction among different neural systems subserving the elementary component in organizing them in a musical dimension is discussed. A critical account of this fruitful approach is found in Carol Krumhansl's thoughtful introduction to the round table discussion.

Music and the Brain

ROBERT J. ZATORRE

Montreal Neurological Institute. McGill University, Montreal, Canada

ABSTRACT: The aim of this paper is to illustrate how studying music from a neuroscience perspective may be a valuable way to probe a variety of complex cognitive functions and their neural substrate. Three different sets of issues are described. First, studies dealing with the brain correlates of musical imagery are discussed. This topic is of interest in that it illustrates how subjective sensations may be studied via objective techniques, and gives insight into neural systems associated with internal phenomena. Second, some findings pertaining to absolute pitch are presented. Absolute pitch is a useful example of a highly specific cognitive skill that is unevenly distributed in the population. Examination of its neural basis helps to understand aspects of memory function and points to ways to explore individual differences in brain organization that underlie differential skills. The final topic, music and emotion, has not been the subject of much systematic research, but it is of great interest because it intersects with a large literature on the neuroscience of affective processing. Findings from some studies indicate that music may engage systems concerned with biological reward, raising interesting but so far unanswered questions about the broader role of music in human experience.

KEYWORDS: music; brain

INTRODUCTION

Why should music interest a neuroscientist as a legitimate object of scientific study? Such a question might be posed in two ways by persons coming from different intellectual traditions. From a scientific perspective, we might wonder whether studying something as complex and "nonessential" as music is likely to yield meaningful insights, the implicit objection being that more fundamental cognitive or perceptual functions are more valid to focus on. From a humanistic perspective, one might harbor suspicions of a perceived reductionist agenda in science, to oversimplify artistic endeavors as mere products of mechanistic organisms. Such points of view may be, to some extent, caricatures of the reactions to the neuroscience of music, but they do represent important critiques that should not be ignored. Happily for those within this field, as exemplified by this volume of proceedings, these questions are increasingly being addressed to the satisfaction of both scientific and artistic communities. Our hope is that the contents of these papers will demonstrate the value of music as a window onto complex brain functions, while at the same time illustrating how a scientific understanding of music can yield deep insights into the nature of human thought and expression.

Address for correspondence: Dr. Robert J. Zatorre, Montreal Neurological Institute, 3801 University St., Montreal, Quebec H3A 2B4, Canada. Voice: 1-514-398-8903; fax: 1-514-398-1338.

Ann. N.Y. Acad. Sci. 999: 4–14 (2003). © 2003 New York Academy of Sciences.
doi: 10.1196/annals.1284.001

In this essay, my aim is to highlight the different approaches that can be taken to the study of music and brain function and how they throw light on three distinct aspects of the neural substrates of cognition, rather than to provide a detailed review of any one research question. Because music encompasses so many aspects of human cognition, it provides a rich source of materials for psychologists and neuroscientists to investigate. This very complexity also renders the study of music inherently complicated, however. It is therefore of great importance to define the specific aspect of musical function to be studied and, wherever possible, to identify the cognitive components associated with those functions. The research to be described in this paper comes from three separate threads of work that have been carried out in my laboratory over the last few years. The first set of studies deal with musical imagery; the second, with absolute pitch; and the third, with music and emotion. Each of these topics touches upon separate questions, yet they have in common that musical processes are central to them. This work also needs to be put in perspective as part of a much broader research agenda, which is sketched out in part by the other papers in this volume. It is therefore important to remember, first, that any one set of studies cannot be interpreted without the context of other information from many other domains; second, that the study of music and its neural correlates is clearly still in its early years and consequently that many more questions than answers are apparent at this point. The latter point is precisely what makes this field of inquiry dynamic; it is therefore a good choice for an investigator who values stimulating debate over clear-cut conclusions.

MUSICAL IMAGERY

The first set of studies to be discussed deal with a cognitive function that is inherently difficult to study because it is entirely internal and does not necessarily have observable concomitants. This work, done over the last few years in collaboration with Andrea Halpern of Bucknell University, has benefited from prior research in cognitive psychology dealing with visual imagery.[1,2] Until the 1960s, mental imagery was generally considered suspect as an object of serious research precisely because of its subjective, internal nature. The behaviorist model eschewed unobservable processes as not subject to scientific scrutiny, and rightly so, for earlier attempts at characterizing mental processes too often devolved into nothing more than "armchair" psychology. Starting with the work of Shepard,[3] however, it was shown that by clever experimental manipulations, it was indeed possible to measure objective behavioral indices of what otherwise would be inscrutable internal events. Thus, for instance, the demonstration that the time taken to make a decision on a rotated letter was strictly proportional to the degree of rotation indicated that a visual imagery process could be isolated and measured. This and other findings quickly became classics in the field of cognition, because they powerfully showed how we could probe internal processes by studying their external correlates. In this sense, cognitive psychology was shown to be no different from other fields of science, where unobservable forces or events are examined indirectly via the traces they leave (think particle physics).

Most work in mental imagery has involved vision. Yet, both introspection and experimentation reveal that most people experience auditory imagery,[4] in particular musical imagery, which might be described intuitively as simply "hearing music in one's mind." The phenomenology of this effect suggests that musical imagery retains many of the properties of real perception (as with visual images), particularly its temporal characteristics, but that it is not confusable with real hearing. Based on this, Halpern designed a series of studies showing that judgments of imagined music generally followed similar patterns to judgments of heard music. For example, in an important set of experiments,[5] it was shown that if persons are asked to judge the pitch change corresponding to two syllables in a well-known song, they can do so fairly accurately, demonstrating that they have access to an internal representation of the pitch structure. That, in itself, is useful because it serves to verify a person's response, because there is an objectively correct answer. But more importantly, Halpern showed that the time taken to make the response was a near-linear function of the time between probed items in the real music. This demonstration of chronometry indicates that the imagery event unfolds in something like real time, much as visual imagery processes are also time dependent. This is important, because it suggests an analogous mental process in imagery and perception. (It is important to note that the alternative, a non–time-dependent process, would be easy to posit: it could certainly be the case, for instance, that the pitches of a tune are stored in memory in something like a look-up table, where it would take no longer to search for two items close together than two items far apart. That this is not how imagery proceeds tells us that the imaginal event unfolds over time just as real perception does.)

The behavioral similarity between imagery and perception leads naturally to the hypothesis that these two functions share an underlying neural representation. This hypothesis was put to the test in a series of studies that can briefly be summarized. The first attempt to address this question made use of a classic paradigm in neuropsychology: the lesion-deficit approach. The basic idea is simply that damage to certain areas of the brain leads to specific disturbances, which yield clues to the function of the damaged areas. In the context of the imagery hypothesis, the prediction would be that if both processes depend on shared neural territory, then a lesion that disrupts perception should also disrupt imagery. It had previously been established that damage to the superior temporal gyrus, a brain area in the temporal lobe that contains the auditory cortices, leads to perceptual deficits on musical tasks, especially when that damage is in the right cerebral hemisphere. (For a review, see Ref. 6.) When we tested patients with cortical excision within these areas, we found that perception and imagery suffered a common fate after damage to the right temporal lobe: performance was poorer than normal on a pitch-judgment task whether performed on a heard tune or one that was merely imagined. By contrast, patients with similar lesions on the left side performed as well as controls on both tasks.[7]

This association between the two functions was tested further in two studies using functional neuroimaging. These techniques have become the mainstay of a great deal of work in cognitive neuroscience and have added immeasurably to our ability to make inferences about human brain function by allowing scientists to probe how regional brain activity patterns change as a function of stimulus characteristics or task demands. In the first of several experiments on musical imagery, listeners were scanned while performing a version of the pitch-judgment task just described.[8] The principal result confirmed earlier findings that activity was observed within the superior temporal gyri on both sides, not only when listening and judging heard tunes, but also when simply imagining the tunes without overt input. This demonstration of auditory cortex activity in the absence of an external stimulus is important in supporting the contention that neural events within sensory areas are responsible for the subjective experience associated with imagery, a conclusion that is also supported in other modalities. To explore further the issue of lateralization of response, a second neuroimaging experiment was carried out in which the listener was asked to imagine the continuation of a familiar musical excerpt after having heard its first few notes.[9] The data again showed activity within the superior temporal gyrus when imagery was involved, as compared to a condition in which tones were perceived without imagery. This time, the pattern showed strong lateralization to the right side, most likely because the tunes were selected to avoid all verbal associations, adding additional support for the predominance of right-hemisphere mechanisms in melodic processing.

In more recent research (Halpern *et al.*, submitted), we approached the problem from a somewhat different perspective. The aim was to test the generality of the musical imagery phenomenon using materials other than songs. Songs have both a sequential component as well as semantic associations in long-term memory, which may give rise to complex patterns of brain activity; we wanted to avoid those factors. We therefore selected timbre as the dimension to focus on. During fMRI scanning, listeners with moderate musical training were asked to imagine the timbres of two instruments cued by a visual word or they listened to the sounds of musical instruments in the perceptual control condition. To obtain a behavioral index of successful imagery, subjects were asked to make a dissimilarity rating of the two sounds in question. Comparison of these ratings using a multidimensional scaling approach showed significant similarities in both perceived and imagined ratings, indicating that subjects can indeed generate internal representations of timbre. The fMRI data were consistent with prior findings in that brain activity elicited during imagery was observed in portions of the auditory cortex and overlapped with the activity elicited by real perception. In addition, there was again a tendency for a greater response in the right than in the left auditory cortex.

Thus, this additional study confirms and extends the general conclusion that imagery does depend on a partially shared neural substrate that forms part of the perceptual analysis mechanism. When we experience an imaginal event, this reflects an engagement of neural systems that are involved in perception. More generally, these studies are of interest because they illustrate one way to probe the complex inner workings of conscious thought processes. It is now possible to state with a certain degree of confidence that we have the tools to begin to unravel what earlier generations of scientists had thought closed to scientific inquiry. However, much work lies ahead, and it should be emphasized that no one technique will be sufficient to give us more than a small piece of the puzzle. Among the many questions that remain to be explained are how the neural traces in sensory cortex that seem to correspond to imagery are activated—what are the pathways and sources of this process? Another question that begs to be studied is how individuals differ in their degree of imagery ability and what in their brains might explain the variability. We need think no further than Beethoven's celebrated ability to compose while utterly deaf to realize the extent to which some people can generate internal representations not only of previously heard sounds, but also of new ones. How does this happen? An inquiry into this process would undoubtedly tell us something interesting about creativity, that is, how the mind can recombine known elements into novel patterns.

ABSOLUTE PITCH

A consideration of how individuals differ from one another in any domain inevitably leads to the question of nature versus nurture. This is such a general issue in all of psychology and biology that we can only touch upon it here. But one circumstance in which these issues come into particularly clear focus concerns the phenomenon of absolute pitch. A review of the many threads related to this phenomenon is beyond our scope (for reviews see Refs. 10 and 11), but what is of relevance for the present discussion is the link between absolute pitch and brain function. Indeed, this phenomenon has begun to generate considerable interest in some scientific circles, because it may yield some clues to brain organization associated with special skills and also because of growing hints that there may be a genetic component to its development (for a review, see Ref. 12). Moreover, it is well-established that the development of absolute pitch is highly dependent on the age at which musical training is begun.[13,14] Thus, absolute pitch may serve as a useful paradigm for understanding the interactions between genetic mechanisms, their role in brain development, and how environmental input modifies and influences behavior by shaping or tuning neural processes that depend on these genetic factors.

The contribution made by functional imaging to this domain has so far been modest, but it seems destined to grow. One of the first questions addressed in our own laboratory concerning absolute pitch was simply to understand the pattern of brain activity that distinguished a possessor of absolute pitch from musicians with similar training but relative pitch.[15] To this end we measured regional cerebral blood flow among possessors and non-possessors under two conditions: while listening to pairs of tones with no explicit instructions and while listening to the same tones and determining the musical interval formed by them. We reasoned that musicians with absolute pitch would engage their ability automatically even when listening "passively" and that this condition would reveal the greatest difference across groups. Making relative judgments of musical intervals, however, should yield more similar patterns of brain activity, although musicians with absolute pitch might also use pitch labels in this task.

The principal finding of this study was that subjects with absolute pitch demonstrated significant brain activity in a region of the frontal cortex while listening to tones without explicit instructions, whereas nonpossessors did not. We might be tempted to conclude from this that the area in question is the "absolute pitch" region. But the pattern of results from the interval-judgment task, in fact, showed a similar pattern of activity in this frontal area for both absolute and relative pitch musicians. Thus, the most parsimonious explanation is that the activity in this region seems to be an index of a neural process related to labeling. Under conditions in which labeling is only possible by absolute pitch subjects, only these subjects show the effect, whereas when labeling of an interval is possible, both groups show similar activity in this area. The particular frontal cortical region in question is known from other research to be concerned with establishing and maintaining conditional associations in memory,[16] that is, the ability to make a specific response in relation to a given stimulus. Thus, it constitutes a logical candidate area for the link between a pitch and its label in absolute pitch subjects. Additional support for this explanation is that the same region of frontal cortex can be recruited, even in nonmusicians, once they are taught to identify chords with arbitrary labels.[17] Thus, it seems that for whatever as yet unknown reason, the associative function of this part of the frontal lobe is somehow facilitated in these individuals so that they form tone-label associations more readily. It should be pointed out that the labeling, while usually of a verbal nature, need not be limited to a note name, because absolute pitch may also be demonstrated by any number of other, nonverbal codes;[18] in particular, auditory imagery and sensorimotor responses such as those associated with playing a tone on a given instrument are often reported among musicians who have absolute pitch.

So what brain mechanism(s) do distinguish people with absolute pitch? At present, the answer to this question remains unknown, but an interesting anatomical correlate has been found that bears mention. The picture emerging from several studies is that there are significant differences in the degree of lateral asymmetry of cortical structures concerned with auditory processing (in the posterior portion of the superior temporal gyrus). Persons with absolute pitch tend to show a greater degree of anatomical hemispheric asymmetry of this area of the brain as compared to either musicians without absolute pitch or nonmusicians.[15,19] However, how to interpret this effect remains uncertain; whereas initial indications were that the enhanced asymmetry might reflect greater growth of left-hemisphere auditory cortical structures, the latest evidence suggests instead that it may be a reduced volume of right-hemisphere auditory cortical areas that accounts for the differences.[20] Thus, perhaps it is not simply the case that "bigger is better," but instead that there may be more complex interactions between gross morphological features and the underlying cortical function. Regardless of the details of this line of investigation, anatomical differences are important to the extent that they may serve as an indication of possible genetic or epigenetic factors. Because asymmetries in this part of the brain are reported to exist even prior to birth,[21] we need to know if they are also somehow markers of a propensity for developing absolute pitch under appropriate environmental conditions. Longitudinal studies to unravel this issue are therefore of great interest. More generally, therefore, the study of this otherwise somewhat arcane phenomenon may reveal important knowledge about how specific cognitive and perceptual skills depend on neural development and its interaction with the environment, a problem that is at the heart of essentially all of cognitive neuroscience.

MUSIC AND EMOTION

The third topic selected to illustrate the value and interest of studying music from a neuroscience perspective is emotion. Studying music and emotion is inherently tricky for a number of reasons, and only relatively recently are efforts being made to achieve a more systematic understanding of this complex phenomenon. (For a recent overview of this field, see Ref. 22.) Ignoring the affective aspects of musical experience (which music cognition has often been guilty of) may be a dangerous approach, however, because we may be missing some of the most salient and important aspects of human response to music. Once again, the contribution of brain imaging to the knowledge base is so far rather limited, and what little has been done tends to offer more questions than answers. Yet, as mentioned at the outset of this essay, at this early stage in our understanding of these phenomena, it may be more valuable to generate appropriate questions than to provide premature answers. Work on these questions done in my laboratory by Anne Blood represents our first attempt to explore the neural substrates of music and emotion, and we hope that it will serve to spur others to address the topic as well.

Among the factors that contribute to the difficulty of studying emotions and music in a scientific context is that emotional responses to music tend to be idiosyncratic and heterogeneous, and depend on a variety of complex and difficult-to-control individual sociocultural, historical, educational, and contextual variables. This presents a particular problem for examining brain responses to music-elicited emotion, because if reactions vary widely for unknown reasons, it will be difficult to set up a consistent experimental protocol. However, there are approaches that we can use to mitigate these problems. In the first study from our group on this topic,[23] we reasoned that one way to elicit consistent responses would be to make use of dissonance. Whereas people often disagree on what music they like, there is a certain amount of agreement that highly dissonant music tends to be unpleasant. We need to determine what is meant by dissonance, because in the appropriate tonal context dissonance serves a critical role in creating esthetically pleasant music; an entire literature exists in music cognition and psychoacoustics dealing with these issues. (For a general review, see Ref. 24.) But intuitively, it is clear that if we listen to music with many "wrong notes," it simply does not sound very good!

Based on this straightforward, albeit potentially simplistic consideration, we designed a stimulus set in which a tonal melody was accompanied by chords that varied in their degree of dissonance. The advantage of this approach is that the degree of dissonance could be varied parametrically to measure changes in brain activity that correlate with the variable of interest; this method avoids having to choose an appropriate baseline condition (often a problem in functional neuro-imaging studies),[25] because any changes observed must be related to the parameter varied. Subjects were scanned while listening to these varying levels of dissonant music and making judgments about their pleasantness, which generally confirmed that the more dissonant the accompaniment, the less pleasant the experience.

The pattern of brain activity associated with increasing and decreasing dissonance revealed distinct sets of correlations in parahippocampal and orbitofrontal regions, respectively. These sites are typically considered paralimbic cortical areas: cortex that stands in an intermediate anatomical relationship between association areas and the limbic system. One way to think of these areas, therefore, is that they mediate between perceptual and cognitive representations, on the one hand, and emotion, on the other. Many prior studies of emotion in the context of both neuro-imaging and pathology have identified orbitofrontal and adjacent frontal areas as well as parahippocampal areas as being related to various aspects of mood and pleasant versus unpleasant emotional responses.[26–29] It therefore seems that the emotional effects of music are mediated via the same areas that mediate emotion elicited by other stimuli or situations. In addition, reciprocal regional interactions were found between increasing versus decreasing dissonance conditions. That is, increasing activity in regions active during negative emotions was observed to be associated with a corresponding decrease in activity in regions that are active during positive emotions. This type of effect suggests a functional interaction between regions mediating opposite emotions, which in turn may indicate that music exerts its effects not only by evoking one reaction, but also by inhibiting incompatible ones.

Having explored aspects of negative affect elicited by music, we turned our attention to positive responses. Here we were once again faced not only with the issue of heterogeneity of responses, but also with the problem of subjectivity. How do we know whether one person's reaction to music is fundamentally similar to another's, even if they both say that they enjoy what they are listening to? This problem is a variant of the issue brought up in the context of imagery, and there is no one solution to exploring subjective states with objective methods (this is, after all, psychology's principal task). However, one phenomenon associated with music that helps greatly in dealing with this problem is the so-called "musical chills" or "shivers" that many people report experiencing on hearing certain musical passages. This effect has been documented by behavioral measures.[30] It occurs in a reasonably large proportion of the population, including persons with no formal musical training; it also has the advantage of being elicited relatively reliably. Moreover, emotions elicited by music can be associated with objective physiological markers, which therefore serve as an external index of the internal phenomenon.[31,32] The chills effect is experienced as a very positive emotion and is sometimes described as ecstatic or euphoric by many individuals. Thus, it seems to capture one of the most intense aspects of the affective response to music.

But what is the neural basis for this response? To address this question we measured the brain activity associated with musically elicited chills using positron emission tomography.[33] First, we had the test subjects select passages of instrumental music that consistently gave them this feeling at a certain specific point in time. We carefully selected both the subjects and their music, such that we found stimuli that elicited a strong chills response in one person, while leaving another person unaffected. The latter served as our control condition, in that any given stimulus served as the experimental condition for one person and the control for another; thus, across the entire subject sample the chills-inducing and neutral musical samples were completely balanced. We obtained subjective numerical ratings on the intensity of the chills experience from each subject after scanning and also measured a number of psychophysiological responses during scanning, including respiration, heart rate, and muscle tension.

The participants reported feeling the chills sensation on an average of 77% of the trials, which is sufficient for purposes of analysis and a reasonable value considering that the listening conditions inside a brain scanner tend not to be as conducive to positive emotional responses as one's favorite easy chair for music listening! Importantly, the psychophysiological measures were consistent with the subjective ratings of chills, so that when subjects reported feeling chills, there were significant increases in all three variables.

Several complementary approaches were taken to analyze the data. One of the findings that emerged most consistently was that several brain areas were significantly engaged as a function of increasing subjective intensity of the chills response. These effects could not be attributed to differences in familiarity among the self-selected and control stimuli (a potential confound), because they were largely preserved even when data from the self-selected music conditions alone were analyzed. Of greatest interest in the complex neural response observed were brain areas previously implicated by other types of studies in response to highly rewarding or motivationally important stimuli. The circuitry implicated, which includes areas in the dorsal midbrain, ventral striatum (which contains the nucleus accumbens), insula, and orbitofrontal cortex, among others, was correlated with the subjective experience of chills. In addition, and reminiscent of the previous study on dissonance, reciprocal interactions were observed such that portions of the amygdala, a brain region implicated in fear and negative emotion,[34,35] showed activity decreases as the chills response became more intense. The amygdala receives inhibitory presynaptic input from cholinergic neurons intrinsic to the nucleus accumbens,[36] suggesting a possible mechanism for decreased activity in these regions as a consequence of activity increases in ventral striatum.

The pattern of activity observed in correlation with music-induced chills is similar to that observed in other brain imaging studies of euphoria and/or pleasant emotion. For example, activity in the nucleus accumbens, dorsal midbrain, and insula has been reported to increase, and that in the left amygdala and ventromedial prefrontal cortex to decrease in response to cocaine administration in cocaine-dependent subjects.[37] In addition, animal studies support a critical role for ventral striatum, several midbrain areas, amygdala, and medial prefrontal cortex in circuitry underlying reward processes, including hedonic impact, reward learning, and motivation.[36,38] Animal studies of endogenous reward in response to natural stimuli, such as food and sexual stimulation, also show involvement of brain activity in similar regions.[39,40] Activity in these regions in relation to reward processes isknown to involve dopamine and opioid systems as well as other neurotransmitters. The pleasant experience of chocolate consumption in humans has also been correlated with activity in midbrain, insula, and ventromedial and orbitofrontal cortex.[41]

In essence, the findings indicate that under some particular conditions, the pleasurable feeling elicited by listening to music seems to depend on a neural substrate that is primarily associated with biologically significant environmental events and is present in many other species. It is easy to understand, in an evolutionary framework, why such reward mechanisms exist for certain stimuli. Intake of food is required for survival of the individual, and sexual reproduction is required for survival of the species; hence, both types of stimuli induce pleasure, and most animals are therefore strongly motivated to seek them out. Certain drugs, such as cocaine, appear to mimic the rewarding features of natural stimuli, no doubt leading to addiction and other negative consequences. But music is not a substance with pharmacological properties. Neither can music be said to be truly essential for survival; if deprived of music, people will neither die of starvation nor fail to reproduce. Yet, we see evidence that it recruits some of the same brain systems associated with these more basic biologically driven stimuli.

Whether music is unique in this respect remains to be seen; it may be one of a class of human constructs that elicit pleasure by coopting ancient neural systems via inputs from the neocortex. We know from much other research that most cognitive aspects of music (pattern processing, working memory, and so forth) depend on neocortical systems in auditory areas of the brain and in frontal and parietal cortex. (For an overview, see Ref. 42.) In fact, damage to cortical areas that results in loss of perceptual capacity also abolishes affective reactions.[43] But music thus presents an interesting puzzle. Appreciation and understanding of music clearly depends on learning, cultural, and even social factors (part of the reason why people differ, sometimes radically, in their preferences). Thus, research into music and emotion opens up some interesting questions not only for the neuroscientist, but also perhaps for the musicologist and even the philosopher. It may prove to be a paradigm for the understanding of interactions between higher-level cognition and subcortically mediated affective responses.

CONCLUSION

This highly selective overview highlights how music can illuminate a variety of distinct aspects of complex human cognitive functions and their neural correlates. The rest of this volume gives many more examples of the important issues that have been tackled by others in this domain. The field is now beginning to show some evidence of maturity, in that there is a larger body of literature on which to draw and also in that investigators are increasingly able to integrate the knowledge gained with more established findings in psychology, neuroscience, or other fields. It is hoped that this volume will contribute towards this evolution by summarizing the state of knowledge and, more importantly perhaps, by indicating the still large gaps in our understanding that future researchers will attempt to cover. Finally, as the theme of the meeting underscored, the continued interaction of musicians and scientists will be important, as the study of music and neuroscience is *mutually* revealing.

REFERENCES

1. FARAH, M.J. 1988. Is visual imagery really visual? Overlooked evidence from neuropsychology. Psychol. Rev. **95:** 307–317.
2. KOSSLYN, S.M., G. GANIS & W.L. THOMPSON. 2001. Neural foundations of imagery. Nature Rev. Neurosci. **2:** 635–642.
3. SHEPARD, R.N. & J. METZLER. 1971. Mental rotation of three-dimensional objects. Science **171:** 791–793.
4. REISBERG, D., Ed. 1992. Auditory Imagery. Lawrence Erlbaum. Hillsdale, NJ.
5. HALPERN, A.R. 1988. Mental scanning in auditory imagery for tunes. J. Exp. Psychol. LMC **14:** 434–443.
6. ZATORRE, R.J., P. BELIN & V.B. PENHUNE. 2002. Structure and function of auditory cortex: music and speech. Trends Cognit. Sci. **6:** 37–46.
7. ZATORRE, R.J. & A.R. HALPERN. 1993. Effect of unilateral temporal-lobe excision on perception and imagery of songs. Neuropsychologia **31:** 221–232.
8. ZATORRE, R.J., A.R. HALPERN, D.W. PERRY, *et al.* 1996. Hearing in the mind's ear: a PET investigation of musical imagery and perception. J. Cognit. Neurosci. **8:** 29–46.
9. HALPERN, A.R. & R.J. ZATORRE. 1999. When that tune runs through your head: a PET investigation of auditory imagery for familiar melodies. Cereb. Cortex **9:** 697–704.
10. WARD, W.D. 1999. Absolute pitch. *In* The Psychology of Music, 2nd Ed. D. Deutsch, Ed. :265–298. Academic Press. San Diego.
11. TAKEUCHI, A. & S. HULSE. 1993. Absolute pitch. Psychol. Bull. **113:** 345–361.
12. ZATORRE, R.J. 2003. Absolute pitch: a paradigm for understanding the influence of genes and development on neural and cognitive function. Nature Neurosci. **6:** 692–695.
13. BAHARLOO, S., S. SERVICE, N. RISCH, *et al.* 2000. Familial aggregation of absolute pitch. Am. J. Hum. Genet. **67:** 755–758.
14. MIYAZAKI, K. 1988. Musical pitch identification by absolute pitch possessors. Percept. Psychophysiol. **44:** 501–512.
15. ZATORRE, R.J., D.W. PERRY, C.A. BECKETT, *et al.* 1998. Functional anatomy of musical processing in listeners with absolute pitch and relative pitch. Proc. Natl. Acad. Sci. USA **95:** 3172–3177.
16. PETRIDES, M. 1995. Functional organization of the human frontal cortex for mnemonic processing: evidence from neuroimaging studies. Ann. N.Y. Acad. Sci. **769:** 85–96.
17. BERMUDEZ, P. & R.J. ZATORRE. 2002. Conditional associative memory for musical stimuli in non-musicians: relationship to absolute pitch. Soc. Neurosci. Abstr.: Abstract Viewer/Itinerary Planner. Society for Neuroscience, Online. Washington, DC.
18. ZATORRE, R.J. & C. BECKETT. 1989. Multiple coding strategies in the retention of musical tones by possessors of absolute pitch. Mem. Cognit. **17:** 582–589.

19. SCHLAUG, G., L. JÄNCKE, Y. HUANG, *et al.* 1995. In-vivo evidence of structural brain asymmetry in musicians. Science **267:** 699–701.
20. KEENAN, J., V. THANGARAJ, A. HALPERN, *et al.* 2001. Absolute pitch and planum temporale. NeuroImage **14:** 1402–1408.
21. WITELSON, S. & W. PALLIE. 1973. Left hemisphere specialization for language in the newborn: neuroanatomical evidence of asymmetry. Brain **96:** 641–646.
22. JUSLIN, P.N. & J.A. SLOBODA. 2001. Music and Emotion: Theory and Research. Oxford University Press. Oxford, UK.
23. BLOOD, A.J., R.J. ZATORRE, P. BERMUDEZ, *et al.* 1999. Emotional responses to pleasant and unpleasant music correlate with activity in paralimbic brain regions. Nature Neurosci. **2:** 382–387.
24. HANDEL, S. 1989. Listening. MIT Press. Cambridge, MA.
25. FRISTON, K., C. PRICE, P. FLETCHER, *et al.* 1996. The trouble with cognitive subtraction. NeuroImage **4:** 97–104.
26. LANE, R.D., E.M. REIMAN, M.M. BRADLEY, *et al.* 1997. Neuroanatomical correlates of pleasant and unpleasant emotion. Neuropsychologia **35:** 1437–1444.
27. DIAS, R., T.W. ROBBINS & A.C. ROBERTS. 1996. Dissociation in prefrontal cortex of affective and attentional shifts. Nature **380:** 69–72.
28. HORNAK, J. & E.T. ROLLS. 1996. Face and voice expression identification in patients with emotional and behavioural changes following ventral frontal lobe damage. Neuropsychologia **34:** 247–261.
29. DAMASIO, A.R. 1996. The somatic marker hypothesis and the possible functions of the prefrontal cortex. Phil. Trans. Roy. Soc. Lond. **351:** 1413–1420.
30. PANKSEPP, J. 1995. The emotional sources of "chills" induced by music. Mus. Percept. **13:** 171–207.
31. KRUMHANSL, C.L. 1997. An exploratory study of musical emotions and psychophysiology. Canad. J. Exp. Psychol. **51:** 336–353.
32. GOLDSTEIN, A. 1980. Thrills in response to music and other stimuli. Physiol. Psychol. **8:** 126–129.
33. BLOOD, A.J. & R.J. ZATORRE. 2001. Intensely pleasurable responses to music correlate with activity in brain regions implicated in reward and emotion. Proc. Natl. Acad. Sci. USA **98:** 11818–11823.
34. ZALD, D.H. & J.V. PARDO. 1997. Emotion, olfaction, and the human amygdala: amygdala activation during aversive olfactory stimulation. Proc. Natl. Acad. Sci. USA **94:** 4119–4124.
35. MORRIS, J.S., C.D. FRITH, D.I. PERRETT, *et al.* 1996. A differential neural response in the human amygdala to fearful and happy facial expressions. Nature **383:** 812.
36. BARDO, M.T. 1998. Neuropharmacological mechanisms of drug reward: beyond dopamine in the nucleus accumbens. Crit. Rev. Neurobiol. **12:** 37–67.
37. KESTER, D.B., A.J. SAYKIN, M.R. SPERLING, *et al.* 1991. Acute effect of anterior temporal lobectomy on musical processing. Neuropsychologia **29:** 703–708.
38. BERRIDGE, K.C. & T.E. ROBINSON. 1998. What is the role of dopamine in reward: hedonic impact, reward learning, or incentive salience? Brain Res. Rev. **28:** 309–369.
39. SCHILSTROM, B., H.M. SVENSSON, T.H. SVENSSON, *et al.* 1998. Nicotine and food induced dopamine release in the nucleus accumbens of the rat: putative role of alpha7 nicotinic receptors in the ventral tegmental area. Neuroscience **85:** 1005–1009.
40. PFAUS, J.G., G. DAMSMA, D. WENKSTERN, *et al.* 1995. Sexual activity increases dopamine transmission in the nucleus accumbens and striatum of female rats. Brain Res. **693:** 21–30.
41. SMALL, D., R.J. ZATORRE, A. DAGHER, *et al.* 2001. Changes in brain activity related to eating chocolate: from pleasure to aversion. Brain **124:** 1720–1733.
42. ZATORRE, R.J. & I. PERETZ. 2001. The Biological Foundations of Music. Ann. N.Y. Acad. Sci. **930:** 1–462.
43. PERETZ, I., A.J. BLOOD, V.B. PENHUNE, *et al.* 2001. Cortical deafness to dissonance. Brain **124:** 928–940.

Toward the Neural Basis of Processing Structure in Music

Comparative Results of Different Neurophysiological Investigation Methods

STEFAN KOELSCH AND ANGELA D. FRIEDERICI

Max Planck Institute of Cognitive Neuroscience, Leipzig, Germany

ABSTRACT: In major-minor tonal music, chord functions are arranged according to certain regularities. The dominant-tonic progression, known as an authentic cadence, is often used as a marker of the end of a harmonic progression and has been considered a basic syntactic structure of major-minor tonal music by several music theorists and music psychologists. We review data from studies in which brain responses to an authentic cadence were compared to those elicited by music-syntactically inappropriate endings. In event-related electric brain potentials (recorded with EEG), the inappropriate endings elicit early right anterior negativity (ERAN), which is maximal around 200 ms after the presentation of an inappropriate chord. The ERAN is reminiscent of early anterior negativities elicited by syntactic incongruities during the perception of language. Magnetoencephalographic (MEG) data suggest that the ERAN is generated in the inferior frontolateral cortex, an area known to be crucially involved in the processing of (linguistic) syntax. Interestingly, the ERAN can be recorded in nonmusicians and in children, indicating that the ability to acquire (implicit) knowledge about musical regularities and to process musical information according to this knowledge is a general ability of the human brain. This ability is probably of great importance for the acquisition of language in infants and children.

KEYWORDS: music; EEG; event-related potential (ERP); structure; syntax; ERAN; music-syntactic mismatch negativity (MMN); MEG; function magnetic resonance imaging (fMRI)

INTRODUCTION

Investigations of the neural correlates of the processing of musical structure are reviewed in this paper. In recent years, several studies investigated the neurocognition of music with electroencephalography (EEG).[1–7] In these studies, event-related brain potentials (ERPs) were calculated from the EEG. ERPs provide information about electric activity of cortical neurons as a response to an experimental stimulus,

Address for correspondence: Dr. Stefan Koelsch, Max Planck Institute of Cognitive Neuroscience, Stephanstr. 1a, 04103 Leipzig, Germany. Voice: +49 (0)341 99 40 121; fax: +49 (0)341 99 40 113.

koelsch@cns.mpg.de

Ann. N.Y. Acad. Sci. 999: 15–28 (2003). © 2003 New York Academy of Sciences.
doi: 10.1196/annals.1284.002

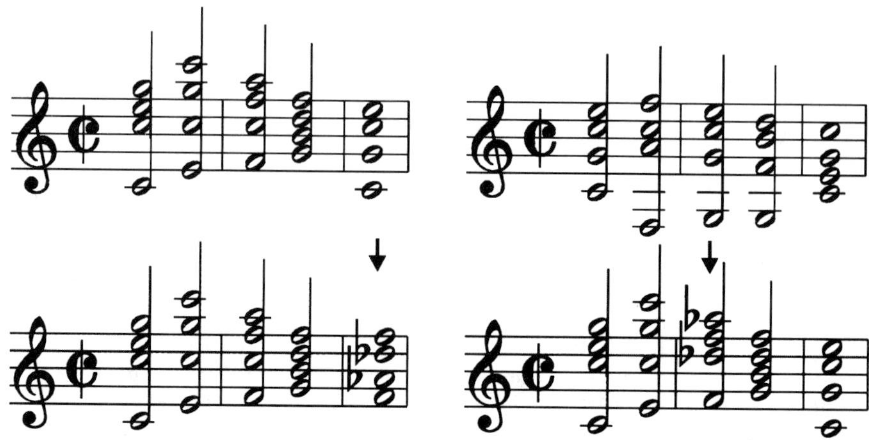

FIGURE 1. Chord sequences exclusively consisting of in-key chords (**a**), containing a Neapolitan sixth chord at the third position (**b**) and at the fifth position (**c**). To exclude the possibility that effects were due to the different interval structure of Neapolitan chords, Neapolitan and tonic chords were presented counterbalanced in root position, as sixth chords, and as six-four chords (in a strict sense, Neapolitans are sixth chords).

such as the presentation of a musical event. The temporal resolution of ERPs is usually in the millisecond range.[8]

In some of the aforementioned studies,[4,6,7] ERPs elicited by harmonically inappropriate (Neapolitan sixth) chords were compared to those elicited by harmonically appropriate tonic chords (FIG. 1). In these studies, chord sequences were presented to the participants, each sequence consisting of five chords, one sequence directly following the other. Sequences consisted mainly of in-key chords and were composed in a way that a musical context was built up towards the end of each sequence (FIG. 1, for examples). Twenty-five percent of the chord sequences contained a Neapolitan chord at the third position, and 25% a Neapolitan chord at the fifth position of the sequence. (For experimental reasons, Neapolitan chords were not always presented as sixth chords.) Neapolitan chords are a prominent stylistic element of major-minor tonal music in both classical and popular music (although Neapolitan chords usually occur in minor tonal environments).

The regularities according to which chord functions and harmonic relations are arranged within a major-minor tonal context have also been denoted as part of a *musical syntax;*[3,4,9–17] the dominant-tonic progression at the end of a harmonic sequence has especially been considered a basic syntactic structure of major-minor tonal music.[9,10,18] The term "musical syntax" does not imply that it is a linguistic syntax in musical terms; rather it points to the fact that music is structured according to complex regularities, a feature reminiscent of language. At the end of a harmonic progression, a tonic chord is the most regular chord function, especially when preceded by a dominant seventh chord. Thus, a Neapolitan chord presented instead of a tonic (after the presentation of a dominant seventh chord) violates the musical regularities, or musical syntax. For listeners familiar with the regularities of major-minor tonal music, the sound of a chord that violates musical regularities is perceived as unexpected.[12–15,18] Note that Neapolitan chords are classically used as a subdominant variation and that they often precede a dominant seventh chord. Thus, Neapolitan chords presented instead of a subdominant at the third position of the sequences (FIG. 1) violate the musical regularities to a smaller degree than do Neapolitans presented instead of a tonic at the end of the chord sequences.

BRAIN SIGNATURES OF MUSICAL SYNTAX

Several experiments have shown that Neapolitan chords elicit a relatively early electric brain response: a negative electric effect with right anterior scalp distribution (FIG. 2), the *early right anterior negativity* (ERAN). Like mismatch negativity (MMN,[19] which is known to reflect the automatic detection of irregularities on a sensory level in a stream of auditory information), the ERAN inverts polarity at mastoidal electrode sites when nose reference is used.[7] The ERAN is usually followed by a late frontal negativity, denoted as the *N5*. The amplitude of the ERAN is maximal around 190–250 ms; the N5 usually peaks around 500–550 ms. ERAN and N5 have been observed in both nonmusicians[4,7] and musicians.[6]

The ERAN is taken as a reflection of the violation of a musical sound expectancy. As just described, such an expectancy is generated according to the complex regularities of major-minor tonal music. We assume that nonmusicians acquire these regularities during exposure to major-minor tonal music in everyday life.[18] With respect to this, ERP effects suggest that even nonmusicians have a sophisticated (implicit) knowledge about the complex regularities of major-minor tonal music and that the acquisition of musical regularities as well as the processing of musical information according to these regularities is a general ability of the human brain.[4] The N5 is taken to reflect processes of subsequent musical integration,[4,6,7,17] an effect that will not be discussed in detail in this paper. (For a further description, see Koelsch *et al.*[20])

Interestingly, the degree of music-structural violation is reflected in an amplitude modulation of the ERAN (FIG. 2): Neapolitan chords presented at the fifth position (a position at which they are structurally highly inappropriate) elicit a larger ERAN compared to Neapolitans presented at the third position (where Neapolitans are less inappropriate).[4] This difference of the degree of music-structural violation is also reflected in behavioral discrimination performance: when participants are instructed to detect the Neapolitan chords, hit rates are higher for Neapolitans that are highly inappropriate than for Neapolitans that are fairly appropriate.[4]

FIGURE 2. Effects of processing musical violations. **(a)** Grand-average ERPs of Neapolitan and tonic chords at the fifth position. Neapolitan chords elicited an early right anterior negativity (ERAN) and a late bilateral negativity (N5). It is suggested that the ERAN reflects the violation of a musical sound expectancy and the N5 integration processes. **(b)** Grand-average ERPs of chords at the third (0–600 ms) and fourth position (600–1200 ms). Chords at the third position were either Neapolitan chords or in-key chord functions. Both ERAN N5 elicited by the Neapolitan chords were considerably smaller than those elicited by Neapolitans at the fifth position, indicating a processing of Neapolitan chords that was dependent on regularities of musical structure. (Reprinted with permission from Koelsch *et al.*[4])

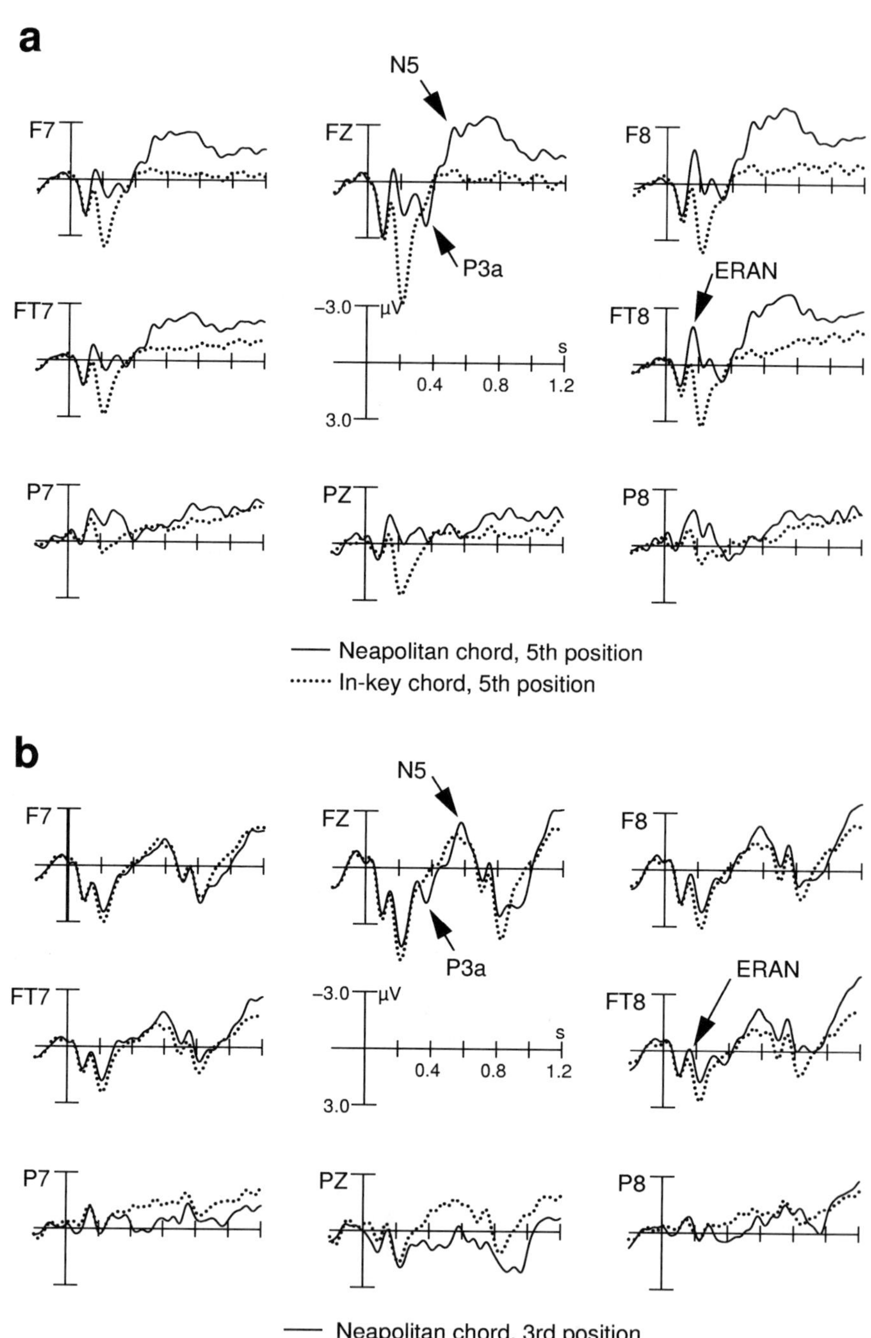

FIGURE 2. *See previous page for legend.*

MUSICAL VERSUS PHYSICAL IRREGULARITY DETECTION

Because Neapolitan chords introduce tones that have not been presented in the directly preceding chords (in C major, the out-of-key notes *a flat* and *d flat*), the structural inappropriateness of Neapolitans confounds with physical deviance. However, note that the physical deviance of the Neapolitans is virtually the same at positions three and five of the sequences, whereas the degree of music-structural violation is different between positions three and five of the chord sequences. Therefore, it seems plausible that the ERAN is partly overlapped by processes of physical irregularity detection (which operate independent of knowledge of music-syntactic regularities), but it is highly likely that the main contribution of the ERAN effect (especially when elicited at the fifth position) originates from neural processes connected to music-syntactic processing. A previous study[17] suggests that the amount of physical irregularity detection contributing to the ERAN is not larger than part of the early effect elicited by Neapolitan chords at the third position. Interestingly, frequency deviants presented with identical time course and probabilities as the Neapolitan chords elicit an MMN that does not differ in amplitude between positions three and five. This finding supports the notion that the amplitude difference of the ERAN elicited by Neapolitan chords at positions three and five is due to the difference in music-structural (in)appropriateness (see above).

To test the hypothesis that the ERAN can be elicited without the presence of a physical deviance, we constructed five-chord sequences (FIG. 3) in which the fourth chord was a dominant and the terminal chord either the tonic (i.e., a structurally regular chord function) or the dominant to the dominant (i.e., a structurally irregular chord function). Note, therefore, that in one sequence type the fourth chord was the dominant of the fifth chord, whereas in the other sequence type, the fifth chord was the dominant of the fourth chord. (The harmonic distance between the last two chords was, thus, always one fifth.) Moreover, the terminal tonic chord had no notes in common with the immediately preceding chord and three notes in common with the other three chords; the terminal dominant to the dominant had one note in common with the immediately preceding chord and four notes in common with the other three chords. If expectancies were generated on the basis of local feature processing and note (pitch) repetition effects, the terminal tonic would violate expectancies more than the terminal dominant to the dominant. On the other hand, if expectancies were generated on the basis of music-syntactic regularities, the dominant to the dominant would violate expectancies more than would the tonic.

Using this stimulus material, a study was conducted in which participants (none had participated in extracurricular music lessons or performances) were asked to press one button in response to the tonic and another in response to the dominant-to-dominant. Each sequence was presented randomly in another key and ended randomly with equal probability on either tonic or dominant-to-dominant (FIG. 3). Compared to the regular terminal chords, the music-structurally irregular chords elicited an ERAN (peaking around 200 ms), which was followed by an N2b-P3 complex (FIG. 3). Note that the N2b does not invert polarity at mastoidal sites, ruling out the possibility that the negativity around 200 ms is (purely) an N2b. Because the terminal dominant to the dominant represented a structural but not a physical irregularity, the data indicate that the ERAN can be elicited even without the presence of a physical deviance (and, thus, reflects the processing of a music-structural or music-

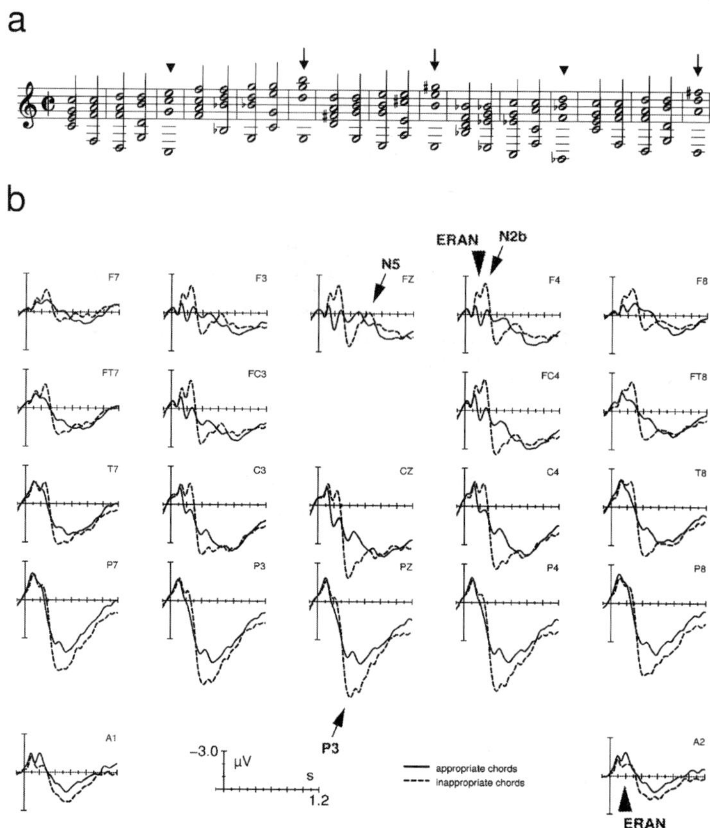

FIGURE 3. (**a**) Examples of chord sequences. *Short arrows* indicate the harmonically appropriate sequence endings (tonic chord), *long arrows* the inappropriate endings (dominant to dominant). Sequences were presented in direct succession, each sequence was presented randomly in a different tonal key, appropriate and inappropriate endings occurred with $P = 0.5$, and the order of the two sequence types was randomized. (**b**) Grand-average of ERP waveforms elicited by sequence endings (*solid line*, harmonically appropriate tonic; *dashed line*, harmonically inappropriate dominant to dominant). Compared to the appropriate chord functions, the inappropriate chord functions elicited an ERAN which inverted polarity at mastoidal sites (*short arrows*), followed by an N2b, a P3, and probably an N5.

syntactic irregularity). Notably, the data also indicate that the ERAN is not specific to the processing of Neapolitan chords.

Note, however, that both Neapolitan and dominant-to-dominant chords represent irregularities of (1) musical structure and (2) tonality (both chord types are out-of-key chords). It is therefore not yet clear if the ERAN can be elicited even without the presence of a violation of tonality. We addressed this issue in recent studies (unpublished data) and found that even in-key chord functions (such as submediants) elicit an ERAN, indicating that the ERAN is also present when a structural violation does not coincide with a violation of tonality.

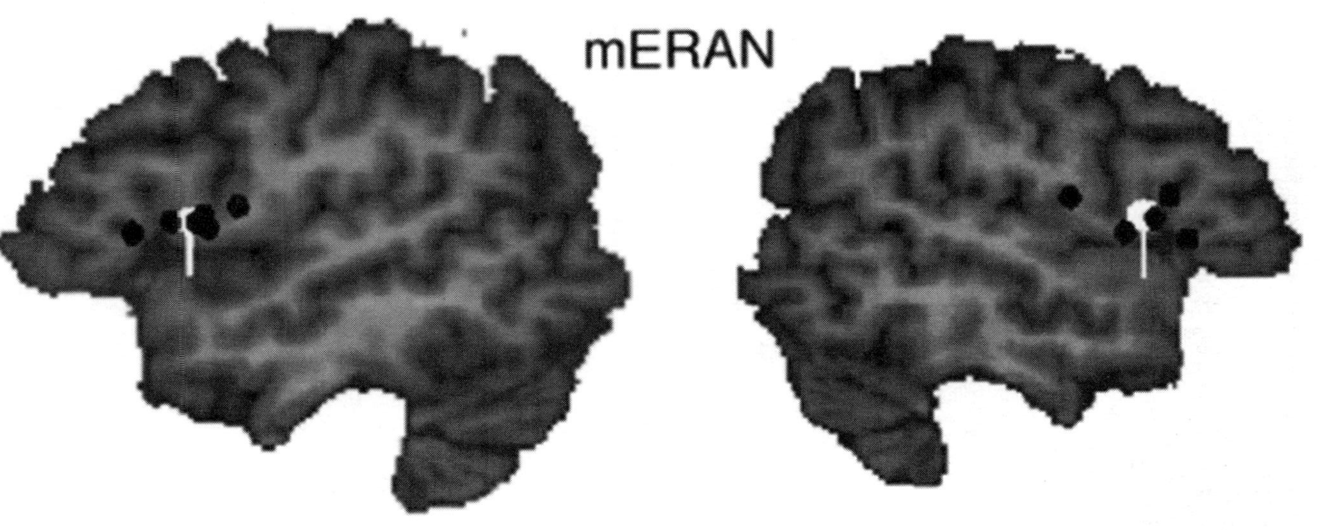

FIGURE 4. Grand average dipole solution (*white*) of the mERAN (from six subjects), from left and right sagittal views. Dipole solutions for the mERAN-effect (difference signals: tonic subtracted from deviants) refer to two dipole configurations (one dipole in each hemisphere). *Black disks* represent single subject solutions. In each hemisphere, a source was located in the frontal opercular cortex, areas known to be crucially involved in the processing of syntactic structure during language comprehension. (Adapted from Maess *et al.*[16])

ATTENTION AND INFLUENCE OF MUSICAL EXPERTISE

The ERAN can be elicited under preattentive listening conditions[7]: when participants read a book while chord sequences were presented, both ERAN and N5 were clearly present. In the second part of the same experiment, participants attended the musical stimulus and detected the inappropriate harmonies. In that part, the ERAN was only marginally larger than when the musical stimulus was ignored.

Recently, we found that the ERAN is larger in musicians than in nonmusicians,[6] reflecting that musicians, presumably because of their explicit knowledge about musical regularities and harmonic relations, are more sensitive to violations of musical structure.

SYNTAX IN MUSIC AND LANGUAGE

With magnetoencephalography (MEG), the main neural generators of the ERAN were localized in the inferior frontolateral cortex (inferior pars opercularis, in the left hemisphere part of *Broca's area*; FIG. 4).[16,21,22]

Interestingly, the ERAN is reminiscent of early anterior negativities that correlate with the early detection of an error in the syntactic structure of a sentence (usually observed with a maximum over the left hemisphere).[23,24] For example, the early left anterior negativity (ELAN)[24,25] has been observed in response to words with unexpected syntactic properties in sentences (phrase structure violations). Thus, both ERAN and ELAN are sensitive to violations of an expected structure; moreover, both ERAN and ELAN are negativities that are maximal around 200 ms, have a lateralized anterior scalp distribution, reflect fairly automatic processes, and receive contributions from generators located in the inferior frontolateral cortex.[22,24–26] The ELAN receives additional contributions from the planum polare, but the dipole solution of the MEG localizing the ERAN in the inferior frontolateral cortex[16] does not exclude that the ERAN also receives contributions from the (anterior) supratemporal cortex.

The similarities between ERAN and ELAN suggest that structural information of both music and language is processed by similar brain mechanisms. With respect to the similarities between ERAN, ELAN, and MMN, it was previously suggested that these ERP effects belong to a family of perisylvian negativities that mediate the processing of irregularities of auditory input[17] and that the neural generators of ELAN, ERAN, and MMN are part of a highly adaptive, perisylvian system of auditory information-processing that mediates the processing of single tones,[19,27] acoustic patterns,[28] phonemes,[29] tonal music,[4] and speech.[30]

Note that in a study by Patel *et al.*,[3] in which music and language processing was directly compared, both linguistic and musical structural incongruities elicited the same P600 (or LPC). Thus, it was suggested that the P600 probably reflects general knowledge-based structural integration processes during the perception of rule-governed sequences. Moreover, in that study the harmonically inappropriate chords elicited an ERP effect similar to the ERAN, denoted by Patel *et al.* as right anterotemporal negativity (RATN). This component was taken to reflect the application of music-specific rules, or probably music-specific working memory processes. In contrast to the ERAN, the RATN had a longer latency (around 350 ms), was

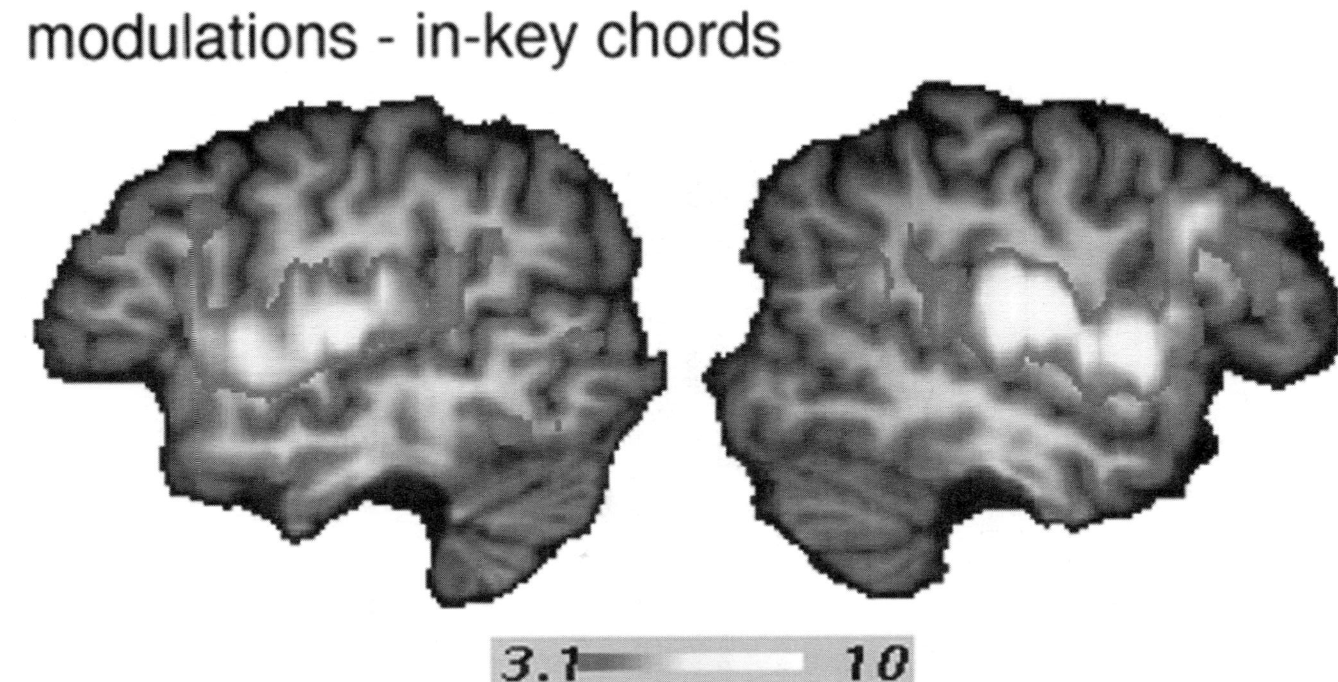

FIGURE 5. Statistical parametric maps (z-scores are indicated in the scale) of the contrast modulations versus in-key chords, mapped onto an individual brain. The panel shows views from left (Talairach coordinate x = −50) and right sagittal (Talairach coordinate x = 46).

observed only over the right hemisphere, and has so far only reportedly been elicited in musicians.

The MEG results (FIG. 4) have been supported by a study with functional magnetic resonance imaging (fMRI).[31] As in the MEG study, the inferior frontolateral cortex was activated during the processing of irregular chord functions (FIG. 5). Interestingly, besides inferior frontolateral cortical areas, posterior temporal areas (in the left hemisphere often referred to as Wernicke's area) are activated bilaterally by the processing of structurally inappropriate musical events. Note that the inferior frontolateral and the posterior temporal cortex are crucially involved in the perception of language.[32–34] The fMRI results concur with findings that indicate considerable overlap of neural structures and processes underlying the perception of music and language.[3,4,16,21,35–39]

In a recent fMRI experiment with a sparse temporal sampling technique we presented Neapolitan chords as structurally inappropriate musical events (as in the aforementioned previous EEG and MEG studies).[40] In that study, activations of the inferior frontolateral cortex as well as some posterior temporal activations were replicated, but no activation of Heschl's gyrus was found. This finding indicates that possible contributions of physical irregularity detection (as reflected electrically in the MMN) to the ERAN are only marginal and that the ERAN originates from different cortical regions than the physical MMN (e.g., frequency MMN): whereas the ERAN appears to receive main contributions from areas located in the inferior frontolateral cortex (and probably the anterior part of the superior temporal gyrus, with no contribution from primary auditory areas), the (physical) MMN receives its main contributions from areas located within or in the close vicinity of primary auditory areas, with additional contributions from anterior frontal cortical areas.

However, given the broadening of the MMN framework, which summarizes perisylvian negativities that mediate the fast and automatic processing of auditory irregularity detection, the ERAN may be regarded as an MMN elicited by the highly abstract feature "music-syntactic appropriateness." With respect to this, we suggested that the ERAN may be understood as a "music-syntactic MMN."[6,7,41]

The localization of (language-) syntactic processes using fMRI suggests involvement of the anterior portion of the left superior temporal gyrus and the frontal operculum adjacent to BA44.[26,42] In a recent fMRI experiment that used the same auditory sentence material as in a previous MEG study,[22] some areas increased their activation as a function of syntactic violation. These are the anterior portion and the posterior portion of the left superior temporal gyrus and the frontal operculum.[43] The anterior portion of the superior temporal gyrus and the frontal opercular activation was also observed in an fMRI study presenting sentences in which all content words were replaced by pseudowords.[44] Thus, it appears that the posterior superior temporal gyrus activation reflects the integration of semantic and syntactic information and that the anterior portion of the superior temporal gyrus together with the inferior frontal region supports syntactic processes as such.

PROCESSING STUCTURAL INFORMATION IN CHILDREN

To investigate developmental aspects of music processing, we conducted the EEG experiment with Neapolitan and in-key chords,[4,6] measuring two groups of children:

9- and 5-year-olds (children who had not received any formal musical training). The harmonies were task irrelevant, and children were asked to respond to infrequently occurring chords played with an instrumental timbre deviating from the standard piano timbre. As in adults, Neapolitan chords presented at the fifth position elicited distinct early and late ERP effects in both 5- and 9-year-old children: an earlier negativity, which was maximal around 320 ms and therefore around 120 ms later than that in adults, and a later negativity, the N5, maximal around 550 ms.

The finding that in both age groups the ERPs of Neapolitan chords differed from those elicited by structurally appropriate chords indicates that already the 5-year-old children processed the chords according to their harmonic appropriateness. As did adults, children showed a remarkable amplitude reduction of ERP effects for Neapolitan chords at the third position, suggesting that already 5-year-old children processed the musical stimulus according to regularities of "classical" major-minor tonal music, which are described by the theory of harmony.

As these regularities have previously been denoted as part of a musical syntax (see above), the findings suggest that already 5- and 9-year-old children have an (implicit) knowledge of musical syntax, that is, the complex regularities of tonal music.

Because children had no special musical expertise, the findings support the hypothesis that the acquisition of knowledge about musical regularities and the processing of musical information according to this knowledge is a general ability of the human brain, which is already present at the age of 5 (if not earlier). At this age, children participating in the study had been exposed in everyday life to major-minor tonal music, that is, the regularities underlying the music of their culture.

The described effects were present although the Neapolitan chords were task irrelevant, suggesting that children process harmonic relations and musical structure fairly automatically, that is, even when they do not specifically attend harmonies. (This finding concurs with the discovery that the ERAN can be elicited preattentively in adults.[7])

The observation of sophisticated musical abilities in children corresponds with the observation that musical elements of speech play an important role in the acquisition of language; it has been hypothesized that music and speech are intimately connected in early life,[45,46] that musical elements pave the way to linguistic capacities earlier than phonetic elements,[47] and that melodic aspects of adult speech to infants represent the infant's earliest associations between sound patterns and meaning[48] as well as between sound patterns and syntactic structure.[49–51] With this respect, it is also interesting to note that the ERAN can already be observed at the age of 5 (presumably even earlier), whereas the fast and automatic processing of syntactic violations in language as reflected in the ELAN (see above) becomes observable around the age of 9–10 (unpublished data from A. Hahne and A.D. Friederici). Thus, the fast and automatic processes of structure analysis during the comprehension of complex regularity-based auditory information (as reflected in early anterior negativity) appear to be employed earlier for musical than for linguistic information. That is, it appears that children do not learn language before music, but music before language, supporting the notion that musical elements of speech (such as prosody) play a crucial role in the infant's earliest associations between sound patterns and syntactic structure,[49–51] and thus for the acquisition of language.[45–48,50,51]

In conclusion, data obtained with different investigation methods and from different subject groups indicate that musical structure can be processed by brain mechanisms that operate (a) fairly automatically, (b) with a latency around 180 ms, and that (c) mainly originate from the frontal cortex (probably with contributions from anterior temporal lobe structures). These brain mechanisms can be observed in nonmusicians, musicians, adults, and children, and these mechanisms appear to be similar, and partly overlapping, with the mechanisms that process structure during language comprehension. Results indicate that the ability to acquire (implicit) knowledge about musical regularities and the ability to process musical information according to these regularities is a general ability of the human brain. This ability appears to be important for the acquisition of language.

NOTE: Examples of the stimuli are available at http://www.stefan-koelsch.de

REFERENCES

1. JANATA, P. 1995. ERP measures assay the degree of expectancy violation of harmonic contexts in music. J. Cognit. Neurosci. **7:** 153–164.
2. BESSON, M. & F. FAITA. 1995. An event-related potential (ERP) study of musical expectancy: comparison of musicians with nonmusicians. J. Exp. Psychol.: Human Percept. Perf. **21:** 1278–1296.
3. PATEL, A.D., E. GIBSON, J. RATNER, et al. 1998. Processing syntactic relations in language and music: an event-related potential study. J. Cognit. Neurosci. **10:** 717–733.
4. KOELSCH, S., T. GUNTER, A.D. FRIEDERICI, et al. 2000. Brain indices of music processing: 'nonmusicians' are musical. J. Cognit. Neurosci. **12:** 520–541.
5. REGNAULT, P., E. BIGAND & M. BESSON. 2001. Different brain mechanisms mediate sensitivity to sensory consonance and harmonic context: evidence from auditory event-related brain potentials. J. Cognit. Neurosci. **13:** 241–255.
6. KOELSCH, S., B. SCHMIDT & J. KANSOK. 2002. Influences of musical expertise on the ERAN: an ERP-study. Psychophysiology **39:** 657–663.
7. KOELSCH, S., E. SCHRÖGER & T. GUNTER. 2002. Music matters: preattentive musicality of the human brain. Psychophysiology **39:** 1–11.
8. RUGG, M. & M. COLES. 1995. Electrophysiology of Mind. Event-Related Brain Potentials and Cognition. Oxford University Press. Oxford.
9. SLOBODA, J. 1985. The Musical Mind: The Cognitive Psychology of Music. Oxford University Press. New York.
10. DELIÈGE, C. 1984, Les Fondements de la Musique Tonale. J.C. Latès. Paris.
11. SWAIN, J. 1997. Musical Languages. Norton. UK.
12. KRUMHANSL, C. & E. KESSLER. 1982, Tracing the dynamic changes in perceived tonal organization in a spatial representation of musical keys. Psychol. Rev. **89:** 334–368.
13. BHARUCHA, J. & C. KRUMHANSL. 1983, The representation of harmonic structure in music: hierarchies of stability as a function of context. Cognition **13:** 63–102.
14. BHARUCHA, J. & K. STOECKIG. 1987. Priming of chords: spreading activation or overlapping frequency spectra? Percept. Psychophysiol. **41:** 519–524.
15. BIGAND, E., F. MADURELL, B. TILLMANN, et al. 1999. Effect of global structure and temporal organization on chord processing. J. Exp. Psychol.: Human Percept. Perf. **25:** 184–197.
16. MAESS, B., S. KOELSCH, T. GUNTER, et al. 2001, 'Musical syntax' is processed in the area of Broca: an MEG-study. Nature Neurosci. **4:** 540–545.
17. KOELSCH, S., T. GUNTER, E. SCHRÖGER, et al. 2001. Differentiating ERAN and MMN: an ERP-study. NeuroReport **12:** 1385–1389.
18. TILLMANN, B., J. BHARUCHA & E. BIGAND. 2000. Implicit learning of tonality: a self-organized approach. Psychol. Rev. **107:** 885–913.
19. NÄÄTÄNEN, R. 1992. Attention and Brain Function. Erlbaum. Hillsdale, NJ.

20. KOELSCH, S., T. GUNTER, E. SCHRÖGER, *et al.* 2003. Processing tonal modulations: an ERP-study. J. Cognit. Neurosci. In press.
21. KOELSCH, S., B. MAESS & A.D. FRIEDERICI. 2000. Musical syntax is processed in the area of Broca: an MEG study. NeuroImage **11:** 56.
22. FRIEDERICI, A.D., Y. WANG, C. HERRMANN, *et al.* 2000. Localisation of early syntactic processes in frontal and temporal cortical areas: an MEG study. Human Brain Mapp. **11:** 1–11.
23. NEVILLE, H., J. NICOL, A. BARSS, *et al.* 1991. Syntactically based sentence processing classes: evidence from event-related brain potentials. J. Cognit. Neurosci. **3:** 151–165.
24. FRIEDERICI, A.D., E. PFEIFER & A. HAHNE. 1993. Event-related brain potentials during natural speech processing: effects of semantic, morphological and syntactic violations. Cognit. Brain Res. **1:** 183–192.
25. HAHNE, A. & A. FRIEDERICI. 1999. Rule-application during language comprehension in the adult and the child. *In* Learning: Rule Extraction and Representation. A.D. Friederici & R. Menzel, Eds. Walter de Gruyter. New York.
26. FRIEDERICI, A. 2002. Towards a neural basis of auditory sentence processing. Trends Cognit. Sci. **6:** 78–84.
27. SCHRÖGER, E. 1998. Measurement and interpretation of the mismatch negativity (MMN). Behav. Res. Meth. Instr. & Comp. **30:** 131–145.
28. SCHRÖGER, E. 1994, An event-related potential study on sensory representations of unfamiliar tonal patterns. Psychophysiology **31:** 17–181.
29. NÄÄTÄNEN, R., A. LEHTOKOSKI, M. LENNES, *et al.* 1997. Language-specific phoneme representations revealed by magnetic brain responses. Nature **385:** 432–434.
30. FRIEDERICI, A.D. 1995. The time course of syntactic activation during language processing: a model based on neuropsychological and neurophysiological data. Brain Lang. **50:** 259–281.
31. KOELSCH, S., T. GUNTER, D.Y. CRAMON, v. *et al.* 2002. Bach speaks: a cortical 'language-network' serves the processing of music. NeuroImage **17:** 956–966.
32. KNECHT, S., A. FLÖEL, B. DRÄGER, *et al.* 2002. Degree of language lateralization determines susceptibility to unilateral brain lesions. Nature Neurosci. **5:** 695–699.
33. JUST, M., P. CARPENTER, T. KELLER, *et al.* 1996. Brain activation modulated by sentence comprehension. Science **274:** 114–116.
34. FRIEDERICI, A.D., Ed. 1998. Language Comprehension: A Biological Perspective. Springer. Berlin.
35. PLATEL, H., C. PRICE, J. BARON, *et al.* 1997. The structural components of music perception, a functional anatomical study. Brain **120:** 229–243.
36. LIEGEOIS-CHAUVEL, C., I. PERETZ. '. BABAIE, *et al.* 1998. Contribution of different cortical areas in the temporal lobes to music processing. Brain **121:** 1853–1867.
37. PERETZ, I., R. KOLINSKY, M. TRAMO, *et al.* 1994. Functional dissociations following bilateral lesions of auditory cortex. Brain **117:** 1283–1301.
38. ZATORRE, R., A. EVANS & E. MEYER. 1994. Neural mechanisms underlying melodic perception and memory for pitch. J. Neurosci. **14:** 1908–1919.
39. WALLIN, N.L., B. MERKER & S. BROWN, Eds. 2000. The Origins of Music. MIT Press. Cambridge, MA.
40. KOELSCH, S. & G. SCHLAUG. 2003. An fMRI study on music perception in children and adults. In preparation.
41. KOELSCH, S., B. MAESS, T. GROSSMANN, *et al.* 2003. Sex difference in music-syntactic processing. NeuroReport **14:** 709–713.
42. KAAN, E. & T. SWAAB. 2002. The brain circuitry of syntactic comprehension. Trends Cognit. Sci. **6:** 350–356.
43. FRIEDERICI, A., S.A. RÜSCHEMEYER, A. HAHNE, *et al.* 2003. The role of the left inferior frontal and superior temporal cortex in sentence comprehension: localizing syntactic and semantic processes. Cereb. Cortex **13:** 170–177.
44. FRIEDERICI, A.D., M. MEYER & D.Y. CRAMON. 2000. Auditory language comprehension: an event-related fMRI study on the processing of syntactic and lexical information. Brain Lang. **74:** 289–300.

45. TRAINOR, L. & S. TREHUB. 1992. A comparison of infants' and adults' sensitivity to western musical structure. J. Exp . Psychol.: Human Percept. Perf. **18:** 394–402.
46. TREHUB, S., G. SCHELLENBERG & D. HILL. 2000. The Origins of Music: Perception and Cognition: A Developmental Perspective. N.L. Wallin, B. Merker & S. Brown, Eds. MIT Press. Cambridge, MA.
47. PAPOUSEK, H. 1996. Musicality in infancy research. *In* Musical Beginnings. J. Sloboda & I. Deliège, Eds. Oxford University Press. Oxford.
48. FERNALD, A. 1989. Intonation and communicative intent in mothers' speech to infants: is the melody the message? Child Dev. **60:** 1497–1510.
49. KRUMHANSL, C.L. & P.W. JUSCZYK. 1990. Infants' perception of phrase structure in music. Psychol. Sci. **1:** 70–73.
50. JUSCZYK, P., K. HIRSH-PASEK, D. KEMLER-NELSON, *et al.* 1992. Perception of acoustic correlates of major phrasal units by young infants. Cognit. Psychol. **24:** 252–293.
51. JUSCZYK, P.W. & C.L. KRUMHANSL. 1993. Pitch and rhythmic patterns affecting infants' sensitivity to musical phrase structure. J. Exp. Psychol.: Human Percept. Perf. **19:** 627–640.

The Promises of Change-Related Brain Potentials in Cognitive Neuroscience of Music

MARI TERVANIEMI[a] AND MINNA HUOTILAINEN[a,b]

[a]*Cognitive Brain Research Unit, Department of Psychology, University of Helsinki, Helsinki, Finland*

[b]*Biomag Laboratory, Helsinki University Central Hospital, Helsinki, Finland*

ABSTRACT: Even when simultaneously performing a task unrelated to sounds, the human auditory cortex can precisely model the invariances of the acoustic environment. Data acquired in a mismatch negativity (MMN) paradigm have shown that temporally and spectrally complex sounds as well as their relations are automatically represented in the human auditory cortex. Furthermore, MMN data indicate that these neural sound representations are spatially distinct from phonetic and musical sounds within and between the cerebral hemispheres. Most MMN studies were conducted in pitch dimension, but temporal aspects of sound processing are also under increasing experimentation. To some extent, musical expertise is also reflected in sound representation accuracy as indexed by the MMN paradigm.

KEYWORDS: neural sound representations; preattentive processing; musical development; auditory memory; complex sounds

THE MISMATCH NEGATIVITY PARADIGM

The architecture of sound representations in the human brain is a key area of interest in the cognitive neuroscience of audition. Knowing how sounds and their elementary continua are encoded and processed is required to comprehend how higher forms of auditory information, namely, speech and music, are represented by the brain. Additionally, knowing which aspects of auditory information are encoded without the conscious attention of the listener and the accuracy of this automatic information processing is needed to understand the processes of speech and music perception.

To provide a tool for answering these concerns, a mismatch negativity (MMN) paradigm was developed over the last 25 years. The MMN, a component in event-related auditory brain potentials, is an index of similarity between two sound representations. In other words, it reflects the accuracy of neural memory traces in encoding the sound features. In the present review, the findings most relevant to music studies are summarized.

Address for correspondence: Dr. Mari Tervaniemi, Cognitive Brain Research Unit, Department of Psychology, P.O. Box 9, FIN-00014 University of Helsinki, Helsinki, Finland. Voice: +357-9-19129456; fax: +358-9-19129450.

Mari.Tervaniemi@Helsinki.FI

Ann. N.Y. Acad. Sci. 999: 29–39 (2003). © 2003 New York Academy of Sciences.
doi: 10.1196/annals.1284.003

The MMN was originally observed in pitch deviance when two sounds of 1,000 and 1,140 Hz were presented in an oddball paradigm.[1] Since then, the MMN has been recorded to index discrimination of changes in various sound parameters such as pitch,[2,3] intensity,[4,5] duration,[6,7] sound order,[8–10] sound omission,[11] rhythm,[12] abstract rules,[13] chord quality,[14,15] and formant structure.[16,17]

A theoretically and practically important feature of the MMN paradigm is that the MMN can be elicited and recorded even when the subject is not attending to the sounds but is concentrating, for instance, on reading or watching a (silent) movie. Despite such "automatic" elicitation of the MMN, its amplitude and latency sensitively reflect the subject's accuracy in detecting the sound changes behaviorally. This correspondence between MMN parameters and discrimination accuracy has been shown by several indices such as hit rates, reaction times, and d' in both oddball and 2AFC paradigms (e.g., see Refs. 3 and 7), musicality tests,[18,19] and similarity ratings.[20]

Theoretically, the MMN thus provides an index of the neural sound representations, the recordings not being contaminated by activation of the subsequent attentional mechanisms. This aspect is especially important in group-comparison designs in which the subject groups might differ in motivation or ability to concentrate on a given task. An obvious benefit of automatic MMN elicitation is that the recordings can be performed with infants, children, and clinical populations with whom paradigms involving active participation of the subject are often difficult, unreliable, or even impossible to conduct.

In the following, the recent literature on music-sound MMN obtained in adults is summarized. Thereafter, the current view on neural sound representation research is described in children of various ages, laying the groundwork for future investigations of neural representations of music in children.

MUSIC-SOUND PROCESSING

Pitch Discrimination

The first MMN study using synthesized musical instrument sounds was conducted in the early 1990s to compare the pitch discrimination accuracy of musicians with and without absolute pitch (AP).[21] Piano sounds were used because AP is often more accurate, first, when the sounds contain overtones, and, second, when the subjects are familiar with the timbre. The sound frequencies belonged to the Western musical scale (C, C#) or were off the scale (C-, C+, C#-). It was hypothesized that in AP musicians, the pitch MMN to on-scale tones would be greater than that to off-scale notes because the sound representations formed by on-scale tones are more stable. The results did not confirm this hypothesis; the MMN parameters did not differ between the on- and off-scale tones, and the MMN did not differentiate the subject groups (FIG. 1). This result was interpreted as showing the importance of attention mechanisms in determining pitch.

Interestingly, however, in both groups the MMN was significantly greater in amplitude and earlier in latency to piano sounds when compared to sinusoidal tones. To systematically investigate the facilitating effect of overtones on the accuracy of pitch discrimination, three MMN studies were conducted. In the first, subjects were

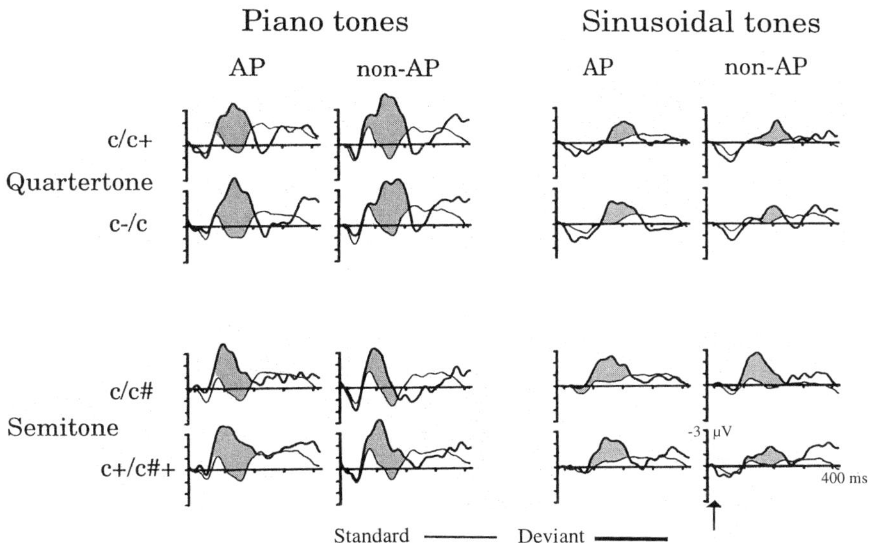

FIGURE 1. Event-related potential (ERP) recordings of absolute-pitch (AP) processors and non–absolute-pitch processors to piano and sinusoidal tones in conditions with different pitch changes, as indicated on the *left*. During the recordings, subjects concentrated on reading a book of their own choice. The *thin line* represents ERPs to standard stimuli; the *thick line,* ERPs to deviant stimuli. *Shadowing* indicates the presence of mismatch negativity (MMN) and the *arrow* the onset of the sound. Altogether, four pitch differences were employed between standard ($P = 0.9$) and deviant ($P = 0.1$) stimuli: 262/269 Hz (C/C+), 254–262 Hz (C–/C), 262–277 Hz (C/C#), and 269–285 Hz (C+/C#+). These differences were equal to quarter tone and semitone intervals on the Western musical scale (262 Hz = C4). The combinations included all possibilities with regard to tone positions either belonging or not belonging to the scale (standard on/off the scale, deviant on/off the scale). The subject's performance in the pitch-naming task was tested after the EEG recordings. Subjects were asked to name the pitch and octave of 50 tape-recorded synthesized piano tones (range: C2–E6). The AP subjects correctly named, on average, 82% and the non-AP subjects 9% of these test sounds.[21]

presented with pitch changes of 2.5, 5, and 10% at 500 Hz sounds during a reading task and a target-detection task.[22] It was found that pitch changes were detected more accurately among sounds that consisted of the three lowest partials than among pure sinusoidal tones. Correspondingly, the pitch-MMN amplitude was greater and latency shorter in spectrally rich sounds than in pure tones. This result suggests that pitch discrimination, under attentional control as well as prior to that, is facilitated by the presence of the harmonic partials.

Consequently, we wanted to determine if adding harmonic partials or prolonging the sound would further facilitate the elicitation of pitch MMN.[23] In separate sequences, sounds with one, three, and five harmonic partials were presented. The fundamental frequency was 500 Hz, and in deviants a 2.5% frequency change of all partials was presented. It was found that the MMN amplitude was enhanced to spectrally rich sounds when compared to pure tones, with no difference between spectrally rich tones having three or five partials. This suggests that relatively few harmonic partials, at least when they fall inside the dominance region of pitch perception, are sufficient to sharpen the neural representations underlying pitch discrimination.

In the third study of this series, the neural and behavioral accuracy of pitch discrimination across a wide frequency range (250–4000 Hz) with pure sinusoidal versus spectrally rich sounds was compared.[24] Also, the magnitude of the frequency change was varied (2.5, 5, 10, and 20%). Confirming the previous findings, the spectrally rich sounds elicited greater MMN across the entire frequency range. Changes at the middle frequencies elicited greater MMN, which, in parallel, had a shorter latency than those at the lowest and highest frequencies. As predicted, MMN amplitude and latency reflected the magnitude of the pitch change.[3]

An alternative approach to the question of the presence of musical context on pitch discrimination was adopted by Brattico *et al.*,[25] who presented reading subjects with a large pitch change in three contexts: isolated sounds, a sequential pattern with tones along the Western musical scale, and a sequential pattern with intervals unfamiliar to the subjects. Data indicated that the MMN amplitude was greater when the pitch change occurred among patterns with a familiar scale than with an unfamiliar scale. Moreover, MMN was greater when the pitch change occurred within an unfamiliar scale than among single tones. This suggests that musical context facilitates pitch discrimination, the facilitation being most effective when the context is familiar to the subject.

Generators of Music-Sound MMN

Alho *et al.*[26] pioneered investigation in the cortical generators of the music-sound MMN. They used sounds with an occasional pitch change of identical magnitude, namely, that of one whole-tone step (G4-A4), in three different contexts: single tones, parallel chords, and sequential chord patterns. The MMN elicited by the deviant pitch in these three contexts did not differ in latency or amplitude. However, comparisons of the pitch-MMN generators revealed that the MMN source was more medial in the auditory cortex with sequential and parallel chords than with single tones. This suggests that within the auditory cortex, there are functionally separate areas for encoding spectrally and/or temporally complex information.

Subsequent studies were conducted to determine if the auditory cortex has spatially distinct areas for encoding musical versus phonetic sounds.[15] Subjects were presented with two phonemes (/e/ *vs* /o/) or chords (A major *vs* A minor). These stimuli were matched in complexity as well as in magnitude of the pitch change between standard and deviant tones. As the main result of the study, the MMNm source for a phoneme change was located superior to that of the chord change. On the contrary, the early exogenous P1m sources, being mainly determined by the physical sound characteristics, did not differ between phonemes and chords. These data thus suggest that in both hemispheres distinct cortical areas specialize in representing phonetic and musical sounds. Furthermore, subsequent data obtained by positron emission tomography (PET) indicate that left and right hemispheres have differential roles in representing these sounds: change-related activity for phonetic sounds was observed in the auditory areas on the left side, whereas chord changes activated the corresponding areas on the left side.[27]

DEVELOPMENT

The development of musical skills during childhood is an extremely interesting combination of innate capabilities, the natural development of these skills, and the effects of musical training. It is still very challenging to study the brain correlates of this multifaceted process, because even the brain correlates of simple sound and sound-change processing in children are relatively poorly known. In the following, a short introduction to the current status of the development of auditory brain responses during childhood is given.

Change-Related Event-Related Potentials in Infants and Children

The full-term newborn's event-related potentials (ERPs) to a repeated sound with constant acoustical properties are characterized by a positive peak at about 250 ms followed by a small negativity at 600–700 ms.[28] Thereafter, two prominent negative peaks (N250 and N450) become apparent in the ERPs, and at the age of 1 year, the response is characterized by the P150-N250-P350-N450 pattern, which remains relatively stable until the age of 10 years.[29] Deviant sounds elicit a response pattern of MMN-P3-LDN (mismatch negativity, P3 or P300 component, and late discriminative negativity) in children of various ages. These components are of particular interest in this context, because they imply the presence of an automatic short-term memory system with storage, comparison, and change-detection capacity, whose existence is required by higher cognitive functions, thus opening a new research era in which questions on innate human musical capabilities can be assessed.

The MMN in children is a frontocentral negativity, peaking at 150–300 ms after the onset of deviation, and it is surprisingly stable in amplitude, latency, and distribution across all ages studied so far.[30–32] It has been found in full-term newborns,[33] premature infants,[34] and even near-term fetuses[35] (FIG. 2).

For speech sounds, the MMN has been shown to selectively increase to native-language vowels between the ages of 6 and 12 months[36] and index speech-sound learning even in newborns.[37] Comparable to the data obtained in adults,[16] the speech-sound MMN in infants of 8 months has been shown to be more mature in the left hemisphere.[38]

The MMN studies in children between the ages of 2 and 6 years are relatively scarce.[39] Typically, the MMN latency has been found to decrease[30] and the memory-trace duration to increase[40] during school age. The MMN has been used as an index of inadequate accuracy of auditory perception in dyslexia and learning problems.[41,42]

The P3 or P300 is a positive response, peaking at 300–400 ms after stimulus onset, which in adults is claimed to index attention switching towards the deviant or target stimulus.[43] The first part of P3, the P3a, appears in conditions in which the subject is engaged in a task completely unrelated to the auditory stimulation stream, but the attention of the subject unintentionally slips to the extremely different and potentially relevant "novel" sounds appearing randomly in the stream of repeated

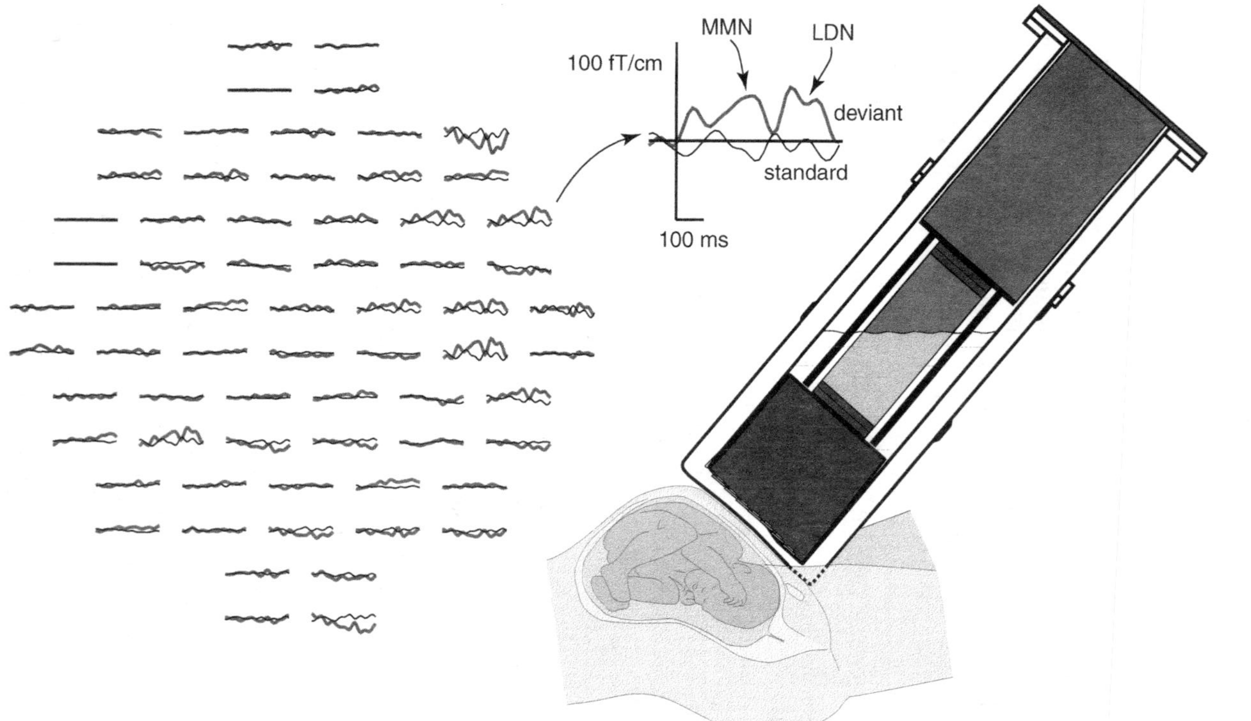

FIGURE 2. Magnetoencephalographic (MEG) recording of a healthy human fetus is performed by placing the instrument tightly against the mother's abdomen so that the surface of the instrument is 4–7 cm from the brain of the fetus (*right*). The 99-channel magnetometer of the BioMag Laboratory, Helsinki University Central Hospital, records data from 33 positions with one magnetometer and two gradiometers in each location. Responses from the gradiometer channels to standard (*black*) tones of 500 Hz and deviant (*gray line*) tones of 750 Hz are shown on the *left*. The *rightmost* channels have recorded the data above the brain of the fetus. One channel is enlarged, showing possible correlates of the fetal mismatch negativity (MMN) and the late discriminative negativity (LDN) responses.

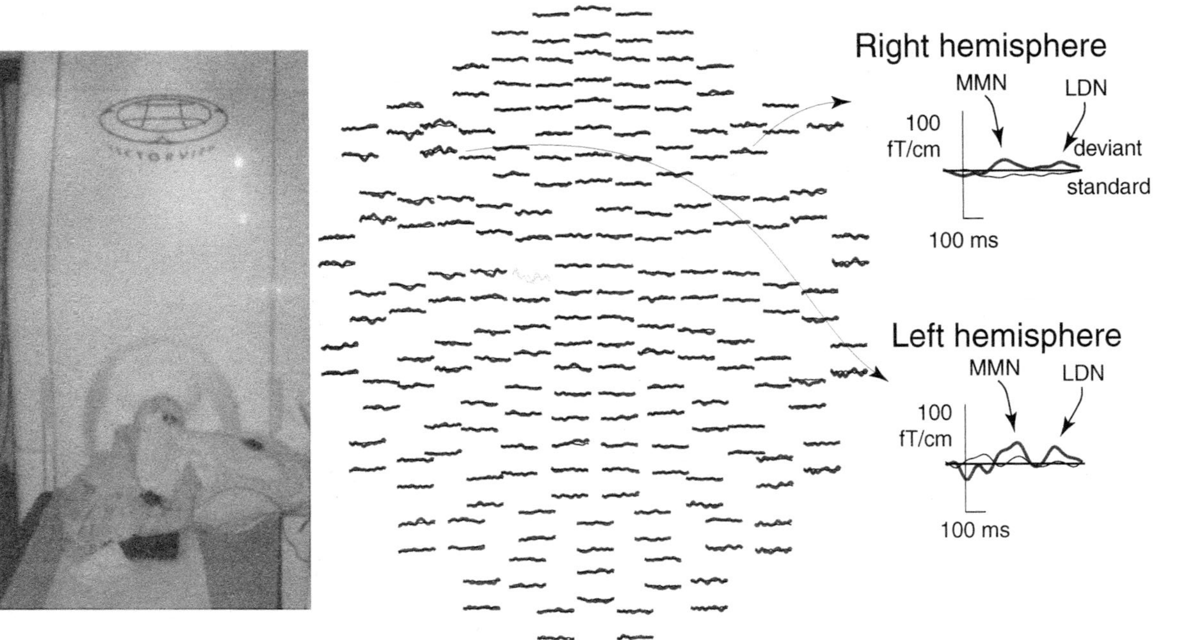

FIGURE 3. Magnetoencephalographic (MEG) recording of the same subject as in FIGURE 2 performed 3 days after birth using a 306-channel Vectorview magnetometer of the BioMag Laboratory (*left*). The infant's head is placed inside the helmet-shaped instrument, and the distance to the infant's brain is 2–5 cm from the surface of the helmet. The instrument records from 102 positions with one magnetometer and two gradiometers in each location. Responses from the gradiometer channels to the same stimuli as in FIGure 2 are shown with the nose up, left hemisphere to the left and right hemisphere to the right side. One channel from each hemisphere is enlarged, showing correlates of the mismatch negativity (MMN) and the late discriminative negativity (LDN) responses.

sounds. This response has been reported in children in active and passive conditions. The P3 also seems to be connected to the involuntary allocation of attention in children.[44]

Late discriminative negativity (LDN), typical in children but occasionally reported in adults,[45,46] is a long-lasting negativity, peaking at around 400–600 ms. In contrast to MMN, it changes markedly during childhood. The latency of the LDN remains relatively stable, whereas the amplitude increases during childhood and diminishes towards adulthood.[47,48] The functional role of the LDN is still unclear, but it has been found to correlate with the magnitude of change only in early childhood, indicating that it may be related to higher cognitive functions, a suggestion also supported by its substantial and long-lasting developmental changes. In language learning, the role of the P3-LDN complex seems to be different from that of the MMN: the MMN indexes the development of long-term traces for foreign vowels,[49] whereas the P3-LDN responses develop even after the vowel learning process, possibly indexing the attention allocation related to the relevancy of the stimulation.[50]

Our recent data in healthy newborns, recorded with magnetoencephalography (MEG, using event-related magnetic fields; FIG. 3), show that the correlates of the MMN-P3-LDN complex can be found at the auditory cortices. As in adults, the newborn's MMN response seems to be generated mainly in the auditory cortex, whereas the P3 and LDN seem to have several sources in newborns (as found in adults for the P3[51]). In the future, the MEG recordings will hopefully reveal the locations and functional relevance of these discriminative components. Thereafter, the use of these responses in studies of musical sound processing will become possible.

DISCUSSION

The present review summarizes the role of preattentively activated memory functions of human auditory memory specifically in music-sound contexts by introducing the stimulation paradigms in which those have been studied so far (for more extensive reviews, see Refs. 13 and 52). In the present context, evidence of facilitated preattentive functions in professional musicians compared with amateur musicians or non-musicians is also of upmost relevance. To summarize, the importance of auditory cortex functions that do not need conscious activation of the listener for their activation is highlighted by current evidence also in the context of music-sound perception, cognition, and musical expertise.

The value of the change-detection paradigm in the neuroscience of music may be its suitability in probing the auditory neurocognition of subjects with highly variable performance in traditional behavioral paradigms in a comparable way. As illustrated in FIGURES 2 and 3, the method is useful in newborns and even in unborn fetuses in probing their automatic change-detection mechanisms, accuracy and duration of their short-term memory traces, and features of sound that they can disentangle. At the same time, this paradigm can be adjusted to study adult experts (e.g., musicians) and to compare their neural memory functions to those of non-experts, without being contaminated by differences in motivation or vigilance that might trouble the interpretation of the results obtained in a behavioral task.

Data obtained thus far suggest that music-sound encoding in the auditory cortex fundamentally differs from that of pure tones and speech sounds in both functional[21–24] and anatomical[15,27] terms. Such specialization of the auditory cortex as a function of the relevant sound features, originally shown in monkeys in intracranial recordings,[53] opens fascinating views to the accuracy of the human brain in representing our auditory environment. Moreover, it also confirms that the present method of cognitive neuroscience can unveil neurocognitive functions that were previously accessible only in nonhuman primates via invasive techniques.

Moreover, there is also evidence of long-term expertise in facilitating the neural encoding of the most relevant sound features during the most rudimentary sound analysis, which takes place without attentional control in sensory memory.[53,54] This suggests a dynamic transfer of information between sensory and long-term memory systems in audition. In other words, while the original memory models assumed information flowed merely from sensory (short-term) stores towards the long-term store (e.g., Ref. 55), the present views also emphasize the role of long-term memory in modulating the earliest processing stages. By showing that personal experience modulates sensory-level sound analysis, such a dynamic view highlights the need for studies on the preattentive functions in the field of music cognition.

Adopting such a view might easily lead us to conclude that most aspects of music perception and cognition are attributed to explicit learning of those skills. However, in our opinion, the specific roles of inherited versus learned processes in musical expertise are far from clear. Even if the evidence provided by infant and fetal recordings (FIGS. 2 and 3) is taken into account, it is highly problematic that it can specify the role of innate skills and learning. One must bear in mind that some learning already occurs during the fetal period, and the nature of this learning with respect to music depends of the musical sound environment before birth, which, in turn, depends on the musical skills and interests of the mother. These complex issues may partly be investigated via change-related brain responses in the future.

REFERENCES

1. NÄÄTÄNEN, R. *et al.* 1978. Early selective-attention effect on evoked potential reinterpreted. Acta Psychol. **42**: 313–329.
2. SAMS, M. *et al.* 1985. Auditory frequency discrimination and event-related potentials. Electroencephal. Clin. Neurophysiol. **62**: 437–448.
3. TIITINEN, H. *et al.* 1994. Attentive novelty detection in humans is governed by pre-attentive sensory memory. Nature **372**: 90–92.
4. NÄÄTÄNEN, R. *et al.* 1987. The mismatch negativity to intensity changes in an auditory stimulus sequence. Electroencephal. Clin. Neurophysiol. Suppl. **40**: 125–131.
5. SCHRÖGER, E. *et al.* 1995. Time course of loudness in tone patterns is automatically represented by the human brain. Neurosci. Lett. **202**: 117–120.
6. KAUKORANTA, E. *et al.* 1989. Reactions of human auditory cortex to a change in tone duration. Hearing Res. **41**: 15–21.
7. AMENEDO E. & C. ESCERA. 2000. The accuracy of sound duration representation in the human brain determines the accuracy of behavioural perception. Eur. J. Neurosci. **12**: 2570–2574.
8. CSEPE, V. *et al.* 1995. Mismatch field to minor pitch changes in single and paired tones: neuromagnetic source localization from derived fields. Electroencephal. Clin. Neurophysiol. **44**: 204–210.
9. Hari, R. *et al.* 1992. Neuromagnetic mismatch fields to single and paired tones. Electroencephal. Clin. Neurophysiol. **82**: 152–154.

10. TERVANIEMI, M. *et al.* 1999. Pre-attentive discriminability of sound order as a function of tone duration and interstimulus interval: a mismatch negativity study. Audiol. Neuro-Otol. **4:** 303–310.

11. YABE, H. *et al.* 1997. Temporal window of integration revealed by MMN to sound omission. NeuroReport **8:** 1971–1974.

12. RÜSSELER, J. *et al.* 2001. Event-related brain potentials to sound omissions differ in musicians and non-musicians. Neurosci. Lett. **308:** 33–36.

13. NÄÄTÄNEN, R. *et al.* 2001. "Primitive intelligence" in the auditory cortex. TINS **24:** 283–288.

14. KOELSCH, S. *et al.* 1999. Superior attentive and pre-attentive auditory processing in musicians. NeuroReport **10:** 1309–1313.

15. TERVANIEMI, M. *et al.* 1999. Functional specialization of the human auditory cortex in processing phonetic and musical sounds: a magnetoencephalographic study. NeuroImage **9:** 330–336.

16. NÄÄTÄNEN, R. *et al.* 1997. Language-specific phoneme representations revealed by the electric and magnetic brain responses. Nature **385:** 432–434.

17. SHESTAKOVA, A. *et al.* 2002. Abstract phoneme representations in the left temporal cortex: magnetic mismatch negativity study. NeuroReport **13:** 1813–1816.

18. LANG, H. *et al.* 1990. Pitch discrimination performance and auditory event-related potentials. *In* Psychophysiological Brain Research. Vol. 1: 294–298. C.H.M. Brunia *et al.*, Eds. Tilburg. Tilburg University Press.

19. TERVANIEMI, M. *et al.* 1997. The musical brain: brain waves reveal the neurophysiological basis of musicality. Neurosci. Lett. **226:** 1–4.

20. TOIVIAINEN, P. *et al.* 1998. Timbre similarity: convergence of neural, behavioral, and computational approaches. Music Percept. **16:** 223–241.

21. TERVANIEMI, M. *et al.* 1993. Absolute pitch and event-related brain potentials. Music Percept. **10:** 305–316.

22. TERVANIEMI, M. *et al.* 2000. Harmonic partials facilitate pitch discrimination in humans: electrophysiological and behavioral evidence. Neurosci. Lett. **279:** 29–32.

23. TERVANIEMI, M. *et al.* 2000. Effects of spectral complexity and sound duration in complex-sound pitch processing in humans: a mismatch negativity study. Neurosci. Lett. **290:** 66–70.

24. NOVITSKI, N. *et al.* 2003. Frequency discrimination at different frequency levels as reflected by electrophysiological and behavioral indices. Submitted.

25. BRATTICO, E. *et al.* 2001. Context effects on pitch perception in musicians and non-musicians: evidence from ERP recordings. Music Percept. **19:** 1–24

26. ALHO, K. *et al.* 1996. Processing of complex sounds in the human auditory cortex as revealed by magnetic brain responses. Psychophysiology **33:** 369–375.

27. TERVANIEMI, M. *et al.* 2000. Lateralized automatic auditory processing of phonetic versus musical information: a PET study. Hum. Brain Mapp. **10:** 74–79.

28. KUSHNERENKO, E. *et al.* 2002. Maturation of the auditory event-related potentials during the first year of life. NeuroReport **13:** 47–51.

29. NEVILLE, H. *et al.* 1993. The neurobiology of sensory and language processing in language-impaired children. J. Cogn. Neurosci. **5:** 235–253.

30. SHAFER, V.L. *et al.* 2000. Maturation of mismatch negativity in school-age children. Ear Hear. **21:** 242–251.

31. KRAUS, N. *et al.* 1999. Speech-sound discrimination in school-age children: psychophysical and neurophysiologic measures. J. Speech Lang. Hear. Res. **42:** 1042–1060.

32. CHEOUR, M. *et al.* 1998. Maturation of mismatch negativity in infants. Int. J. Psychophysiol. **29:** 217–226.

33. ALHO, K. *et al.* 1990. Event-related brain potential of human newborns to pitch change of an acoustic stimulus. Electroencephal. Clin. Neurophysiol. **77:** 151–155.

34. CHEOUR-LUHTANEN, M. *et al.* 1996. The ontogenetically earliest discriminative response of the human brain. Psychophysiology **33:** 478–481.

35. HUOTILAINEN, M. *et al.* 2003. Memory-related responses from the human fetus. In preparation.

36. CHEOUR, M. *et al.* 1998. Development of language-specific phoneme representations in the infant brain. Nature Neurosci. **1:** 351–353.

37. CHEOUR, M. *et al.* 2002. Speech sounds learned by sleeping newborns. Nature **415:** 599–600.
38. PANG, E.W. *et al.* 1998. Mismatch negativity to speech stimuli in 8-month-old infants and adults. Int. J. Psychophysiol. **29:** 227–236.
39. MORR, M.L. *et al.* 2002. Maturation of mismatch negativity in typically developing infants and preschool children. Ear Hear. **23:** 118–136.
40. Gomes, H. *et al.* 1999. Electrophysiological evidence of developmental changes in the duration of auditory sensory memory. Dev. Psychol. **35:** 294–302.
41. KUJALA, T. & R. NÄÄTÄNEN. 2001. The mismatch negativity in evaluating central auditory dysfunction in dyslexia. Neurosci. Biobehav. Rev. **25:** 535–543.
42. KRAUS, N. 2001. Auditory pathway encoding and neural plasticity in children with learning problems. Audiol. Neuro-Otol. **6:** 221–227.
43. ESCERA, C. *et al.* 2000. Involuntary attention and distractibility as evaluated with event-related brain potentials. Audiol. Neuro-Otol. **5:** 151–166.
44. GUMENYUK, V. *et al.* 2001. Brain activity index of distractibility in normal school-age children. Neurosci. Lett. **314:** 147–150.
45. ALHO, K. *et al.* 1992. Intermodal selective attention II: effects of attentional load on processing of auditory and visual stimuli in the central space. Electroencephal. Clin. Neurophysiol. **82:** 356–368.
46. TREJO, L.J. *et al.* 1995. Attentional modulation of the mismatch negativity elicited by frequency differences between binaurally presented tone bursts. Psychophysiology **32:** 319–328.
47. KRAUS, N. *et al.* 1993. Mismatch negativity in the neurophysiologic/behavioral evaluation of auditory processing deficits: a case study. Ear Hear. **14:** 223–234.
48. KUSHNERENKO, E. *et al.* 2001. Central auditory processing of durational changes in complex speech patterns by newborns: an event-related brain potential study. Dev. Neuropsychol. **19:** 85–99.
49. CHEOUR, M. *et al.* 2002. Mismatch negativity shows that 3-6-year-old children can learn to discriminate non-native speech sounds within two months. Neurosci. Lett. **325:** 187–190.
50. SHESTAKOVA, A. *et al.* 2003. Event-related potentials associated with second language learning in children. Clin. Neurophys. **114:** 1507–1512.
51. HUOTILAINEN, M. *et al.* 1998. Combined mapping of human auditory EEG and MEG responses. Electroencephal. Clin. Neurophysiol. **108:** 370–379.
52. TERVANIEMI, M. 2003. Cortical representations for music sounds–EEG and MEG evidence. *In* Cognitive Neuroscience of Music. I. Peretz & R.J. Zatorre, Eds. :294–309. Oxford University Press. Oxford. In press.
53. RAUSCHECKER, J.P. 1998. Cortical processing of complex sounds. Curr. Opin. Neurobiol. **4:** 516–521.
54. SCHRÖGER, E., M. TERVANIEMI & M HUOTILAINEN. 2003. Bottom-up and top-down flows of information within auditory memory: electrophysiological evidence. *In* Psychophysics beyond Sensation: Laws and Invariants of Human Cognition. C. Kaernbach, E. Schröger & H. Müller, Eds. Erlbaum. Hillsdale, NJ.
55. ATKINSON, R.C. & R.M. SHIFFRIN. 1968. Human memory: a proposed system and its control processes. *In* The Psychology of Learning and Motivation: Advances in Research and Theory, 2nd Ed. K.W. Spence & J.T. Spence, Eds. Academic Press. New York.

Functional Imaging of Pitch Analysis

TIMOTHY D. GRIFFITHS

Auditory Group, Newcastle University, Newcastle upon Tyne, UK;

Wellcome Department of Imaging Neuroscience, University College, London, UK; and

Centre for the Neural Basis of Hearing, Cambridge University, UK

ABSTRACT: This work addresses the brain basis for the analysis of pitch and pitch patterns required for normal musical perception. Recent functional imaging experiments are consistent with a hierarchical scheme for the analysis of pitch. Mechanisms in the ascending auditory pathway to the primary auditory cortex allow the representation of the spectral and temporal features of individual notes required for the perception of their pitch. Converging experiments where pitch strength is manipulated in different ways suggest that there may be a "pitch center" in the lateral part of Heschl's gyrus, adjacent to the primary auditory area. The suggestion is that there is a representation in this area that correlates with the perception of pitch rather than a simple mapping of physical stimulus characteristics. The analysis of patterns of pitch such as melodies, as opposed to the pitch of individual notes, involves much more distributed processing in the superior temporal lobes and frontal lobes. Involvement of the frontal lobe in pitch pattern analysis may in part reflect whether subjects analyze the pitch patterns in order to carry out an output task.

KEYWORDS: pitch analysis; functional imaging; music; brain; PET; fMRI

INTRODUCTION

The processing of pitch and pitch patterns is a critical aspect of musical cognition, and recent functional imaging studies using positron emission tomography (PET) and functional MRI (fMRI) have shed considerable light on the neural bases for this. Other types of functional imaging using electroencephalography and magnetoencephalography are considered elsewhere in this volume. Functional imaging has begun to establish a fundamental framework that allows us to understand how the normal brain processes pitch patterns. Additionally, this framework provides a basis from which to consider what happens when the system goes wrong in patients with congenital and acquired disorders of musical cognition.

Address for correpondence: Professor T.D. Griffiths, School of Neurology, Neurobiology and Psychiatry, Newcastle University Medical School, Framlington Place, Newcastle upon Tyne, NE2 4HH, UK. Fax: 44 191 222 5227.

t.d.griffiths@ncl.ac.uk

Ann. N.Y. Acad. Sci. 999: 40–49 (2003). © 2003 New York Academy of Sciences.
doi: 10.1196/annals.1284.004

FUNCTIONAL IMAGING OF HUMAN PITCH ANALYSIS

Theoretical Considerations

Functional imaging experiments considered here assess the hemodynamic response to different types of auditory stimuli; they measure brain signals related to regional blood flow. The techniques allow measurement of activity in the whole brain with a spatial precision that can be less than 1 cm, but cannot be used to follow brain activity in the form of rapid temporal patterns where brain activity changes occur over a timescale of less than 1 second. In PET, regional cerebral blood flow is measured directly using a radioactive tracer, whereas in fMRI the blood oxygenation level dependent (BOLD) response is measured. Only recently has direct evidence demonstrated a direct link between the hemodynamic response and local brain activity. Logothetis[1] measured both the hemodynamic response using BOLD and local neural brain activity in the macaque in response to a visual stimulus. The BOLD response in macaque visual cortex is actually very similar to the response of human auditory cortex to sound stimulation.[2] Local brain activity was assessed by both local field activity, a measure of dendritic activity, and multiunit activity, a measure of axonal activity. The Logothetis work directly demonstrated a link between the BOLD response and neuronal activity. The best correlation was found between BOLD and the local field potential, suggesting that BOLD reflects the dendritic input to neurons rather than their axonal output. Recent work suggests a particular importance of glial cells in this coupling.[3] Regardless of the exact mechanism, the work is therefore of importance in demonstrating that the activity measured in functional imaging reflects the local rate of firing in populations of neurons. The hemodynamic techniques are not well suited to the demonstration of other types of neural code, especially temporal codes (although it is possible to gain some insight into any stages of processing where temporal firing patterns are converted into a rate code; see below). Much debate in pitch processing centers on the relative importance of different types of neural codes,[4] and it is important to realize this limitation.

Pitch is the name given to a percept rather than a particular stimulus structure; we perceive a given pitch whether the pitch is associated with a pure tone or a harmonic stimulus such as a musical instrument or voice. A critical aspect of musical appreciation is the analysis of perceptual patterns where the individual percepts are combined simultaneously to form chords and harmonic structure and sequentially to form melodies. In functional imaging, it is important to consider whether the hemodynamic responses that we observe correspond to the encoding of the stimulus properties or whether they correspond to neural correlates of the conscious perception[5] that is called pitch. In many experiments, it is very difficult to tell. For example, in a recent series of experiments we have varied the temporal regularity of noise in order to vary its pitch[6–8] and measured the corresponding change in the hemodynamic response. In these experiments there is no absolute way of interpreting the measure of neural activity as a correlate of the stimulus properties or as a correlate of the percept that is generated. All that can be said is that the mean local firing rate increases in certain areas in response to the stimulus manipulation. Auditory neuroscience is a little behind visual neuroscience in this respect. In visual neuroscience, for example, a number of functional imaging experiments have looked at the brain response to bistable percepts such as binocular rivalry,[9] where fixed stimulus properties can lead to a varying percept. These influential studies allow inference about the neural correlates of perception. In the case of pitch, the development of such stimuli for imaging experiments could lead to important insights in the future.

Processing of Stimuli Associated with Pitch in the Ascending Auditory Pathway

Recent experiments have identified structures in the ascending auditory pathway that respond to systematic variation in the regularity of the stimulus,[7] which produces a change in pitch. This work was made possible by the technical advances of sparse imaging and cardiac triggering. Sparse imaging[2] is a way of overcoming the considerable noise that is produced by the MRI scanner during measurement of the BOLD response.[10] The technique uses the slow build-up of the BOLD response to advantage. Essentially, infrequent scans are carried out at intervals of the order of 10 seconds. Because of the slow build-up of the BOLD response (typically 10 s to peak in primary auditory cortex), brain activity measured in this way will correspond to the period before scanning, when the test stimuli were presented in quiet (without "contaminating" scanner noise). Apart from the improvement in measurement of the brain signal, there are good biological reasons to measure brain responses without the additional noise produced during scanning. The presence of the additional noise will change the nature of the listening task into one for which the substrate may not be the same as the detection of the stimulus in silence. The second advance, cardiac triggering, is a technique that is of particular benefit for imaging of the brain stem.[11] The technique is a way of overcoming the considerable movement of the brain stem caused by pulsation of the basilar artery by always scanning at the same point in the cardiac cycle.

Using sparse imaging and cardiac gating, functional activation can be demonstrated in the whole of the human auditory brain including the brain stem.[7] Activation due to sound was demonstrated in the cochlear nucleus, lateral lemniscus, and inferior colliculus within the brain stem, the medial geniculate body, and auditory cortex. The sound stimuli used in this study were regular-interval sounds in the form of iterated rippled noise.[12] These noises are created by using a delay-and-add algorithm that produces regularity in the stimulus and a pitch. The strength of the pitch corresponds to the regularity of the stimulus. Analysis was carried out on each of the structures of the ascending auditory pathway to test the hypothesis that there is a relation between local brain activity, measured indirectly by the BOLD signal, and temporal regularity in the stimulus. The contrast between the regular-interval sound and noise matched in intensity and passband was significant in both cochlear nuclei at the $P < 0.05$ level, whereas the same contrast was significant in both inferior colliculi at the $P < 0.005$ level. The study therefore represents a demonstration of an increase in the local mean firing rate as a function of the stimulus regularity as early as the cochlear nucleus, with a more significant relation in the inferior colliculus. This can be interpreted in two possible ways. One possibility is that there may be a subpopulation of cells in the cochlear nucleus that increase their mean firing rate in response to particular temporal regularities corresponding to particular pitches. However, neurophysiologic studies in the guinea pig[13] have not demonstrated such a selective response; responses to temporal regularity in the onset-chopper cell type are selective, but the selectivity is demonstrated in the temporal firing pattern rather than the mean rate. A second possibility in the cochlear nucleus is that the mean firing rate in a larger population of neurons increases as a function of *synchronization* of the local networks of neurons due to the regularity of the stimulus. Modeling studies[14] were motivated by a need to study cortical processing, but used networks of excitatory and inhibitory neurons with conventional Hodgkin-Huxley neural dynamics and a pattern of interconnections that could plausibly be applied to brainstem nuclei. The studies demonstrated tight coupling of mean activity levels and synchronization that was not sensitive to large changes in the model parameters. On the basis of the absence of a candidate cell in the cochlear nucleus with a tuned rate response to regular interval sound, the synchronization mechanism for increasing the local BOLD response would seem more plausible. In the case of either possible mechanism in the cochlear nucleus, the more significant relation in the inferior colliculus points to a stabilized neural representation of regularity at that level that is based on a local rate code. Such a representation is predicted by physiologic models such as that of Langner[15] and the psychophysical auditory image model (AIM) of Patterson *et al.*[16] However, the brain-stem fMRI studies lack the spatial precision to define maps of temporal stimulus properties predicted by the Langner model.

The brain-stem study demonstrated the coding of a temporal stimulus property as a local neural rate code. This is not to dismiss the relevance of spectral encoding to the perception of pitch, especially at low frequencies, where the presence of the resolved lower harmonics in harmonic sounds increases pitch salience. In the brain stem, demonstration of frequency mapping (tonotopy) suffers from the same problem as demonstration of the mapping of regularity: the lack of anatomic resolution. Melcher and colleagues at Massachusetts General Hospital have seen trends consistent with tonotopic organization in the inferior colliculus (Melcher, unpublished observation), but no published study to date has demonstrated any systematic mapping. The brain-stem mapping of tonotopy in mammalian neurophysiologic studies by Langner and others is one form of indirect argument for its existence in humans. A much stronger argument is the tonotopy that has been demonstrated in the human cortex. This could not occur without preservation of tonotopic mapping in the human brain stem.

Processing of Pitch in the Auditory Cortex

In terms of the spectral structure of sound relevant to pitch, a number of studies have addressed the question of tonotopic mapping in the auditory cortex. In humans, the auditory cortex is located in the superior temporal plane, the superior surface of the temporal lobe within the Sylvian fissure (FIG. 1). The superior temporal plane is conveniently shown in a tilted axial section like that in FIGURE 1. The primary auditory cortex is located in the region of Heschl's gyrus (HG), running laterally and anteriorly in the superior temporal plane. There is a degree of macroscopic variability in that there might be one, two, or even three HG, and the number can differ on either side. Detailed human anatomic studies confirm that primary auditory cortex, defined on the basis of microscopic structure (cytoarchitectonics), is related to the medial part of HG (or the most anterior HG if there are more than one). However, there is considerable anatomic variation of the cytoarchitechtonic area with respect to the macroscopic boundaries, whereby as little as 30% or as much as 80% of the gyrus can be taken up by primary cortex.[17] It is important to bear this in mind when considering imaging studies of the cortex. When these are able to look at activation in

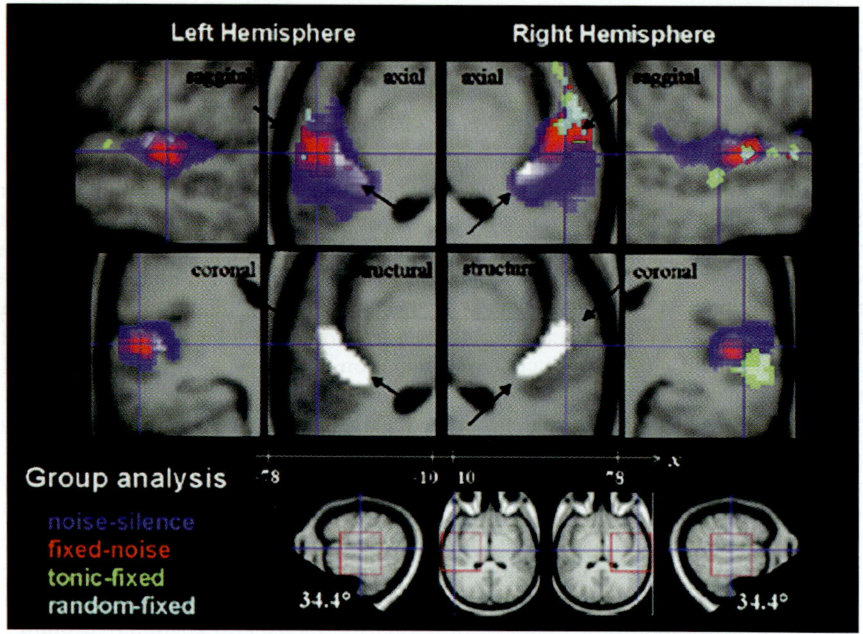

FIGURE 1. fMRI activation for contrasts between noise stimuli with different temporal regularity and pitch strength and with different pitch patterns. Group data for nine subjects. The contrasts are rendered onto the average structural image of the group. *Blue*: activation in response to noise bursts (vs silence); *Red*: differential activation in response to notes with fixed pitch (vs noise bursts); *Green*: differential activation in response to tonic melodies (vs fixed pitch); *Cyan*: differential activation in response to random melodies. The *white area* shows the mean position of Heschl's gyrus for the group. *Arrows* show the midline of Heschl's gyrus separately in each hemisphere. The position and orientation of the sections are illustrated in the bottom panels of the figure. The "axial" section is tilted by 0.6 radians (or 34.4 degrees) relative to the horizontal plane to show the entire surface of the temporal lobe in one plane. The other sections are sagittal and coronal with respect to the surface of the temporal lobe. The sagittal sections show front to the left for the left hemisphere and front to the right for the right hemisphere, that is, they are being viewed from outside the brain volume. Reproduced from Ref. 8 with permission.

individual subjects, lack of consistency between subjects may partly reflect the fact that such consistency can only be defined with respect to macroscopic landmarks, rather than with respect to the functional auditory areas that are not accurately delineated by such landmarks.

Lauter *et al.*[18] carried out the first study to examine tonotopic organization in the human cortex using PET. A comparison was made in a group study between activation based on pure tones at 500 Hz and 4 kHz. Both tones produced activation in the superior temporal plane, with more lateral activation being produced by the lower-frequency tone. In a later PET study, Lockwood and colleagues[19] found two foci of activation in the left hemisphere (in medial and lateral HG) for a 4-kHz pure tone presented to the right ear at the 90-dB hearing level. For a 500-Hz tone, a single focus of activation was demonstrated in the lateral part of HG, at the same point as the lateral focus for the 4-kHz tone. The distinct patterns produced by the two tones provide evidence for a tonotopic mapping in the superior temporal plane. However, the precise pattern of mapping is difficult to demonstrate in PET experiments based on group data where spatial smoothing rarely exceeds a filter width at half maximum of 10 mm.

Some studies have employed fMRI to investigate tonotopic mapping in the cortex (e.g., Refs. 20 and 21) where increased spatial resolution and the ability to carry out individual analyses are a particular advantage. A disadvantage of fMRI is the scanner noise; the studies of both Wessinger *et al.* and Talavage *et al.* used "epoch mode" designs where there is continuous acquisition of data (and therefore scanner noise) during presentation of the stimuli of interest. In the Wessinger *et al.* study, harmonic tones with most spectral energy at 55 Hz or 880 Hz were presented to both ears of subjects. A consistent pattern of activation was demonstrated in the left hemispheres of subjects, where the activation due to the high frequency tone was more medial in the superior temporal plane than the activation due to the low frequency tone, but the same consistency was not observed in the right hemispheres of the subjects. Talavage *et al.* used a variety of stimuli where the spectral distribution could be varied (pure tones at 650 Hz and 2.5 kHz, 10-Hz amplitude-modulated tones with same carrier frequencies, and broadband stimuli [AM noise and music] that were low- or high-pass filtered). For each type of stimulus type there was a low- and a high-frequency condition. Areas were defined where there was a greater response to either the high or the low frequency. A low frequency area was identified in HG, whereas high-frequency areas on HG were identified both lateral and medial to it. Talavage *et al.* proposed that the organization of responses along HG is consistent with having two adjacent tonotopic maps with mirror reversal between them at the low frequency point. Furthermore, they proposed that the medial and lateral areas correspond, respectively, to areas A1 and R in the macaque.[22,23] A similar mirror reversal of tonotopy at low frequency is seen between A1 and R in the macaque studies.

In terms of the representation of the temporal regularity of the stimuli, an early study with regular-interval sounds used PET to identify cortical areas showing a relation between stimulus regularity and local brain activity.[6] Activation was shown in HG as a function of the temporal regularity of the stimulus. On the basis of that study it is not possible to say whether the peak activation as a function of stimulus regularity was in primary cortex in medial HG or secondary cortex in lateral HG, and PET did not allow examination of individual data. A recent fMRI group analysis of the cortical BOLD response to temporal regularity[8] has demonstrated bilateral peak activation as a function of temporal regularity in secondary auditory cortex in lateral HG (FIG. 1). This was demonstrated in both group analyses and in eight of the nine individual analyses.

Evidence for a "Pitch Center" in the Secondary Auditory Cortex

The studies of the effect of temporal regularity on cortical responses beg the question: what is represented in lateral HG when the temporal regularity of the stimulus is varied? As argued above, it is not possible to make a definitive statement on the basis of the data. The activation corresponds to a population rate code that might be related to temporal regularity in the stimulus or to the percept of pitch. Evidence for a perceptual "pitch center" comes from another experiment in which pitch salience is varied in a different manner. In an fMRI experiment, Penagos and colleagues[24] showed that activation in lateral HG due to harmonic stimuli within a particular passband increased when the harmonics were resolved and the pitch salience was greater, than when the harmonics were unresolved and the pitch salience was weaker. Stimulus regularity was the same in both cases, as both were exactly periodic stimuli.

An argument about a human homologue of area R in the macaque in lateral HG has already been made on the basis of tonotopic data.[21] Patterson and colleagues[8] also speculate about the existence of such a homologue and suggest that a neural correlate of the percept of pitch may exist in this center.

Distributed Temporal Lobe Mechanisms for the Processing of Pitch Patterns

The work considered so far suggests that processing of the pitch of individual sounds depends on hierarchical mechanisms. Stabilized representations of stimulus properties relevant to pitch are constructed in the brain stem and are used to form a cortical representation corresponding to the percept of pitch. Musical cognition depends on the processing of patterns of pitch presented either simultaneously or sequentially. Melodies represent sequential pitch patterns within a temporal window of seconds or tens of seconds rather than the temporal window of milliseconds relevant to the processing of temporal regularity within individual sounds. This section considers such pitch patterns.

In the work using regular-interval sounds with associated pitch as just described,[8] the individual sounds containing regularity were presented in sequences with both fixed pitch and as two types of pitch pattern: a random variation of pitch and a tonal melody. Comparison between the sequences containing the pitch variation produced a pattern of cortical activation that was distinct from the pattern produced by the fixed pitch. A striking negative finding here was the absence of any significant activation in the brain-stem nuclei. The processing of pitch patterns within this long temporal window involves the cortex, and the network of areas activated is distinct from the region of the primary auditory cortex. Bilateral activation was demonstrated in the posterior superior temporal lobe (in the region of the planum temporale, PT, and adjacent superior temporal gyrus) and in the anterior superior temporal lobe in the planum polare, PP (FIG. 1). This is the first level of pitch processing at which asymmetries emerge, with a greater extent and significance for the right-sided activations. The other feature to emerge at this level of pitch processing is a much greater variation between individuals. The four areas demonstrated in the group analysis in this experiment were similar to four areas demonstrated in a previous interaction analysis in a PET study that was designed to demonstrate areas interested in "long-term" temporal structure.[6] However, the previous study did not demonstrate the asymmetry that was apparent in the recent fMRI study. The right lateralization during the processing of pitch pattern would be consistent with studies of melody perception and imagery.[25,26]

From a musical perspective, a surprising aspect of the fMRI cortical study was the absence of any marked differences between the random pitch and melody conditions. Although the melodies were novel and would not have semantic associations, they were tonal melodies with a contour or long-term structure not present in the random pitch condition. Cognitive neuropsychology suggests that the processing of the "local" structure of pitch sequences has a different basis from the processing of contour.[27] Of possible relevance here is the absence of any task in the fMRI experiment that was primarily concerned with pitch perception. In experiments where comparison tasks for pitch sequences are employed,[25,28] frontal activation is seen that does not occur in the experiments considered above. It is conceivable that differences between the cortical processing of different types of pitch pattern may only emerge when the brain has to make use of them.

The Processing of Distinct Pitch Dimensions in the Cerebral Cortex

A recent experiment has addressed the question of whether the dimensions of pitch have distinct representations in the human brain (Warren *et al.*, this volume). Much of neuroscience treats pitch as a unitary concept that increases along a single dimension. This experiment was based on the manipulation of two pitch dimensions according to the helix model of pitch perception.[29] On a keyboard, the dimension of pitch chroma corresponds to note letter, which repeats in a cycle every octave. The other dimension of pitch, pitch height, corresponds to the distance towards the right on the keyboard. Together, these dimensions result in a perceptual helix that is able to represent, among other phenomena, the perceptual closeness of notes with the same chroma that are an octave apart in pitch height. The experiment used harmonic stimuli. The height of notes was manipulated by changing the fine structure of the spectrum of the stimulus, whereas chroma was changed independently by manipulating the fundamental frequency. As in previous experiments using regular-interval sounds, areas were demonstrated in lateral HG that responded to stimuli with associated pitch compared to noise that was matched in spectrum. Additionally, specific areas were demonstrated in PT posterior to HG that were responsive to changes in pitch height (but not to changes in chroma), whereas areas were demonstrated in the PP anterior to HG that were responsive to changes in pitch chroma (but not to changes in height). The data therefore support the idea that the musical concepts of height and chroma have distinct representations in the human brain. Pitch height is an important aspect of object separation recognition (for example, a soprano or tenor singing the same melody will differ in pitch height), whereas pitch chroma is important for the formation of melody. Therefore, the data also suggest the possibility that early stages in object separation are instantiated in the posterior temporal lobe, whie the anterior temporal lobes are engaged in the analysis of auditory informational streams. A number of other experiments (reviewed in Ref. 30) also point to early acoustic-object processing in the posterior temporal lobe.

CONCLUSIONS

These studies suggest that the analysis of pitch in music depends on a number of stages. The spectral and temporal features of sounds relevant to pitch are encoded in the brain stem, whereas a neural correlate of the conscious perception of pitch exists in areas of auditory cortex distinct from the primary auditory cortex. Longer time-scale patterns of pitch are processed in networks including areas in the temporal lobes (distinct from the primary and secondary areas) and in the frontal lobes.

ACKNOWLEDGMENTS

My work is supported by the Wellcome Trust. The pitch helix work involved Jason D. Warren and members of the Centre for the Neural Basis of Hearing (CNBH), Cambridge (Roy Patterson and Stefan Uppenkamp). The work using regular interval sounds was also carried out with members of the CNBH in addition to Ingrid Johnstrude, Cambridge University.

REFERENCES

1. LOGOTHETIS, N.K. *et al.* 2001. Neurophysiological investigation of the basis of the fMRI signal. Nature **412:** 150–157.
2. HALL, D.A. *et al.* 1999. "Sparse" temporal sampling in auditory fMRI. Hum. Brain Mapp. **7:** 213–223.
3. PARRI, R. & V. CRUNELLI. 2003. An astrocyte bridge from synapse to blood flow. Nature Neurosci. **6:** 5–6.
4. PLACK, C.J. *et al.* 2003. Pitch Perception. Springer Handbook of Auditory Research. In press.
5. FRITH, C., R. PERRY & E. LUMER. 1999. The neural correlates of conscious experience: an experimental framework. Trends Cognit. Sci. **3:** 105–114.
6. GRIFFITHS, T.D. *et al.* 1998. Analysis of temporal structure in sound by the human brain. Nature Neurosci. **1:** 421–427.
7. GRIFFITHS, T.D. *et al.* 2001. Encoding of the temporal regularity of sound in the human brainstem. Nature Neurosci. **4:** 633–637.
8. PATTERSON, R.D. *et al.* 2002. The processing of temporal pitch and melody information in auditory cortex. Neuron **36:** 767–776.
9. LUMER, E.D., K.J. FRISTON & G. REES. 1998. Neural correlates of perceptual rivalry in the human brain. Science **280:** 1930–1934.
10. RAVICZ, M.E., J.R. MELCHER & N.Y. KIANG. 2000. Acoustic noise during functional magnetic resonance imaging. J. Acoust. Soc. Am. **108:** 1683–1696.
11. GUIMARES, A.R. *et al.* 1998. Imaging subcortical activity in humans. Hum. Brain Mapp. **6:** 33–41.
12. YOST, W.A., R. PATTERSON & S. SHEFT. 1996. A time domain description for the pitch strength of iterated rippled noise. J. Acoust. Soc. Am. **99:** 1066–1078.
13. WIEGREBE, L. & I.M. WINTER. 2001. Temporal representation of iterated rippled noise as a function of delay and sound level in the ventral cochlear nucleus. J. Neurophysiol. **85:** 1206–1219.
14. CHAWLA, D., E.D. LUMER & K.J. FRISTON. 1999. The relationship between synchronization among neuronal populations and their mean activity levels. Neural Comput. **11:** 1389–411.
15. LANGNER, G. 1992. Periodicity encoding in the auditory system. Hear. Res. **60:** 115–142.
16. PATTERSON, R.D., M.H. ALLERHAND & C. GIGUERRE. 1995. Time-domain modeling of peripheral auditory processing: a modular architecture and a software platform. J. Acoust. Soc. Am. **98:** 1890–1894.
17. MOROSAN, P. *et al.* 2001. Human primary auditory cortex: cytoarchitechtonic subdivisions and mapping into a spatial reference system. NeuroImage **13:** 684–701.
18. LAUTER, J.L. *et al.* 1985. Tonotopic organisation in the human auditory cortex revealed by positron emission tomography. Hear. Res. **20:** 199–205.
19. LOCKWOOD, A.H. *et al.* 1999. The functional anatomy of the normal human auditory system: responses to 0.5 and 4.0kHz tones at varied intensities. Cereb. Cortex. **9:** 65–76.
20. WESSINGER, C.M. *et al.* 1997. Tonotopy in human auditory cortex examined with functional magnetic resonance imaging. Hum. Brain Mapp. **5:** 18–25.
21. TALAVAGE, T.M. *et al.* 2000. Frequency-dependent responses exhibited by multiple regions in human auditory cortex. Hear. Res. **150:** 225–244.

22. KAAS, J.H. & T.A. HACKETT. 1998. Subdivisions of auditory cortex and levels of processing in primates. Audiol. NeuroOtol. **3:** 73–85.
23. MERZENICH, M.M. & J.F. BRUGGE. 1973. Representation of the cochlear partition on the suprior temporal plane of the macaque monkey. J. Neurophysiol. **24:** 193–202.
24. PENAGOS, H., A.J. OXENHAM & J. MELCHER. 2003. Effects of harmonic resolvability on the cortical activity produced by complex tones. Assoc. Res. Otolaryngol.
25. ZATORRE, R.J., A.C. EVANS & E. MEYER. 1994. Neural mechanisms underlying melodic perception and memory for pitch. J. Neurosci. **14:** 1908–1919.
26. ZATORRE, R.J. *et al.* 1996. Hearing in the minds ear- a PET investigation of musical imagery and perception. J. Cognit. Neurosci. **8:** 29–46.
27. LIEGEOIS-CHAUVEL, C. *et al.* 1998. Contribution of different cortical areas in the temporal lobes to music processing. Brain **121:** 1853–1867.
28. GRIFFITHS, T.D. *et al.* 1999. A common neural substrate for the analysis of pitch and duration pattern in segmented sound? NeuroReport **18:** 3825–3830.
29. KRUMHANSL, C.L. 1990. Cognitive Foundations of Musical Pitch. Oxford University Press. New York.
30. GRIFFITHS, T.D. & J.D. WARREN. 2002. The planum temporale as a computational hub. Trend Neurosci. **25:** 348–353.

Music Agnosia and Auditory Agnosia

Dissociations in Stroke Patients

LUIGI A. VIGNOLO

Neurology Department, University of Brescia, 25125 Brescia, Italy

ABSTRACT: A review and an experimantal study were carried out in search of dissociations between the recognition of music (music agnosia) and that of environmental sounds (auditory agnosia) in stroke patients. The review focused on 45 adequately studied cases published since 1883. The experimental study consisted of administering standard tests of music and environmental sound recognition to 40 unselected patients with unilateral stroke. Among case reports, music was selectively impaired more frequently than environmental sounds, whereas the reverse occurred in the experimantal study. In this, right hemisphere lesions tended either to disrupt the apperception of environmental sounds, sparing music entirely, or to disrupt both environmental sounds and melody, sparing rhythm, whereas left hemisphere lesions tended to spare melody and to disrupt rhythm, either selectively or in association with the semantic identification of environmental sounds.

KEYWORDS: music agnosia; auditory agnosia; sound recognition disorders; hemispheric lateralization; brain damage; stroke

INTRODUCTION

The auditory receptive aspect of amusia, music agnosia, has been studied over the years essentially in those rare cases of musicians who happened to be so unlucky as to incur both a brain lesion and a music-minded neurologist. However, music recognition has also been examined, although in a somewhat ancillary fashion, in a number of non-musicians in whom damage to the brain brought about other clinically striking auditory imperception disorders such as central deafness, pure word deafness, and nonverbal auditory agnosia. The aim of clinicians confronted with these patients was to define the quality of the disorder more precisely, that is, to determine whether the auditory imperception was confined to one specific set of stimuli (spoken language, environmental sounds, music) or extended to more than one set. Dissociations between music and the other categories of auditory stimuli were detected in some cases. This is an important finding, as it may help to understand to what extent music recognition should be viewed as an autonomous ability within the auditory cognitive domain. The independence of music with respect to recognition of spoken language, a matter of debate in the past, seems well established now, particularly after the research of the Montreal group.[1–3] By contrast, the specificity of mu-

Address for correspondence: Dr. Luigi A. Vignolo, Neurology Department, University of Brescia, Spedali Civili, 25125 Brescia, Italy. Voice: +39.030.3995632; fax: +39.030.394313. vignolo@numerica.it

Ann. N.Y. Acad. Sci. 999: 50–57 (2003). © 2003 New York Academy of Sciences. doi: 10.1196/annals.1284.005

sic with respect to other nonverbal auditory stimuli, such as environmental sounds, is still open to investigation. Evidence of such specificity rests on cases showing a dissociation between these two types of stimuli, music and environmental sounds. This paper reports the findings of a review and of an experimental study carried out in search of such dissociation. The review included the clinical case reports of auditory perceptual disorders due to cerebrovascular lesions of the cerebral hemispheres published since 1883. The experimental study consisted of administering a battery of standard tests of music and environmental sound recognition to a series of patients with unilateral stroke.

REVIEW OF CASE REPORTS FROM 1883–2001

We reviewed 175 case reports of auditory perceptual disorders due to acquired lesions (of all etiologies) of the cerebral hemispheres published from 1883–2001.[a] Three cases of lesions of the brain stem and one of section of the corpus callosum were not included. Of the total number of 175 case reports under scrutiny, 130 were excluded due to left handedness (14), etiology other than stroke (35), lack of post-mortem and/or imaging data (18), and inadequate examination of music and/or environmental sounds (61). We were left with 45 cases of circumscribed vascular damage to one or both hemispheres with sufficient evidence concerning both music and environmental sound recognition. Of these, 32 were not dissociated: both music and environmental sounds were spared in 5 and impaired in 27. The remaining 13 cases (28.9%) were dissociated. Music was found to be selectively impaired in 11 cases and environmental sounds in 2, as shown in TABLE 1.

In a general sense, the dissociations (TABLE 1) suggest some specificity for music recognition with respect to environmental sounds. However, they do not allow us to draw any definite conclusions. Apart from the inevitable quota of subjective arbitrariness inherent in a retrospective analysis, clinical case reports are often objectively heterogeneous. Individual differences in examination and evaluation make comparisons difficult. In the reviewed series, musical stimuli varied from familiar tunes to more complex melodic patterns; preserved enjoyment of music meant intact recognition to some authors, but not to others; inability to recall the name of a familiar song was interpreted as a recognition defect by some investigators, but not by others; etc. Moreover, the neuroradiologic or anatomic evidence of these cases yielded little information about the relations between each particular auditory disorder and the damaged hemisphere, because lesions were mostly bilateral. The need for further research, carried out with well-defined behavioral criteria, led to the following experimental study.

EXPERIMENTAL STUDY

This study was undertaken to search for test-wise dissociations between music agnosia and auditory agnosia in a prospective series of unilateral stroke patients.

[a]In 1883, Wernicke and Friedlander[4] published their case of "deafness following bilateral temporal damage" with autopsy.

TABLE 1. Published case reports of stroke patients (n = 13) showing dissociation between music agnosia and auditory agnosia, divided according to damaged hemisphere and clinical diagnosis

	Damaged Hemisphere		
	Left	Right	Both
M+ ES−			Motomura *et al.*, 1986 (AA)[12]
			Godefroy *et al.*, 1995 (AA)[13]
M− ES+	Eustache *et al.*, 1990 case 1 (WD)[14]	Mazzucchi *et al.*, 1982 (MA)[15]	Kanshepolsky *et al.*, 1973 (WD)[18]
		Steinke *et al.*, 2001 (MA)[16]	Shindo *et al.*, 1981 (WD)[19]
		Griffiths *et al.*, 1997 (MA)[17]	Coslett *et al.*, 1984 (WD)[20]
			Tanaka *et al.*, 1987 (WD)[21]
			Peretz *et al.*, 1994 case 1 (MA)[22]
			Peretz *et al.*, 1994 case 2 (MA)[22]
			Peretz *et al.*, 1997 (MA)[3]

M, music; ES, environmental sounds; +/-, unimpaired/impaired.
Clinical diagnosis: AA, auditory agnosia; MA, music agnosia; WD, word deafness.

Subjects

We examined 40 unselected right-handed patients hospitalized in our Neurology Department after their first unilateral hemispheric stroke and harboring a small or medium-sized (diameter ≤30 mm) lesion on computerized tomograpy (CT) scan. Testing took place 3–12 weeks poststroke. Twenty patients had a lesion in the right hemisphere and 20 in the left. Six of these scored under cut-off on the Token Test. Mean age was 61.0 years for the right brain-damaged group and 65.9 years for the left (SD 13.02 vs 7.26; range 40–49 vs 49–79); average schooling was 7.8 vs 6.2 years (range 5–18 vs 5–13). There were 13 males and 7 females in both groups. None of the patients was a professional musician. Musical education, evaluated by Grison's 6-point scale[5] modified by Benton,[6] was not significantly different in the right and left groups (overall mean 2.8, range 1–5). No patient had clinically evident amusia, auditory agnosia, or word deafness. None complained of either hypoacusis in everyday life or auditory problems during the testing sessions. All patients could hear the ticking of a stopwatch held at about 2 cm from either ear. All of them gave informed consent to participate in this study.

Procedures

Auditory stimuli were presented at a comfortable volume through earphones. Two sets of stimuli, music and environmental sounds, were administered.

Music Recognition

The melodic and temporal organization subtests of the Montreal Musical Battery, as adopted by Ayotte *et al.*,[7] were used. In this battery, recognition of melodic organization is evaluated by three subtests, that is, Scale, Contour, and Interval, each of which is characterized by a particular type of pitch variation. The first variation consists of creating a scale-violated comparison melody by modifying the pitch of one note of the original melody, so as to bring it out of scale. This change is particularly salient because the pitch sounds out of tune. The second variation consists of creating a contour-violated melody, changing the pitch of one note, while maintaining the original key; the pitch direction of the adjacent intervals is thus modified. The third variation consists of creating a contour-preserved or interval-violated comparison melody by modifying the pitch of a note to the same extent (in terms of semi-tone distance), in keeping with the original key and contour.

Each subtest consists of 30 pairs of melodies, a target one followed by a comparison one. An equal number of identical melodies and manipulated melodies is included in the three subtests. The task is same-different recognition: the subject has to judge if the comparison melody is the same as or different from the target melody.

Recognition of temporal organization is evaluated by two subtests, that is, Rhythm and Meter. The Rhythm subtest is made-up like the melodic organization subtests, except that manipulation of the comparison melody consists of changing the duration values of two adjacent tones, keeping the meter and total number of sounds identical. The Meter subtest consists of 30 sequences recorded in a random order, half of them written in double meter and half in triple meter. The subject is informed that he/she will be hearing waltzes and marches that he/she has to discriminate along this dimension. In all tests, each correct answer scores 1 point (maximum score 30).

Environmental Sound Recognition

The Environmental Sound Recognition Test, developed by Schnider *et al.*,[8] was employed. The test consists of 25 items from 5 semantic categories: human non-verbal sounds (e.g., crying baby), man-made non-living sounds (e.g., hammer), animal sounds (e.g., cat meowing), natural non-living sounds (e.g., thunder), and technical sounds (e.g., car engine). While hearing a sound sequence (e.g., a crying baby), the subject is shown four vertically arranged photographs: the actual source of the sound, an acoustic foil (would produce a similar sound but is semantically different, e.g., a cat), a semantic foil (which would produce a different sound, but belongs to the same semantic category, e.g., a laughing baby), and an unrelated foil (e.g., a car). The subject is requested to point to the picture showing the source of the sound being heard. Each correct answer scores 1 point (maximum score 25). The types of error are considered to reflect impairment of different recognition processes: acoustic errors are assumed to reflect an apperceptive-discriminative disorder, whereas semantic errors are assumed to reflect an associative-semantic defect.

Normative values with cut-off scores for all tests were available for 40 control subjects matched for age, years of schooling, and Grison scores. The statistical significance of comparisons was analyzed by the Mann-Whitney U test.

Results

Group performances, shown in FIGURES 1 and 2, indicate that right brain-damaged subjects, as a group, perceived the melodic aspects of music significantly worse than did the left brain-damaged ones, whereas the reverse occurred with the rhythmic aspects. (In the Interval and Meter subtests, this difference fell short of significance.) The environmental sound test was performed less well by right brain-damaged patients than by the left, and the types of error were different: right brain-damaged subjects made significantly more acoustic errors, whereas left brain-damaged subjects tended to make more semantic errors.

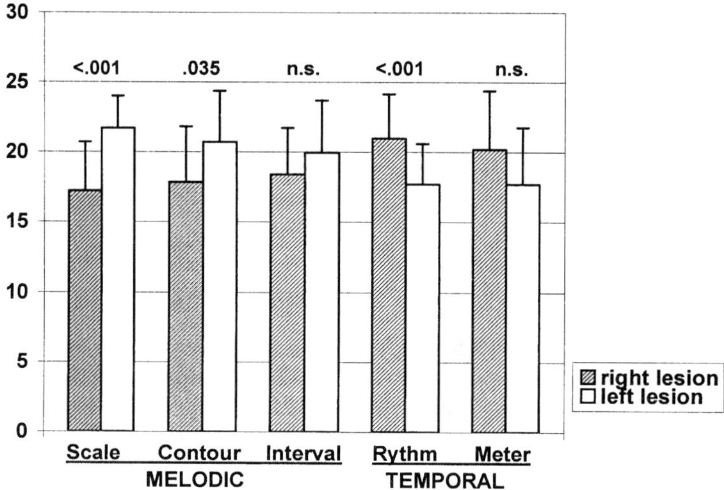

FIGURE 1. Mean scores (and SD) obtained by right and left brain-damaged patients on the Melodic and Temporal subtests of the Montreal Musical Battery.

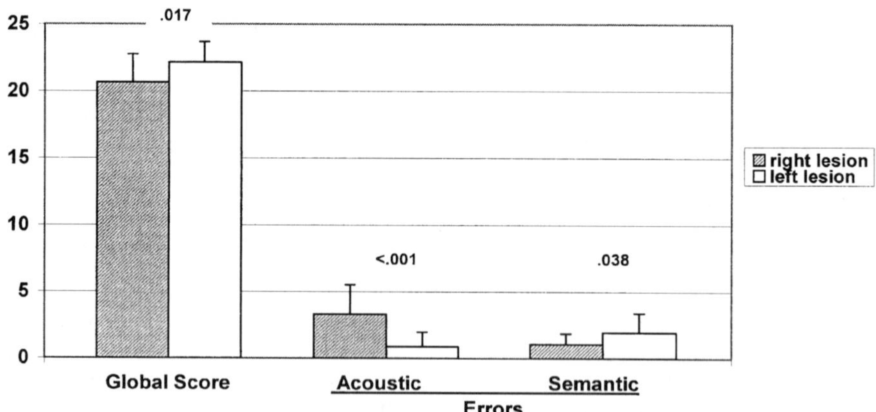

FIGURE 2. Mean global scores and mean types of errors of right and left hemisphere-damaged patients on the Environmental Sound Recognition Test.

TABLE 2. Unilateral stroke patients ($n = 19$) showing dissociation between music and environmental sound recognition tests

Type of dissociation	Number of patients	Music		Environmental sounds		
		Melodic aspects	Temporal aspects	Score	Prevalent error type[a]	Prevalent side of lesion
I	8	Good	Good	**Poor**	A	Right[b]
II	2	**Poor**	Good	**Poor**	A	Right
III	3	Good	**Poor**	Good	—	Left
IV	6	Good	**Poor**	**Poor**	S	Left[c]

good = above cut-off; **poor** = under cut-off.
[a]A, acoustic; S, semantic.
[b]7 patients with right lesion; 1 patient with left lesion.
[c]5 patients with left lesion; 1 patient with right lesion.

Analysis of performances by individual subjects indicated some kind of dissociation in 19 of 40 patients. TABLE 2 discloses that in 8 patients music was entirely spared, in contrast to impairment of environmental sounds.

Seven of these patients had a lesion in the right hemisphere and only one in the left. Six of 7 patients with right-sided lesions made more acoustic than semantic errors in sound recognition, whereas the one patient with a left-sided lesion made more semantic errors.

Contrary to what we expected on the basis of the review of single cases, the reverse clear-cut dissociation, that is, music impaired *in toto* and environmental sounds spared, was not found. Dissociations appeared only when the distinction between the melodic aspect (including Scale, Contour, and Interval subtests) and the temporal aspect (including Rhythm and Meter) of music was taken into account.

Impairment on melodic tasks (at least 2 of 3) and on the environmental sound test (with predominantly acoustic errors) in contrast to unimpaired temporal tasks was found in 2 patients, both with a lesion of the right hemisphere. Temporal tasks (Rhythm, in particular) were selectively impaired, whereas both melodic tasks and environmental sounds were spared in 3 patients, all with left hemisphere lesions. Temporal tasks were impaired in association with environmental sounds (with predominantly semantic errors), in contrast to unimpaired melodic tasks in 6 patients, all of whom had a lesion of the left hemisphere.

DISCUSSION

When both music and environmental sound agnosias were investigated by behavioral tests, a double dissociation between the type of auditory stimulus and the hemispheric side of the lesion was found in almost half (19 of 40, 47.5%) an unselected series of unilateral stroke patients. However, in most of them (11 of 19), the dissociation did not involve music in its entirety, but it did involve either its melodic or its temporal aspect. Likewise, the apperceptive-discriminative aspect of sound recognition behaved differently from the associative-semantic one in the presence of the brain damage.

In a general way, lateralization of the lesion underlying impairment (TABLE 1) was as expected from previous experimental studies in brain-damaged patients focusing separately on either environmental sounds[8–10] or music.[2,7,11] In the dissociated cases (TABLE 2), right hemisphere lesions tended either to disrupt the apperception of environmental sounds, sparing music *in toto*, or to disrupt both the apperception of environmental sounds and melody, sparing rhythm. By contrast, left hemisphere lesions tended to disrupt rhythm either selectively or in association with the semantic identification of environmental sounds, sparing melody.

How do these findings compare with evidence stemming from the review of single case reports? Dissociations were found less frequently in the literature than in the experimental study (28.9 vs 47.5%), music being selectively impaired more frequently than environmental sounds in single case reports and less frequently in the experimental study. A possible reason is that most clinicians evaluated the environmental sounds and sometimes also the music in a global way, whereas the standard tests allowed significant qualitative distinctions to be drawn within both domains.

Results confirmed that neither music agnosia nor environmental sound agnosia is a unitary disorder, reflecting the disruption of unitary abilities. Rather, both should be conceived as compound disorders consisting of different defects, each of which selectively affects a specific structural dimension of the stimulus or a particular type of brain process. In music agnosia, it has been suggested that the melodic (pitch) dimension of music should be distinguished from the temporal dimension,[2] whereas in auditory agnosia, it is useful to distinguish between two levels of the recognition process, that is, the acoustic discrimination from the semantic identification of environmental sounds.[9] In the present study, the first element of each pair (melodic; acoustic discrimination) was found to be more vulnerable to lesions of the right hemisphere, whereas the second element (temporal; semantic identification) was more vulnerable to lesions of the left hemisphere. The dissociations found in our sample seem to reflect such selective vulnerability of distinct aspects of music and of sound processing to the lateralization of the lesion.

ACKNOWLEDGMENTS

The assistance of Eleonora Cattaneo, Elisabetta Del Zotto, and Sara Ugazio is gratefully acknowledged.

REFERENCES

1. ZATORRE, R.J. 1985 Discrimination and recognition of tonal melodies after unilateral cerebral excisions. Neuropsychologia **23:** 31–41.
2. PERETZ, I. 1990. Processing of local and global musical information by unilateral brain-damaged patients. Brain **113:** 1185–1205.
3. PERETZ, I., S. BELLEVILLE, S. FONTAINE, *et al.* 1997. Dissociations entre musique et langage après atteinte cérébrale: un nouveau cas d'amusie sans aphasie. Rev. Canad. Psychol. Exp. **51:** 354–367.

4. WERNICKE, C. & C. FRIEDLANDER. 1883. Ein Fall von Taubheit in Folge von doppelseitiger Laesionen des Schlafenlappens. Fortschr. Med. **1:** 177–185.
5. GRISON, B. 1972. Une Étude sur les Altérations Musicales au Cours des Lésions Hémisphériques. M.D. Thesis, Paris.
6. BENTON, A.L. 1977. The amusias. *In* Music and the Brain: Studies in the Neurology of Music. M. Critchley & R.A. Henson, Eds. :378–397. Heinemann Medical Books. London.
7. AYOTTE, J., I. PERETZ, I. ROUSSEAU, *et al.* 2000. Patterns of music agnosia associated with middle cerebral artery infarcts. Brain **123:** 1926–1938.
8. SCHNIDER, D., D.F. BENSON, D.N. ALEXANDER, *et al.* 1994. Nonverbal environmental sound recognition after unilateral hemispheric stroke. Brain **117:** 281–287.
9. VIGNOLO, L.A. 1969. Auditory agnosia: a review and report of recent evidence. *In* Contributions to Clinical Neuropsychology. A.L. Benton, Ed. :172–208. Aldine Publishing Company. Chicago, IL.
10. VIGNOLO, L.A. 1982. Auditory agnosia. Philos. Trans. Roy. Soc. Lond., series B **298:** 49–57.
11. LIÉGEOIS-CHAUVEL, C., I. PERETZ, M. BABAï, *et al.* 1998. Contribution of different cortical areas in the temporal lobes to music processing. Brain **121:** 1853–1867.
12. MOTOMURA, N., A. YAMADORI, E. MORI, *et al.* 1986. Auditory agnosia. Analysis of a case with bilateral subcortical lesions. Brain **109:** 379–391.
13. GODEFROY, O., D. LEYS, A. FURBY, *et al.* 1995. Psychoacoustical deficits related to bilateral subcortical hemorrages. A case with apperceptive auditory agnosia. Cortex **31:** 324–332.
14. EUSTACHE, F., B. LECHEVALIER, F. VIADER, *et al.* 1990. Identification and discrimination disorders in auditory perception: a report on two cases. Neuropsychologia **28:** 257–270.
15. MAZZUCCHI, A., C. MARCHINI, R. BUDAI, *et al.* 1982. A case of receptive amusia with prominent timbre perception defect. J. Neurol. Neurosurg. Psychiatry **45:** 644–647.
16. STEINKE, W.R., L. CUDDY & L.S. JACOBSON. 2001. Dissociations among functional subsystems governing recognition following right-hemisphere damage. Cognit. Neuropsychol. **18:** 411–437.
17. GRIFFITHS, TD., A. REES, C. WITTON, *et al.* 1997. Spatial and temporal auditory processing deficits following right hemisphere infarction. A psychophysical study. Brain **120:** 785–794.
18. KANSHEPOLSKY, J., J. KELLEY, J.D. WAGGENER, *et al.* 1973. A cortical auditory disorder: clinical, audiologic and pathologic aspects. Neurology **23:** 699–705.
19. SHINDO, M., K. KAGA & Y. TANAKA. 1981. Auditory agnosia folllowing bilateral temporal lobe lesions. Report of a case. Brain Nerve **33:** 139–147.
20. COSLETT, H.B., H.R. BRASHEAR & K.M. HEILMAN. 1984. Pure word deafness after bilateral primary auditory cortex infarcts. Neurology **34:** 347–352.
21. TANAKA, Y., T. KAMO, M. YOSHIDA, *et al.* 1991. So-called cortical deafness. Clinical, neurophysiological and radiological observations. Brain **114:** 2385–2401.
22. PERETZ, I., R. KOLINSKY, M. TRAMO, *et al.* 1994. Functional dissociations following bilateral lesions of auditory cortex. Brain **117:** 1283-1301.

Varieties of Musical Disorders

The Montreal Battery of Evaluation of Amusia

ISABELLE PERETZ, ANNE SOPHIE CHAMPOD, AND KRISTA HYDE

University of Montreal, Montreal, Canada

ABSTRACT: Multiple disorders of musical abilities can occur after brain damage. Conversely, early brain anomalies or vast brain injuries may sometimes spare ordinary musical skills in individuals who experience severe cognitive losses. To document these incidences, comprehensive behavioral testing is required. We propose to use the Montreal Battery of Evaluation of Amusia (MBEA) because it is arguably the best tool currently available. Over the last decade, this battery was developed and validated in populations with brain damage of various etiologies. Furthermore, the MBEA is theoretically motivated and satisfies important psychometric properties. It is sensitive, normally distributed, reliable on test-retest, and correlates with Gordon's *Musical Aptitude Profile*, another more widely used battery of tests. To promote its wide usage, the MBEA is now available upon request. In addition, individual MBEA data of 160 normal participants of variable age and education have been made available to all via the internet.

KEYWORDS: musical disorders; amusia evaluation; tests of musical abilities; disorders of music perception and memory; learning deficits

INTRODUCTION

The study of musical disorders dates back to the origins of neuropsychology. In 1865, soon after Broca[1] reported the first case of language disorder as a result of a lesion in the frontal area of the left hemisphere, Bouillaud[2] described the first series of cases in which various musical abilities were lost consequent to brain insult. These are commonly referred to as *amusias*. In addition, Bouillaud reported that music was selectively spared in a man who could no longer speak or write, but retained the ability to compose and write music. Bouillaud comments: "Et l'on nierait encore la pluralité et la spécialité des facultés!!!" ("And we would deny the multiplicity and specialization of faculties!!!", p 755). Since that time, similar cases and conclusions have been drawn over various facets of musical abilities (see Ref. 3 for a recent review).

This long-standing interest in musical disorders results from the fascinating observation that musical functions can be impaired or spared in a highly selective fashion. In particular, brain damage can selectively interfere with musical abilities while the rest of the cognitive system, including language, remains essentially intact

Address for correspondence: Isabelle Peretz, Department of Psychology, University of Montreal, C.P. 6128 succ. Centre-ville, Montreal, Quebec, H3C 3J7, Canada. Fax: (514) 343-5787.
Isabelle.Peretz@umontreal.ca

Ann. N.Y. Acad. Sci. 999: 58–75 (2003). © 2003 New York Academy of Sciences.
doi: 10.1196/annals.1284.006

(e.g., Ref. 4). Moreover, not all musical abilities are equally affected. Specific musical disorders can occur as a result of damage to particular processing components. Sustained interest in these musical impairments is motivated by the fact that these disorders are spectacular expressions of the principles that underlie the organization of musical functions in the normal brain. As clearly stated by McCloskey,[5] "Complex systems often reveal their inner working more clearly when they are malfunctioning than when they are running smoothly" (p.594). Thus, despite the development of neuroimaging techniques, which provide a means to explore the normal brain, the study of musical disorders continues to provide key information for the understanding of the musical brain.

While the study of patients with accidental brain damage is the oldest and hence the best known method in neuropsychology, it is not the only setting in which examination of musical disorders is worthwhile. One important area in which evaluation of residual musical skills would be highly valuable is dementia. It is widely acknowledged that demented patients maintain a level of musical achievement and enjoyment that cannot be matched by other activities, even at the most advanced stages of the disease. To our knowledge, no systematic study of their musical abilities has yet been undertaken. Similarly, musical skill appears to be one of the few areas in which autistic individuals appear to function normally.[6] Once again, we do not know of any systematic study of their musical abilities. Finally, about 10% of the general population have learning disabilities in particular domains, such as reading (dyslexia) and speaking (language-specific impairments). It would certainly be of interest to know whether these learning deficits generalize to the musical domain. In these special circumstances, the assessment of musical abilities would be informative from both a clinical and a theoretical perspective.

To progress in such diverse neuropsychological settings, it is necessary to use a standardized evaluation of musical abilities. The goal of this paper is to present such an instrument. Since 1987, we have developed and adjusted a battery of musical tests for the evaluation needs of the perceptual and memory skills of the ordinary adult listener. We refer to this battery as the Montreal Battery of Evaluation of Amusias (MBEA). Since the most recent version of the MBEA appears to meet the criteria that make it suitable for the screening of amusic cases, we consider this a prime opportunity to share this tool with neurosciencists who have an interest in music processing.

MONTREAL BATTERY OF EVALUATION OF AMUSIA[a]

Music perception and memory are the most investigated musical functions in both the cognitive and neuropsychological domains. Indeed, discrimination and recognition abilities implicate basic and common functions that can be assessed across all normal listeners, non-musicians and musicians alike. Yet, evaluation of these skills is not a simple matter. These basic skills depend on the adequate functioning of multiple components. Therefore, it is essential to have a model that specifies the processing components that are involved and their likely interactions as well as a set of tests that allow their respective assessment. One of us[7] has proposed a model that

---

[a]A copy of the MBEA can be obtained by sending a request to Isabelle.Peretz@umontreal.ca

has both guided the development of the MBEA and proved its neuropsychological adequacy. This model along with the relevant neuropsychological findings is outlined here (for a more detailed review of the literature, see Ref. 8). Its brief description here is necessary to introduce the tests of the MBEA, which will be described in the following section. The suitability and validity of the tests for diagnostic purposes are then presented and discussed.

The Model

In our model of music perception and memory, musical input must be processed along the *melodic dimension* (defined by sequential variations in pitch) and the *temporal dimension* (defined by sequential variations in duration) and then mapped onto a stored long-term representation if there is one that matches the input. This model, represented in FIGURE 1, only considers monophonic music, that is, music that contains a single voice. This may correspond to the tune component of a song. The lyrics component of a song is assumed to be processed in parallel in a different system, the language processing system sketched at the right of the figure. Some aspect, yet to be defined, in the auditory stream is supposed to trigger the intervention of the music or the language processing system. The musical input is then analyzed by two parallel and largely independent subsystems whose functions are to specify, respectively, the melodic content (i.e., representing the melodic contour and the tonal functions of the successive pitches) and the temporal content (representing the metric organization as well as the rhythmic structure of the successive durations). To simplify, the melodic route represents *what* and the temporal route represents *when* events occur in the auditory musical input. Both routes, defining the musical analysis

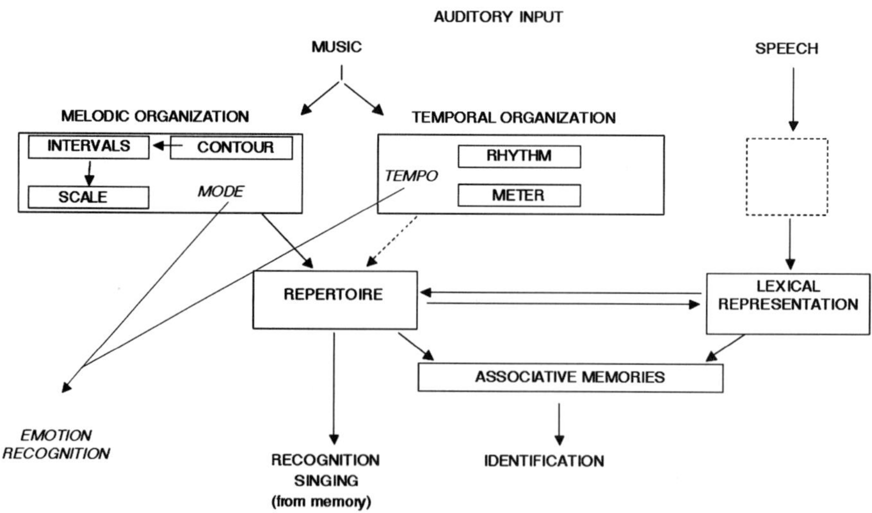

FIGURE 1. Schema of the processing components involved in music recognition. The processing components involved in emotion recognition are represented in *italics*.

components, send their respective outputs—or perhaps a combination of the two—to the *repertoire*. The *repertoire* is conceived of as a perceptual representation system, available to all listeners, that contains all the representations of the specific musical phrases to which one has been exposed during one's lifetime. In turn, the repertoire output can activate stored representations in other systems, such as the lexical representations for the retrieval of the accompanying lyrics, if any, or the associative memories for retrieving and pronouncing the title (leading to identification) of the musical excerpt and for retrieving all sorts of nonmusical information (such as an episode related to the first hearing of the music concerned). Naming a musical excerpt via associative memories, however, is not necessary for recognition. Successful activation (or selection) of one particular musical candidate in the *repertoire* will evoke a sense of familiarity and hence lead to recognition.

The model is supposed to apply to the recognition of both familiar and unfamiliar musical input. The main processing difference is that unfamiliar sequences do not yet have a mental representation stored in the *repertoire* when first presented. However, the same repertoire component seems to be involved in the recording and storage of both familiar and novel music. Support for this view comes from different neuropsychological findings. For example, when the *repertoire* component is damaged, the patient can no longer learn novel music or relearn familiar ones.[9] Similarly, when the lesion interrupts the melodic route more severely than the temporal route, its behavioral effects are similar for familiar and unfamiliar material. In particular, we found a similar pattern of depressed scores with the MBEA that involves novel musical excerpts (see FIGURE 2 for an example of stimulus) and their equivalent versions constructed with familiar tunes. (See Ref. 10 for results on the unfamiliar set and Ref. 11 for results on the familiar set.)

In fact, each processing component considered in the model has been isolated by studying the musical disorders experienced by brain-damaged patients. Hence, we briefly review here the relevant work, beginning with the main division between melodic and temporal organization. As posited in FIGURE 1, the musical input is conceived of as undergoing transformations along two parallel perceptual routes: melodic and temporal. Support for this dual route comes from the observation of double dissociations between the processing of melodic and temporal information in music perception. The temporal route may be spared when the melodic one is severely compromised by the lesion,[12–16] and, vice versa, temporal discrimination can be impaired when melodic discrimination is preserved.[12,14,17] Thus, brain damage can produce a selective loss of the "what" or the "when" in music perception. This explains why we assume the operation of separable perceptual subsystems for analyzing pitch and temporal variations, respectively.

Similarly, neuropsychological support for the existence of two separate codes, one best described in terms of contour and the other in terms of interval sizes, comes from the study of the effects of a unilateral lesion on melody discrimination. We[12,14,15] have repeatedly observed that a lesion in the left hemisphere can spare the ability of representing melodies in terms of their contour, but it interferes with an interval-based procedure, whereas a lesion in the right hemisphere disrupts both procedures. This pattern of results suggests the intervention of two distinct mechanisms, one for extracting pitch directions (i.e., contour) and one for abstracting pitch interval sizes, that are serially organized. According to this principle, a right hemisphere lesion, by disrupting the procedures required for representing the melody contour, deprives the intact left hemispheric structures of the anchorage points necessary for encoding interval information. This is expressed by a unidirectional arrow between the contour and the interval-processing component in FIGURE 1.

The interval representation, in turn, is essential for perceiving *musical* pitch. Intervals allow the emergence of scale structures, musical keys, and tonal functions. The use of this tonal knowledge, represented here by the term "scale," is an important part of the model. For instance, we have argued that this processing component is subserved by a modular organization and hence is music specific.[18,19] Nevertheless, neuropsychological support for its isolation remains partial. To our knowledge, the only supportive evidence for its distinct existence comes from the case study of G.L.,[20] a patient who became amusic as a result of bilateral damage to the auditory cortex. This lesion produced an isolated disturbance in tonal interpretation of pitch. Pitch was no longer encoded in terms of its tonal function, but rather it was perceived into a unidimensional continuum of pitch height (i.e., intervals) and an ordinal dimension of pitch direction (i.e., contour). As a consequence, the patient encountered specific difficulties with melodic as opposed to temporal patterns, experienced memory difficulties for pitch material, and was relatively insensitive to tonal structure in perception (in singing, though, residual tonal knowledge was noted). The observation of an opposite pattern in another patient[21] who exhibits signs of preserved tonal knowledge despite severe difficulties in pitch discrimination provides converging support for the existence of a distinct network subserving the component named "scale." The localization of this network was recently ascribed to the the rostromedial prefrontal cortex.[22]

The temporal route has been less studied than the melodic route. Yet, neuropsychological evidence points to a dissociation between two types of temporal organization mechanisms. The first type of temporal organization, to which we refer as "rhythm," corresponds to the tendency to group events according to temporal proximity without regard to periodicity. The second, to which we refer as "meter," is concerned with the extraction of an underlying temporal regularity or beat. Selective disruption of the rhythmic structure relative to spared metric organization has been observed in several studies.[12,14,23] Conversely, the inability to hear or tap the beat while maintaining intact rhythmic perception and production has also been reported.[23,24] This double dissociation between the treatment of rhythm and meter justifies their distinct representation, as shown in FIGURE 1.

The Tests

The MBEA contains six tests that allow for the assessment of the functioning of each of the musical components depicted in FIGURE 1. Accordingly, these tests are referred to as the contour, interval, scale, rhythm, meter, and memory tests. All six tests use the same pool of 30 novel musical phrases that were composed according to the rules of the Western tonal system by Irène Deliège. An example is provided in FIGURE 2. The musical material was intentionally written with sufficient complexity to guarantee its processing as a meaningful structure rather than as a simple sequence of tones. The selections lasted 3.8 to 6.4 seconds (mean: 5.1 s) in all but the metric test in which the stimuli lasted twice as long (mean: 11 s). The computer-generated versions were created on a microcomputer running a MIDI sequencing program (Sequencer Plus Gold) and controlling a sample playback digital synthesizer (Rolland Sound Canevas SC 50). The scores were manually entered so that each tone occupied its precise value in terms of pitch and duration, keeping intensity and velocity constant, and was delivered with a piano sound.

The tests were constructed with this pool of stimuli as follows. In the melodic organization tests, three types of manipulations were applied to the same tone in 15 sequences. One manipulation consists of a scale-violated alternate melody created by modifying the pitch to be out of scale, while retaining the original melodic contour. This change is particularly salient because the changed pitch sounds out of tune. The second manipulation consists of a contour-violated alternate melody created by modifying the critical pitch so as to change the pitch direction of the surrounding intervals, while maintaining the original key. The third manipulation consists of a contour-preserved or interval-violated alternate melody that is created by modifying the same critical pitch to the same extent (in terms of semi-tone distance), while maintaining the original contour and scale (see FIG. 2 for an example). The serial position of the modified pitch varies across melodies; half the pitch changes fall in the beginning of the melody and half at the end, while avoiding the first and

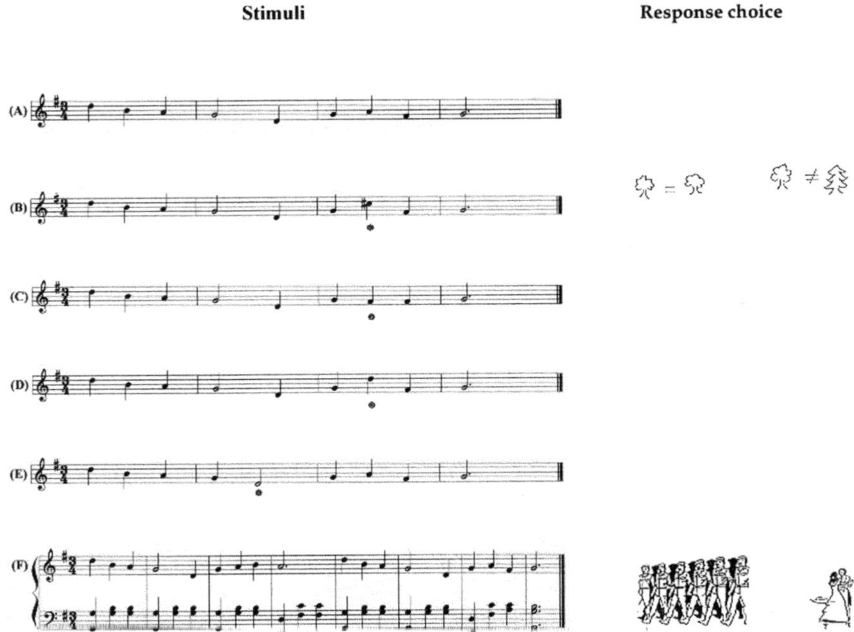

FIGURE 2. Example of one musical stimulus as used in the six tests of the Montreal Battery of Evaluation of Amusia (MBEA). The standard stimulus is represented in **A**, its scale alternate in **B**, its contour alternate in **C**, its interval alternate in **D**, and its rhythm alternate in **E**. *Asterisk* indicates the changed note. In **F** is represented the full accompanied phrase that serves for the metric test. On the *right* are presented the symbols provided on the response sheet.

last tone positions. Average pitch interval changes are made equivalent across the three conditions, with a mean of 4.3, 4.3, and 4.2 semitones apart from the original pitch, in the scale-violated, contour-violated, and interval-violated condition, respectively.

Three sets of stimuli, each comprised of 2 practice trials and 30 experimental trials, were constructed with these melodies. Each trial is preceded by a warning tone and consists of a target melody and comparison melody separated by a 2-second silent interval and with a 5-second intertrial interval. The first set, the scale-violated condition, was constructed so that 15 trials contained identical melodies and 15 trials included a different scale-violated comparison melody. The second and third set, serving for the contour-violated and the interval-violated condition, respectively, are similar to the scale-violated condition set in that they use the same target melodies. The only modification is that each comparison melody is replaced by its contour-violated alternate or its preserved-contour alternate. Melody pairs are presented in each set in random order. These three sets are referred to as the scale, contour, and interval tests, respectively. In each test, a catch trial has been inserted to ensure that the participant is paying attention. The catch is that the comparison melody has all its pitches set at random. Subjects are required to perform a "same-different" classification task. They have to judge, on each trial, whether the target and the comparison sequence are the same or not.

The *temporal organization conditions* involve the rhythmic and metric tests. For the rhythmic test, the stimuli are the same as those used in the melodic organization tests. To create different comparison patterns, the manipulation involves changing the duration values of two adjacent tones so as to change rhythmic grouping by temporal proximity, while retaining the same meter and total number of sounds. This can be done either by changing two quarter notes to a dotted quarter and an eighth note or by interchanging the order of two successive notes of different duration values (e.g., a half note followed by a quarter note becomes a quarter note followed by a half note, as illustrated in FIGURE 2). The serial positions of these changes vary across patterns. A set of two practice trials, one catch trial and 30 experimental trials, was constructed with these stimuli. The task also requires a "same-different" classification. For the metric test, two-phrase sequences are used (rather than the one-phrase sequences used in the previous tests). Moreover, in its most recent version, these melodies are presented in a harmonized version, where chords are added to accentuate the binary and ternary structure of the melodies (see FIG. 2 for an illustration). Half of these sequences are written in duple meter and half in triple meter; they are recorded in random order with a 5-second intertrial interval. Subjects are informed that they will hear a series of melodies that they have to categorize as either a waltz or a march. They are encouraged to tap along with what they perceive to be the underlying beat of each sequence. Four practice trials precede 30 experimental trials.

The last test of the musical battery is a *memory recognition test*. From the initial set of 30 single-phrase melodies, 15 were selected for the recognition part of this study. Each had been presented at least five times in the same format (including the one embedded in the two-phrase harmonized sequences). In addition to these "old" selections, a set of 15 recognition foils was prepared. The "new" melodies were constructed along the same principles, but differed from the "old" ones in their exact temporal and pitch pattern. The 30 sequences were then recorded in random order with a 5-second intertrial interval. The subjects were requested to respond "yes" if they recognized a melody as having been presented earlier during the session and otherwise to respond "no." This last test comes as an incidental memory test, because the subjects are not informed in advance that their memorization of the tunes would later be tested.

The six subtests are presented in a single session that lasts approximately 1 hour and a half. The order of presentation is generally fixed in single case studies beginning with the scale, contour, and interval tests followed by the rhythmic, meter, and memory tests. In group studies, the order of tests is counterbalanced across participants, except for the incidental memory test, which is always the final test of the session. Participants may ask for as many breaks between tests as required to ensure vigilance during testing.

Psychometric Properties of the MBEA

As summarized previously, the MBEA has been useful in detecting a variety of disorders in music perception and memory after brain damage. The brain lesions were the result of either a stroke,[12,25] the surgical clipping of an aneurysm,[4,9,15] or the resection of epileptic tissue.[14] Therefore, a large pool of adults of various ages and levels of education as well as their matched controls have been assessed with the MBEA over the years. Recently, a homogeneous group of 68 fire workers (in training) were tested with the latest version of the MBEA for validation purposes. Since the results obtained by this group of young adults (mean: 21 years) with average education (mean: 13 years) did not differ significantly from those obtained with previous versions of the battery (except for the metric test reported in TABLE 1), we now have at our disposal the results of 160 neurologically intact adults, 62 women and 98 men, aged 14–79, with 7–21 years of education. Their scores are summarized in TABLE 1. The individual scores can be found and downloaded from our web site (http://www.fas.umontreal.ca/psy/iperetz.html).

Sensitivity. As can be seen in TABLE 1, levels of performance are similar across tests with average normal performance being around 90% correct in each test (i.e., 27 of 30 correct responses). Despite the generally high level of performance, the tests are sensitive, since more than 80% of the participants do not obtain a perfect score for each individual test. However, the data for each subtest are skewed towards the higher scores and thus violate normality (by the Kolmogorov-Smirnov test). When individual scores are averaged over the six tests to produce a single global value, these composite scores follow a normal distribution, as shown in FIGURE 3. At the upper extreme, only 3% of the normal participants achieved a perfect score. At the lower extreme, we found that three individuals, representing less than 2% of the population, obtained scores that were below 2 standard deviations from the mean. Note that these outliers obtained performance higher than 70% correct. Therefore, the global score obtained in the MBEA can serve as a good index of the functioning of the music perceptual and mnesic system. Therefore, we propose that this index should be used to distinguish between normal from abnormal performance. We discuss this issue in more detail when we discuss the diagnostic value of the MBEA.

Test-Retest Reliability. We also assessed the test-retest reliability of the MBEA with a subgroup of 28 fire workers (in training). The retest took place 4 months after the initial testing session. Their individual global scores (averaged across the six

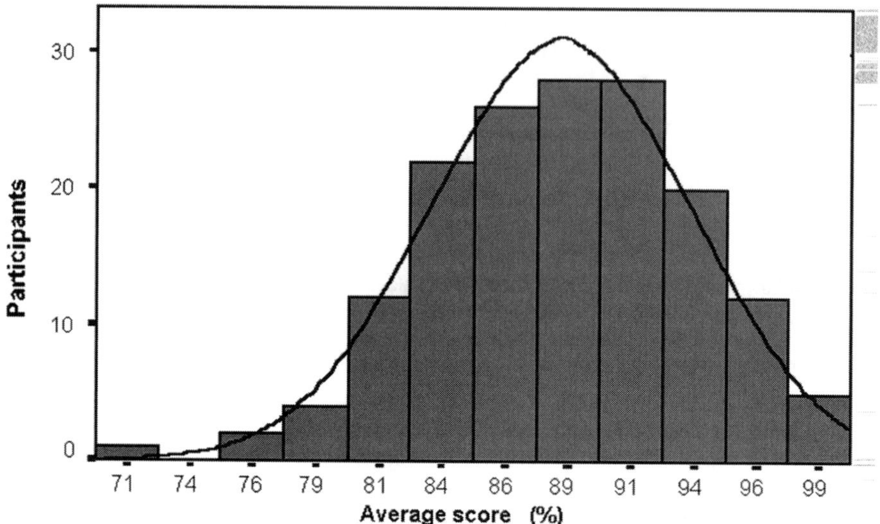

FIGURE 3. Distribution of global composite scores obtained on the Montreal Battery of Evaluation of Amusia (MBEA) for the 160 normal participants. The mean corresponds to 88% correct and the standard deviation to 5.2.

TABLE 1. Mean correct responses and standard deviations on the 30 experimental trials for each test obtained by 160 normal participants

	Scale	Contour	Interval	Rhythm	Meter	Memory	A
Mean	27	27	26	27	26^a (25)	27	27
SD	2.3	2.2	2.4	2.1	2.9^a(3.5)	2.3	1.6
% of individuals with PF	17	9	7	15	14^a (10)	10	3
Cut-off score	22	22	21	23	20^a (18)	22	23
% of N below cut-off	3	1	1	1	1^a (2)	1	2

Note: N, normal participants; PF, perfect score; A, average. Cut-off scores correspond to 2 standard deviations below the mean.

[a]Latest, harmonized, version of the test. Improvement over the prior nonharmonized version scores (shown in *parentheses*) is significant ($t_{158} = 2.332$; $P <0.02$).

tests) are presented in FIGURE 4. As can be seen, participants improved their performance on retest, although none obtained a perfect score. This suggests that the MBEA remains sensitive on retest. This is particularly important in neuropsychology, where recovery of functions is likely to occur 1 year posttrauma (see Ref. 9 for an illustration of the usefulness of the MBEA in assessing recovery). These data also demonstrate the usefulness of the battery for assessing the effects of training and/or therapy on basic musical abilities.

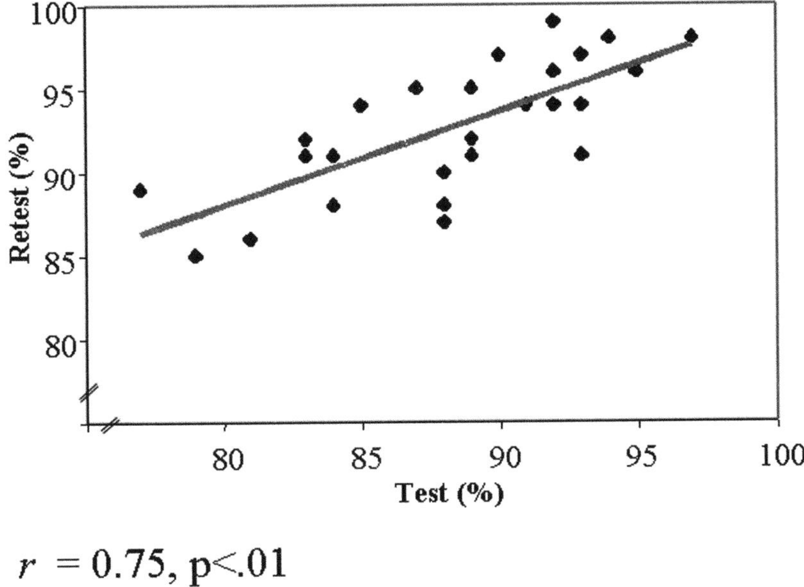

$$r = 0.75, p<.01$$

FIGURE 4. Plot of individual composite scores obtained on the Montreal Battery of Evaluation of Amusia (MBEA) by the 28 fire workers (in training) on test and retest.

Validation with Gordon's Musical Aptitude Profile. The overall validation of the MBEA as a tool for the evaluation of musical abilities was assessed with the *Musical Aptitude Profile* tests of Gordon.[26] The choice of the latter battery was motivated by the fact that it is currently the most commonly used test of musical abilities in North America and hence has the largest norms available, with 12,809 subjects from grade 4 to grade 12. From the seven subtests of Gordon's Musical Aptitude Profile battery, we selected the two subtests that were the closest in format and content to the tests of the MBEA. These were the melody and the meter imagery subtests. Each subtest is comprised of 40 trials in which a musical phrase followed by a musical answer is presented. The duration of the musical stimuli is similar to those used in the MBEA (lasting 3–13 s, with a mean of 6 s). The task of the subjects is to judge whether the two successive stimuli sound the same or different. In the melodic imagery subtest, the musical phrase is embedded in ornamental pitches, as in a hidden figure test. Thus, the musical characteristics that are relevant for discrimination are undetermined. The musical features that seem relevant in the meter imagery subtest are meter, tempo, and rhythm.

These melodic and meter imagery subtests of Gordon's battery were submitted to the group of 68 fire workers (in training) at the same time as they were tested with the MBEA. Half of them were first tested with Gordon's two subtests and then with the MBEA; the other half were tested in the reverse order. Within each group, defined by battery order, care was taken to counterbalance the order of presentation of each subtest. The whole testing session lasted approximately 200 minutes.

The group of fire workers obtained similar levels of performance on Gordon's subtests and the MBEA, with 82% and 89% correct, respectively. More importantly, their scores are positively correlated ($r = 0.53$, $P < 0.001$). That is, overall, the participants were found to achieve similar levels of performance on the MBEA and on Gordon's subtests. Thus, the two batteries appear to tap, at least in part, a common pool of abilities. Incidentally, the two sets of tests that appear to have more in common are the melodic tests for which the correlation was higher ($r = 0.41$, $P < 0.001$) than the meter tests ($r = 0.23$, $P < 0.05$). In fact, Gordon's meter test correlated best with the rhythm test of the MBEA (with $r = 0.43$, $P < 0.001$). These differences highlight the fact that these two batteries are not equivalent even if they tap some common resources or abilities.

To summarize, the MBEA possesses several qualities that make it suitable for detecting the presence of a musical disorder in perception and memory. The MBEA provides an index of functioning that is normally distributed in the population and that is both sensitive and reliable. Moreover, the usefulness of the MBEA has been demonstrated in the past with brain-damaged populations and its validity as a test of musical abilities has now been validated with Gordon's *musical aptitude profile*. Thus, the MBEA can serve as a diagnostic tool for the presence of amusia in the general population, and not solely in brain-damaged patients.

Diagnostic Value

We were recently confronted with the question of the diagnostic power of the MBEA. This question is not as relevant in the case of brain damage given that the neurological pathology provides independent evidence for the presence of an anomaly. This was the case in all populations that we have studied so far. Now, the critical point is to assess to what extent the MBEA may diagnose the presence of a disorder on its own. That is, we need to determine the diagnostic value of the MBEA when there is no known evidence of brain damage. We were faced with this diagnostic issue in our recent studies of congenital amusia,[27,28] a learning disability specific to music, that is not associated with any known neurological anomaly. Yet, the condition can be very severe and interrupt the normal development of musical competence in a highly selective manner. In principle, the MBEA should be able to identify these congenitally affected individuals.

In our initial studies of congenital amusia,[27–30] we used the MBEA to verify whether or not self-declared amusic individuals were indeed impaired musically. These persons declared themselves "musically impaired" in response to newspaper advertisements or volunteered themselves after having read or heard about the condition by way of the media. Those who reported having difficulties in recognizing music and tunes (in the absence of lyrics) and who were told that they sang out of tune (but could not hear it) were considered potential amusic subjects. Among these, only participants with a high degree of education (mean: 17 years) were invited to come to our laboratory. This latter criterion is not essential, but it guarantees that the person does not suffer from a general cognitive deficiency. To the time of the writing, we have tested 27 such self-declared amusic adults with the MBEA. The group is comprised of 21 women and 6 men aged 20–89 years (mean: 55 years). Their results are presented in TABLE 2.

TABLE 2. Mean correct responses and standard deviations in 30 experimental trials for each test obtained by 27 self-declared amusic participants

	Scale	Contour	Interval	Rhythm	Meter	Memory	A
Mean	19	19	19	22	20	20	20
SD	4.4	3.5	3.3	4.2	3.6	4.9	2.9

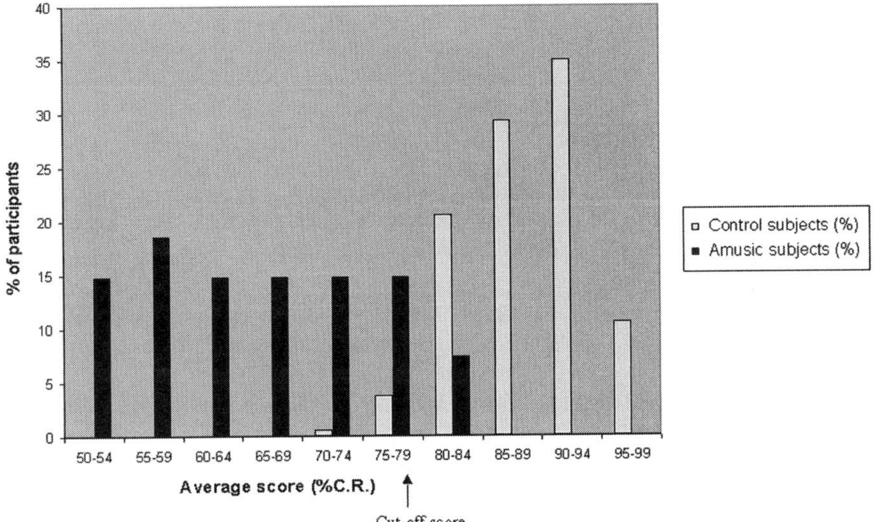

FIGURE 5. Distribution of composite scores obtained on the Montreal Battery of Evaluation of Amusia (MBEA) by self-declared amusic volunteers and by normal participants. Distributions are expressed in terms of percentages of participants per group. The cut-off score of 78% represents 2 standard deviations below the normals' mean.

As a group, the self-declared amusics do perform below controls in each test, with $F(1,185) = 315.1$, $P < 0.001$, hence confirming objectively their subjective experience. As can be seen in TABLE 2, the self-declared amusic participants obtain scores that, on average, lie below the cut-off scores (that correspond to 2 standard deviations below the mean of the control participants; TABLE 1) with the exception of the metric test. On the latter test, amusic subjects' results are significantly lower than normal performance ($F(1,185) = 59.39$, $P < 0.001$). However, as the size of the standard deviations indicates in TABLE 2, there appear to be a few false alarms. Three of the 27 self-declared amusic individuals obtained normal scores on the MBEA. This is best illustrated in the distribution of the normal and amusic scores presented in FIGURE 5.

The slight discrepancy between the self-reported and normal behavioral performance on the MBEA for some of the self-declared amusic subjects deserves further discussion. The first idea that comes to mind is that these self-declared amusic participants are normal and simply underestimate their musical competence. Although a few individuals might indeed demean their musical abilities, this may not apply to the three false-alarm cases. First, all potential amusics were carefully screened with questionnaires. Secondly, we find that non-musicians have some insight regarding their level of performance. For example, the group of 28 fire workers (in training) who participated in the retest session with the MBEA were invited to evaluate their musical skills on a rating scale, with 1 as very poor and 10 as excellent. Their rating was found to correlate with their overall performance on the MBEA ($r = 0.51$, $P < 0.001$). Therefore, we are more inclined to infer that most self-declared amusic participants who perform in the high range on the MBEA have difficulties in musical spheres other than perception and memory. For example, it is quite common to find individuals who are highly competent in music analysis and yet sing very poorly. This form of amusia cannot be detected with the MBEA. In other words, it is likely that the false alarms obtained with the MBEA are not due to false declaration of some volunteers, but rather reflect the plausible heterogeneity of the disorders experienced by individuals afflicted with a music-learning disability. Identification of the sources of this variability is theoretically important and should be the goal of future studies.

The MBEA, nevertheless, succeeded in confirming objectively the musical defect experienced by 89% of the self-declared amusic volunteers. This observation is a good departure point. It is important to emphasize that this is not the only diagnostic conclusion that can be reached with the MBEA. In addition, the MBEA allows the formulation of hypotheses regarding the faulty mechanisms that may be involved. For example, we noted in our prior studies that confirmed self-declared amusic individuals, hereafter referred to as congenitally amusic individuals, have a deficit on the melodic dimension. This is illustrated in FIGURE 6, which shows the scores of the 24 amusic individuals whose global composite score lies below the cut-off score of 78%. As can be seen, all but one amusic participant score below the normal range in the discrimination of musical stimuli that differ on the melodic dimension, whereas about half of them succeed in discriminating the same stimuli when these differ in rhythm. This pattern of performance led us to argue that the core deficit in congenital amusia is related to a basic pitch perception defect.[12,33] Yet, there might be other forms of deficits, as suggested by the recent discovery of a congenitally amusic person with an isolated difficulty in rhythm discrimination (FIG. 6).

It is also interesting (FIG. 6) that individuals with confirmed congenital amusia do not exhibit depressed performance across the board. Some perform normally on some tests. This is reflected in their scores obtained on the six tests of the MBEA that do not all correlate with each other, as can be seen in TABLE 3. For example, all three melodic tests correlate with each other, whereas none of them correlates with the metric test, hence supporting the distinction between melodic and temporal processes. This highlights the idea that the MBEA taps relatively distinct abilities. The correlations also show that congenital amusia does not simply represent the lower tail of a normal distribution. If this were the case, amusic individuals would be expected to obtain similar scores in all six tests of the MBEA. Rather, there seem to be different forms of congenital amusia that differentially affect the mechanisms involved in music processing much in the same way that there are a variety of acquired amusias that can occur after brain damage.

TABLE 3. Pearson correlations between scores obtained by all 24 congenital amusic participants whose global score on the MBEA lies below 2 standard deviations from the mean

	Scale	Contour	Interval	Rhythm	Meter	Memory
Scale	1.00	.50[a]	.44[a]	.15	.18	.42[a]
Contour		1.00	.56[a]	.27	−.03	.54[a]
Interval			1.00	.64[a]	.16	.58[a]
Rhythm				1.00	.38	.53[a]
Meter					1.00	.15
Memory						1.00

[a]$P < 0.05$.

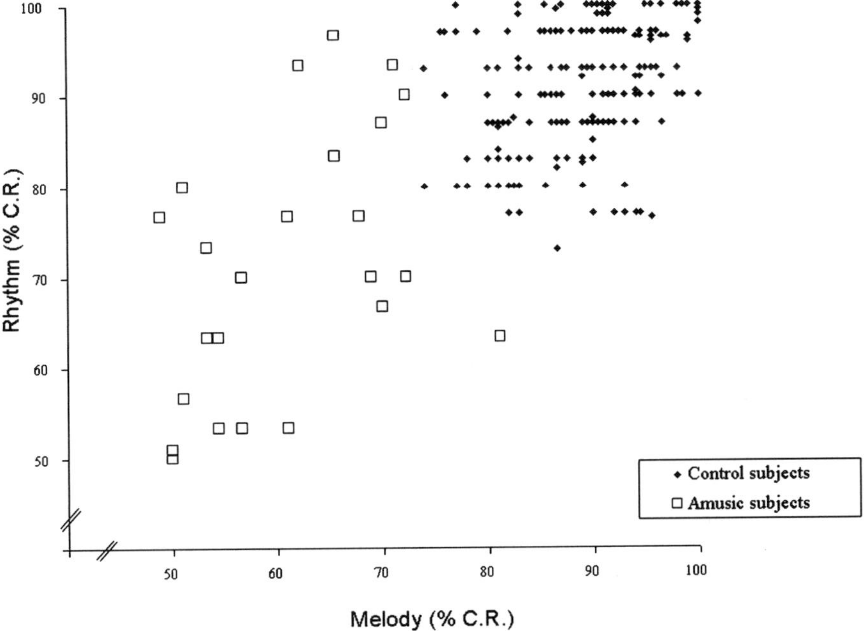

FIGURE 6. Percentage of correct responses obtained by 24 persons with confirmed congenital amusia relative to the normal distribution in the melodic organization tests (abscissa) and the rhythm test (ordinate axis) of the Montreal Battery of Evaluation of Amusia (MBEA).

In summary, the MBEA presents a number of advantages for diagnosis. We can certainly use it as a measure of the prevalence of congenital amusia in the general population, with the realization that the MBEA may, in fact, underestimate the true proportion of affected individuals. We may also exploit the MBEA for educational purposes. For example, we are interested in assessing the adequacy of the MBEA for testing children. Such an extension would provide an opportunity to conduct prospective studies of learning disabilities and thereby assess remediation strategies. In such future developments, it would be important to make the information available to all interested investigators so as to promote comparisons across laboratories and to update normative data. This is what we plan to do via our web site.

Advantages and Limits of the MBEA

The MBEA is more suitable for the evaluation of musical abilities in neuropsychological settings than are any other batteries that we are aware of, including Gordon's Musical Aptitude Profile[26] mentioned earlier and Seashore's tests of musical ability.[31] Because the MBEA was developed more recently, the battery reflects better current concepts of music perception and cognition. The MBEA is better motivated theoretically. This indicates that the battery is more suitable for identifying a problem in music perception and memory than are the other batteries. As described previously, the MBEA allows the pinpointing of which music recognition-processing component is likely at fault in a given individual, whereas Gordon's Musical Aptitude Profile battery may not do so because multiple structural aspects are all tested at once. While Seashore's tests of musical ability are more apt than Gordon's battery to distinguish the different mechanisms that come into play in music perception, the tests remain less selective than the MBEA. For example, the Seashore's tonal memory subtest involves the discrimination of two short sequences which may differ by a single pitch, just as in the MBEA. However, unlike the MBEA, the change of pitch is not systematically manipulated for pitch direction, pitch distance, and key. Therefore, the test may overlook the presence of a theoretically interesting deficit. Furthermore, Seashore's tonal memory subtest requires the participant to localize the pitch change in the comparison melody. This meta-analytic demand may recruit processes that are unrelated to melody discrimination or to ordinary musical abilities. Thus, while Seashore's tests of musical ability were at one time the best available tool for neuropsychologists, they are no longer adequate for detecting the presence of a deficit in a meaningful way. This is because Seashore's battery was created before the development of cognitive psychology and the subsequent explosion of research on music cognition.

Another advantage of the MBEA is that it focuses on the skills of the ordinary listener so as to uncover underlying deficits. Most batteries aim at evaluating exactly the opposite. The explicit goal is to assess musical aptitudes in order to identify talented children who might benefit from instrumental lessons.[32] This difference in perspective explains why most batteries incorporate tests that do not tap specific and basic mechanisms but rather probe the ability to integrate multiple musical aspects in complex tasks. Gordon's Musical Aptitude Profile is an example of this type of battery that aims at identifying talent and not disability. Thus, the MBEA fills a gap in the assessment of musical abilities by allowing the evaluation of deficits rather than high levels of achievement. However, the battery does not respond to all testing needs. For example, it is probably too easy for professional musicians and hence may lack sensitivity to detect a potential loss of musical skills as a consequence of brain damage in this particular expert group. Nevertheless, the MBEA is not limited to nonmusicians as its adequacy in testing amateur choir singers recently demonstrated.[25]

Like any other behavioral test, the MBEA is not a fixed tool. It is a dynamic instrument that depends on the progress made in research. For example, we recently realized a shortcoming of the battery regarding emotions. One brain-damaged amusic patient (I.R.) who has the most severe case that we have been able to assess with the MBEA[10] was able to recognize the emotional tone of music.[11] More precisely, like normal subjects, I.R. was able to use the mode and tempo in which musical excerpts were played in order to judge whether these were happy or sad.[33] This spectacular dissociation between emotion and recognition does not necessarily imply that emotion and recognition follow distinct pathways in the brain as some posit.[34] Rather, the structural determinants of emotions, such as mode and tempo, are different from those involved in recognition. (See FIGURE 1 for a plausible addition to the model.) Hence, brain damage may spare the processing of these emotion-specific characteristics, while compromising the functioning of the rest of the recognition system. In other words, the MBEA does not encompass all musical perceptual abilities; in particular, it does not evaluate emotional appreciation.

This caveat can be complemented by the use of additional batteries.[35] However, we should bear in mind that we should conceive of additions rather than modifications to the existing MBEA. The battery needs to be maintained in the same form for standardization purposes. Thus, the MBEA should only be revised if it is ever shown to no longer meet research and clinical standards. In the meantime, other diagnostic tools should be devised for the evaluation of other basic musical abilities that the MBEA does not assess. Singing and beat extraction are two such common musical skills; yet, there are no standard tests for the evaluation of these skills. The development of such tools is currently in progress in our laboratory.[36]

CONCLUSIONS

We have presented and discussed the suitability of the MBEA in identifying disorders in music recognition abilities in most Western listeners. The MBEA has been shown to have good sensitivity, reliability and validity. Moreover, it can detect the presence of a deficit in at least six distinct processing components. This specificity in assessment allows, in turn, the formulation of hypotheses on the functional origin of the putative problem in a given individual.

The usefulness of the MBEA is probably at its best in single case studies, particularly now that norms are available on the internet. However, the MBEA is also instructive in group studies of special populations. As mentioned previously, it can be used to document a preserved area of functioning in pervasive disorders of development such as autism. The battery can be used to explore the domain specificity of developmental disorders such as language-specific impairments. Finally, the MBEA can be used to measure evolution of performance as after recovery or therapy. As long as the cognitive decline is not too severe, as in advanced stages of Alzheimer's type of dementia or in individuals affected with Williams' syndrome, the MBEA can be very informative. Therefore, we would like to offer the MBEA as a standard battery for music assessment in neuropsychological settings. The widespread use of the MBEA should facilitate progress in the discipline by allowing comparisons across periods of testing of the same individual as well as by allowing comparisons across studies of different individuals in the same laboratory. Above all, the wide distribution of the MBEA should facilitate exchanges across laboratories in Western or Westernized countries. This joint effort should, in turn, strengthen the scientific rigor and vigor of our discipline.

ACKNOWLEDGMENTS

The paper is based on studies supported by grants from the Canadian Institute of Health Research and the Canadian Natural Science and Engineering Research Council to the first author. A.-S. Champod was supported by a summer studentship from the Canadian Natural Science and Engineering Research Council at the time of study and Krista Hyde by a fellowship from the Fonds de la Recherche en Santé du Québec. We wish to thank Julie Ayotte for her help in collecting a large proportion of the data and Séverine Samson for insightful comments on the paper.

REFERENCES

1. BROCA, P. 1861. Remarques sur le siège de la faculté du langage articulé, suivies d'une observation d'aphémie. Bull. Mem. Soc. Anat. Paris **2:** 330–357.
2. BOUILLAUD, J. 1865. Sur la faculté du langage articulé. Bull. Acad. Natl. Med. **30:** 752–768.
3. MARIN, O.S.M. & D.W. PERRY. 1999. Neurological aspects of music perception and performance. *In* The Psychology of Music. D. Deutch, Ed. :653–724. Academic Press. New York.
4. PERETZ, I. *et al.* 1994. Functional dissociations following bilateral lesions of auditory cortex. Brain **117:** 1283–1302.
5. MCCLOSKEY, M. 2001. The future of cognitive neuropsychology. *In* Handbook of Cognitive Neuropsychology. B. Rapp, Ed. :593–610. Psychology Press. New York.
6. MOTTRON, L., I. PERETZ & E. MENARD. 2000. Local and global processing of music in high-functioning persons with autism: beyond central coherence? J. Child Psychol. Psychiatry **41:** 1057–1065.
7. PERETZ, I. 1993. Auditory agnosia: a functional analysis. *In* Thinking in Sound. The Cognitive Psychology of Human Audition. S. McAdams & E. Bigand, Eds. :199–230. Oxford University Press. New York.
8. PERETZ, I. 2001. Music perception and recognition. *In* The Handbook of Cognitive Neuropsychology. B. Rapp, Ed. :519–540. Psychology Press. Hove.
9. PERETZ, I. 1996. Can we lose memories for music? The case of music agnosia in a non-musician. J. Cognit. Neurosci. **8:** 481–496.
10. PERETZ, I., S. BELLEVILLE & F.S. FONTAINE. 1997. Dissociations entre musique et langage après atteinte cérébrale: un nouveau cas d'amusie sans aphasie. Rev. Canad. Psychol. Exp. **51:** 354–367.
11. PERETZ, I. & L. GAGNON. 1999. Dissociation between recognition and emotional judgment for melodies. Neurocase **5:** 21–30.
12. PERETZ, I. 1990. Processing of local and global musical information in unilateral brain-damaged patients. Brain **113:** 1185–1205.
13. PERETZ, I. & R. KOLINSKY. 1993. Boundaries of separability between melody and rhythm in music discrimination: a neuropsychological perspective. Q. J. Exp. Psychol. **46:** 301–325.
14. LIÉGEOIS-CHAUVEL, C. *et al.* 1998. Contribution of different cortical areas in the temporal lobes to music processing. Brain **121:** 1853–1867.
15. AYOTTE, J. *et al.* 2000. Patterns of music agnosia associated with middle cerebral artery artefact. Brain **123:** 1926–1938.
16. PICCIRILLI, M., T. SCIARMA & S. LUZZI. 2000. Modularity of music: evidence from a case of pure amusia. J. Neurol. Neurosurg. Psychiatry **69:** 541–545.

17. MAVLOV, L. 1980. Amusia due to rhythm agnosia in a musician with left hemisphere damage: a non-auditory supramodal defect. Cortex **16:** 331–338.
18. PERETZ, I. & J. MORAIS. 1989. Music and modularity. Contemp. Music Rev. **4:** 277–291.
19. PERETZ, I. & J. MORAIS. 1993. Specificity for music. *In* Handbook of Neuropsychology. Vol. 8 : 373-390. F. Boller & J. Grafman, Eds. Elsevier. Amsterdam.
20. PERETZ, I. 1993. Auditory atonalia for melodies. Cognit. Neuropsychol. **10:** 21–56.
21. TRAMO, M., J. BHARUCHA & F. MUSIEK. 1990. Music perception and cognition following bilateral lesions of auditory cortex. J. Cognit. Neurosci. **2:** 195–212.
22. JANATA, P. *et al.* 2002 The cortical topography of tonal structures underlying western music. Science **298:** 2167–2170.
23. POLK, M. & A. KERTESZ. 1993. Music and language in degenerative disease of the brain. Brain Cognit. **22:** 98–117.
24. FRIES, W. & A. SWIHART. 1990. Disturbance of rhythm sense following right hemisphere damage. Neuropsychologia **28:** 1317–1323.
25. STEINKE, W.R., L.L. CUDDY & L.S. JAKOBSON. 2001. Dissociations among functional subsystems governing melody recognition after right-hemisphere damage. Cognit. Neuropsychol. **18:** 411–437.
26. GORDON, E. 1965. Musical Aptitude Profile. GIA. Chicago.
27. AYOTTE, I., I. PERETZ & K. HYDE. 2002. Congenital amusia: a group study of adults afflicted with a music-specific disorder. Brain **125:** 1–14.
28. PERETZ, I. 2001. Brain specialization for music: new evidence from congenital amusia. Ann. N.Y. Acad. Sci. **930:** 153–165.
29. PERETZ, I. *et al.* 2002. Congenital amusia: a disorder of fine-grained pitch discrimination. Neuron **33:** 185–191.
30. PERETZ, I. 2003. Brain specialization for music: new evidence from congenital amusia. *In* The Neuroscience of Music. I. Peretz & R. Zatorre, Eds. :192–203. Oxford University Press. Oxford.
31. SEASHORE, C.E. *et al.* 1960. Seashore Measures of Musical Talents. Psychological Corporation. New York.
32. SHUTER-DYSON, R. 1999. Musical ability. The Psychology of Music. 2nd Edition. D. Deutsch, Ed. :627-651. Academic Press. San Diego.
33. PERETZ, I., L. GAGNON & B. BOUCHARD. 1998. Music and emotion: perceptual determinants, immediacy and isolation after brain damage. Cognition **68:** 111–141.
34. LEDOUX, J. 2000. Cognitive-emotional interactions: listen to the brain. *In* Cognitive Neuroscience of Emotion. R. Lane & L. Nadel, Eds. :129–155. Oxford University Press. New York.
35. PERETZ, I., N. GOSSELIN, S. KHALFA & B. BOUCHARD. 2003. Musical clips for research on emotion. In preparation.
36. DALLA BELLA, S. & I. PERETZ. 2003. Congenital amusia interferes with the ability to synchronize with music. Ann. N.Y. Acad. Sci., this volume.

Music and the Brain

Lessons from Brain Diseases and Some Reflections on the "Emotional" Brain

HEINZ GREGOR WIESER

*Department of Neurology, Epileptology and Electroencephalography,
University Hospital Zurich, Zurich, Switzerland*

ABSTRACT: Studies are reviewed from the perspective of a neurologist and epileptologist interested in "music and the brain." At the neurocognitive level, deficits in pitch discrimination of patients with brain lesions and those during the intracarotid amobarbital test are outlined, because they show that the temporal lobe and, in particular, the right acoustic cortex are crucial. Hallucinations of music during epileptic seizures as well as the analysis of musicogenic epilepsy point to the same gross localization and lateralization. At the esthetic level, music theoretical concepts on the consonance-dissonance dichotomy and related EEG examinations are reported, which illustrate the importance of mesiolimbic temporal lobe structures for the pleasure that we might experience when listening to music. The complex interaction of many neuronal circuits and assemblies of both hemispheres in musical perception and performance is illustrated by musical analysis of a recording by an organ player who experienced a right temporal lobe seizure. This analysis revealed that the seizure-induced errors of the left hand were compensated with the right hand in a musically meaningful way.

KEYWORDS: music; brain; tonotopy; neurocognition; esthetics; brain map; music abilities; pitch discrimination; brain lesions; amusia; hallucination; musicogenic epilepsy; intracarotid amobarbital test; right temporal lobe seizure; EEG; consonance; dissonance; music intervals; dichotomy; limbic brain

INTRODUCTION

The human brain has several unique properties. It is: (1) procrustean, that is, it reduces information of the 'outer world' to its own categories; (2) deterministic, that is, it seeks security and expediency (usefulness); (3) habituative, that is, it suppresses repetitive and expected information; (4) synthetic, that is, it seeks a gestalt even when there is none; (5) active, (6) predictive, (7) hierarchical, (8) rhythmic, (9) self-rewarding, and (10) self-reflecting, that is, it is able by introspection to examine its own operations (thereby the hardware of the brain can serve as its input and even as

Address for correspondence: Prof. Dr. med. H.G. Wieser, Abteilung für Epileptologie & Elektroenzephalographie, Neurologische Klinik, Universitätsspital, CH-8091 Zürich, Switzerland. Voice: 0041/1 255 55 30 (or 31); fax: 0041/1 255 44 29.
hgwepi@neurol.unizh,ch

Ann. N.Y. Acad. Sci. 999: 76–94 (2003). © 2003 New York Academy of Sciences.
doi: 10.1196/annals.1284.007

its own program; (11) social, that is, the operations of the brain are tightly connected with culture; (12) hemispherically specialized; and finally, (13) kalogenetic (a term introduced by Turner and derived from kalos (Gr.) beauty, the good and right; genesis (Gr.), the productive reason), that is, it seeks grace; and (14) monocausotaxophile (a term introduced by Pöppel to indicate that the brain prefers a single reason that explains all).[1]

The theme, "music and the brain," is usually dealt with at three levels: acoustic, cognitive, and esthetic.[2–7] At the acoustic level, the spatiotemporal pattern of sound waves, the specific excitation of sensory cells, the acoustic pathway with its relays and functions, and the acoustic cortices should be dealt with. At the cognitive level, acoustic memory, spatiotemporal analysis of sounds and tones, pitch, timbre, rhythm, and the apprehension of melodies are the center of interest. Here, analysis often relies on the study of circumscript and more diffuse brain diseases and alterations of musical perception in the context of other pathologic conditions. At the esthetic level, we have to ask ourselves why we like music and why, in most instances, we like only certain types of music. This category encompasses musical preferences, music education, and sociocultural aspects of music.

THE NEUROCOGNITIVE LEVEL

Some examples of the neurocognitive and esthetic aspects are as follows. Tonotopy, that is, the spatial organization of neuronal tissue according to the frequency of an acoustic stimulus, is present in the cochlea,[8] acoustic nuclei,[9] and acoustic cortex. In the acoustic cortex, it has been demonstrated with evoked potential studies,[10] particularly magnetoencephalograpy,[11] but also with electrical intracerebral stimulation.[12] Crottaz-Herbette and Ragot[13] found that in the perception of complex sounds, the N1 latency codes pitch and topography codes spectra.

Relations between Music Processing and Neuroanatomical Regions of the Brain

Support for the existence of music-specific neural networks is found in various pathologic conditions that isolate musical abilities from the rest of the cognitive system.[14–16] Brain damage and congenital brain anomalies can lead to selective disorders of music processing, and epilepsy can reveal the autonomous functioning and selectivity, respectively, of the neural networks that subserve music.[17] Based on several studies, brain maps have been constructed that indicate brain areas important for musical abilities. It has been claimed that for the perception of (consonant) music the orbitofrontal cortex and the right frontal lobe are particularly important, whereas for the perception of dissonant sounds the right G. parahippocampalis may play an important role. Pitch discrimination has been linked to the planum temporale (in musicians and persons with so-called absolute pitch, the left planum temporale is much larger than the right), and melodic properties of music and harmony to the temporal lobe in general. Rhythm was associated with the cerebellum. According to Penhune *et al.*,[18] the lateral cerebellar cortex and cerebellar vermis exert a supramodal contribution to the production of timed motor response, particularly when it is complex and/or novel. Musical education/training also has been linked to the cerebellum, because it was found that the cerebellum is about 5% larger in musicians

TABLE 1. Categorization of amusias (Benton[27])

Amusia type	Characterization and localization	Case reports (literature)
Oral-expressive or vocal amusia	If "pure," R T, R F	Mann, 1898; Jossmann, 1926, 1927
	If with other symptoms: bilateral T, L F	Pötzl & Uiberall, 1937 Jellinek, 1933
Instrumental apraxia	R F (accordion player)	Botez & Wertheim, 1959
Musical agraphia	Inability to make notes of heard tones or to copy notes from a paper = visual perceptive problem. Nearly always associated with verbal agraphia (Gerstmann Syndrome)	
Musical amnesia	Inability to identify or to sing generally known melodies, whereas the person can reproduce tone by tone	von Grison, 1972
Musical alexia	Inability to read musical notations; words or other symbols can be read; often associated with verbal memory deficits, other deficits of musical perception and language deficits	
Perturbations of rhythm	Not well studied; usually both sides, perceptive and expressive, affected	
Receptive amusia	Inability to discriminate pitch, loudness, duration, timbre, and rhythm; often associated with word deafness and other auditory agnostic deficits	

than in nonmusicians. Enlargement of the hand motor area in association with the amount of musical training of musicians (piano, string instruments) was shown. Such brain maps are certainly only useful if these locations are seen as centers of gravity, because many other studies revealed that in almost all aspects of music perception and performance the whole brain is engaged in a certain manner. Multiple interconnected neural networks are engaged, some of which may capture the essence of brain specialization for music, such as encoding of pitch along musical scales.

Brain Diseases and Amusias

Broca and Wernicke pioneered the analysis of aphasias, and Henschen,[19] Feuchtwanger,[20] Kleist,[21] Ustvedt,[22] Wertheim,[23,24] Alajouanine,[25] who described Ravel, and Luria,[26] who described Shebalin, pioneered the "amusias." Benton[27] categorized the amusias into (1) oral-expressive or vocal amusia, (2) instrumental apraxia, (3) musical agraphia, (4) musical amnesia, (5) musical alexia, (6) perturbations of rhythm, and (7) receptive amusia (TABLE 1). Peretz *et al.*[28] and Ayotte *et al.*[29] studied well-documented cases of congenital amusia to find that this condition is related to severe deficiencies in processing fine-grained pitch variations. The deficit extends to impairments in music memory and recognition as well as in singing and the ability to tap in time to music. Because the 11 adult congenital amusical individuals that were studied processed and recognized speech, including speech prosody, common environmental sounds, and human voices, this disorder appears specific to the musical domain.

Alajouanine[25] reported on Ravel's severe receptive aphasia associated with complete loss of the ability to compose, because he had lost the knowledge of the composition rules. Ravel, however, could play many of his compositions on the piano if someone gave him the first measures. Previous to his disease, Ravel had characterized himself "as a musical hydrant," because "music flowed out of him like water."

Hallucinations of Music

Hallucinations of music as correlates of "spontaneously" occurring epileptic discharges or induced by electrical stimulation of certain brain areas have been studied in detail by Penfield and Perot.[30] In 520 epilepsy patients who underwent electrical stimulation of the brain in the context of epilepsy surgery, they studied the evoked auditory symptoms: elementary auditory hallucinations were evoked 66 times in 24 patients: music (17/11), voices (46/16), and sounds (3/3). Evoked music was familiar in 13/7 and unfamiliar in 4/4, whereas voices were familiar in 17/8 und unfamiliar in 29/12. Right hemispheric was more often accompanied by auditory experiential responses than was left hemispheric stimulation (n positive stimulations, right vs left: all auditory 45:21; music 13:4; voices 31:15; and sounds 1:2). Analyzing the location of the stimulation in the temporal lobe (TL), auditory responses were evoked more frequently in the mid- and posterior TL than in the other locations: anterior TL 11, middle TL 22, posterior TL 19, and opercular insular 10.

In the classic study of Penfield and Perot[30] on experiential responses elicited by electrical brain stimulation (66 positive auditory responses in 24 patients), we found on reanalysis that music hallucinations elicited 3.3 times more often from the right than from the left side (17 positive responses in 11 patients; right:left = 13:4). Whereas electrically induced hallucinations of voices (46 positive responses in 16 patients) had a right over left preponderance of 31:15 (2.1:1), all experiential responses ($n = 137$) showed a right over left preponderance of 94:43 (2.2:1). Although the investigators themselves cautiously concluded that due to the uneven number of stimulations in both hemispheres no conclusions with regard to hemisphere lateralization should be drawn, we believe that their data speak for at least a tendency for lateralization of musical hallucinations as experiential responses to the right side.

We published a case entering psychomotor status epilepticus during presurgical evaluation with depth electrodes for seizure monitoring.[31] This young Portuguese woman exhibited epileptic discharges that originated in the right auditory cortex

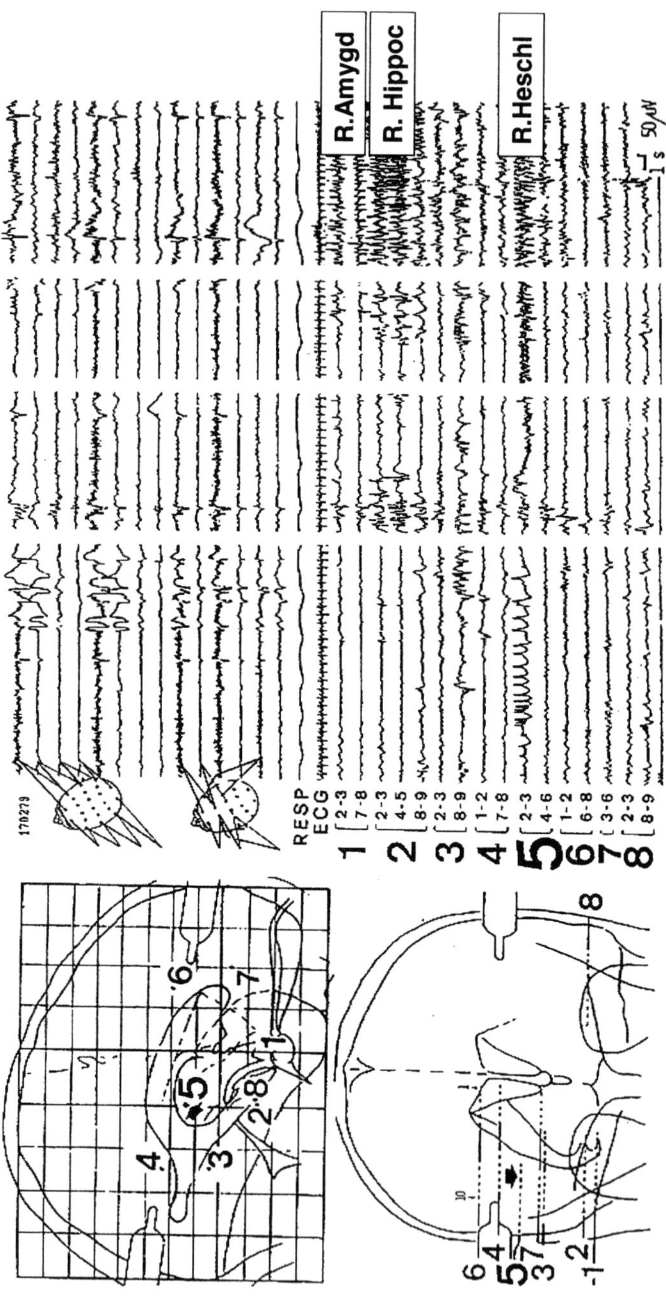

FIGURE 1. Depth EEG recorded onset of prolonged right temporal lobe seizure discharges originating in the right acoustic area (right Heschl gyrus) and spreading to the ipsilateral mesial limbic temporal areas (amygdala and hippocampal formation). During these prolonged discharges the patient hallucinated a song, "Santa Maria," very familiar to her. (Modified from Ref. 31, with kind permission of the publisher.)

(right Heschl gyrus) but spread to the right mesiobasal temporal lobe structures (amygdala and hippocampus). Associated with the epileptic discharge (FIG. 1), the patient hallucinated a Portuguese song, "Santa Maria," very familiar to her. She had heard this song for the first time when she fell in love with a college student. This has been interpreted as evidence of the role of the limbic system (the "emotional brain") in the sense of a "familiarity experience" (déjà entendu).

Insights from the Intracarotid Amobarbital Test

A unique method to study hemispheric lateralization for musical abilities is the so-called intracarotid amobarbital test (IAT, also known as the Wada test). This test serves for preoperative hemispheric language lateralization and assessment of memory functions.[32] Borchgrevink,[33] using this test to study pitch ability, found that after right hemispheric anesthesia four right-handed persons abruptly lost control of pitch and tonality – counting monotonously with preserved musical rhythm. Tonal control was gradually regained during recovery. First regained were the gross characteristics of the melodic envelope, the accuracy of pitch being the last element to be restored. Consciousness and normal speech comprehension and production, including the local dialect (prosody, intonation, and stress), were preserved throughout. Left hemispheric anesthesia in these four right-handed persons produced abrupt loss of speech comprehension, speech production, and singing ability. Normal functions were regained suddenly and were perfectly controlled at once. Borchgrevink therefore concluded that for the "normal" right-handed person, the left hemisphere controls speech perception, speech production, prosody, musical rhythm, and the act of singing, whereas the right hemisphere controls pitch and tonality in singing (but not in speech). In early extensive damage to one hemisphere, the "healthy" hemisphere might control all of these functions. Prosody, the musical element of the language, might be linked to the right hemisphere in childhood, but it gradually shifts to the left hemisphere at the end of puberty. This was postulated because children usually take over the dialect of a new surrounding, whereas adults usually keep going with their original dialect.

Pitch Discrimination in Patients with Brain Lesions

Together with Wittlieb-Verpoort,[34] we studied pitch discrimination in patients with various brain lesions (left temporal [$n = 11$], right temporal [$n = 13$]); left frontal [$n = 10$]; right frontal [$n = 9$)]; and a control group ($n = 20$).

The study design included three levels of difficulty. At the first level (Condition D T; Introduction T D), one target (T = 500 Hz) and one distractor (D = 600 Hz) tone were presented in a random series (…-D-T-D-T-T-D-T-D-… etc).

At the second level (Condition D T T T D; Introduction D1 T1 T2 T3 D2), three target (T1 = 400 Hz, T2 = 500 Hz, T3 = 600 Hz) and two distractor (D1 = 300 Hz and D2 = 700 Hz) tones were presented in a random series (…-D1-T3-D2-T1-T3-D1-T1-D2-, etc).

At the third level (Condition D T D T; Introduction D1 T1 D2 T2), two target (T1 = 500 Hz, T2 = 700 Hz) and two distractor (D1 = 400 Hz and D2 = 600 Hz) tones were presented in a random series (…- D1-T2-D2-T1-T2-D1-T2-D2-, etc).

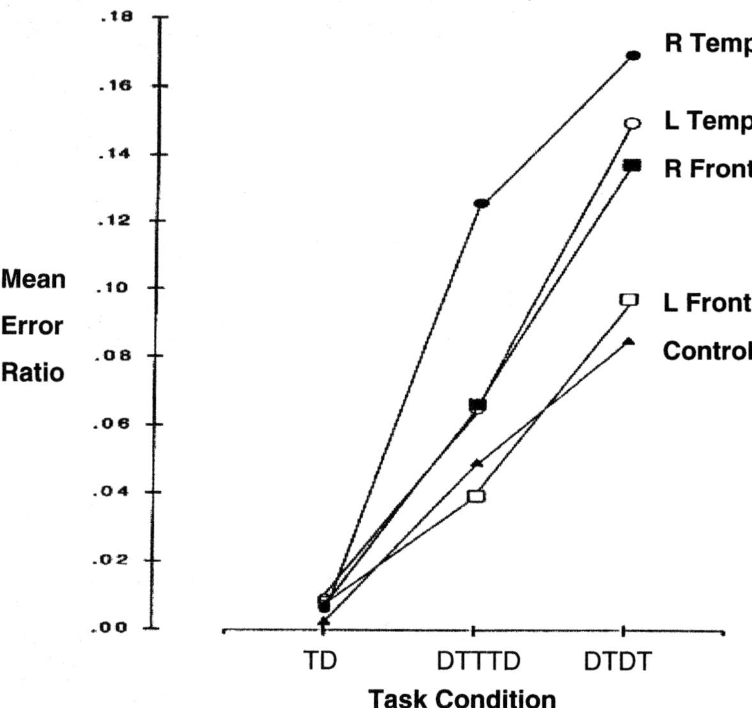

FIGURE 2. Discrimination of auditory signals in brain-damaged patients and control subjects. Mean error ratios of controls and patient groups with lesions right (R) and left (L), temporal (Temp), and frontal (Front) for the three task conditions with increasing demands from TD (*left*) to DTDT (*right*). T, target tone, D, distractor tone; see text for details. (Modified from Ref. 34, with kind permission of the editor and publisher.)

The results (FIG. 2) were as follows. There was no difference between groups at level 1. At level 2 the right temporal lobe (TL) group did significantly worse than did all other patient groups and controls. At level 3 the differences between right TL and left TL and right frontal lobe groups were less obvious

Performance of Patients on the Seashore Test after Selective Amygdalohippocampectomy

Ferrara[35] examined 22 patients who underwent selective amygdalohippocampectomy (AHE; 11 right, 11 left) because of mesial temporal lobe epilepsy with unilateral hippocampal sclerosis), who were right-handed, postoperatively completely seizure- and aura-free, and without any psychiatric or clinically relevant neurological-neuropsychological deficits. Patient groups were compared with 11 age- and sex-matched controls. We used the Seashore Test because Milner[36] has used this test in pre- and postoperative studies on the effects of temporal lobectomy. This test was developed by Seashore[37,38] around 1922 and is still used to examine musical abili-

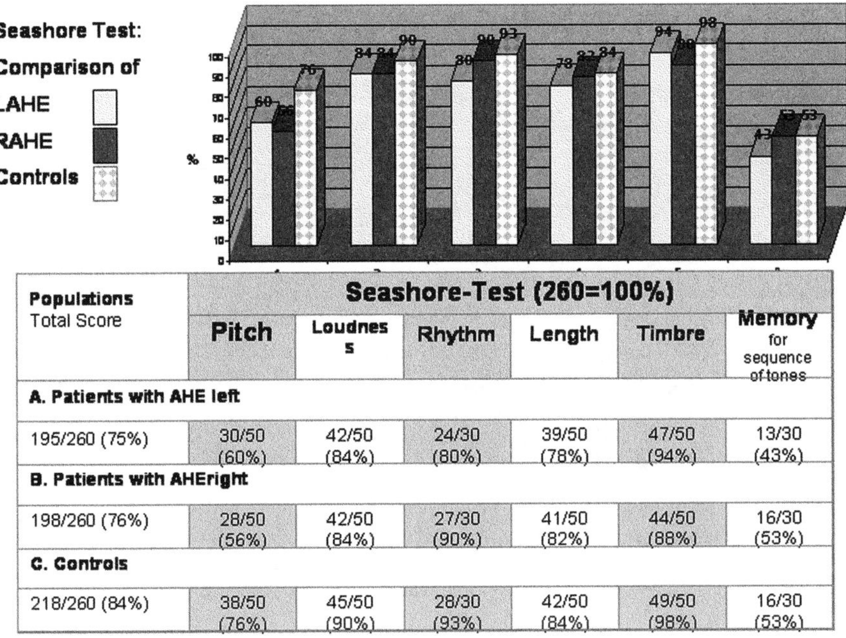

Populations Total Score	Seashore-Test (260=100%)					
	Pitch	**Loudness**	**Rhythm**	**Length**	**Timbre**	**Memory** for sequence of tones
A. Patients with AHE left						
195/260 (75%)	30/50 (60%)	42/50 (84%)	24/30 (80%)	39/50 (78%)	47/50 (94%)	13/30 (43%)
B. Patients with AHEright						
198/260 (76%)	28/50 (56%)	42/50 (84%)	27/30 (90%)	41/50 (82%)	44/50 (88%)	16/30 (53%)
C. Controls						
218/260 (84%)	38/50 (76%)	45/50 (90%)	28/30 (93%)	42/50 (84%)	49/50 (98%)	16/30 (53%)

FIGURE 3. Performance of control persons ($n = 11$) and patient groups after left (LAHE; $n = 11$) and right-sided selective amygdalohippocampectomy (RAHE, $n = 11$) in the Seashore test. Results are displayed as % of the maximum obtainable points. (Modified from Ferrara, L. 2003. Musikperzeption bei Epilepsiepatienten nach selektiver Amygdala-Hippocampektomie, Thesis, University Zurich. In preparation. Reproduced with kind permission of the author.)

ties. The Seashore test consists of a total of 260 tasks, allowing for a maximum of 260 points to be gained. It examines discriminative abilities for pitch, loudness, rhythm, length, timbre, and memory for sequences of tones.

The mean age of left-sided AHE patients (LAHE) was 48, for right-handed AHE (RAHE) patients it was 45, and for controls it was 50 years. The gender distribution was 8/3 for LAHE, 5/6 for RAHE, and 6/5 for controls. All 11 patients with LAHE were on antiepileptic medication, whereas all patients with RAHE were not. The musical standard was medium; there were no music professionals in either the patient groups or the control group.

The results, stratified for side of operation and gender, are displayed in FIGURES 3 and 4. As can be seen in FIGURE 3, for all groups together performance was worst in the category memory, and LAHE patients did worse than RAHE patients. Pitch discrimination also was reduced in AHE patients compared to controls. In the whole patient population it was worse in RAHE patients than in LAHE patients. Analyzing gender differences in this population, female controls did worse than did male controls in the categories "memory for tone sequences" and "pitch." Female patients with LAHE performed poorly in pitch discrimination.

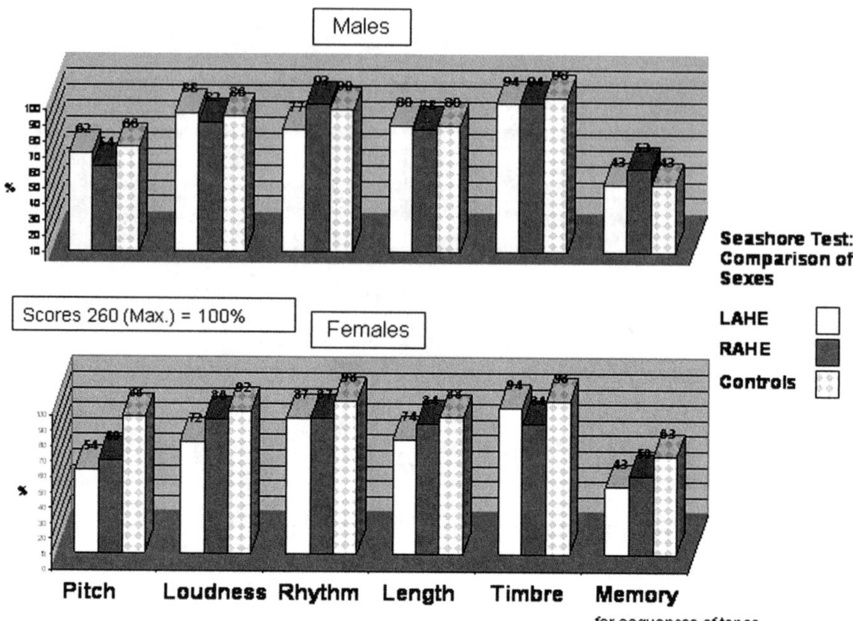

FIGURE 4. Performance of control persons ($n = 11$) and patient groups after left- (LAHE; $n = 11$) and right-sided selective amygdalohippocampectomy (RAHE, $n = 11$) in the Seashore test, stratified for gender. Results are displayed as % of the maximum obtainable points. (Modified from Ferrara L. Musikperzeption bei Epilepsiepatienten nach selektiver Amygdala-Hippocampektomie, Thesis, University Zurich; in preparation, with kind permission of the author.)

Musical Abilities after Temporal Lobe Cortectomy

Liegeois-Chauvel *et al.*[39] studied the ability to recognize changes in note intervals and to distinguish between different rhythms and meters in 65 right-handed patients who had undergone unilateral temporal cortectomy for the relief of intractable epilepsy. Their results showed that right temporal cortectomy impaired the use of both contour and interval information in the discrimination of melodies and left temporal cortectomy impaired only the use of interval information. They underlined the importance of the superior temporal gyrus in melody processing. The excision of the posterior part of the superior temporal gyrus was most detrimental for pitch and temporal variation processing. In the temporal dimension, they observed dissociation between meter and rhythm and the critical involvement of the anterior part of the superior temporal gyrus in metric processing.

Musicogenic Epilepsy

Musicogenic epilepsy is a rare disorder in which seizures are induced by hearing certain sounds, typically music.[40] An affective component of the stimulus is evident in some patients, yet nonmusical sounds, such as whirring machinery, can be effective triggers in others. The seizures are of simple or complex partial type, with inter-ictal and ictal epileptiform activity recorded from either temporal region.

There is often a latency of several minutes during which the patient must be continually exposed to the stimulus for the seizure to occur, and autonomic symptoms and signs may be noted before overt ictal behavior begins. Patients with musicogenic attacks may also have spontaneous seizures.

In 1997 we reported a case of musicogenic epilepsy with ictal SPECT study and discussed the findings of this patient in the context of 76 cases with musicogenic epilepsy found in the literature and an additional 7 cases followed in Zurich.[41] We analyzed these 83 patients according to precipitating musical factors, type of epilepsy, presumed localization of seizure onset, and demographic data. We concluded that musicogenic epilepsy is a particular form of epilepsy with a strong correlation to the temporal lobe and a right-sided preponderance and that a high musical standard might predispose to musicogenic epilepsy.

In this review, we found 34 patients in whom the investigators had commented on the characteristics of the musical stimulus in terms of musical category, familiarity, and instruments. In 17 patients, the stimulus was rated as "specific" and in 17 patients as "unspecific." In 15 patients, the "familiarity and/or the affective content" of the stimulus was found to be very important, whereas in 1 patient "novelty" of the musical stimulus was important.

Concerning the presumed "specificity" of instruments, the following rank order appeared: piano and organ 11 times, "jazz-instruments" (not specified) 2 times, string instruments 1, and wind instruments 1.

For 35 patients their music standard was reasonably well described: 4 patients (11%) were professional and 11 (31%) were amateur musicians. Five patients (14%) were "music-fans" and 7 (20%) had an above average interest in music. Only 8 patients (23%) with musicogenic epilepsy declared that they were not especially interested in music. Although no comparison with the level of general musical education of the relevant population is possible, these data suggest that a high music standard might predispose to musicogenic epilepsy.

Hemispheric Specialization and "Compensation:" How the Left Helps the Right Hemisphere

We studied the recorded organ performance of a professional musician with right temporal lobe epilepsy during a right temporal lobe seizure.[42] While playing an organ concert (John Stanley's Voluntary VIII, Op. 5), he suffered a complex partial seizure. Music analysis of the recorded concert performance during the seizure and comparison with other available exercise recordings and with the composition itself indicated seizure-induced variations. At the beginning of the seizure, the left hand started to become imprecise in time and deviated from the score, whereas the right hand remained faultless at this time. With increasing duration of the seizure discharge, the dissociation of both hands from the score increased, but the right hand compensated for the errors of the left hand in a musically meaningful way, that is, aiming to compensate for the seizure-induced errors of the left hand. The case illustrates untroubled musical judgment during epileptic activity in the right temporal lobe at the onset of the seizure. Whereas the temporal formation of the performance was markedly impaired, the ability of improvisation, in the sense of a "perfect mu-

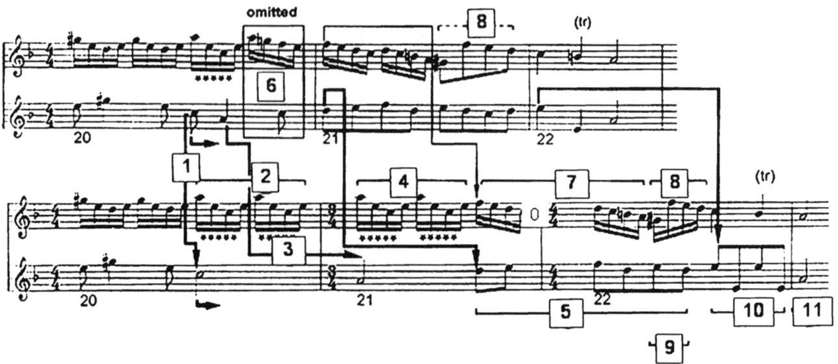

FIGURE 5. Musical analysis of a recording of an organ player who experienced an epileptic seizure with right temporal lobe onset. The seizure slowly progressed during a performance. The upper notation is the composition (original scores bars 20–22 of the Voluntary VII Op. 5 for organ by John Stanley (1713–1786); the lower notation, the distorted part at the beginning of the seizure. Note the errors of the left hand (l.h.), which are compensated by the right hand (r.h.) in a "perfect musical" way. To facilitate reading of the musical analysis, the main elements of seizure-induced variations are indicated by *numbers in boxes*. Main findings are: In the middle of the bar, the C is played four times longer as a minim (half-note) instead of a quaver (eighth note); see [1]. The r.h. playing the *top row* continues to play the sixth-motif normally, but twice; see [2]. The l.h. plays the following crotchet (quarter note) A as a minim, thereby doubling the normal duration of this note; see [3]. Again the r.h. repeats the sixth-motif twice; see [4]. In that way, the sixth-motif is expanded but remains musically meaningful. The following six quavers of the l.h. correspond exactly to the first six quavers of bar 21 of the original score and are played correctly in the normal tempo; see [5]. The last quaver of bar 20 (a C) is omitted; see [6]. The r.h. reacts to this omission meaningfully, and the corresponding semiquavers (sixteenth notes) of the original bar 21 are played; see [7]. The last four quavers of bar 21 (see [8]). are played in double speed by the r.h. as a sequence of semiquavers, synchronously with the last two quavers of the group of six mentioned above; see [9]. The organist seems to realize the waste of time and tries to compensate for it by doubling the tempo. The l.h. follows this idea by playing the octave-theme on E of bar 22 twice as quavers in arpeggiated fashion; see [10]. The r.h. finishes bar 22 in accordance with the original score, and this divergent part ends correctly in unison on A; see [11]. The second, stronger effect of the seizure is on bars 40–43. Beginning at bar 40, the tempo is continuously slowed down without any variation in the score until the performance is completely stopped in the middle of bar 43 and remains interrupted for 5 minutes. (Slightly modified from Ref. 42 with kind permission of the publisher.)

sical solution" to errors of the left hand, remained intact. As seen in FIGURE 5, the left hand performs with imprecise tone lengths, whereas the right hand performs perfectly. Deviations from the notation are evident in both hands, but the right hand (directed by the "healthy" = unaffected left hemisphere) compensates for the errors of the left (affected) hand in a musically meaningful way.

ESTHETIC LEVEL: CONSONANCE-DISSONANCE DICHOTOMY

A good example of what is called "Anmutungsqualität" in German and probably best translated into "pleasantness" is given in FIGURE 6, where the painting of

FIGURE 6. "Anmutungsqualität." The effect of lateral inclination of the head exemplified by Pablo Picasso, Les Amoureux, 1917 (Private property, New York). Taken from Frey S. and republished with kind permission of the author and the publisher Hans Huber, Bern, from the article: H.G. Wieser. 1982. Zur Frage der lokalisatorischen Bedeutung epileptischer Halluzinationen. *In* Halluzinationen bei Epilepsien und ihre Differentialdiagnose, K. Karbowski, Ed. Huber. Bern.

Picasso was modified by changing the inclination of the head relative to the body axis.[43] Whereas on the left the female gives the impression of being a devoted listener or of taking a liking to the young man, on the right side she looks somewhat proud and reserved.

In Western tonal-harmonic music, we traditionally speak of consonant and dissonant accords (intervals), although the consonance/dissonance dichotomy of the 12 half-tones is a difficult concept because it depends on sociocultural aspects and on the level of musical education and experience.

Western tonal music relies on this concept of pitch structure as a formal geometric structure that determines distance relationships within a harmonic or tonal space, explaining musical scales, harmonics, and relations between musical scales. In the Western tonal-harmonic concept, music is organized around one or more stable reference tones/(the tonic). Geometric models that conform to descriptions in music theory have been computed.[44–46]

Together with Mazzola, we undertook several studies to examine whether the human brain knows and operates with the mathematical transfer formula, which might be viewed as the basic mathematical principle of the classical consonance-dissonance dichotomy which has been used in the classical Western music tradition since Palestrina and Bach.[47,48]

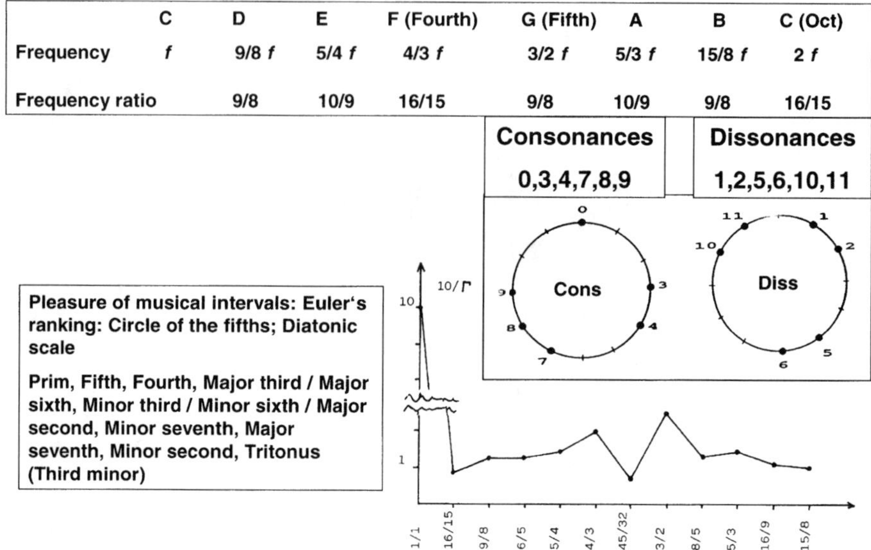

FIGURE 7. Euler's ranking of pleasantness of musical intervals, their frequency ratios, and the consonance-dissonance dichotomy.

Let me quickly introduce this consonance-dissonance dichotomy. According to the classical music theory in consonances, the frequency ratio is small. FIGURE 7 gives these frequency ratios and depicts Euler's ranking which is: prime, fifth, fourth, major third/major sixth, minor third/minor sixth/major second, minor seventh, major seventh, minor second, and tritonus (third minor).

If one gives the following task:
- divide the 12 half-tones into 2 groups of 6 in such a way that
  - one half can be transformed into the other half and that
  - the "transfer formula" is as simple as possible,

the result is:

- Of 462 mathematically possible solutions, only 2 solutions are a linear function.
- One of them is the classical "consonance-dissonance dichotomy," as used by Palestrina. It is "strong" and "autocomplementary."

Now make the test with the "transfer function" $y = (f)x = 5x + 2$ and calculate the remainder from 12. You will transfer the intervals on the left (consonances) to that written on the right (dissonances); for example, take the 3: $5 \times 3 = 15 + 2 = 17 - 12 = 5$.

Consonance $\rightarrow$ Dissonance	
0	2
3	5
4	10
7	1
8	6
9	11

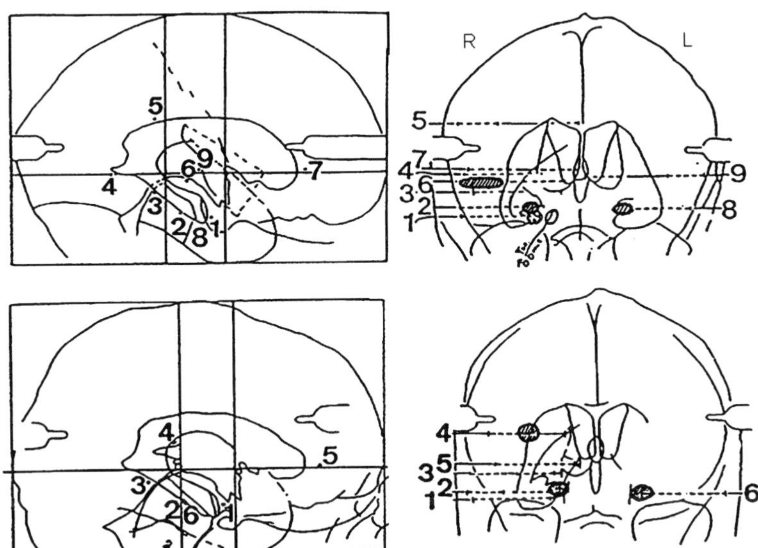

FIGURE 8. Graph of position of depth electrodes in two patients who volunteered in the Musical Consonance/Dissonance Test of Guerino Mazzola during their presurgical evaluation (long-term seizure monitoring). The EEG of the *hatched areas* was analyzed during the presentation of tones and musical accords (intervals). Parts of this figure are taken from Ref. 47 with kind permission of the publisher.

We then examined whether there is a brain representation of these rules of music theory, that is, we asked ourselves if the brain "knows and operates" with the "transfer function" $y = (f)x = 5x + 2$. To this end we examined (with informed consent) patients who underwent presurgical evaluation with depth electrodes (stereoelectroencephalography) and who had electrodes in the acoustic areas and in both mesiobasal limbic temporal lobe regions (FIG. 8). The examined patients were not especially interested in music but experienced pleasure when listening to simple folkloric ("harmonic") music. The special "music test" (G. Mazzola) was presented through earphones from a music computer, as described in detail elsewhere.[47,48] The consonant or dissonant tones were presented either simultaneously (as accords) or in sequence.

We then analyzed the EEG recorded directly from the hippocampal formation and from the acoustic cortex quantitatively per musical interval. The depth-recorded EEG during the presentation of an acoustic event was processed using FFT. Spectral participation vectors were calculated and used for statistical calculations (FIG. 9).

Results are depicted in FIGURE 10 and can be summarized as follows: The calculation of spectral participation values from EEG revealed that consonance or dissonance of musical intervals is reflected in the electrical activity (depth EEG) of the human brain. The reactions shown by the depth EEG of the limbic system (the spectral participation vectors) corresponded in precise and distinctly differing patterns to the consonances and dissonances presented. The main findings are: the EEG of the

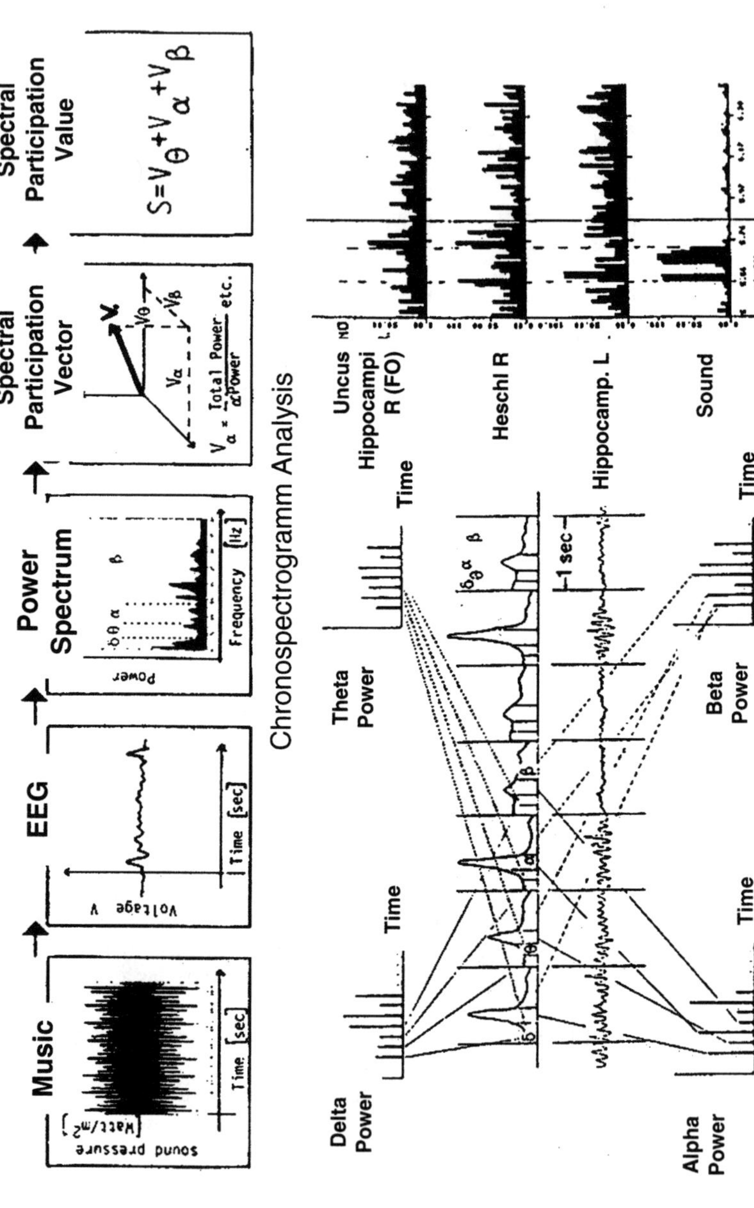

FIGURE 9. Illustration of the processing of EEG, recorded with depth electrodes, with calculation of Spectral Participation Values (SPVs) for each presented musical accord (consonances and dissonances) either in succession or simultaneously.

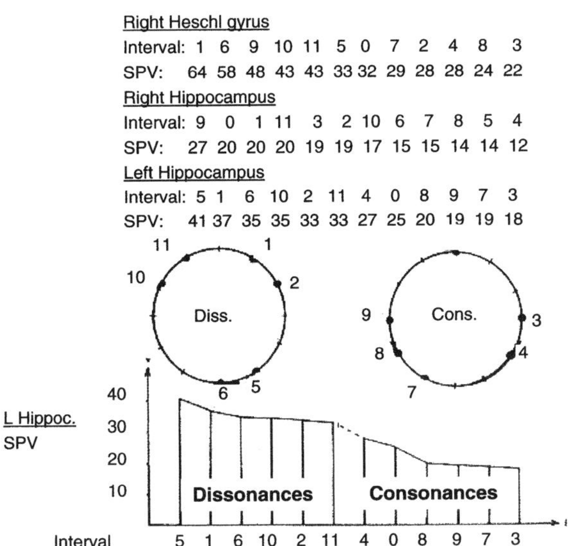

FIGURE 10. Spectral Participation Values (SPVs) for the EEG of the right acoustic cortex (Heschl gyrus) and the right and left hippocampal formation during listening to the Musical Consonance/Dissonance Test of Guerino Mazzola. The ranking order of the SPVs of the left hippocampal formation is graphically displayed and indicates that the EEG of the left hippocampus reacts differently to consonances compared to dissonances. This is an indication that the left hippocampus operates with the transfer function $y = (f)x = 5x + 2$. (Modified from Ref. 49 with kind permission of the editor and publisher.)

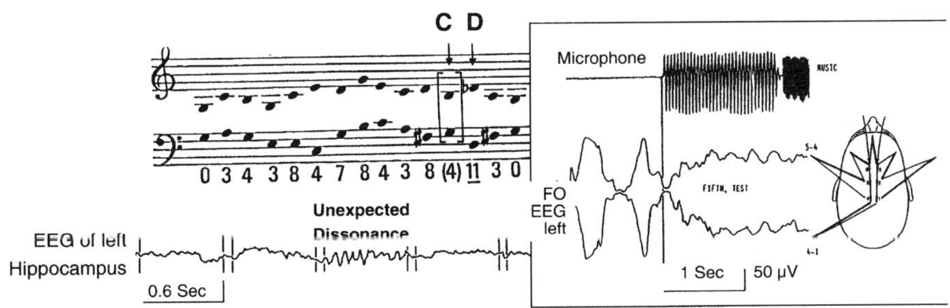

FIGURE 11. (*Left*) EEG of the left hippocampus during an unexpected dissonance in the context of consonant intervals. (*Right*) The foramen ovale electrode recorded EEG of the left mesial temporal lobe changed abruptly with the onset of a consonant tone displayed through earphones and recorded simultaneously with the EEG as a "microphone."

hippocampus reflects the consonance-dissonance dichotomy; the limbic brain thus knows and operates with the "transfer function" $y = (f)x = 5x + 2$. The EEG activity of the right planum temporale (close to Heschl's gyrus; the left Heschl's gyrus was not recorded) responded differently to a given succession of intervals (FIG. 10).

Consonant intervals were then substituted by a dissonance [D] in Fux's *Gradus ad Parnassum*, and the EEG reaction of the hippocampal formation was observed. In the example depicted in FIGURE 11, among 12 consonant intervals, we changed the twelfth into a dissonant major seventh (between bars). In most instances, the EEG activity of the left hippocampus was affected by a sudden unexpected dissonance.

These results may help to understand the different emotional responses which usually arise in response to consonant or dissonant musical intervals.

REFERENCES

1. PETSCHE, H., Ed. 1989. Musik-Gehirn-Spiel (Beiträge zum 4. Herbert von Karajan Symposium) Birkhäuser. Basel.
2. CRITCHLEY, M. & R.A. HENSON. 1977. Music and the Brain. W. Heinemann. London.
3. CLYNES, M., Ed. 1982. Music, Mind, and Brain. Plenum Press. New York.
4. BERENDT, J.E. 1983. Nada Brahma, Die Welt ist Klang. Insel-Verlag. Frankfurt a. Main.
5. DEUTSCH, D. 1984. The Psychology of Music. Academic Press. New York.
6. PEERY, J.C., P. WEISS., T.W. DRAPER & B. YOUNG, Eds. 1987. Music and Child Development. Springer. Berlin-Heidelberg.
7. STEINBERG, R., Ed. 1995. Music and the Mind Machine. Springer. Berlin-Heidelberg.
8. VON BEKESY, G. 1960. Experiments in Hearing. McGraw-Hill. New York.
9. ROUILLER, E.M. & D.K. RYUGO. 1984. Intracellular marking of physiologically characterized cells in the ventral cochlear nucleus of the cat. J. Comp. Neurol. **225:** 167–186.
10. VAUGHAN, H.G., JR. & W. RITTER. 1970. The sources of auditory evoked responses recorded from the human scalp. Electroencephalogr. Clin. Neurophysiol. **28:** 360–367.
11. HARI, R., K. AITTONIEMI, M.L. JARVINEN, *et al.* 1980. Auditory evoked transient and sustained magnetic fields of the human brain. Localization of neural generators. Exp. Brain Res. **40:** 237–240.
12. LIEGEOIS-CHAUVEL, C., A. MUSOLINO & P. CHAUVEL. 1991. Localization of the primary auditory area in man. Brain **114:** 139–151.
13. CROTTAZ-HERBETTE, S. & R. RAGOT. 2000. Perception of complex sounds: N1 latency codes pitch and topography codes spectra. Clin. Neurophysiol. **111:** 1759–1766.
14. GORDON, H.W. 1975. Hemispheric asymmetry and musical performance. Science **189:** 68–69.
15. HOWELL, P., I. CROSS & R. WEST, Eds. 1985. Musical Structure and Cognition. Academic Press. New York.
16. SUNDBERG, J. 1991. The Science of Musical Sounds. Academic Press Series: Cognition and Perception. Academic Press. San Diego.
17. PERETZ, I. 2002. Brain specialization for music. Neuroscientist **8:** 372–380.
18. PENHUNE, V.B., R.J. ZATTORE & A.C. EVANS. 1998. Cerebellar contributions to motor timing: a PET study of auditory and visual rhythm reproduction. J. Cogn. Neurosci. **10:** 752–765.
19. HENSCHEN, S.E. 1920. Aphasie, Amusie und Akalkulie. *In* Klinische und anatomische Beiträge zur Pathologie des Gehirns. Vol. 5. Nordiska Bokhandeln. Stockholm.
20. FEUCHTWANGER, E. 1930. Amusie. Studien zur pathologischen Psychologie der akustischen Wahrnehmung und Vorstellung und ihrer Strukturgebiete besonders in Musik und Sprache. Springer-Verlag. Berlin.
21. KLEIST, K. 1934. Grosshirnpathologie. Barth. Leipzig.

22. USTVEDT, H.J. 1937. Ueber die Untersuchung der musikalischen Funktionen bei Patienten mit Gehirnleiden, besonders bei Patienten mit Aphasie. Acta Med. Scand. (Suppl.) **96:** 1–737.
23. WERTHEIM, N. 1963. Disturbances of the musical functions. *In* Problems of Dynamic Neurology. L. Halpern, Ed. :162–180. Rothschild Haddassah University Hospital, Jerusalem.
24. WERTHEIM, N. 1969. The amusias. *In* Handbook of Clinical Neurology. P.J. Vinken & G.W. Bruyn, Eds. Vol. **4:** 195–206. North-Holland Publ. Co. Amsterdam.
25. ALAJOUANINE, T. 1948. Aphasia and artistic realisation. Brain **71:** 229–241.
26. LURIA, A.R., L.S. TSVETKOVA & D.S. FUTER. 1965. Aphasia in a composer (V.G. Shebalin). J. Neurol. Sci. **2:** 288–292.
27. BENTON, A.L. 1977. The amusias. *In* Music and the Brain. M. Critchley & R.A. Henson, Eds. Heinemann. London.
28. PERETZ, I., J. AYOTTE, R.J. ZATORRE, *et al.* 2002. Congenital amusia: a disorder of fine-grained pitch discrimination. Neuron **33:** 185–191.
29. AYOTTE, J., I. PERETZ & K. HYDE. 2002. Congenital amusia: a group study of adults afflicted with a music-specific disorder. Brain **125:** 238–251.
30. PENFIELD, W. & P. PEROT. 1963. The brain's record of auditory and visual experience: a final summary and discussion. Brain **86:** 595–696.
31. WIESER, H.G. 1980. Temporal lobe or psychomotor status epilepticus. A case report. Electroencephal. Clin. Neurophysiol. **48:** 558–572.
32. JONES-GOTMAN, M., M.L. SMITH & H.G. WIESER. 1997. Intra-arterial amobarbital procedures. *In* Epilepsy. A Comprehensive Textbook. J. Engel, Jr. & T.A. Pedley, Eds. :1767–1775. Raven Press. New York.
33. BORCHGREVINK, H.M. 1982. Prosody and musical rhythm are controlled by the speech hemisphere (incl. sound example). *In* Music, Mind, and Brain. M. Clynes, Ed. :151–157. Plenum Press. New York.
34. WIESER, H.G. & E. WITTLIEB-VERPOORT. 1995. Tone discrimination in patients with temporal lobe lesions. *In* Music and the Mind Machine. The Psychophysiology and Psychopathology of the Sense of Music. R. Steinberg, Ed. :115–126. Springer. Berlin.
35. WIESER, H.G., J. ENGEL, JR., P.D. WILLIAMSON, *et al.* 1993. Surgically remediable temporal lobe syndromes. *In* Surgical Treatment of the Epilepsies, 2nd Ed. J. Engel, Jr., Ed. :49–63. Raven Press. New York.
36. MILNER, B. 1962. Laterality effects in audition. *In* Interhemispheric Relations and Cerebral Dominance. V.B. Mountcastle, Ed. Hopkins. Baltimore.
37. CARY, H. 1923. Are you a musician? Professor Seashore's specific psychological tests for specific musical abilities. Scientific Am. **12:** 326–327.
38. LINTON, S.J. 1980. The man and the mind: Carl Emil Seashore, a pioneer on two frontiers. Swed. Pioneer Hist. Q. **31:** 122–128.
39. LIEGEOIS-CHAUVEL, C., I. PERETZ, M. BABAI, *et al.* 1998. Contribution of different cortical areas in the temporal lobes to music processing. Brain **121:** 1853–1867.
40. CRITCHLEY, M. 1937. Musicogenic epilepsy. Brain **60:** 13–27.
41. WIESER, H.G., H. HUNGERBUHLER, A.M. SIEGEL & A. BUCK. 1997. Musicogenic epilepsy: review of the literature and case report with ictal single photon emission computed tomography. Epilepsia **38:** 200–207.
42. WIESER, H.G. & R. WALTER. 1997. Untroubled musical judgement of a performing organist during early epileptic seizure of the right temporal lobe. Neuropsychologia **35:** 45–51.
43. FREY, S. 1977. Jahresbericht 1977: Schweizerischer Nationalfonds zur Förderung wissenschaftlicher Forschung. Bern.
44. MAZZOLA, G. 1990. Geometrie der Töne: Elemente der mathematischen Musiktheorie. Birkhäuser. Basel.
45. JANATA, P., J.L. BIRK, J.D. VAN HORN, *et al.* 2002. The cortical topography of tonal structures underlying Western music. Science **298:** 2167–2170.
46. ZATORRE, R.J. & C.L. KRUMHANSL. 2002. Mental models and musical minds. Science **298:** 2138–2139.

47. WIESER, H.G. & G. MAZZOLA. 1986. Musical consonances and dissonances: are they distinguished independently by the right and left hippocampi? Neuropsychologia **24:** 805–812.
48. MAZZOLA, G., H.G. WIESER, V. BRUNNER & D MUZZULINI. 1989. A symmetry-oriented mathematical model of classical counterpoint and related neurophysiological investigations by depth-EEG. Computers Math. Applic. (Pergamon Press) **17:** 539–594.
49. WEISER, H.G., & G. MAZZOLA. 1987. EEG responses to music in limbic and auditory cortices. *In* Fundamental Mechanisms of Human Brain Function. J. Engel, Jr. *et al.*, Eds. :131–143. Raven Press. New York.

Musicogenic Seizures

GIULIANO AVANZINI

Istituto Nazionale Neurologico C. Besta, Milan, Italy

ABSTRACT: Eighty-seven reports of patients with seizures induced by listening and/or playing music and one personal observation are reviewed. Music-induced (or musicogenic) seizures are currently classified among the reflex seizures precipitated by complex stimuli. According to the available information, they are defined as focal seizures due to a discharge involving lateral and mesial temporal and orbitofrontal areas. The specific musical component responsible for seizure precipitation is still undetermined. An important role is attributed to the emotional aspect of music. The existence of this rare disorder should be borne in mind by neurologists, who should also be aware of the existing musical test batteries that may help in understanding better the nature of triggering mechanisms responsible for this unique pathological condition. The implementations of the results of ongoing investigations on brain processing of musical information will advance our understanding of the mechanisms responsible for the transition from interictal to ictal phases of epilepsy.

KEYWORDS: seizures; music; epilepsy

INTRODUCTION

In most epilepsies, the transition from interictal to ictal phase seems to occur spontaneously due to homeostatic changes that are still poorly understood.[1] In a relatively small number of cases, however, external seizure-precipitating factors can be identified that provide the opportunity to investigate ictogenic mechanisms. Epilepsies presenting with seizures precipitated by exogenous stimuli are often designated as reflex epilepsies. Such a definition is unsatisfactory for two reasons. In many instances, stimulus-induced seizures may coexist with spontaneous seizures, thus making it inappropriate to use the precipitating factor as the main criterion for defining the syndromic entity. Moreover, the precipitating mechanism (e.g., reading) may involve high integratory cerebral functions, making problematic the boundaries between such "complex reflex epilepsies" and other types of epilepsy in which the relevant precipitating event (thinking, arithmetic calculation) seems independent of any external influence. Some clarifications are expected from the ongoing revision of the classification system of Seizures and Epilepsies by the *ad hoc* task force of the International League against Epilepsies[2] based on the scheme

Address for correspondence: Giuliano Avanzini, Department of Experimental Research and Diagnostics, National Neurologic Institute "Carlo Besta," via Celoria 11, 20133 Milan, Italy. Voice: (39) 02 2394253; fax: (39) 02 70600775.

avanzini@istituto-besta.it

Ann. N.Y. Acad. Sci. 999: 95–102 (2003). © 2003 New York Academy of Sciences.
doi: 10.1196/annals.1284.008

TABLE 1. Classification of reflex seizures

Precipitated by simple stimuli

Photosensitive and television-induced seizures

Eye closure sensitivity

Fixation-off sensitivity

    Precipitation by somatosensory stimuli

    Precipitation by proprioceptive stimuli (movements, startle)

Miscellaneous (olfactory, gustatory, audiogenic, vestibular)

Precipitated by complex stimuli

Reading epilepsy (primary, secondary)

Practice-induced seizures

Seizures precipitated by eating

Other complex sensorimotor triggers

    Hot water epilepsy, water immersion epilepsy

    Seizures precipitated by tooth brushing

Complex auditory stimuli

Musicogenic seizures

Other (musical pitch, voices of radio speakers)

Complex visual stimuli (own hand, safety pin, etc.)

Emotional precipitation, psychogenic epileptic seizures

reported in TABLE 1 that deals with reflex seizures and does not take into account the still unresolved problem of syndromic classification.

Currently, seizures in which a precipitating factor (either exogenous or endogenous) is detectable are referred to as provoked seizures, bearing in mind that other apparently spontaneous seizures may, in the future, be found to be provoked by a hitherto unknown precipitating mechanism.

Among provoked seizures as just defined, seizures induced by music or musicogenic seizures were the object of special interest after the seminal paper of Critchley[3] who reported clinical findings in 11 patients in whom seizures were precipitated by listening or playing various types of music. In this and in subsequent reports,[4,5] Critchley acknowledged previous observations published in medical journals beginning with Merzheevski's paper[6] in 1884 and suggested that unrecognized mentions of seizures induced by music could be found earlier in classic literature (namely, Shakespeare's "The Merchant of Venice"). After the attention of the neurological community had been alerted to this peculiar disorder, further clinical observations were reported that, since 1947,[7] were consistently documented by EEG recording.

Excellent reviews have been reported by Poskanzer et al.,[8] Titeca,[9] Critchley,[5] Scott,[10] Wieser,[11] and Zifkin and Zatorre.[12] Further observations published in recent years prompted me to reconsider this fascinating subject.

DEFINITION AND FREQUENCY

Audiogenic seizures, that is, seizures induced by acoustic stimulation, are among the most frequently observed stimulus-induced seizures in genetically predisposed rats, mice, and rabbits.[13] In humans, sudden noises can precipitate seizures in patients with cerebral palsy or metabolic disorders.

These sound-induced seizures are attributed to a peculiar susceptibility of subcortical structures and are by no means related to seizures induced by music, a complex stimulus whose processing involves cortical integratory activity. Thus, according to Zifkin and Zatorre,[12] after the Oxford English Dictionary's definition of music, musicogenic seizures are defined as seizures induced by sounds in melodic or harmonic combination. This definition includes every type of music, independent of any stylistic and cultural constraint and excludes other nonstartling acoustic stimulations that are exceptionally observed to elicit epileptic seizures or epileptiform EEG patterns.

Based on the literature, musicogenic seizures are a rare phenomenon with a prevalence as low as 1 case per 10,000,000 population,[5] whereas the prevalence of epilepsy is 6–8 per 1,000 persons and that of provoked seizures is estimated to range from 1–3 per 10,000 persons.[14] However, the prevalence of music-induced seizures may be underestimated because they can be unnoticed due to the long latency between stimulus onset and elicited clinical and EEG events and because of the limited musical education of people and physicians.

In the following sections, the characteristics of the effective musical stimuli and the clinical and EEG findings of the music-triggered seizures are summarized, based on 88 informative observations, 83 of which were previously reviewed by Wieser *et al.*[11] Since that time, four cases have been reported by Nakano *et al.*,[15] Trevathan *et al.*,[16] Genç *et al.*,[17] and Mórocz *et al.*,[18] and one is a personal observation.

CHARACTERISTICS OF THE MUSICAL STIMULUS

Analysis of 67 published observations in which music was the only known precipitating stimulus shows that different types of music can be effective. In most cases, however, seizures were reported to be consistently evoked by a specific type of music or a specific piece of music. The available information on the effective musical stimulation is summarized in TABLE 2.

Further specifications on the type of effective stimulation are available in 46 patients, as reported in TABLE 3. As for the type of instrument, the available information is reported in TABLE 4. Information on the level of musical training is available in 38 cases and is reported in TABLE 5.

CLINICAL CHARACTERISTICS OF MUSIC-INDUCED SEIZURES

Common to all cases that have been analyzed is the finding of a latency of up to several minutes between stimulus and seizure onset. A build-up of an emotional feeling during this latency is inconsistently reported. From personal observations, the seizures were invariably precipitated by playing the piano and occurred after an

TABLE 2. Effective musical stimulus in 67 patients in whom music was the only seizure-provoking stimulus

Classic	5
Predominant melodic	11
Predominant rhythmic	6
Melodic and rhythmic	23
Songs (text may be important)	10
Uncertain	14

TABLE 3. Type of effective stimulation available in 46 patients

Familiarity/affective content	15
Novelty	1
Specific musical piece	8
Specific musical style	6
Playing instruments	6

[a]"Arabesque" Turkish music;[17] folkloric Sardinian;[21] National hymns;[23] sad songs where violin predominates;[27] choral or sentimental music;[28] and romantic, popular, and waltz music.[29]

TABLE 4. Effective musical instruments

Piano and organ	12
"Jazz instruments" (not specified)	2
String instruments	1
Wind instruments	1

TABLE 5. Level of musical training

Professional	4
Amateur	12
"Music fan"	5
Interested in music	7
No special interest in music	12

induction time of several minutes. During this time the patient experienced such an unpleasant feeling that he completely stopped his beloved activity of performing music and was very reluctant to undergo a test while playing the piano during the EEG recording.

In several cases, namely, those reported in earlier publications, the information provided is not sufficient to define the type of seizures. In all of the observations that are suitably documented according to the clinical initial symptoms and their EEG correlates (when available), musicogenic seizures can be classified as focal seizures that may subsequently evolve into generalized tonic–clonic seizures. Genuine generalized seizures (that is, seizures in which the initial signs/symptoms suggest simultaneous involvement of both cerebral hemispheres) have not been reported in any case.

According to Wieser et al.[11] in a review of 44 patients with a suitable description of localization of the seizure focus, right lateralization was defined in 61% of the cases and temporal lobe epilepsy was diagnosed in 75%. A right temporal origin of the ictal discharge was subsequently confirmed by ictal recording performed by Nakano et al.,[15] Trevathan et al.,[16] and Genç et al.[17] In addition, ictal SPECT performed by Wieser et al.,[11] Nakano et al.,[15] and Genç et al.[17] pointed again to the right temporal area. Furthermore, in Trevathan et al.'s[16] patients, the MRI scan demonstrated a subtle area of cortical thickening in the right superior temporal gyrus, which was the site of ictal onset according to subdural EEG recordings. On the basis of these findings, the patients underwent surgical resection of the right superior temporal gyrus with complete disappearance of musicogenic seizures without a neurological deficit.

On the basis of the prominent role of the mesial temporal structures in the pathophysiology of temporal lobe epilepsy, the emotional and autonomic component observed in several music-induced seizures, and the results of available electrophysiological and SPECT data, the primary responsibility of mesial temporal structures in musicogenic seizures has been inferred. However, to confirm this possibility requires depth electrode ictal recording, which has never been performed. On the other hand, Mórocz et al.,[18] using fMRI measurements, reported in their case that activation of the right frontoorbital area corresponding to gyrus rectus occurred at an early phase of epileptogenic music before seizure-related activity was recorded from the left temporal lobe. The investigators suggested that the observed activity changes in the frontoorbital area reflected emotional arousal and memory related to the music rather than to seizure activity *per se*.

Overall, according to the foregoing data, musicogenic seizures fall into the category of focal (or partial) seizures whose generating discharge in most cases arises in the right hemisphere. The first temporal gyrus seems to be consistently involved in the induction process; other areas (namely, frontal ones), however, are reported to be coactivated or even to lead the sequence of activation of cortical areas.

TRIGGERING MECHANISMS FOR MUSICOGENIC SEIZURES

From analysis of reported observations, it is difficult to identify a critical factor common to all types of musical stimuli that is effective in inducing seizures. This lack of specificity of the stimulus and long duration of the induction time led to a dispute over the place of musicogenic seizures among reflex seizures, which in a strict sense are defined as those that are reliably provoked by natural or artificial stimulation of a certain receptor or group of receptors.[14] These characters, however, are common to other types of event-related seizures that are triggered by integration of higher cortical activities such as reading and thinking and that are included under the heading of complex reflex epilepsies (TABLE 1).

Whatever the conclusion of the *ad hoc* task force of the International League Against Epilepsy,[19] it is clear that the cause-effect relation in music-induced seizures is not simple, and it postulates the occurrence of some mediating process, leading several investigators to stress the putative role of music-associated emotions. According to Critchley,[5] Scott,[10] Gastaut and Tassinari,[20] Vizioli,[21] and Wieser,[11] music may create an emotional state that *per se* is responsible for seizure precipitation. This view is supported by the observation that the presentation of a musical trigger can cause increasing agitation and autonomic signs such as tachycardia and hyperventilation before the clinical attack begins and it fits the reports of primary or secondary involvement of frontal and temporal limbic structures that have a crucial role in our emotional life (Refs 11,16–18 and personal observation). Although the affective aspect of musical experience is, without a doubt, a crucial component of the human response to music, as exhaustively elaborated by Zatorre elsewhere in this volume, its role in seizure induction is not always obvious. For instance, the patient reported on by Brien and Murray[22] identified as effective stimulus a "metallic" quality of the singer's voice, and one of Jallon *et al.*'s[23] patients was particularly sensitive to very rhythmic musical pieces that did not seem to elicit a particular emotional response. However, even patients in whom music-associated emotion was considered a crucial precipitating factor did not report the occurrence of seizures in other, nonmusical, emotional situations. Patients may have a consistent emotional feeling associated with the seizure-inducing music, because that specific music was listened to by him or her in a highly emotional situation, as in the case studied by Gastaut and Tassinari.[20] A conditioned mechanism has been hypothesized to intervene by Shaw and Hill[7] according to a view that it is currently no longer accepted.

Caution is needed in interpreting the role of emotion that is associated with listening to music as the main trigger of musicogenic seizures because of the risk of improperly including musicogenic seizures in the general category of seizure precipitated by emotion (or psychogenic seizures) (TABLE 1), thus disregarding the specifity of the musical trigger.

The nature of music-associated emotions and the way in which they are processed in the brain was investigated by Krumhansl,[24] Blood *et al.*,[25] and Blood and Zatorre[26] and is further discussed in Zatorre's paper in this volume. A differential activation of limbic structures was observed to depend on the pleasant or unpleasant emotional feeling associated with the type of musical stimulus, but the role of the specific affective aspects of musical experience needs to be further characterized.

CONCLUSIONS

Although rare, music-induced seizures are of considerable interest to students of brain activity involved in musical perception and in mechanisms underlying seizure precipitation. Their analysis may:

- Advance our understanding of the mechanisms responsible for the transition from interictal to ictal states, and

- Contribute to defining the anatomic functional organization of neural systems dedicated to processing musically relevant acoustic features.

The available data indicate that definite neural systems in the temporal lobe are consistently involved in triggering music-induced seizures. They include not only the areas that are known to have a primary role in processing musically relevant acoustic features (namely, the superior temporal gyrus), but also the mesial temporal structures that are part of the limbic system. Moreover, the paralimbic frontoorbital structures have been reported to be activated in an early phase of exposure to epileptogenic musical stimulation.[18] The involvement of limbic and paralimbic systems that are known to participate in processing the emotional aspects of music perception stresses the role of emotion in triggering musicogenic seizures that had been repeatedly suggested on the basis of several clinical observations.

The neurologist who observes a patient presenting seizures precipitated by music should take advantage of existing musical test batteries to analyze the specific role of the different musical components in seizure precipitation. Standard stimulation procedures for eliciting music-induced EEG and/or behavioral changes in music-susceptible (but not only) patients should be defined, and guidelines on how to manipulate the effective musical stimulus and to identify its relevant epileptogenic components should be made available to clinicians.

Results of ongoing investigations aimed at defining the specific nature of musical emotion may be implemented to provide us with better tools to test patients with musicogenic seizures.

ACKNOWLEDGMENTS

The assistance of Professor Heinz Gregor Wieser in making available the original material on which his 1997 review was based is gratefully acknowledged.

REFERENCES

1. DE CURTIS, M. & G. AVANZINI. 2001. Interictal spikes in focal epileptogenesis. Prog. Neurobiol. **63:** 541–567.
2. ENGEL, J., JR. 2001. A proposed diagnostic scheme for people with epileptic seizures and with epilepsy: report of the ILAE task force on classification and terminology. Epilepsia **42:** 796–803.
3. CRITCHLEY, M. 1937. Musicogenic epilepsy. Brain **60:** 13–27.
4. CRITCHLEY, M. 1942. Musicogenic epilepsy: two cases. J.R.N. Med. Ser. **28:** 182–184.
5. CRITCHLEY, M. 1977. Musicogenic epilepsy. *In* Music and the Brain. M. Critchley & R.A. Hensen, Eds. :344–353. Heinemann. London.
6. MERZHEEVSKI, I.P. 1884. Slochai epliepsi pripedki kotorsi vizibayootsya kekotopemi misikelnimi tonomi. Minutes of the Meeting of the S. Petersbourg Soc. Psichiatr. (Quoted by Titeca, 1965).
7. SHAW, D. & D. HILL. 1947. A case of musicogenic epilepsy. J. Neurol. Neurosurg. Psychiatry **10:** 107–117.
8. POSKANZER, D.C., A.E. BROWN & H. MILLER. 1962. Musicogenic epilepsy caused only by a discrete frequency band of church bells. Brain **85:** 77–82.
9. TITECA, J. 1965. L'épilepsie musicogénique. Revue générale à propos d'un cas personnel suivi pendant quatorze ans. Acta Neurol. Belgica **65:** 598–648.
10. SCOTT, D. 1977. Musicogenic epilepsy. *In* Music and the Brain. M. Critchley & R.A. Hensan, Eds. :354–364. Heinemann. London.
11. WIESER, H.G., H. HUNGERBÜHLER, A.M. SIEGEL & A. BUCK. 1997. Musicogenic epilepsy: review of the literature and case report with ictal single photon emission computed tomography. Epilepsia **38:** 200–207.
12. ZIFKIN, B.G. & R.J. ZATORRE. 1998. Musicogenic epilepsy. *In* Reflex Epilepsies and Reflex Seizures: Advances in Neurology, vol. 75. B.G. Zifkin, F. Andermann, A.

Beaumanoir & A.J. Rowan, Eds. :273–381. Lippincott-Raven Publishers. Philadelphia.

13. AVANZINI, G. 1995. Animal models relevant to human epilepsies. Ital. J. Neurol. Sci. **16:** 5–8.

14. GASTAUT, H. 1989. Synopsis and conclusions of the International Colloquium on reflex seizures and epilepsies, Geneva 1988. *In* Reflex Seizures and Reflex Epilepsies. A. Beaumanoir, H. Gastaut & E. Naquet, Eds. :497–507. Editions Médicine & Igiene. Genève.

15. NAKANO, M., Y. TAKASE & C. TATSUMI. 1998. A case of musicogenic epilepsy induced by listening to an American pop music. [Article in Japanese]. Rinsho Shinkeigaku **38:** 1067–1069.

16. TREVATHAN, E., R.J. GEWIRTZ, J.E. CIBULA, *et al.* 1999. Musicogenic seizures of right superior temporal gyrus origin precipitated by the theme song from "The X-Files." Epilepsia **40** (Suppl 2): 23.

17. GENÇ, B.O., E. GENÇ, N. TASTEKIN & N. IIHAN. 2001. Musicogenic epilepsy with ictal single photon emission computed tomography (SPECT): could these cases contribute to our knowledge of music processing? Eur. J. Neurol. **8:** 191–194.

18. MOROCZ, I.A, A. KARNI, S. HAUT, *et al.* 2003. fMRI of triggerable aurae in musicogenic epilepsy. Neurology **60:** 705–709.

19. COMMISSION ON CLASSIFICATION AND TERMINOLOGY OF THE INTERNATIONAL LEAGUE AGAINST EPILEPSY. 1989. Proposal for revised classification of epilepsies and epileptic syndromes. Epilepsia **30:** 389–399.

20. GASTAUT, H. & C.A. TASSINARI. 1966. Triggering mechanisms in epilepsy. The electroclinical point of view. Epilepsia **7:** 85–138.

21. VIZIOLI, R. 1989. Musicogenic epilepsy. Int. J. Neurosci. **47:** 159–164.

22. BRIEN, S.E. & T.J. MURRAY. 1984. Musicogenic epilepsy. Can. Med. Assoc. J. **131:** 1255–1258.

23. JALLON, P., L.A. HERAUT & J.M. VANELLE. 1989. Musicogenic epilepsy. *In* Reflex Seizures and Reflex Epilepsies. A. Beaumanoir, H. Gastaut & E. Naquet, Eds. :269–274. Editions Médicine & Igiene. Genève.

24. KRUMHANSL, C.L. 1997. An exploratory study of musical emotions and psychophysiology. Canad. J. Exp. Psychol. **51:** 336–353.

25. BLOOD, A.J., R.J. ZATORRE, P. BERMUDEZ, *et al.* 1999. Emotional responses to pleasant and unpleasant music correlate with activity in paralimbic brain regions. Nature Neurosci. **2:** 382–387.

26. BLOOD, A.J. & R.J. ZATORRE. 2001. Intensely pleasurable responses to music correlate with activity in brain regions implicated in reward and emotion. Proc. Natl. Acad. Sci. USA **98:** 11818–11823.

27. HAMOIR, R. & J. TITECA. 1948. Étude éléctroéncephalographique d'un cas d'épilepsie musicogénique. Rev. Neurol. **80:** 635.

28. REESE, H.H. 1948. The relation of music to diseases of the brain. Occup. Ther. Rehab. **27:** 12–18.

29. WEBER, R. 1956. Musikogene epilepsie. Nervenartz **27:** 337–340.

Roundtable I: Dissecting the Perceptual Components of Music

Introduction

CAROL L. KRUMHANSL

Department of Psychology, Uris Hall, Cornell University, Ithaca, New York 14853, USA

As initially conceived by Dr. Giuliano Avanzini, the roundtable was designed to bring together neuroscientists and musicians to discuss the elementary components of music. The idea was to challenge the traditional elements upon which musical education is based. The musical curriculum traditionally divides the study of Western music into such components as rhythm, melody, and harmony, with less attention to musical timbre and orchestration. Historical and analytic studies are distinguished from music theory, which is treated somewhat separately from music composition.

In the science of music, these divisions have suited the scientific habit of reducing complex problems to simpler ones. Thus, music research has tended to focus on one or another of the components: melody versus harmony, meter versus rhythm, and so on. The disciplinary subdivisions have introduced further distinctions, such as those between perception and performance, psychoacoustics and cognition, and effects of development as opposed to effects of musical training or expertise. Empirical laboratory research and computational modeling have developed along parallel but somewhat independent lines, and the neuroscience of music has appeared more recently as a largely separate strand of research.

One possible outcome anticipated in Dr. Avanzini's initial planning was that the discussion might bring to our attention the utility of further dissecting music into yet more elementary components. However, another, more intriguing possibility was proposed: "On the other hand, some musical abilities as defined by neuropsychological investigations may be relevant for more than one of the competencies as traditionally identified by music theorists." This invitation to challenge traditional assumptions and approaches was richly rewarded. Despite the rather open-ended nature of the charge given to the roundtable participants, a remarkably strong consensus emerged on several overarching themes.

In various ways, the contributions show an increased respect for the complexity of music than was generally evident in earlier research. Implicitly, the investigators acknowledge the advantage of employing experimental materials that are as close to real music as allowed by the question of interest. Thus, a certain amount of "experimental control" is sacrificed for the musicality, or "ecological validity," of the materials employed. One example of this, in Séverine Samson's precise studies of

Address for correspondence: Department of Psychology, Uris Hall, Cornell University, Ithaca, NY 14853, USA. Voice: 1.607.255.6351; fax: 1.607.255.8433.

clk4@cornell.edu

Ann. N.Y. Acad. Sci. 999: 103–105 (2003). © 2003 New York Academy of Sciences.
doi: 10.1196/annals.1284.009

the neuropsychological basis of timbre perception, is the emphasis on the multi-dimensional aspect of timbre and the extension of timbre judgment tasks to melodies, not just single tones. Although most research has pointed to right temporal lobe involvement in timbre discrimination tasks, these innovations have pointed to additional left temporal lobe involvement when processing tone sequences with tones varying in duration and frequency.

To cite another example of the concern with musical realism, the studies of Thomas Münte and collaborators include an experiment that focuses on event-related brain potentials to rhythmic sequences. Despite the fact that this method requires precise registration of brain activity to events in time, the investigation used real drum sequences taken from a drum school. These sequences contained micro-variations in the range of 15–20 ms. As a consequence, brain responses to deviations in these contexts were found to require a longer offset than had previously been found with artificial stimuli, probably representing a more realistic finding for music. Similarly, Luisa Lopez and collaborators have included in their electrophysiological studies of pitch deviations tuning deviations in block and arpeggiated chords as well as more extended passages of music.

A related development is the inclusion of materials outside the tradition of Western tonal music. Ian Cross argues, "…an inclusive delineation of music is an absolute prerequisite if cognitive neuroscience is to aspire to an adequate degree of generalisability in respect of the theories that underpin empirical research and the findings that it generates in respect to music." Given the wide range of materials in the various empirical contributions presented in the roundtable, the cognitive neuroscience of music appears well on its way to achieving this kind of generality. Represented, for example, is African-American music in the performance expression study by Giovanni De Poli. Interestingly, this program of research has found that the framework developed in the context of Western classical music for communication between performer and listener of the intended expression, with suitable modifications, extends well to this and other musical styles.

The issue of cultural familiarity is addressed perhaps most directly in the fMRI study by Steven Demorest and Steven Morrison. Their study measured brain responses in Western listeners as they heard Western and Chinese music. Behaviorally, differences appeared between the two styles of music in that listeners were more able to recognize the music subsequent to the brain scanning session. However, the fMRI results did not show differences between the two styles of music. That is, similar areas were activated whether listeners heard a familiar or an unfamiliar style. The investigators discuss various theoretical and methodological possibilities for this lack of difference, but this finding may signal the difficulty of finding localized brain responses as a function of the cultural origin of music. Instead, the fMRI results may reflect common processing mechanisms applied to a range of musical styles.

The concern with cultural influences also raises interesting questions apart from neuropsychological correlates. Among them is the possible influence that a culture's language has on its music. In his contribution, Aniruddh Patel addresses this by considering rhythmic differences between the French and English language, and tests whether a similar difference might be found in the musical rhythms of those cultures. Spoken English exhibits a greater variability of vowel duration than does spoken French. Analysis of a large number of musical themes from English and French classical music discloses a similar difference in the variability of tone duration. This study, then, finds some cultural (language) imprint on the music.

Although the findings concerning culture presented in the roundtable were on the whole somewhat mixed, influences of musical training were strong and consistent. As argued by Thomas Münte, trained musicians provide a "model-organism for neuroplasticity" given the sustained and extensive training that musicians undergo. In the experiments that he and collaborators have done, specific effects of musical training are found in event-related potentials. One experiment, focusing on pitch, found a stronger pattern of results for string players (especially trained on intonation) than for nonmusicians. Another experiment, focusing on processing of spatial location, found differences between pianists and nonmusicians, on the one hand, and conductors, on the other. Conductors need to combine two skills: focusing attention on a particular spatial location, while monitoring the entire auditory space. Finally, in the study mentioned earlier, drummers had a stronger and more sustained response to rhythmic deviations than did woodwind players and nonmusicians.

A similar strategy of comparing musicians and nonmusicians was followed in the study conducted by Luisa Lopez and collaborators on pitch discrimination using EEG. The amplitudes and latencies of both early and late responses differed as a function of the subject's musicality score, especially for the more musical sequences. Although Steven Demorest and Steven Morrison did not find fMRI differences as a function of cultural familiarity, their study did uncover differences between musicians and nonmusicians. Finally, Séverine Samson reviews a number of studies finding musical training effects in physiological measures for tasks involving musical timbre. Thus, musical training appears to be emerging as one of the most promising strategies in the cognitive neuroscience of music, and its effects appear to cut across the various capacities that are often covered separately in traditional music education.

Music as a Biocultural Phenomenon

IAN CROSS

Faculty of Music, University of Cambridge, Cambridge CB3 9DP, UK

ABSTRACT: There is a need to clarify the domain of music as an object of cognitive and neuroscientific research. This paper explores some ramifications of an inclusive delineation of the domain of music for such research.

KEYWORDS: music; culture; ethnomusicology; language; evolution

There is now a vast amount of cognitive-scientific research on music. Most of that research has focused on music as organized sound and on the human experience of organized sound. The question this paper raises is, "Is this enough?" Is a cognitive-scientific understanding of organized sound and the human experience of organization in sound adequate to account for the human experience of music?

At a first approximation, the answer may seem to be "yes." Indeed, although musical practices are hugely diverse and some remain virtually unexplored from the perspective of cognitive science, what is known about the human experience of musical sound can at least suggest hypotheses about how the cognition of this diverse and cognitively underexplored music can be understood.

For example, within the *campesino* culture of Northern Potosí in Bolivia, there is a marked preference for the timbre of many musical sounds (particularly those on wind instruments) to be severely inharmonic, a quality referred to as *tara*.[1] This stands in sharp distinction to "Western" preferences for euphonious and harmonic timbres, but it is at least comprehensible. Timbral preference does appear to be one of the more culturally variable dimensions of music, and in this case it would appear to relate to qualities of sounds in the natural environment that have significant survival value—here, sounds made by rutting male llamas!

A somewhat more complex situation seems to pertain in the several Central African cultures in which Simha Arom[2] found that the transposition of a melody that changed the interval size between equivalent pairs of notes was judged to be "the same" melody as the original; here, it seems that that judgments about identity between transposed melodies are based not on preservation of exact interval sizes but on what he refers to as "permutational" identity, involving the preservation of interval in terms of number of scale steps, irrespective of the relative size of the scale steps.

Address for correspondence: Dr. Ian Cross, Faculty of Music, University of Cambridge, West Road, Cambridge CB3 9DP, UK. Voice: +44 (0)1223 335185; fax: +44 (0)1223 335067.
ic108@cus.cam.ac.uk
http://www.mus.cam.ac.uk/~cross/

Ann. N.Y. Acad. Sci. 999: 106–111 (2003). © 2003 New York Academy of Sciences.
doi: 10.1196/annals.1284.010

MUSIC AS MORE THAN SOUND

However, despite the fact that an understanding of the human experience of organization in sound can explain much about music cognition, music has more dimensions than the sonic. There are instances in which the "foundational" categories of theorized western musical practice that form the basis for most perceptual and cognitive theory and experimentation about music do not appear adequate to explain the practice and experience of music. For example, in certain Southern African cultures, Blacking[3] showed that the structure of melodies on kalimba (thumb-piano) can on occasion depend more on the sequence of movements involved in production of the melody than on the properties of the pitch patterns produced. More recent work[4] demonstrates that the pitch structures of blues guitar solos over the last century are better explained as consequences of constraints on dynamic patterns of hand and finger movements on the guitar fretboard than in terms of abstract principles of melodic or harmonic organization.

To explore perception and cognition in terms of these instances of musical practice requires a reformulation of the dimensions of music and musicality, a redefinition of music in terms of both sound and movement and of the relationships between them. This notion is given added weight by the existence of cultures and societies in which no term cognate with the conventional western term "music" appears to exist, particularly evidenced in some African societies as, for example, in the Igbo notion of *nkwa* as embracing singing, making music, and dancing.[5] Here, it seems that music-making is conceived of as a component of a broader suite of practices, all of which involve overt action.

Indeed, just as with language, music is properly not a characteristic of individuals but of communities, evidenced not just in individual cognitions and behaviors but in inter-individual interactions that may especially involve entrainment of action and interaction,[6,7] based on the abstraction of temporal regularities from sound sequences[8] and action patterns in time.[9]

It appears that the human experience of music is most adequately conceived of as having a social and interactive dimension,[10] although this may not be immediately evident if we take the paradigm case of music in cognition to be represented by the solitary listener. However, the postulate that *the* predominant mode of engagement with music lies in the listening experience appears to apply only in certain (largely, Western) cultural contexts, and although participation in Western art-music as a mature listener may appear not to involve overt action or interaction, the *acquisition* of that capacity almost invariably does.[11,12]

In fact, the notion of music as best represented by individual engagement with the "autonomous artwork" seems to be a product of the last 200 years of Western practice and theory. Music construed in a broader context than that which has been described as the "élite, canonic practices in the West" seems to be embedded in more extensive suites of social practices in most cultures, gaining and returning meaning from and to the contexts within which it is embedded. As Bohlman[13] notes, "Both ethnomusicologists and music sociologists insist that all human beings produce music and that expressive practices do not divide into those that produce music and those that produce something else, say ritual or dance. Music accumulates its identities… from the ways in which it participates in other activities…."

In other words, music is not only sonic, embodied, and interactive; it is bound to its contexts of occurrence in ways that enable it to derive meaning from, and interactively to confer meaning on, the experiential contexts in which it occurs, these meanings being variable and transposable. Some instances from non-Western sources might clarify the range of contexts within which music can be embedded and hint at the diverse significances it can possess and confer. Music has functioned as a medium for communication with the dead for the Kaluli of Papua New Guinea,[14,15] forming part of a nexus that binds birds, souls, places, and people at a time of emotional and numinal transformation; music has been a mechanism for restructuring social relations, as in the *domba* initiation of the Venda;[16] for shamans of Outer Mongolia, music helps constitute the path that they travel in the course of a healing process, while at the same time assisting to render their actions liminal for their patients;[17] and music provides a flexible and ambiguous medium for the complex social, ritual, and sexual dialogues that make up the "flower songs" – *hua'er* – of northwest China.[18,19]

Indeed, music in globalized Western culture is also embedded in a broad range of activities and contexts: just think of the embeddedness of music in religious and liturgical ceremonies, in theater, television, and film, in dance, and of course in the ubiquitous "background" music encountered as the soundtrack to life in supermarkets, malls, lifts, restaurants, and bars. In each of these very different circumstances, music's meaning is rarely, if ever, explicit; the music can appear to be *about* something, but its *'aboutness'* can vary from context to context, within a context, and from individual to individual.

TOWARD AN INCLUSIVE DEFINITION OF MUSIC

To sum up, as I've suggested elsewhere,[20] *"music embodies, entrains and transposably intentionalises time in sound and action."*

Music's embodied nature is evident in the fact that, as just noted, many cultures do not differentiate between activities that in Western and westernized societies might be separately categorized as "music" and "dance." In other words, the Western notion of music as, in essence, "organized sound" might constitute a culturally specific and partial demarcation of the correlates of human musicality. Moreover, the developmental precursors of music in infancy (in the form of proto-musical behaviors) and through early childhood are exploratory and kinesthetically embedded, being closely bound to vocal play and to whole body movement.

Music's embodied characteristics may provide the basis for music's capacities to coordinate and entrain action in time. While types of interactional synchrony may be observed in conversational interactions, music, unfolding within periodic and hierarchically structured temporal frameworks that are grounded in bodily periodicities, enables patterned interaction in time that may involve entrainment to an isochronous repetitive pulse.[21]

Music's capacity to entrain overlaps with, but is distinct from, its capacity to mean; music is embedded in social action, deriving meaning from that action and in turn endowing it with significance—"intentionalizing" it—for its participants. Music's significances may arise from association, or by means of the evocation of emotional body-state, or through some mimetic mapping whereby attributes of musical patterns of sound and movement share or track temporal characteristics of events as experienced.[22] However, the significances of one and the same musical activity can be interpreted quite differently by participants without undermining the apparent integrity of collective musical behaviors. Such behaviors have an ambiguity of reference, but an unambiguous referentiality, that may afford cost-free modes of engaging in and rehearsing social interaction.

This model of music has various implications for an understanding of the neuroscience of music. In the first place, depending on the particular forms and situations in which music may be manifested for particular cultures, we would expect a greater or lesser likelihood that neurophysiological correlates of sonic pattern and of patterned movement are more or less likely to co-occur. Certainly, the notion of exploring music by examining the neurophysiological correlates of the acoustic signal that partially constitutes it must be recognized as severely culture specific and as likely to afford only partial access to an understanding of the neurophysiology of music.

Similarly, it can be proposed that the capacity of joint musical behaviors to entrain collective behaviors might result in joint neurophysiological dynamics at various levels of specificity; in the absence of neuroscientific tools and methods that can address this question, it appears that the present focus on the individual within the neurosciences of music might again yield a severely incomplete account.

Moreover, given music's intentional status we would expect a significant overlap between the neurophysiological correlates of speech and music; it is conceivable that speech and music are best considered as constituting poles of a continuum rather than existing as categorically discrete phenomena. Indeed, neurophysiologically speaking, it is likely that "music" in this model will overlap with many, if not most other, domains of human activity and behavior, particularly those concerned with social interaction.

What this implies is that an inclusive delineation of music is an absolute prerequisite if cognitive neuroscience is to aspire to an adequate degree of generalizability with respect to the theories that underpin empirical research and the findings that it generates in respect of music. The cognitive sciences need operational definition(s) of music that connect biology to culture. This contrasts markedly with the situation in the study of language, where current thinking and research have produced delineations of the language faculty in broad (concerned with sensorimotor, conceptual-intentional, and computational systems and their interactions) and narrow (concerned solely with the computational system) senses[23] that are grounded in both biology and culture and that are likely to be productive with respect to empirical research. It seems probable that the empirical study of the cognitive and neural bases of music will require the development of similar broad and narrow delineations of the domain of research; the definition of music given here is only a minimal step towards achieving a broad delineation, while those theories that might form the substrate of a narrow delineation[24] are restricted in their applicability to specific types of western tonal music (although computational theories have been developed for a few non-Western musics[25]).

Given the diversity of musical practices and the variability in the ways in which music's meanings appear bound to its uses, this implies that a pluralistic and multidisciplinary approach in which there is a continual and ongoing dialogue between cognitive neuroscientists, musicians, musicologists, and ethnomusicologists is necessary. There is an imperative need for the cognitive sciences of music to "triangulate" music as both a biological and a cultural phenomenon (indeed, as a peculiarly human phenomenon, as soon as we penetrate beyond music as organized sound, music appears to be quite outside the repertoire of behaviors of other species).

CONCLUSION

To close, does any of this matter? I suggest that there are at least two reasons why it should: (1) If we are to establish cognitive-scientific understandings of musics, those understandings must be as generalizable as those that are applicable in, say, the domain of language. (2) The sciences have political consequences in the real world for their objects of study. Given this, it would be reassuring to be able to believe that the sciences in question are as well founded as they should be to qualify as science.

REFERENCES

1. STOBART, H.F. 1996. *Tara* and *Q'iwa*: worlds of sound and meaning. *In* Cosmología y Música en los Andes (Music and Cosmology in the Andes). M.P. Baumann, Ed. :67–81. Biblioteca Iberoamericana and Vervuert Verlag. Madrid and Frankfurt.
2. AROM, S. 1997. Le "syndrome" du pentatonisme africain. Musicæ Scientiæ **1**: 139–163.
3. BLACKING, J. 1961. Patterns of Nsenga *kalimba* music. African Music **2**: 3–20.
4. NELSON, S. 2002. Melodic improvisation on a twelve-bar blues model: an investigation of physical and historical aspects, and their contribution to performance. Ph.D thesis, Department of Music, City University London, London, UK.
5. GOURLAY, K.A. 1984. The non-universality of music and the universality of non-music. World of Music **26**: 25–36.
6. MERKER, B. 2000. Synchronous chorusing and human origins. *In* The Origins of Music. N. Wallin, B. Merker & S. Brown, Eds. :315–328. MIT Press. Cambridge, MA.
7. AYOTTE, J., I. PERETZ & K. HYDE. 2001. Congenital amusia: a group study of adults afflicted with a music-specific disorder. Brain **125**: 238–251.
8. DRAKE, C. 1998. Psychological processes involved in the temporal organization of complex auditory sequences: universal and acquired processes. Music Percept. **16**: 11–26.
9. STOBART, H.F. & I. CROSS. 2000. The Andean anacrusis? Rhythmic structure and perception in Easter songs of Northern Potosí, Bolivia. Br. J. Ethnomusicol. **9**: 63–94.
10. MAGRINI, T. 1989. The group dimension in traditional music. World of Music **31**: 52–79.
11. PAPOUSEK, H. 1996. Musicality in infancy research: biological and cultural origins of early musicality. *In* Musical Beginnings. I. Deliège & J.A. Sloboda, Eds. :37–55. Oxford University Press. Oxford.
12. PAPOUSEK, M. 1996. Intuitive parenting: a hidden source of musical stimulation in infancy. *In* Musical Beginnings. I. Deliège & J.A. Sloboda, Eds. :88–112. Oxford University Press. Oxford.
13. BOHLMAN, P. 2000. Ethnomusicology and music sociology. *In* Musicology and Sister Disciplines. D. Greer, Ed. :288–298. Oxford University Press. Oxford.
14. SCHEIFFLIN, E.L. 1993. Performance and the cultural construction of reality: a New Guinea example. *In* Creativity/Anthropology. R. Rosaldo, S. Lavie & K. Narayan, Eds. :270–295. Cornell University Press. London.
15. FELD, S. 1982. Sound and Sentiment: Birds, Weeping, Poetics and Song in Kaluli Expression. University of Pennsylvania Press. Philadelphia.
16. BLACKING, J. 1976. How Musical is Man? Faber. London.
17. PEGG, C.A. 2001. Mongolian Music, Dance and Oral Narrative: Performing Diverse Identities. University of Washington Press. London.

18. TUOHY, S. 1999. The social life of genre: the dynamics of folksong in China. Asian music. J. Soc. Asian Music **30**: 39–86.
19. YANG, M. 1994. On the hua'er songs of north-western China. Yearbook for Traditional Music **26**: 100–116.
20. CROSS, I. 2003. Music and biocultural evolution. *In* The Cultural Study of Music: A Critical Introduction. M. Clayton, T. Herbert & R. Middleton, Eds. :19–30. Routledge. London.
21. MERKER, B. 2002. Principles of interactive behavioral timing. *In* Proceedings of the 7th ICMPC, Sydney. C. Stevens, D. Burnham, G. McPherson, E. Schubert & J. Renwick, Eds. :149–152. Causal Productions. Adelaide.
22. CROSS, I. 1999. Is music the most important thing we ever did? Music, development and evolution. *In* Music, Mind and Science. S.W. Yi, Ed. :10–39. Seoul National University Press. Seoul.
23. HAUSER, M.D., N. Chomsky & W.T. Fitch. 2002. The faculty of language: what is it, who has it and how did it evolve? Science **298**: 1569–1579.
24. LERDAHL, F. & R. JACKENDOFF. 1983. A Generative Theory of Tonal Music. MIT Press. Cambridge, MA.
25. HUGHES, D. 1991. Grammars in non-Western musics: a selective survey. *In* Representing Musical Structure. P. Howell, R. West & I. Cross, Eds. :327–362. Academic Press. London.

Exploring the Influence of Cultural Familiarity and Expertise on Neurological Responses to Music

STEVEN M. DEMOREST AND STEVEN J. MORRISON

University of Washington School of Music, Seattle, Washington 98195-3450, USA

ABSTRACT: Contemporary music education in many countries has begun to incorporate not only the dominant music of the culture, but also a variety of music from around the world. Although the desirability of such a broadened curriculum is virtually unquestioned, the specific function of these musical encounters and their potential role in children's cognitive development remain unclear. We do not know if studying a variety of world music traditions involves the acquisition of new skills or an extension and refinement of traditional skills long addressed by music teachers. Is a student's familiarity with a variety of musical traditions a manifestation of a single overarching "musicianship" or is knowledge of these various musical styles more similar to a collection of discrete skills much like learning a second language? Research on the comprehension of spoken language has disclosed a neurologically distinct response among subjects listening to their native language rather than an unfamiliar language. In a recent study comparing Western subjects' responses to music of their native culture and music of an unfamiliar culture, we found that subjects' activation did not differ on the basis of the cultural familiarity of the music, but on the basis of musical expertise. We discuss possible interpretations of these findings in relation to the concept of musical universals, cross-cultural stimulus characteristics, cross-cultural judgment tasks, and the influence of musical expertise. We conclude with suggestions for future research.

KEYWORDS: music; musical expertise; magnetic resonance imaging; culture; brain mapping; auditory comprehension

INTRODUCTION

In recent years, music education in the United States and elsewhere has undergone a revolution with increased emphasis on the study of music of other cultures and increased sensitivity to the diverse cultural backgrounds of the student population.[1–3] This attention to multicultural music education raises questions about the nature of musicianship and its development. We do not know if studying a variety of world music traditions involves the acquisition of new skills or an extension and refinement of traditional skills long addressed by music teachers. In other words, is

Address for correspondence; Dr. Steven M. Demorest. University of Washington School of Music, Box 353450, Seattle, WA 98195-3450, USA. Voice: 206-543-7587; fax: 206-616-4098. demorest@u.washington.edu

Ann. N.Y. Acad. Sci. 999: 112–117 (2003). © 2003 New York Academy of Sciences.
doi: 10.1196/annals.1284.011

a student's familiarity with a variety of musical traditions a manifestation of a single overarching "musicianship" or is knowledge of the various musical styles similar to a collection of discrete skills much like learning a second language?

Research on the comprehension of spoken language has disclosed a neurologically distinct response among subjects listening to their native language compared to an unfamiliar language.[4,5] In a recent study, we hypothesized that an analogous distinction might be observed when comparing subjects' responses to music of their native culture and music of an unfamiliar culture.[6] The few neurological studies that have involved musical culture as a variable supported our hypothesis of a distinct "first music" response. Genç and associates reported a case in which epileptic seizures in a 48-year-old Turkish woman with no musical training were precipitated by both familiar and unfamiliar Turkish arabesques but not by other musical styles or by musical stimuli presented out of context.[7] Similarly, event-related potential (ERP) data have identified an increase in P3 amplitude, interpreted as indicating attention allocation and memory updating, among Turkish listeners when hearing a familiar instrument (ney) rather than an unfamiliar instrument (cello).[8]

AN fMRI EXPLORATION OF CULTURAL FAMILIARITY IN MUSIC

We compared the responses of six professional string players and six untrained control subjects to music from a familiar culture (Western) and an unfamiliar culture (Chinese traditional) while undergoing a functional magnetic resonance scan.[6] Our hypothesis was that activation would be associated with a comprehension response present for culturally familiar music and absent for culturally unfamiliar music. In addition, we were interested in differences related to musical training.

Contrary to our hypothesis, no overall differences in activation were observed for either subject group when activation responses to the two musical cultures were compared, although significant differences in later recall performance were based on cultural familiarity. There were also activation differences based on training. In a direct comparison between trained and untrained listeners' responses to music versus rest, significant activation in the right superior temporal gyrus was noted in professional musicians only. This difference was present regardless of the familiarity of the music. Prior research has identified areas in Heschl's gyrus adjacent and proximal to the right primary auditory cortex as critical to the processing of tonal information beyond basic pitch perception.[9] Musicians apparently involved these areas during listening to a greater degree than did untrained listeners. Musicians also demonstrated additional activation in the right and left midfrontal regions for Western music and Chinese music, respectively, when compared to rest.

This initial study provided no clear evidence of a distinct pattern of activation for culturally familiar versus unfamiliar music regardless of the degree of musical training. Although no distinct activation differences were observed, subjects could more accurately identify excerpts from the culturally familiar music in a post-scan recognition task. The conflict between activation and behavioral data leaves open the question of what processes may have contributed to superior recall performance and raises questions about what was or was not measured by the fMRI scans. Perhaps the task was too broad to elicit culturally based differences in activation discernible by functional imaging. It is interesting, however, that the nature of the task during and after scanning was identical in many ways to studies of the comprehension of spoken language.[5] Analysis of an English/Cantonese speech comparison of these same subjects revealed language-based differences in activation. If we assume that our results are valid, that no difference in activation exists in subjects' responses while listening to culturally familiar and unfamiliar music, and that activation does differ by expertise, we may consider possible interpretations.

INTERPRETATIONS

Musical Semantics

The most obvious interpretation suggested by a lack of difference in activation is best expressed by the popular notion of music as a universal language without cultural boundaries. We can confidently reject this interpretation based both on results of previous cross-cultural research[10,11] and on differences in our own subjects' recall performance. It appears that cultural familiarity did play a role in music comprehension at some level. Perhaps culture-based differences in activation would have emerged had the recall task been performed in the scanner instead of afterwards or had a more specific musical judgment been required of the subjects. In the absence of a specific judgment task, musical semantics may be sufficiently broad to allow listeners to make sense of what they are hearing without shared comprehension. By this we mean to distinguish between the facile, dynamic comprehension of the encultured insider and the more negotiated comprehension of the cultural outsider. To a listener outside the Chinese musical tradition, the structures of an unfamiliar system of pitch organization may be understood as wrong or out-of-tune notes. The listener has made sense of the music, but on his or her own terms, however inappropriate they may be. In this way, the term "musical accommodation" might be more accurate than "musical comprehension." Rather than music acting as a universal language to all listeners, each listener, depending on his or her enculturation and possibly on training, may universally apply his or her own comprehension strategies to all music, although with varying degrees of success. Such an interpretation would account for both the neurological and the behavioral data we observed.

Stimulus Characteristics

A second interpretation of our results focuses on the stimuli chosen for study. The fundamental differences found in some cross-cultural musical comparisons may be more covert than those in other more extreme pairings. Our selection of Western and Chinese musical examples—two traditions that are culturally distinct—may have resulted in stimuli that in some ways sounded quite similar.

For the purposes of experimental control, we were careful to match our chosen examples for external musical characteristics so that listeners' responses were not based simply on musical surface differences. Specifically, we controlled for instrumental timbre, texture, and tempo in choosing excerpts by Scarlatti and from the traditional piece, *Liu Qin Niang*. Had we chosen differently, Scarlatti contrasted with, say, an example of African drumming, listener responses may have differed simply because of differences in instrument color, de-emphasis of melodic content, or the fact that half our subjects were professional string players rather than professional percussionists.

The arguable similarity among the examples we selected could have accounted for the lack of difference in neurological response, but it would not have predicted differential performance on our post-scan memory task. Despite some outward similarities, it appears that one set of examples, those from the home culture, were easier to recognize.

Musical Expertise

Professional musicians exhibited greater activation in the right superior temporal gyrus than did untrained controls when processing both types of music, suggesting that formal training does influence brain activation. For the Western music versus rest comparison, additional activity was noted in the right midfrontal area. A recent reanalysis by Zatorre[9] suggested that areas of the right frontal lobe are involved in the storage and retrieval of tonal information in working memory. The present results might be interpreted as reflecting the trained listeners' use of tonal information to identify characteristics of the excerpts heard. They may then rely on stored information regarding features, such as melodic structure, to aid in the recall of excerpts during the post-test recognition task. Although prior research in this area was conducted among untrained listeners,[12] the researchers in that study specifically directed subjects to complete pitch-related tasks. In our study, subjects were only engaged in focused attending for later recall, leaving the strategies to be employed up to each subject. Musicians, perhaps as a result of their extensive training in such areas as tonal analysis and aural skills, may have more quickly turned to assessment of tonal relationships as a strategy to facilitate recall. In contrast, the Chinese music versus rest comparison elicited left midfrontal activation for trained listeners. Given the association of this region with verbal working memory, this may reflect subjects' attempts to apply verbal descriptors to less familiar musical stimuli. This interpretation is similar to recent observations of young students who demonstrated left frontal activity after receiving music instruction carried out using verbal, as opposed to performance-based, teaching strategies.[13] In our study, the difference in frontal activation did not appear in the direct comparison between trained and untrained listeners, however, suggesting that any strategic difference may only be one of degree rather than kind.

FUTURE RESEARCH

While speculative at this point, these interpretations may be clarified by further research into cross-cultural responses to music. First, it is vital that researchers compare not only music but also listeners from other cultures. To that end, we are currently investigating responses of native Chinese listeners, again with varied degrees of formal musical training. Preliminary imaging data from untrained control subjects reveal no differences in activation between Western classical and Chinese traditional listening conditions, a finding that parallels that of Western listeners. Results of a post-test memory task, however, also reveal no difference in success at recalling Western or Chinese examples. Such results may reflect the inadvisability of using

Western music as an "unfamiliar" stimulus for non-Western subjects due to its ubiquity throughout many of the world's societies. A better choice would be to use music of a non-Western subject's own culture and music unfamiliar to both Western and non-Western subjects alike.

To minimize the potential for listeners accommodating rather than accurately comprehending musical information, it may be advisable to design specific judgment tasks that tease out or expose cultural differences (see Neuhaus, this volume, for an example). This is not without its perils, as many musical judgments are often culturally derived or require a certain level of expertise. Despite this challenge, there are some judgment tasks that may be amenable to cross-cultural investigation. For example, findings from the cognitive psychology literature propose musical enculturation as a key variable in an individual's construction of tonal hierarchy[14,15] and establishment of tempo synchronization.[16] An examination of more specific, yet culturally neutral tasks could allow us to explore the degree to which culturally derived understandings of pitch and rhythm are manifested neurologically.

The surface similarity of our stimuli raises another interesting question. If subjects continue to demonstrate a consistent neurological response to music even when encountering musical styles or traditions far distant from their own, it would be useful to explore the limits of such a response, exploring the very conditions under which sounds are heard as music. It may be that a stimulus must include particular characteristics or combinations of characteristics before an individual independently codes it as music, but that once coded, the response is similar across cultures. If so, then varying the amount or type of "musical" characteristics in auditory stimuli during scanning might lead to interesting findings about the brain's approach to categorizing and coding auditory information. Perhaps the need for these characteristics varies according to listening context or is subject to manipulation (see Griffiths, this volume). Is the boundary of "music" and "other," neurologically speaking, determined more by culture or by expertise?

CONCLUSION

In today's music classrooms, teachers are choosing to perform songs and listen to recordings that allow students to interact with a wide range of musical styles and traditions. In many countries, classroom populations have become as diverse as the curriculum itself. It remains unclear how this new diversity of music and students will impact our concept of what it is to be musically educated. Drawing on the questions discussed here, we need to better understand the complementary roles played by enculturation and formal training in developing musical skills and concepts. It may be that students will need to develop new strategies to interact with unfamiliar musical styles or that new styles will emerge from the variety of musical systems encountered in the classroom.

As is often the case with the early stages of a research program, our findings raise more questions than they answer. The one conclusion of which we are confident is that those interested in discovering the biological foundations of human musical thought and behavior must employ stimuli, tasks, and subjects that encompass the world of music, not just a single culture. Only in this way can we avoid the perils and pitfalls that characterized some of the earlier research into music psychology and avoid an ethnocentric view of the human musical brain.

REFERENCES

1. ANDERSON, W.M. & P.S. CAMPBELL, Eds. 1996. Multicultural Perspectives in Music Education. Music Educators National Conference. Reston, VA.
2. MARK, M.L. 1998. Multicultural music education in the United States. Bull. Hist. Res. Music Ed. **19:** 177–186.
3. VOLK, T.M. 1998. Music, Education, and Multiculturalism: Foundations and Principles. Oxford University Press. New York.
4. PERANI, D., S. DEHAENE, F. GRASSI, *et al.* 1996. Brain processing of native and foreign languages. Neuroreport 7: 2439–2444.
5. SCHLOSSER, M.J., N. AOYAGI, R.K. FULBRIGHT, *et al.* 1998. Functional MRI studies of auditory comprehension. Hum. Brain Mapp. **6:** 1–13.
6. MORRISON, S.J., S.M. DEMOREST, E.H. AYLWARD, *et al.* 2003. fMRI investigation of cross-cultural music comprehension. NeuroImage 20: 378–384.
7. GENÇ, B.O., E. GENÇ, G. TASTEKIN, *et al.* 2001. Musicogenic epilepsy with ictal single photon emission computed tomography (SPECT): could these cases contribute to our knowledge of music processing? Eur. J. Neurol. **8:** 191–194.
8. ARIKAN, M.K., M. DEVRIM, O. ORAN, *et al.* 1999. Music effects on event-related potentials of humans on the basis of cultural environment. Neurosci. Lett. **268:** 21–24.
9. ZATORRE, R.J. 2001. Neural specializations for tonal processing. Ann. N.Y. Acad. Sci. **930:** 193–210.
10. CARTERETTE, E.C. & R.A. KENDALL. 1999. Comparative music perception and cognition. *In* The Psychology of Music. D. Deutsch, Ed. :725–791. Academic Press. San Diego.
11. NETTL, B. 2001. An ethnomusicologist contemplates universals in musical sound and musical culture. *In* The Origins of Music. N.L. Wallin, B. Merker & S. Brown, Eds. :463–472. MIT Press. Cambridge, MA.
12. ZATORRE, R.J., A.C. EVANS & E. MEYER. 1994. Neural mechanisms underlying melodic perception and memory for pitch. J. Neurosci. **14:** 1908–1919.
13. ALTENMUELLER, E., W. GRUHN, D. PARLITZ, *et al.* 1997. Music learning produces changes in brain activation patterns: a longitudinal DC-EEG study. Int. J. Arts Med. **5:** 28–33.
14. KRUMHANSL, C.L. 2000. Tonality induction: a statistical approach applied cross-culturally (cognitive reference points in Finnish folk hymns and North Sami yoiks). Music Percept. **17:** 461–479.
15. PERLMAN, M.A. & C.L. KRUMHANSL. 1996. An experimental study of internal interval standards in Javanese and Western musicians. Music Percept. **14:** 95–116.
16. DRAKE, C. & J.B. HANI. 2002. Do you hear this music in the same way as me? Intercultural differences in the perception of musical structure. Presented at 7th International Conference on Music Perception & Cognition. Sydney, Australia, July 19.

Analysis and Modeling of Expressive Intentions in Music Performance

GIOVANNI DE POLI

CSC, Department of Information Engineering, University of Padua, 35100 Padua, Italy

ABSTRACT: A performer can convey different expressive intentions when playing a piece of music. Perceptual and acoustic analysis of expressive music performances attempts to understand the musicians' strategies. Moreover, models for rendering different expressive intentions were developed for both analysis and deeper multimedia products fruition.

KEYWORDS: music performance; expressive intentions; perceptual analysis; acoustic analysis

MUSICAL PERFORMANCE AND EXPRESSIVE INTENTIONS

Music is an important means of communication among three participants: the composer, the performer, and the listener. The composer instills into his works his own emotions, feelings, and sensations, and the performer communicates them to the listener. The performer uses his own musical experience and culture to obtain from the score a performance that may convey the composer's intention.

Different musicians, even when referring to the same score, can produce very different performances. The score carries information such as the rhythmic and melodic structure of a certain piece, but as yet there is no notation able to describe precisely the temporal and timbre characteristics of the sound. The conventional score is inadequate to describe the complexity of a musical performance so that a computer might perform it. Whenever the information of a score (essentially notes, pitch, and duration) is stored in a computer, the performance sounds mechanical and unpleasant. The performer introduces microdeviations in the timing of the performance, dynamics, and timbre, following a procedure that correlates to his own experience and is common in instrumental practice. From such measurements, general performance features and principles can be deduced.

However, no musician plays the same piece in the same way on every occasion. Each performance depends on the performer's emotional state at that particular moment as well as his/her hypothetical dialogue with other musicians and subjective artistic choices. Moreover, the same piece of music can be performed to convey different interpretations of the score and emotions, according to different "expressive

Address for correspondence: Giovanni De Poli, CSC, Department of Information Engineering, University of Padua, Via Gradenigo 6/a, 35100 Padua, Italy. Voice: ++39 049 8277631; fax: ++39 049 8277799.

depoli@dei.unipd.it

Ann. N.Y. Acad. Sci. 999: 118–123 (2003). © 2003 New York Academy of Sciences.
doi: 10.1196/annals.1284.012

intentions," which can even be in contrast to the usual performance practice of that particular piece. A textual or musical document can assume different meanings and nuances depending on how it is played.

A major problem of the analysis-by-measurements method is that a specific deviation on one note could originate from several different principles, so the "true" origin may be impossible to trace. It is difficult to identify a multidimensional structure underlying a surface level merely by analyzing this surface level.

Some musical performance studies have attempted to understand how expressiveness is conveyed in music performance using the analysis-by-synthesis approach. For this purpose, some models or rule systems for generating automatic performance were developed. First, a theory is hypothesized, then it is realized in terms of a synthetic performance, and finally it is evaluated by listening. If needed, the hypothesized principle is further modified and the process repeated. Eventually, a new rule is formulated. In other words, the method is to teach the computer how to play more musically. The success of this method is entirely dependent on the formulation of hypotheses and on competent listeners.

For great variety in the performance of a piece, it is difficult to determine a general system of rules for its execution. An important step in this direction was made by Sundberg (KTH) and co-workers. They determined a group of criteria that, once applied to the generic score, can effect a musically correct performance.[1] Further on, the performer operates on the microstructure of the musical piece not only to convey the structure of the text as written by the composer, but also to communicate his/her own feelings or expressive intentions. Many studies have attempted to understand how much the performer's intentions are perceived by the listener, that is to say, how far they share a common code. Gabrielsson,[2] in particular, studied the importance of emotions in the musical message. An increasing number of studies is concerned with how the musician's intentions affect the performance. In this case, the experimental material is obtained by asking the performer either to provide different performances of his own choice, describing the intentions behind them, or to play with a certain intention in mind (see Ref. 2 for a review).

In this context, we began research to understand the way an expressive intention can be communicated to the listener, and we realized a model that can explain how it can modify the performance of a musical piece so that it may convey a certain expressive intention. In the following section, the main aspects of analysis and modeling expressive intentions will be presented with reference to our results.

ANALYSIS OF EXPRESSIVE INTENTIONS

The analysis-by-synthesis method is summarized in FIGURE 1.

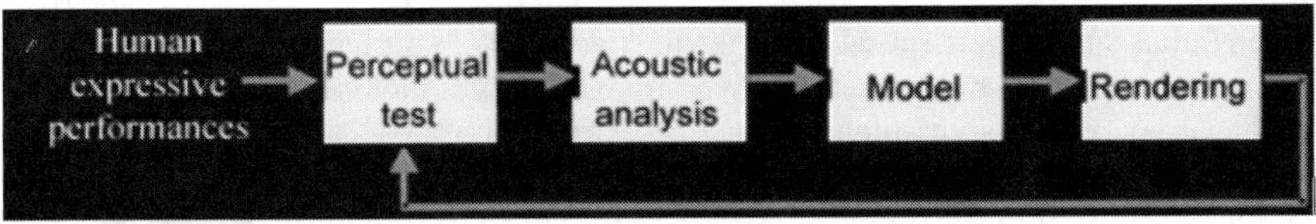

FIGURE 1. Analysis-by-synthesis method.

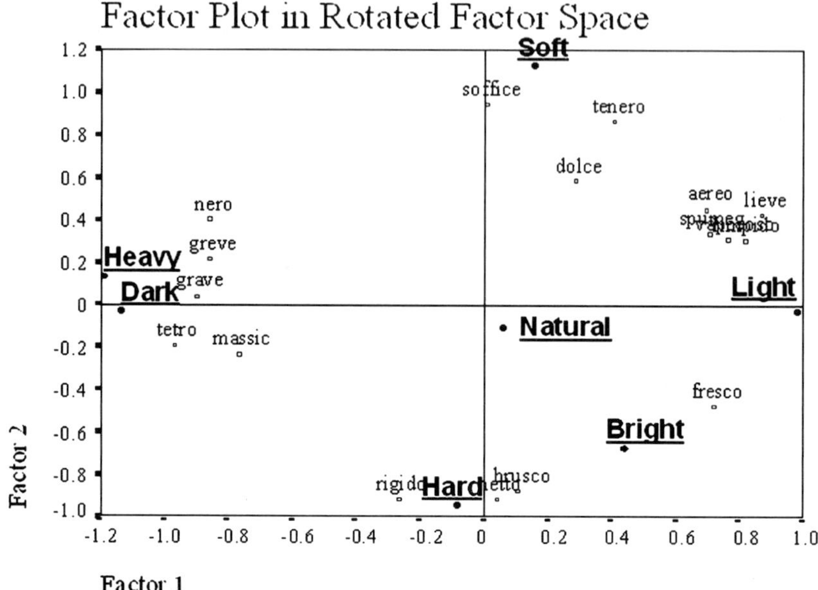

FIGURE 2. Factor analysis on adjectives. *Evaluation adjectives*: black, oppressive, serious, dismal, massive, rigid, mellow, tender, sweet, limpid, airy, gentle, effervescent, vaporous, fresh, abrupt, sharp. *Performances*: neutral, light, bright, hard, dark, heavy, soft. Noted the good recognition of the performer's intentions by the subjects.

Perceptual Analysis

The aim of perceptual analysis is to judge whether or not the performer's intentions are grasped by the listener and to determine the judgment categories used by the listener. We selected a set of scores from Western classical and African-American music. For each score, different performances (correlated with different expressive intentions) were played by professional musicians. Two different factor analyses were made. Factor analysis on adjectives (e.g., see FIG. 2) allowed us to determine a semantic space defined by the adjectives proposed to the listeners. By means of factor scores, it was possible to insert the performances into this space. A comparison between the performance positions and the evaluation adjectives demonstrated good recognition of the performer's intentions by the subject.

The second factor analysis used performances as variables (e.g., see FIG. 3). It showed that the subjects had placed the performances along only two axes. The two-dimensional space (perceptual parametric space, PPS) so obtained represents how subjects arranged the pieces in their own minds. The first factor (expressive intention bright versus expressive intention dark) seems to be closely correlated with the acoustic parameters that regard the kinetics of the music (for instance, tempo). The second factor (expressive intention soft versus expressive intention hard) is connected to the parameters that concern the energy of the sound (intensity, attack time).

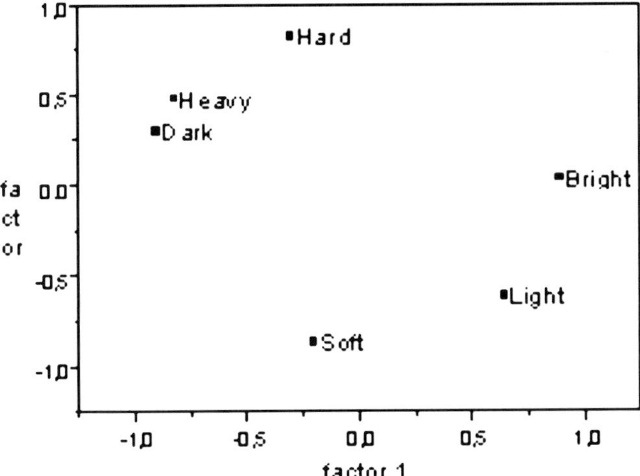

FIGURE 3. Factor analysis using performance as variables. First factor is correlated with kinetics of music; second factor is correlated with energy of the sound.

Other acoustic parameters (e.g., legato, brightness) are related to the PPS axes and were deduced from acoustic analyses.[3]

Acoustic Analysis

Acoustic analysis aims to identify which physical parameters and how many of them are subject to modifications when the expressive intention of the performer is varied.

Every musical instrument has its own expressive resources (vibrato in strings, the tongue in wind instruments, etc.) that are used by the musician to communicate his/her expressive intention. It is inevitable, therefore, that the results of any acoustic measure depend not only on the score, but also on the characteristics of the instrument used and the strategy adopted by the musician. Consequently, it is necessary to compare the data relative to different scores, musicians, and instruments to identify the expressive rules that can be considered valid in a general sense and in specific cases.

Several acoustic analyses of various musical pieces have been carried out using different instruments and performers.[3] The recordings are in either MIDI or audio format. A typical relation among expressive intentions and acoustic parameter variation in terms of a neutral performance (i.e., a performance without any expressive intentions) is shown in TABLE 1.

Modeling Expressive Intentions

According to the analysis-by-synthesis method, using the results of analysis and the expert's experience, some models for producing performances with different expressive intentions have been developed. Combinations of KTH performance

TABLE 1. Relation among expressive intentions and acoustic parameter variation in relation to a neutral performance of Violin Sonata Op. V by Arcangelo Corelli

	Hard	Soft	Heavy	Light	Bright	Dark
Tempo	+		− −	+ +	+ + +	
Legato		+	+	−	− −	
Attack duration	−	+		+	− −	−
Dynamic		+		−	−	+
Upbeat / downbeat		−		+		
Envelope centroid	Beginning	Center	Beginning			Center
Brightness	+ +	− −	+	−	+ +	− −
Vibrato		+ +	+			+ +

rules and of their parameters were used for synthesizing interpretations that differ in emotional quality.

We developed suitable models to compute the expressive deviations necessary to the rendering step in order to synthesize an expressive performance starting from a neutral one. The rendering step can be done in MIDI or by real-time post-processing of a recorded human performance. The system was used to generate performances of different musical repertories. Despite the fact that the models were developed mainly for Western classical music, they showed a general validity in their architecture, even if some tuning of the parameters is needed. Expressive syntheses of pieces belonging to different musical genres (European classical, European ethnic, African-American) verified the generalization of the rules used in the models.[4]

Besides analysis validation, these models allowed more general practical applications. In fact, in multimedia products, textual information is enriched by graphic and audio objects. A correct combination of these elements is extremely effective for the communication between author and user. Usually, attention is placed on the visual rather than on sound, which is merely used as a realistic complement to image or as a musical comment to text and graphics. With increasing interaction, while the visual part has evolved consequently, the paradigm of the use of audio has not changed adequately, resulting in a choice among different objects rather than in a continuous transformation of these. Performance researchers have demonstrated that it is possible to communicate expressive content at an abstract level, so to change the interpretation of a musical piece. A more intensive use of expressive intention control in multimedia systems will allow interactive adaptation of music to different situations. Our models permit a gradual transition (morphing) between different expressive intentions, leading to a deeper fruition of the multimedia product.[5]

REFERENCES

1. FRIBERG, A., L. FRYDÉN, L.-G. BODIN & J. SUNDBERG. 1991. Performance rules for computer-controlled contemporary keyboard music. Computer Music J. **15:** 49–55.
2. GABRIELSSON, A. 1999. The performance of music. *In* The Psychology of Music. 2nd Edition. D. Deutsch, Ed. :501–602. Academic Press. San Diego.

3. DE POLI, G., A. RODA' & A. VIDOLIN. 1998. Note by note analysis of the influence of expressive intentions and musical structure in violin performance. J. New Music Res. **27:** 293–321.
4. CANAZZA, S., G. DE POLI, G. DI SANZO & A. VIDOLIN. 1998. A model to add expressiveness to automatic musical performance. *In* Proceedings of 1998 International Computer Music Conference. :163–169. Computer Music Assoc. Ann Arbor, MI.
5. CANAZZA, S., G. DE POLI, C. DRIOLI, *et al.* 2000. Audio morphing different expressive intentions for multimedia systems. IEEE Multimedia, July–September **7:** 79–83.

Musicians versus Nonmusicians

A Neurophysiological Approach

LUISA LOPEZ,[a,b] REINHART JÜRGENS,[a] VOLKER DIEKMANN,[a]
WOLFGANG BECKER,[a] SIBILLE RIED,[a] BERTA GRÖZINGER,[a] AND
SERGIO NICOLA ERNÉ [b]

[a]*Center for Developmental Disabilities "E. Litta" Grottaferrata, Rome, Italy*

[b]*Sektion Neurophysiologie and [c]Zentralinstitut für BiomedizinischeTechnik,
Universtität Ulm, Ulm, Germany*

ABSTRACT: The ability to perceive sounds and correctly categorize them within a scale is the result of the interaction between inherited capabilities and acquired rules. If a subject listens to a melody, occasional and unexpected endings of the melody typically evoke characteristic auditory evoked responses in the latency range of 300–400 ms (P300). Also, earlier stages of auditory information processing have been exhaustively investigated by means of mismatch negativity (MMN), a deflection that occurs in the auditory evoked response at a latency of about 200 ms, whenever a deviance is randomly inserted in a series of otherwise equal stimuli. Conceivably, perceptual deviations could also be detected against expectancies that are based on abstract rules; introspective experience suggests that such deviations may also elicit fast intuitive responses that typically initiate processes of analytical reasoning for confirmation. In music, the physical features of the stimulus are, in fact, always changing, because the melodic contour consists of a series of notes with different pitch characteristics. In such a condition, a typical mismatch negativity would not be evoked on the basis of physical deviance, but rather of criteria involving the musical contour of the stimulus. In this study, 20 healthy subjects (10 nonmusicians and 10 musicians) underwent auditory stimulation (tone, chord, chord sequence, Mozart and Bach melodies) and both electrical and magnetic recordings. Clear N1 was recorded for all paradigms, in all subjects; MMN and P300 were also recorded, and their amplitudes and latencies were significantly correlated with the musicality score and with the paradigm's difficulty.

KEYWORDS: pitch discrimination; mismatch negativity (MMN); P300; music; auditory evoked potentials; EEG; MEG

INTRODUCTION

Although research in music usually focuses on "musical" abilities such as pitch, timbre, and consonance discrimination, the need for quantifiable data implies the use of experimental protocols that necessarily address only specific aspects of music.

Address for correspondence: Dr. Luisa Lopez, Center for Developmental Disabilities "E. Litta," Via Anagnina Nuova 13, 00046 Grottaferrata, Rome, Italy. Voice: +39 06 94315621; fax: +39 06 9411463.

lopez@uniroma2.it

Ann. N.Y. Acad. Sci. 999: 124–130 (2003). © 2003 New York Academy of Sciences.
doi: 10.1196/annals.1284.013

For example, when exploring pitch discrimination, evoked or event-related potentials may be used, and pitch variations are inserted in a stable temporal pattern. An average is performed to record electrical activity in the brain. This procedure, on the one hand, limits the richness of the stimulus and, on the other, limits the responses to those temporally correlated to the stimulus itself. Yet, the unfavorable signal-to-noise ratio advises against analyzing these data without using averages in the temporal or frequency domain. New methodologies, such as autoregressive models, allow analysis of frequency domains without losing temporal information, but they have shown some limitations in the very fast events such as those likely to occur in musical processing. Conversely, imaging techniques are bound to have even slower temporal resolution (500–1,000 ms).

In our study, increasing levels of complexity in "musical sequences" were compared to very predictable temporal patterns and to a small percentage of mistakes (or deviants). We recorded the ongoing electrical and magnetic activity in the brain in two groups divided by their "musicality." We focused our attention on the auditory responses evoked by a deviance in the stimulus. For example, mismatch negativity (MMN)[1] is a deflection that occurs in the auditory evoked response at a latency of about 200 ms, whenever a deviance in the physical features is randomly inserted in a series of otherwise equal stimuli and is reported as the effect of the joint comparison of several physical aspects with respect to a memory trace. Changes in a pattern of the presented stimuli can also evoke MMN, for example, in complex phonetic stimulation[2] and in complex sound patterns.[3,4] These pattern changes still comply with the prevailing view that MMN occurs when a deviance is detected. Another possibility is that on the basis of acquired rules, a prediction can be made as to the following stimulus. In this case, a feed-forward, rather than a feed-back mechanism would be involved. Moreover, a difference in performance between levels of exposure to the same patterns would be expected. A similar phenomenon is known to occur when an incongruous ending is provided in a linguistic and a nonlinguistic context,[5] although it is thought to involve responses in the 300–400 latency range.

MATERIAL AND METHODS

Twenty healthy subjects, 7 female and 13 male, aged 27.8 ± 9.3 years, half of whom were performing amateur musicians and half nonmusicians, underwent the Bentley test of musical ability[6] and were grouped accordingly (Hi, score >38; Lo, score <38). While the subjects lay in a magnetically shielded room, auditory stimuli were delivered via plastic tubes connected to speakers mounted outside. Sounds were generated by an SB16 board (Creative Labs.) programmed with a MIDI code. Each experimental session consisted of five oddball paradigms, termed NOTE, CHORD, ARPEGGIO, MOZART, and BACH, which were given in five blocks in fixed order. Each block lasted about 10 minutes, including 2 or 3 brief pauses. In the NOTE condition, 300 standard (STD) tones (1310 Hz) and 100 deviant (DEV) tones (1390 Hz) were presented every 1 second. The CHORD/ARPEGGIO paradigms were similar to those of NOTE, except that stimuli consisted of three simultaneous (CHORD) or consecutive (ARPEGGIO) tones, the middle of which had the same characteristics as those in the NOTE condition[7] (1040, 1310, and 1560 Hz for STD and 1040, 1390, and 1560 Hz for DEV). All tones had a sinusoidal waveform of 40 ms' duration and 10 ms' rise and fall time. The three tones in ARPEGGIO were displaced by 80 ms each. The MOZART paradigm consisted of 40 repetitions of the familiar song "Twinkle Twinkle Little Star," which we refer to as MOZART. The BACH paradigm used the two voices in Invention No. 8. There were 15 repetitions of the whole melody. Deviant (DEV) stimuli occurred at an average rate of 25% and started with a wrong tone, a half note above the correct (STD) tone of the melody. Both melodies were played with synthesized piano sound. To avoid predictions of wrong tones, the occurrence of DEVs was different in each repetition and was determined by a pseudo-random list that was identical for all subjects. During the experiment, subjects were instructed to listen carefully to the tones/music, mentally count the number of perceived deviations in each experimental block, and report their counts in the short pauses between the blocks. EEG was recorded from 25 specially designed electrodes combining extra flat Ag/Ag Cl pellets with planar coils, allowing magnetic localization of electrode placement, 19 according to the 10–20 system plus 3 extra electrodes per hemisphere, around positions C3 and C4. The average from the two ear electrodes was used as a reference. Magnetic fields were recorded simultaneously with the EEG, over the right (first session) and left (second session) side of the subjects' head using a 22 software gradiometer (DORNIER MEG system–DC-64 Hz).[8] EEG and MEG were acquired continuously and sampled at 500 Hz together with recordings of eye movements and blinks, respiration, and rectified and filtered sound. The evoked brain responses were averaged offline. Musical stimuli were averaged using the onset of the first tone of each beat as a trigger and the duration of beats as averaging epochs. Statistical comparisons were carried out with a multifactorial analysis of variance (ANOVA).

RESULTS

Overall, the count errors given by our subjects were smaller than the number of presented deviants. The count error was smallest in NOTE and largest in BACH. In all paradigms, the Hi group performed better than the Lo group, with the difference being largest in ARPEGGIO. All paradigms elicited (1) a response of the primary auditory cortex, N1; (2) a frontotemporal negativity at 200 ms, mismatch negativity MMN, to the DEV stimuli; and (3) a late positive component between 350 and 450 ms of latency, which we will refer to as P300, in conjunction with both the detection of deviants and the counting task. While in non-music paradigms and even in MOZART we still obtained discrete N1 peaks, the BACH melody evoked an almost sinusoidal pattern because of the high-tone repetition rate.

The electric and magnetic responses to the STD stimulus (FIG. 1) consisted of rhythmic evoked activity with typical N1-like peaks that are time-locked to the regular stimulation, whereas the DEV stimulus elicited a similar pattern onto which both an MMN and a P300 appeared to be superimposed. For all subjects and for all paradigms, the N1 was generally most prominent over the central electrodes, the MMN over the central or frontal electrode and the P300 over the parietal electrodes. Although in the MEG typical N1M was seen in almost all of our subjects, the shape of the magnetic counterparts of MMN and P300 showed a considerably larger idiosyncratic variability than that in the electric recordings.

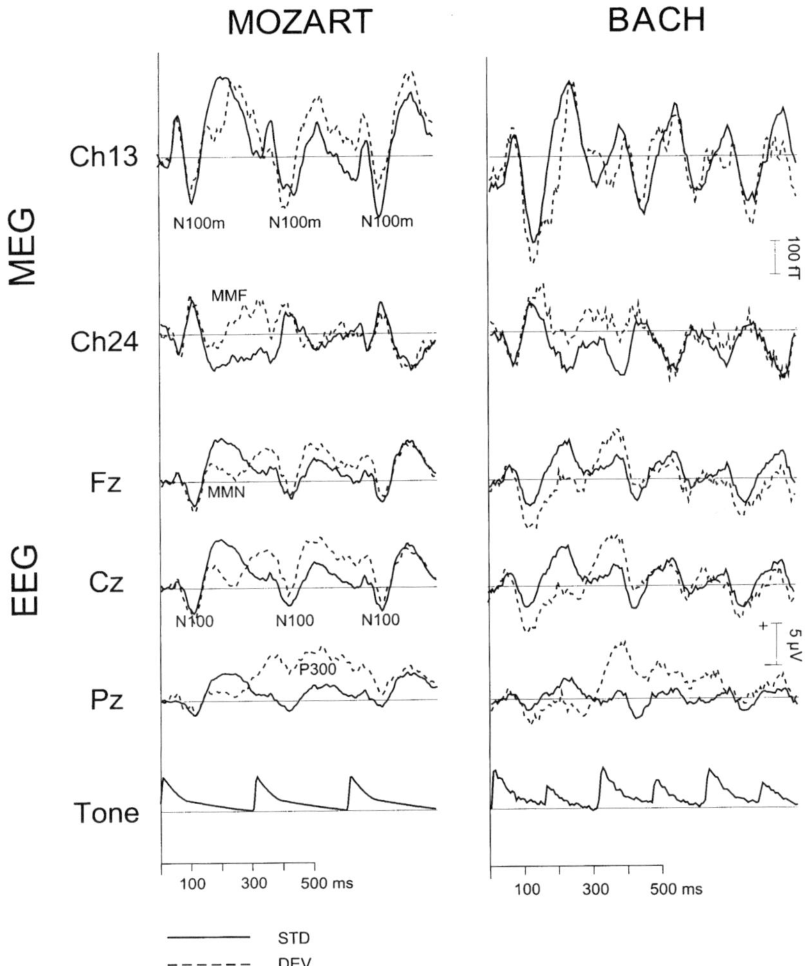

FIGURE 1. Magnetic (traces, 13 posterior and 24 anterior to the acoustical cortex) and electric (Fz, Cz, Pz) brain responses to acoustical stimulation with a simple (MOZART) and a complex (BACH) melody. Ton = rectified sound signal. Data from one subject. *Solid lines*, STD stimulus; *dashed lines*, DEV stimulus.

In FIGURE 2 we compare grand averages of the responses of the subjects with high and low musical ability. Although the N1 responses are similar in both groups, at least for the paradigms NOTE, CHORD, and ARPEGGIO, in the two music paradigms the N1-like deflections were larger in the Hi group. Both groups showed clear MMN in all paradigms, with amplitudes being larger for the Hi group. The influence of musicality was strongest for the P300 in ARPEGGIO and BACH, whereas wrong notes in the MOZART paradigm evoked similar P300 amplitudes in the two groups.

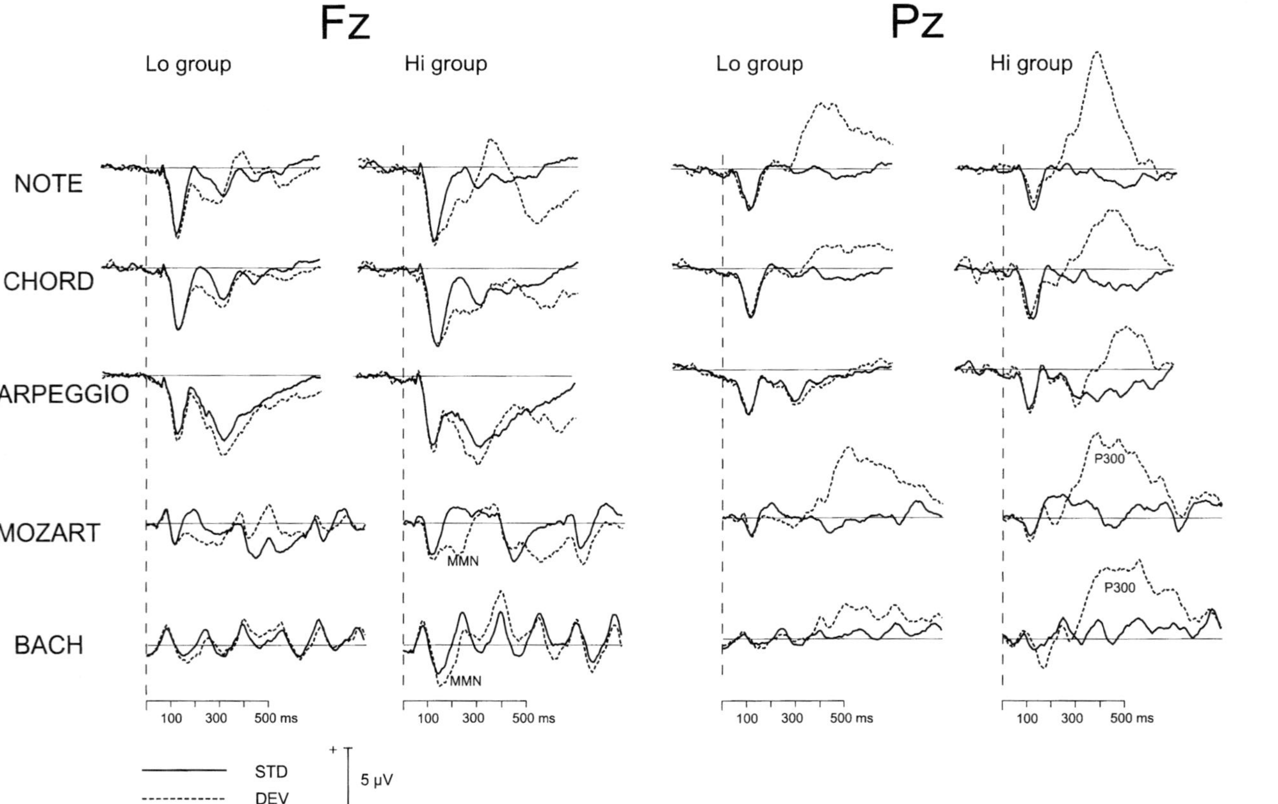

FIGURE 2. Grand averages of electric responses to stimulation with NOTE, CHORD, ARPEGGIO, MOZART, and BACH paradigms. (*Upper panel)* Electrode position Fz; (*lower panel)* Pz. Lo, group of subjects with a Bentley score of 31–36; Hi, group of subjects with a score of 42–48. *Solid lines*, STD stimulus; *dashed lines*, DEV stimulus.

The statistical analysis of variance (ANOVA) confirmed that amplitudes of MMN and P300 were larger for the Hi group than for the Lo group ($P < 0.01$), whereas the N1 peak measurements did not show any significant differences. However, the periodical activity evoked by the BACH melody was further evaluated by fitting the subjects' brain responses with sinusoids of the fundamental tone repetition frequency. This particular analysis revealed a significantly higher amplitude in the Hi group than in the Lo group (1.05 vs 0.72 5V, $P < 0.01$ for STD stimulus, 1.07 vs 0.73 5V, $P < 0.01$ for DEV). The Hi group also had significantly shorter peak latencies for the MMN ($P < 0.05$) and the P300 ($P < 0.01$). Interestingly, however, the Hi group exhibited longer latencies for the N1 component than for the Lo group ($P < 0.01$); this was true for responses to both the STD and the DEV stimuli. The statistical analysis of variance did not show any hemispheric differences for amplitudes or latencies of N1 (electrodes C3/C4), MMN (F3/F4), and P300 (P3/P4), whereas it showed a significant difference for N1 STD (amplitude over C3 = -3.5 μV and C4 = -3.2 μV, $P < 0.05$). The Equivalent Current Dipole locations for the N1 were clustered on the right hemisphere, more medial and posterior in the Hi group with respect to the Lo group, while being scattered on the left hemisphere.

DISCUSSION

As expected from the literature, an early process, termed here "mismatch negativity" (MMN) because of the similarity of its latency, polarity, and distribution with Näätänen's MMN, was observed in all subjects in response to NOTE, CHORD, and ARPEGGIO, with a significant difference between the two groups. There are also similarities with the early right-anterior negativity in its dependence from harmonic context,[9] but not in its scalp distribution. In our paradigm, requiring attention to the deviance, neither one of the two terminologies should probably be used, although MMN is reported to be increased in amplitude by attention manipulation.[10]

Conceivably, incongruence of deviant tones in melodies can be identified if either the whole memory is stored in a memory trace[11] or they are detected because certain tones do not comply with learned rules, for example, schemes of tonality. We hypothesize that each future tone (or set of future tones) is predicted by a parallel process and then offered for comparison with the incoming tones. In this sense, the advantage of "musicianship"[11–13] is more evident in more complex musical sequences, whereas responses to simpler ones do not discriminate between the two groups, probably because of an implicit knowledge of basic tonality rules.[14] In conclusion, the present study of the electric and magnetic evoked responses during the processing of melodic information, and their dependency on musical ability, may provide a new speech-independent protocol for the evaluation of some aspects of fast auditory processing in both physiological and pathological conditions.

REFERENCES

1. Näätänen, R. & C. Escera. 2000 Mismatch negativity: clinical and other applications. Audiol. Neurootol. **5:** 105–110.
2. Kraus, N. & M. Cheour. 2000. Speech sound representation in the brain. Audiol. Neurootol. **5:** 140–150.

3. Lopez, L., A. Mogilner, E. Lado, *et al.* 1991. Auditory magnetic evoked fields in response to an incongruous note within a musical sequence in musicians and non-musicians. Soc. Neurosci. Abstr. **17:** 655.
4. Näätänen, R., E. Schröger, M. Tervaniemi, *et al.* 1993. Development of a memory trace for complex sound patterns in the human brain. Neuroreport **4:** 503–506.
5. Besson, M., F. Faïta & J. Requin. 1994. Brain waves associated with musical incongruities differ for musicians and non-musicians. Neurosci. Lett. **168:** 101–105.
6. Bentley, A. 1963. A Study of Some Aspects of Musical Ability amongst Young Children, Including Those Unable to Sing in Tune. University of Reading. UK Library.
7. Alho, K., M. Huotilainen, H. Tiitinen, *et al.* 1993. Memory-related processing of complex sound patterns in human auditory cortex: an MEG study. Neuroreport **4:** 391–394.
8. Becker, W., V. Diekmann, R. Jurgens & C. Kornhuber. 1993. First experiences with a multichannel software gradiometer recording normal and tangential components of MEG. Physiol. Meas. 14 (Suppl. 4A): A45–A50.
9. Koelsch, S., T.C. Gunter, E. Schroger, *et al.* 2001. Differentiating ERAN and MMN: an ERP study. Neuroreport **12:** 1385–1359.
10. Muller, B.W., C. Achenbach, R.D. Oades, *et al.* 2002. Modulation of mismatch negativity by stimulus deviance and modality of attention. Neuroreport **19:** 1317–1320.
11. Tervaniemi, M., M. Rytkonen, E. Schroger, *et al.* 2001. Superior formation of cortical memory traces for melodic patterns in musicians. Learn Mem. **8:** 295–300.
12. Tervaniemi, M., T. Ilvonen, K. Karma, *et al.* 1997. The musical brain: brain waves reveal the neurophysiological basis of musicality in human subjects. Neurosci. Lett. **226:** 1–4.
13. Russeler, J., E. Altenmuller, W. Nager, *et al.* 2001. Event-related brain potentials to sound omissions differ in musicians and non-musicians. Neurosci. Lett. **308:** 33–36.
14. Tillmann, B., J.J. Bharucha & E. Bigand. 2000. Implicit learning of tonality: a self-organizing approach. Psychol. Rev. **107:** 885–913.

Specialization of the Specialized: Electrophysiological Investigations in Professional Musicians

THOMAS F. MÜNTE,[a] WIDO NAGER,[b] TILLA BEISS,[a] CHRISTINE SCHROEDER,[b,c] AND ECKART ALTENMÜLLER[c]

[a]*Department of Neuropsychology, Otto-von-Guericke-Universität, Magdeburg, Germany*

[b]*Department of Neurology, Medizinische Hochschule, Hannover, Germany*

[c]*Institute for Perfoming Arts Medicine and Music Physiology, Hochschule für Musik und Theater, Hannover, Germany*

ABSTRACT: Several event-related brain potential (ERP) studies examining the processing of auditory stimuli by professional musicians compared with non-musicians are reviewed. In the first study, musicians (string players) and non-musicians attended to one of two streams of auditory stimuli characterized by a specific pitch. Musicians showed a prolonged ERP attention effect, the late portion of which was more frontally distributed than was that of the non-musicians. In the second study, we investigated auditory spatial processing in conductors, pianists, and nonmusicians. Only the conductors showed behavioral selectivity of sound sources located in the peripheral auditory space. In addition, this group showed a negative/positive mismatch response for deviant stimuli occurring outside the focus of spatial attention. Finally, a group of drummers was compared to woodwind players and nonmusicians in a passive listening task. A real continuous drum sequence was manipulated so that some beats were anticipated by 80 ms. The drummers showed a mismatch response not only for the anticipated beats but also for the subsequent beats, suggesting a more complex representation of the temporal aspects stimulus sequence in this subject group. Together, these studies suggest qualitative differences of the neural correlates of auditory processing between musicians and non-musicians. Moreover, these differences appear to be shaped by the specific training of a musician.

KEYWORDS: event-related potentials; conductors; pianists; drummers; attention; music perception; mismatch negativity (MMN)

INTRODUCTION

This volume attests to the growing interest of the neurosciences in the musician as an object for scientific inquiry. This interest goes beyond just simple curiosity; it is motivated by the fact that making music at a professional level is among the most

Address for correspondence: Thomas F. Münte, Department of Neuropsychology, Otto von Guericke Universität, Universitätsplatz 2, Gebäude 24, 39106 Magdeburg, Germany. Fax: +49-391-6711947.

thomas.muente@medizin.uni-magdeburg.de

Ann. N.Y. Acad. Sci. 999: 131–139 (2003). © 2003 New York Academy of Sciences.

doi: 10.1196/annals.1284.014

complex tasks a human can perform. Moreover, the making of music engages multiple brain systems including the motor, auditory, limbic, and executive systems and requires integration of the activity in all of these areas. We have thus argued that the musician's brain might serve as a model organism for neuroplasticity.[1] In fact, the time-on-task in musicians exceeds that of any laboratory animal used for plasticity research by several orders of magnitude, and the complexity of the stimulation is much greater. Although the stimulation conditions in musicians seem ideal to give rise to plastic changes of the brain, the organ of interest can be assessed only non-invasively, necessitating the use of functional neuroimaging techniques such as fMRI, PET, MEG, or event-related brain potentials (ERPs). In the present investigation, we review recent experiments from our laboratory that take advantage of the event-related potential's ability to reveal the temporal dynamics of sensory and cognitive processing.[2,3] In these studies, we were generally concerned with the musician's ability to select stimuli on the basis of some physical cue, that is, pitch, spatial position, or timing. These abilities were tested with respect to both attentive and pre-attentive processing. Moreover, we were interested in the extent to which special training with a particular instrument or musical task would lead to subspecialization within the group of professional musicians.

PITCH (STRING PLAYERS)

Music is a highly organized stimulus along several dimensions, one of them being pitch. In a first study, we therefore asked whether or not trained musicians are better able to attentionally focus on a stream of sounds defined by their pitch. This study thus took advantage of the well-known ERP effects of selective auditory attention that have been studied extensively over the last two decades (for a review, see Ref. 3). Regardless of whether a stream of stimuli is defined by their location or pitch, paying attention to a stimulus stream while ignoring other concurrent streams gives rise to a more negative ERP waveform, termed the Nd (negative displacement), the onset of which often coincides with the exogenously evoked N1 component. As far as attention to pitch is concerned, Hansen and Hillyard[4] in an early study showed that the Nd wave could be subdivided into an earlier phase (150–300 ms, henceforth early Nd) and a more frontally distributed later phase (300–beyond ms, henceforth late Nd). A recent study[5] links the early effect to a more broadly tuned selection taking place over the first 200 ms after the stimulus, whereas the late Nd probably reflects the subsequent more finely tuned stimulus selection. Given that musicians, especially string players, are highly trained auditory discriminators with regard to pitch, we hypothesized that they should exhibit a changed auditory attention effect especially for the late Nd. Twelve musicians (mean age 25 years, 10 with string instruments as their primary instrument), all with extensive musical training since early childhood, and 12 matched nonmusician controls participated. None of the nonmusicians had formal training in music or experience with a musical instrument.

Stimuli were computer-generated, brief, sine-wave tone-pips. Two channels of information, defined by the pitch of the stimuli, were presented via a single speaker standing in front of the subjects. Three different conditions were created on the basis of pitch separation between channels:

50 Hz condition, 800 Hz /. 850 Hz
100 Hz condition, 800 Hz /. 900 Hz
500 Hz condition, 800 Hz /. 1300 Hz

Within each channel, 90% of the stimuli were of 60 ms' duration (standard stimuli), whereas 10% of the stimuli were of slightly longer duration and served as the target stimuli. The duration of the target stimuli (80–120 ms) was adjusted individually so that a target detection rate of about 70% was achieved. For any given run, subjects were instructed to attend selectively to high or low tones and to respond to target stimuli in the "attended channel" by a speeded button response. Stimuli were presented with a randomized interstimulus interval between 150 and 350 ms at about 70 dB(SL). Multichannel ERPs were obtained using standard procedures.[3] To ensure that only ERPs to physically identical stimuli were compared, measurements were taken exclusively on the ERPs to the 800-Hz stimuli. Waveforms were quantified by mean-amplitude measures in time windows 250–300 ms and 600–800 ms.

Grand average event-related potentials are shown for both groups for the frontal midline electrode and the three different pitch separation conditions (FIG. 1). Waveforms are characterized by an initial positivity at around 100 ms followed by a negativity peak at about 180 ms (N1). In the 500-Hz pitch separation condition, the onset of the attention effect appears to coincide with the upward flank of the N1. The attention effect extends to the end of the recording epoch in the musicians, whereas in the control subjects the attention effect lasts up to only about 600 ms in the 100-Hz and 500-Hz conditions. Statistically, a highly significant main effect of attention was observed for both early (250–300 ms, $F_{(1,22)} = 58.14$, $P < 0.0001$) and late (500–750 ms, $F_{(1,22)} = 19.28$, $P < 0.0002$) time windows. Whereas the size of the attention effect was virtually identical during the early time-window (group × attention inter-

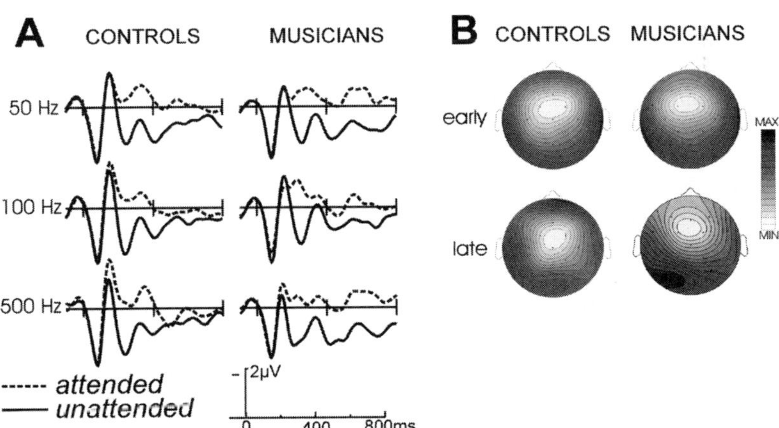

FIGURE 1. Group averages for the musicians and controls for the frontal midline site. From about 150 ms onwards, the ERPs to the attended tones are characterized by a more negative waveform in both groups. This effect is more short-lived in the control group, however. The spline-interpolated isovoltage maps show a similar distribution of the early attention effect (250–300 ms) in musicians and nonmusicians, whereas during the 500–750 ms interval a considerably more frontal distribution is observed in the musician group.

action $F_{(1,22)} = 0.01$, ns), the larger late Nd wave in the musicians was reflected in a group × attention interaction for the late time window ($F_{(1,22)} = 4.42$, $P <0.05$).

The distribution of the early and late attention effects can also be derived from the spline-interpolated isovoltage maps in FIGURE 1. Although the distribution of the early attention effect is virtually identical, the late Nd is considerably more frontally distributed in the musicians. In terms of absolute position, the maximum of the late attention effect of the musicians is shifted frontally by about 3.5 cm. Thus, the present study was successful in revealing differences in attentive processing between the two subject groups. The first difference between musicians and control subjects was an extended late Nd wave in the former group that was absent or much smaller in the latter. This extended negativity, furthermore, had a more frontal distribution in the musicians. Differences between early and late auditory attention effects have been observed in several circumstances. An early interpretation of the late Nd (also termed processing negativity) was that it indicated a match between an attentional trace—that is, a template against which incoming stimuli can be matched—and the eliciting stimulus.[6] In the present circumstances, this would indicate that musicians achieve a better fit between the attentional trace and incoming stimuli, possibly because their attentional trace is more finely tuned and precise. Another interpretation of the prolonged attention effect in musicians can be deduced from a recent study on the spatial gradient of auditory attention.[5] In this study, it was shown that stimuli from locations immediately adjacent to the attended location showed an Nd effect. Moreover, these stimuli had a larger late Nd effect than did the attended stimuli themselves. This finding indicates that the late Nd might reflect the extent and intensity of analysis devoted to the stimulus. Thus, it can be speculated that musicians as highly trained auditory observers analyze incoming auditory stimuli more thoroughly than do less experienced listeners.

It is interesting that the distributions of the late Nd were different for the two groups. Several papers have addressed the localization of auditory attention effects,[7] suggesting that the later attention-related components emanate from frontal cortex. Musicians therefore appear to engage different neuronal populations in this attention task than do nonmusicians. Therefore, a qualitative difference in the processing of pitch seems to exist between the two subject groups.

AUDITORY SPACE PERCEPTION (CONDUCTORS)

Aside from her/his less quantifiable talent to give the orchestra artistic direction, a conductor has to fulfill two apparently contradictory tasks: in order to hone in on a single section or player, she/he has to use selective spatial attention mechanisms, whereas monitoring an entire auditory scene likely requires different cognitive mechanisms such as a comparison of incoming information with some kind of internal model. To assess these two aspects, we recorded ERPs from seven professional classical music conductors (mean age 45 years, mean conducting experience 19 years) in a task that entailed selective attention to one of six sound sources, three located in front of the subjects and three located to the right. Rapid sequences of standard and deviant sounds (random order, interstimulus intervals between 90 and 270 ms) were presented from all of the speakers with either the centermost or the rightmost speaker being relevant at any given time. This experimental setup was very similar to previous studies using ERPs to investigate spatial attention mechanisms in normal listeners[5] and congenitally blind subjects.[8] To distinguish changes in ERPs specific to the conducting experience from general changes related to prolonged professional musicianship, we included a group of professional pianists ($n = 7$) as controls in addition to the nonmusicians ($n = 7$).

With regard to the first task of a conductor, honing in on a certain sound source, ERPs and behavioral indices showed no differences between the conductors and the other two groups for the group of speakers located in front of the subjects. Here, all groups had good selectivity with a comparably low false alarm rate to deviants occurring in the speaker next to the relevant, attended one. Also, the ERPs showed a steep gradient of the amplitude of the so-called Nd attention effect from the relevant to the neighboring speakers for all groups. For the peripheral group of speakers, however, only the conductors were able to distinguish the three sound sources with

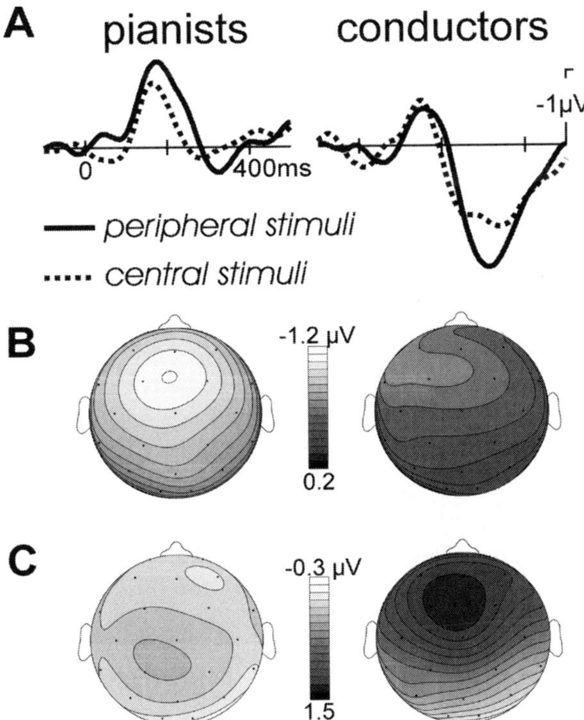

FIGURE 2. (**A**) Deviant minus standard difference waves for stimuli coming from the unattended direction (e.g., from the three speakers located to the right of the subject, when the centermost speaker was attended). In pianists, sizable mismatch negativity can be observed. The mismatch negativity (MMN) is somewhat smaller in conductors. However, it is followed by pronounced positivity in this group. (**B**) Spline-interpolated isovoltage maps illustrating the distribution of the mismatch negativity. A typical frontocentral maximum is observed. (**C**) Distribution of the positive peak following mismatch negativity. A distribution very similar to the MMN is observed.

reasonable precision. Only in this group a decline of the Nd attention effect from the relevant to the adjacent speakers was observed. These results have been discussed in more detail elsewhere.[9]

With regard to the second task, monitoring the entire auditory scene, we focused on the analysis of ERPs to the deviants coming from the unattended direction. As these stimuli occur outside the focus of attention, mismatch negativity (MMN)[10] is expected. In the nonmusician controls, only a very small MMN (less than 0.5 μV) was observed, which is therefore not shown in FIGURE 2. Professional pianists, on the other hand, displayed considerable MMN, whereas in the conductors an intermediate size MMN was followed by a positive deflection. These difference waveforms were quantified by mean amplitude measures in time windows designed to pick up the MMN (140–190 ms, electrodes Fz/Cz) and subsequent positivity (250–350 ms, electrodes Fz/Cz). For the early time window encompassing the MMN, a main effect of the group was found ($F_{(2,18)} = 4.35$, $P < 0.03$), which was followed up by pairwise comparisons, which revealed a significant difference between pianists and nonmusicians ($F_{(1,12)} = 5.54$, $P < 0.04$). The amplitude difference of the MMN between conductors and pianists did not reach significance ($F_{(1,12)} = 3.03$, $P < 0.11$).

For the later time window (250–350 ms), the positivity that was present in the conductor group was statistically reflected in a main effect of group ($F_{(2,18)} = 6.1$, $P < 0.01$), with pairwise comparisons revealing a significant difference in conductors versus nonmusicians ($F_{(1,12)} = 13.55$, $P < 0.004$) and conductors versus pianists ($F_{(1,12)} = 6.16$, $P < 0.03$). Interestingly, a brain-electric source analysis (BESA[11]) showed that mismatch effects in conductors and pianists can be explained by a three-source model with two sources located in the superior temporal lobe and an additional medial source located frontally. The latter source appeared to best explain the positivity observed in the conductors. We propose that this positivity might be an instance of the P3a component, which is found to deviant/novel auditory items.[12,13] This P3a effect maps nicely onto a model proposed by Schröger,[14] who envisions the mismatch detection process as consisting of several stages. The features of an incoming stimulus are initially compared to the established representation of the standard stimulus. The feature-specific mismatch signals are then integrated, giving rise to MMN. If the integrated mismatch signal exceeds a (variable) threshold, conscious detection of the deviant takes place and initiates an involuntary attention shift, which ultimately leads to identification of the stimulus. Initiation of the involuntary attention shift has been associated with the P3a effect.[13] This, in essence, suggests that a qualitatively different mode of processing of deviants occurs outside the attentional focus in conductors compared to pianists.

TIME (DRUMMERS)

Whereas analysis of a conductor's task suggests superior spatial attention abilities, percussionists and drummers, especially from the jazz and rock traditions, can be assumed to possess superior timing abilities. Previous investigations from our group have shown that professional musicians are generally more precise in detecting anticipations of stimuli that occur in an otherwise regular temporal sequence.[15] Stimuli that were anticipated by 20 ms elicited a reliable mismatch negativity in musicians in a passive listening experiment, whereas nonmusicians did not show

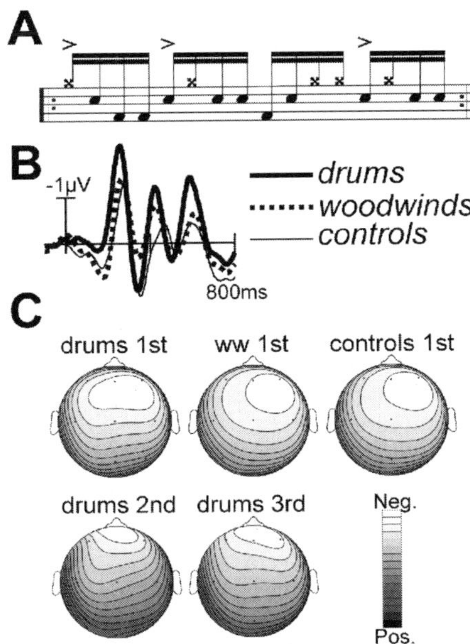

FIGURE 3. Study of the preattentive processing of temporally deviant beats in a real drum sequence. (**A**) Drum sequence (different notes represent the different drums of the drum set; x = hi-hat). The drum sequence was presented in loop mode. Occasionally, one of the beats was anticipated by 30 or 80 ms. (**B**) Deviant minus standard difference wave forms from the frontal midline electrode. In drummers, beats following the anticipated beat gave rise to a mismatch response as well. This effect was less prominent in the woodwind players and in the nonmusician controls. (**C**) Topographic maps show that the second and third mismatch peaks have considerably more frontal distribution than does the first mismatch response showing the typical frontocentral maximum.

MMN. Stimuli that were anticipated by 50 ms led to MMN in both groups, but musicians showed a larger amplitude. In a second experiment, occasionally omitted "sounds" in an otherwise regular tone series evoked a reliable MMN at interstimulus intervals (ISIs) of 100, 120, 180, and 220 ms in musicians. In nonmusicians, the MMN was smaller/absent in the 180 and 220 ms ISIs, respectively.[15]

To follow up these studies, we investigated 10 drummers, 10 woodwind players, and 10 nonmusician controls in a more natural task, which involved the presentation of a real drum sequence taken from a drum school (FIG. 3A). This sequence was presented as a continuous loop with no breaks. Occasionally, a beat was anticipated by 30 or 80 ms. The resulting sequence was presented in the format of a passive listening experiment. ERPs to standard (unaltered) parts of the sequence were subtracted from ERPs to deviant (anticipated) parts of the drum sequence. Beats anticipated by 30 ms did not give rise to a discernible mismatch response in any of the groups. Temporal analysis of the drum sequence revealed that a time jitter of 15–20 ms from beat to beat was present in this sequence. Therefore, it is not surprising that no MMN was observed for the 30-ms condition. The voltage-time diagram of the difference potential for the 80-ms anticipations is shown for a midline frontal electrode in FIGURE 3B. All three subject groups displayed a mismatch response for the first deviant beat of a sequence ($F_{(1,27)}$ = 10.5, P <0.001). This has a typical frontocentral, slightly right preponderant maximum (FIG. 3C). Interestingly, the subsequent beats of the sequence gave rise to additional mismatch responses, manifesting themselves as further negative peaks. These were more prominent in the musicians and especially in the drummers. Also, as illustrated by the spline-interpolated isovoltage maps in FIGURE 3C, the distribution of the second and third mismatch peak was different from that of the first peak. This pattern of results suggests that drummers' preattentive processing of temporally organized sound sequences is more profoundly disturbed. Whereas the amplitude of the mismatch response rapidly decays for the second and third anticipated beats, this is not the case in the drummers. They therefore appear to have a more complex representation of the musical time structure than do nonmusicians and woodwind players.

CONCLUSIONS

The studies reviewed in this brief contribution represent a first step towards the characterization of augmented auditory processing in musicians. All three examples show *qualitative* rather than *quantitative* processing differences between musicians and nonmusicians. Attentive processing of pitch was characterized by longer duration and more frontally distributed late ERP attention effect in musicians, preattentive surveillance of auditory space gave rise to a negative/positive mismatch response in conductors, which was not present in pianists and nonmusicians, and, finally, drummers appear to generate a more complex memory trace of the temporal organization of a stimulus sequence reflected in repetitive mismatch responses.

Further work is needed to clarify the extent of these qualitative processing differences. The fact that several of these differences were present under passive listening conditions or for stimuli in the unattended auditory channel, however, suggests that we are dealing with deeply engraved processing strategies shaped by years of musical experience. Future studies have also to address the question of whether or not these processing strategies are accompanied by neuroplastic changes in the circuitry of the primary and secondary auditory cortices in musicians.

ACKNOWLEDGMENTS

We are indebted to our musician subjects and to the Deutsche Forschungsgemeinschaft for continuing support of our research.

REFERENCES

1. MÜNTE, T.F., E. ALTENMÜLLER & L. JÄNCKE. 2002. The musician's brain as a model of neuroplasticity. Nat. Rev. Neurosci. **3:** 473–478.
2. MÜNTE, T.F., T. URBACH, E. DÜZEL & M. KUTAS. 2000. Event-related brain potentials in the study of human cognition and neuropsychology. *In* Handbook of Neuropsychology, 2nd Ed. F. Boller, J. Grafman & G. Rizzolatti, Eds. :240–321. Elsevier. Amsterdam.

3. HILLYARD, S.A., W.A. TEDER-SALEJARVI & T.F. MÜNTE. 1998. Temporal dynamics of early perceptual processing. Curr. Opin. Neurobiol. **8:** 202–210.

4. HANSEN, J.C. & S.A. HILLYARD. 1980. Endogenous brain potentials associated with selective auditory attention. Electroencephalogr. Clin. Neurophysiol. **49:** 277–290.

5. TEDER-SALEJARVI, W.A. & S.A. HILLYARD. 1998. The gradient of spatial auditory attention in free field: an event-related potential study. Percept. Psychophysiol. **60:** 1228–1242.

6. ALHO, K., M. SAMS, P. PAAVILAINEN, *et al.* 1989. Event-related brain potentials reflecting processing of relevant and irrelevant stimuli during selective listening. Psychophysiology **26:** 514–528.

7. GIARD, M.H., F. PERRIN, J. PERNIER & F. PERRONNET. 1988. Several attention-related wave forms in auditory areas: a topographic study. Electroencephalogr. Clin. Neurophysiol. **69:** 371–384.

8. RODER, B., W. TEDER-SALEJARVI, A. STERR, *et al.* 1999. Improved auditory spatial tuning in blind humans. Nature **400:** 162–166.

9. MÜNTE, T.F., C. KOHLMETZ, W. NAGER & E. ALTENMÜLLER. 2001. Neuroperception. Superior auditory spatial tuning in conductors. Nature **409:** 580.

10. PICTON, T.W., C. ALAIN, L. OTTEN, *et al.* 2000. Mismatch negativity: different water in the same river. Audiol. Neurootol. **5:** 111–139.

11. SCHERG, M. & P. BERG. 1991. Use of prior knowledge in brain electromagnetic source analysis. Brain Topogr. **4:** 143–150.

12. COURCHESNE, E., S.A. HILLYARD & R. GALAMBOS. 1975. Stimulus novelty, task relevance and the visual evoked potential in man. Electroencephalogr. Clin. Neurophysiol. **39:** 131–143.

13. FRIEDMAN, D., Y.M. CYCOWICZ & H. GAETA. 2001. The novelty P3: an event-related brain potential (ERP) sign of the brain's evaluation of novelty. Neurosci. Biobehav. Rev. **25:** 355–373.

14. SCHRÖGER, E. 1997. On the detection of auditory deviations: a pre-attentive activation model. Psychophysiology **34:** 245–257.

15. RÜSSELER, J., E. ALTENMÜLLER, W. NAGER, *et al.* 2001. Event-related brain potentials to sound omissions differ in musicians and non-musicians. Neurosci. Lett. **308:** 33–36.

Rhythm in Language and Music

Parallels and Differences

ANIRUDDH D. PATEL

The Neurosciences Institute, San Diego, California 92121, USA

ABSTRACT: Rhythm is widely acknowledged to be an important feature of both speech and music, yet there is little empirical work comparing rhythmic organization in the two domains. One approach to the empirical comparison of rhythm in language and music is to break rhythm down into subcomponents and compare each component across domains. This approach reveals empirical evidence that rhythmic grouping is an area of overlap between language and music, but no empirical support for the long-held notion that language has periodic structure comparable to that of music. Focusing on the statistical patterning of event duration, new evidence suggests that the linguistic rhythm of a culture leaves an imprint on its musical rhythm. The latter finding suggests that one effective strategy for comparing rhythm in language and music is to determine if differences in linguistic rhythm between cultures are reflected in differences in musical rhythm.

KEYWORDS: rhythm; language; music

INTRODUCTION

Widespread consensus exists among linguists that rhythm is an important aspect of spoken language. That is, just as languages can differ in terms of their inventory of phonemes and the lexicon they make from these sounds, they can also differ in rhythmic organization. Furthermore, the degree of rhythmic similarity is not simply a reflection of the historical relationship of languages. American English and Arabic, for example, are considered to be rhythmically similar, whereas American versus Jamaican English are considered rhythmically different.

What is linguistic rhythm? It is a composite of several aspects of a language that influence how it is organized in time. One component is the pattern of grouping/ phrasing of words within utterances and pausing between utterances. A second component is the durational patterning of syllables. A third component is the "configurational" patterning of stressed versus unstressed syllables. For example, English has a roughly alternating pattern of stressed and unstressed syllables and has mechanisms to avoid too many stressed syllables in close adjacency (stress clashes) or long sequences of unstressed syllables (stress lapses).[1] Greek, in contrast, has a greater tolerance for stress lapses, so that stressed syllables do not recur as regularly as they do in English.[2]

Address for correspondence: Dr. Aniruddh D. Patel, The Neurosciences Institute, 10640 John Jay Hopkins Drive, San Diego, CA 92121. Voice: 858-626-2085; fax: 858-626-2099.
apatel@nsi.edu

Ann. N.Y. Acad. Sci. 999: 140–143 (2003). © 2003 New York Academy of Sciences.
doi: 10.1196/annals.1284.015

What is musical rhythm? As with language, it is not simply one thing but a number of things that influence the way music is organized in time. These include the grouping of tones into phrases, a stable underlying beat (if present), and, if a beat is present, the organization of periodicity on multiple time scales to create musical meter. Just as linguists agree that rhythm is an important component of language, musicologists agree that rhythm is a fundamental aspect of music. This naturally raises the question of how rhythmic organization compares in the two domains in terms of both structure and cognitive processing. Although some cross-fertilization in terms of theoretical descriptions of rhythm has occurred in the two domains (especially regarding metric structure), little empirical work has been done on the similarities and differences of linguistic and musical rhythm.

EMPIRICAL SIMILARITY BETWEEN RHYTHM IN SPEECH AND MUSIC: GROUPING STRUCTURE

In speech perception, the words of a sentence are not perceived as being equally separated from one another. Rather, the boundaries between words vary in the amount of perceived "juncture," so that some words group together to form larger rhythmic chunks or phrases.[3] A similar situation obtains in music, where the tones of a melodic line are usually perceived as being organized into phrases even if there is no physical discontinuity between tones. Some of the acoustic cues to phrase boundaries in language and music, such as pitch drop and durational lengthening, are similar.[4] These cues are so salient that even young infants are sensitive to them.[5] It would be quite interesting to know if any relationship exists between the length of phrases in language and music. For example, do cultures with long linguistic phrases also have long phrases in their music? To my knowledge, no comparative data exist on this issue.

EMPIRICAL DIFFERENCE BETWEEN RHYTHM IN SPEECH AND MUSIC: PERIODICITY

An early and influential theory of speech rhythm posited that languages fell into one of two rhythmic classes. In "stress-timed" languages such as English and German, stressed syllables were claimed to occur at regular temporal intervals, whereas in "syllable-timed" languages such as Italian and French, syllable onsets were claimed to be evenly timed.[6] The underlying claim was that speech is produced in an isochronous way, the only difference being the units that occurred periodically. Subsequent empirical research has failed to confirm this hypothesis,[7] and the search for periodicity in speech has not produced any compelling quantitative data. This stands in sharp contrast to music, where there is abundant evidence for tightly regulated periodic patterns. Indeed, persons often synchronize their movements to a stable perceived beat in music, whereas normal speech does not induce a beat in this sense.

Despite the lack of evidence for isochrony in speech, linguists have maintained the terminology of "stress-timed" and "syllable-timed" languages to capture a perceived difference between the rhythms of different languages. Considerable debate exists, however, over whether two categories are really enough for classifying linguistic rhythms or even whether the categories are simply ends of a continuum.[8]

A NEW APPROACH TO THE EMPIRICAL COMPARISON OF RHYTHM IN LANGUAGE AND MUSIC

The foregoing two sections demonstrate that an effective approach to the empirical comparison of rhythm in language and music is to break rhythm down into subcomponents and compare each in turn. In addition to grouping and periodicity, another component of rhythm is the statistical patterning of event duration, particularly the variability of event duration. This has been the theme of recent research on speech rhythm, which has examined the variability of vowel duration in sentences of stress- versus syllable-timed languages.[9,10] The motivation for this research is the observation that stress-timed languages tend to allow a greater degree of vowel reduction than do syllable-timed languages (i.e., the shortening of vowels in unstressed syllables). A quantitative measure of vowel duration variability, called the "normalized pairwise variability index," or nPVI, was applied to sentences in stress- and syllable-timed languages and was shown to have a greater value in the former languages. This work provided the background for a comparative study of rhythm in language and music that was recently completed in our laboratory. Full details of this study can be found in Ref. 11; here, the main points of the study are described.

We wanted to test the idea that the linguistic rhythm of a culture might leave an imprint on its musical rhythm. This is a popular idea in musicology, but to our knowledge no convincing empirical evidence had been produced to support it. Our approach was to focus on cultures with strong linguistic rhythmic differences and to compare the musical rhythms of these cultures using the same measures that had been applied to speech rhythm. The nPVI allowed us to do this, as it measures the relative variability in a set of durations and can be applied to event durations in language (e.g., vowels) and in music (e.g., musical tones). We chose to focus on the speech and music of England and France, because English and French are canonical examples of languages with different rhythms and because we had access to a large number of English and French classical musical themes from the musicological literature.[12] We examined turn-of-the century composers (such as Elgar and Debussy), as this is considered a time of "musical nationalism" during which the music of different countries took on characteristic styles.

Research on English and French speech using the nPVI had demonstrated that English had a significantly higher value (i.e., a greater variability of vowel duration).[10] We applied the nPVI to the musical themes of 16 composers (6 English, 10 French) and found that, on average, English and French musical themes had significantly different nPVI values. Furthermore, this difference was in the same direction as the difference observed in language, with English having a greater nPVI value than French, meaning that tone durations were more variable in English than in French music.

In conducting this study, we focused exclusively on instrumental music (versus vocal music) in order to examine the "rhythmic thinking" of composers when they were free of a linguistic context. Our data suggest that even when thinking instrumentally, composers can be influenced by the rhythms of their native language. We do not believe that this influence is obligatory. Rather, it is likely to surface at times of musical nationalism, when composers seek to produce music that is particularly representative of their country.

CONCLUSION

Rhythm is a fundamental aspect of language and music, and empirical comparisons between rhythm in speech and music can benefit from careful dissection of rhythm into subcomponents for comparison across domains. This strategy shows that certain aspects of rhythm do show cross-domain similarities, that is, rhythmic grouping and the statistical patterning of event duration. These findings can help guide neural studies of rhythm, which will need to consider the different subcomponents of rhythm and the extent to which the components engage domain-general versus domain-specific processes.

ACKNOWLEDGMENTS

This work was supported by an Esther J. Burnham fellowship and by the Neurosciences Research Foundation as part of its program on music and the brain at The Neurosciences Institute.

REFERENCES

1. NESPOR, M. & I. VOGEL. 1989. On clashes and lapses. Phonology **6:** 69–116.
2. ARVANITI, A. 1994. Acoustic features of Greek rhythmic structure. J. Phonetics **22:** 239–268.
3. WIGHTMAN, C.W., S. SHATTUCK-HUFNAGEL, M. OSTENDORF, *et al.* 1992. Segmental durations in the vicinity of prosodic boundaries. J. Acoust. Soc. Am. **91:** 1707–1717.
4. TODD, N. 1985. A model of expressive timing in tonal music. Mus. Percept. **3:** 33–58.
5. JUSCZYK, R. & C. KRUMHANSL 1993. Pitch and rhythmic patterns affecting infants' sensitivity to musical phrase structure. J. Exp. Psychol. Hum. Percept. Perf. **19:** 627–640
6. PIKE, K.N. 1945. The Intonation of American English. University of Michigan Press. Ann Arbor.
7. ROACH, P. 1982. On the distinction between 'stress-timed' and 'syllable-timed' languages. *In* Linguistic Controversies: Essays in Linguistic Theory and Practice in Honour of F.R. Palmer. D. Crystal, Ed. :73–79. Edward Arnold. London.
8. NESPOR, M. 1990. On the rhythm parameter in phonology. *In* Logical Issues in Language Acquisition. I. Rocca, Ed. :157–175. Foris Publications. Dordrecht.
9. GRABE, E. & E.L. LOW. 2002. Durational variability in speech and the rhythm class hypothesis. *In* Laboratory Phonology 7. C. Gussenhoven & N. Warner, Eds. :515–546. Mouton de Gruyter. Berlin.
10. RAMUS, F. 2002. Acoustic correlates of linguistic rhythm: perspectives. *In* Proceedings of Speech Prosody 2002: 115–120. Université de Provence. Aix-en-Provence, France.
11. PATEL, A.D. & J.R. DANIELE. 2003. An empirical comparison of rhythm in language and music. Cognition **87:** B35–B45.
12. BARLOW, H. & S. MORGENSTERN. 1983. A Dictionary of Musical Themes, Revised Edition. Faber & Faber. London.

Neuropsychological Studies of Musical Timbre

SÉVERINE SAMSON

University of Lille 3, URECA, and Epilepsy Unit, Salpêtrière Hospital, Paris, France

ABSTRACT: Musical timbre is a multidimensional property of sound that allows one to distinguish musical instruments. In this paper, studies that explore the cerebral substrate underlying the processing of musical timbre are discussed. Perceptual asymmetries measured in normal participants, deficits of musical timbre perception obtained in brain-damaged patients, as well as results obtained with various neuroimaging methods are reviewed. The findings obtained in all of these studies generally support the predominant involvement of right temporal lobe areas, and more specifically of its anterior part, in processing spectral and temporal envelopes of musical timbre. However, controversies still exist about the contribution of the left temporal lobe in timbre perception. The necessity of comparing data obtained with different perceptual paradigms (same-different discrimination and similarity judgment) and various types of stimuli (single tones and melodies) was emphasized by reporting lesion studies carried out in patients with unilateral temporal lobe lesions. The few neuroimaging studies published in this domain provided additional and complementary findings. Unlike lesion studies that allow us to infer the cerebral structures that are essential for timbre perception, the latter investigations implicate a more distributed neural network in timbre processing that extends along the superior temporal gyrus to include not only anterior but also posterior temporal regions and possibly frontal areas as well.

KEYWORDS: musical timbre; perception; brain

INTRODUCTION

A growing interest in the psychology of music and in neuropsychology has been devoted to the study of musical timbre, a multidimensional property of sound that allows a person to distinguish musical instruments when pitch, loudness, and perceived duration remain identical.[1] Unlike these latter acoustical attributes that depend on a single physical dimension, timbre is defined as a combination of various acoustical parameters involving stationary (e.g., steady-state spectral energy) and dynamic (e.g., attack and decay times) features of tone.[2–6]

The covariation of these multiple cues leads timbre to be more distinctive than pitch or duration. Indeed, it has been shown that nonmusicians are more sensitive to changes in timbre than to changes in pitch in a sound categorization task.[7] A number of studies also showed that differences in timbre are very effective initiators of

Address for correspondence: Dr. Séverine Samson, Université de Lille 3, URECA, UFR de Psychologie, BP 149, 59653 Villeneuve d'Ascq cedex, France. Voice: +33 3 20 41 64 43; fax: +33 3 20 41 63 24.

samson@univ-lille3.fr

Ann. N.Y. Acad. Sci. 999: 144–151 (2003). © 2003 New York Academy of Sciences.
doi: 10.1196/annals.1284.016

stream segregation,[8,9] being often more powerful than segregation based on pitch differences.[6,10,11] Given the importance of timbre in perception and organization of the auditory scene, it seems relevant to improve our knowledge of the neural network underlying musical timbre.

To explore the cerebral structures involved in timbre, laterality and lesion studies as well as brain imaging investigations were carried out. In the present paper, perceptual asymmetries in normal participants and deficits of musical timbre perception obtained in brain-damaged patients as well as results obtained with various neuroimaging methods are reviewed.

LATERALITY STUDIES

Cerebral lateralization of musical timbre was investigated by interpreting perceptual asymmetries tested with dichotic or monaural stimulation. Generally, ear advantages for nonverbal stimuli appear to be less robust than those for verbal stimuli. Indeed, some studies failed to detect ear asymmetries for timbre processing.[12–15] However, left-ear advantage (LEA), indicative of a right-hemisphere superiority, was reported in several studies using natural instruments,[16–18] synthetic sounds,[19,20] and realistic or unrealistic (played backward) musical instruments.[17] All of these laterality studies strongly support a right-hemisphere priority for timbre perception.

LESION STUDIES

The first study that examined the effect of a unilateral temporal lobe lesion in timbre discrimination was carried out by Milner[21] in 1962. Using the Seashore musical battery,[22] she showed that compared to patients with left temporal lobectomy, patients who had undergone right temporal lobectomy for the relief of intractable epilepsy demonstrated a significant deficit in discriminating spectral information of harmonic tones, suggesting that processing of spectral information involved in timbre relies on the integrity of the right superior temporal lobe. Convergent evidence was reported in several subsequent studies on timbre perception in surgical cases or vascular stroke patients.[23–26] However, a more recent study failed to replicate Milner's finding.[27] No impairment on the Seashore musical battery was obtained in patients tested before or after unilateral temporal lobe surgery for epilepsy. The authors attributed the divergent results to differences between studies in the extent of temporal lobe resection along the Sylvian fissure. Whereas in Milner's study a large portion of the superior temporal gyrus was removed, their resection either spared or was limited to its most anterior part, indicating that the auditory areas implicated in timbre processing were still functional in this latter study.

In only one study, to our knowledge, was a left-hemisphere advantage for timbre processing found in lesion studies.[24] The investigators interpret the left-hemisphere superiority obtained in patients with commissurotomy during identification of musical instruments as evidence of the left-hemispheric semantic processing involved in a verbal identification task. However, this effect may also be attributable, at least in part, to the role of the left hemisphere in processing timing cues reported in several studies.[28–34] Since temporal information given by the initial part of the sound seems to be important for the recognition of musical instruments,[35–37] recognition deficits of musical instruments may be affected by deficits in perception of temporal information involved in timbre.

DISCRIMINATION OF SPECTRAL AND TEMPORAL ENVELOPES OF TIMBRE

To specify the role of each temporal lobe in perceiving the spectral and temporal cues involved in musical timbre, Zatorre and I tested patients with unilateral temporal lobe lesions involving the anterior part of the lateral temporal neocortex as well as various extent of the medial temporal lobe structures in a discrimination task.[38] Tones with varying spectral or time envelopes were created by changing the number of harmonics or the rise time duration, respectively. Half the stimuli consisted of complex tones with 3, 4, or 5 harmonics with a constant temporal envelope (1-ms rise time). The other half had rise times of 1, 50, or 190 ms with a constant spectral content (4 harmonic tones). Tones were presented in pairs, and the subject had to respond whether the two tones were the same or different. The results showed that right but not left temporal lobectomy impaired the discrimination of tones differing in their spectral envelope or their temporal envelope, suggesting that the right temporal lobe is essential for processing the temporal as well as the spectral information involved in musical timbre.

Although right temporal lobe superiority in processing spectral information confirmed previously reported data, the predominant role of the right temporal lobe in processing time-related cues was a new finding. It seems to contradict previous results showing that left hemisphere structures may be predominantly involved in temporal processing.[30,32,39–41] However, the temporal features manipulated in these other studies involve rapid temporal coding presented sequentially. In our investigation, the discrimination of temporal information is generally based on slower temporal changes that holistically modified tone perception. From our findings, we can conclude that timbre discrimination of spectral energy and of temporal variations depends on the integrity of auditory areas from the right temporal neocortex.

DISSIMILARITY JUDGMENT OF MUSICAL TIMBRE

Recently, we have used another paradigm involving similarity judgments to study musical timbre perception in single tones as well as in melodies.[42] Similarity judgment is based on the extraction of particular features that play an important role in recognition by requiring higher order perception than discrimination or detection of differences.[43] The participants included in this study were also tested with a conventional "same-different" task,[38] thereby allowing the comparison of two perceptual procedures. For this purpose, synthetic sounds were created by selecting three levels of spectral changes (1, 4, and 8 harmonics) and three levels of temporal changes (1, 100, and 190-ms rise time duration). All of these tones were difficult to label, therefore precluding the use of verbal strategies. Pairs of different stimuli were presented to patients with unilateral temporal lobe excisions and normal control participants for judgment of their similarity on a rating scale. This task, already validated in normal participants,[44] provides an opportunity to examine the dependence of timbre on both spectral and temporal envelopes of tones. The method used to analyze the data was based on a multidimensional scaling (MDS) technique that allows us to represent the dissimilarities between a set of physically different stimuli as distances between points in an n-dimensional space.[45–47] MDS reveals the extent to which physical distances between multidimensional acoustic information are actually perceived by a subject, instead of investigating only the ability to distinguish same from different pairs of tones.

In a previous study,[44] we showed that normal participants could recover in their perceptual judgments the unambiguous acoustic dimensions that were physically inherent to the stimuli. More specifically, normal listeners could perceive the number of harmonics and rise time duration in a gradual and ordered way and to use this information to make perceptual judgments of musical timbre in single tones as well as in melodies. By contrast, perceptual ratings obtained by patients with right temporal lobe lesions revealed a disturbed perceptual space in single tones and melodic conditions. The most distorted results were obtained with single tones, in which the temporal parameter was less prominent. Tones were grouped according to their spectral content, but the results did not reflect a coherent underlying perceptual dimension. The results in patients with left temporal lobe lesions suggest that they could use the spectral and the temporal envelopes of tones independently in making perceptual judgments of single tones. In the melodic condition, their results were significantly different from those of normal participants, suggesting that left temporal lesions do affect subtle aspects of timbre perception despite these patients' preserved ability to make discrimination judgments using traditional paradigms. In general, data from both patient groups showed that extraction of temporal cues was easier in the melodic than in the single tone condition, suggesting that the different durations and frequencies heard within a musical context facilitate timbre perception. The findings of the present study replicate and extend previous results showing that timbre perception mainly depends on the integrity of right neocortical structures. However, the contribution of left temporal regions is also apparent, indicating that multidimensional techniques are sensitive to more subtle perceptual disturbances that may not be revealed by "same-different" discrimination paradigms.[38]

Lesion studies converge to ascribe a crucial role to the superior right temporal lobe regions in timbre perception. This right hemisphere bias seems to result from the dysfunction of the anterior temporal lobe structures rather than from the dysfunction of posterior temporal lobe areas, because these later regions are generally spared in surgical excision of the temporal lobe. The question regarding the possible involvement of left hemisphere structures in timbre processing remains open.

BRAIN IMAGING STUDIES

The cerebral substrate associated with timbre processing has also been examined using various neuroimaging techniques. However, the goals of these studies were very different, making comparison between studies difficult. In electroencephalography studies, cortical auditory evoked potentials in response to different timbres were explored. The results showed that larger electrical activities were recorded over the right than over the left hemisphere at temporal electrodes[48–50] and, to a lesser degree, at frontal electrodes,[48] suggesting the existence of a predominant right hemisphere neural network responding to timbre changes.

In another study[5] using magnetoencephalography, the effect of musical training on auditory cortical representations in response to tones of different instruments (trumpet and violin) was examined. In this study, neuromagnetic measures were recorded in skilled musicians (trumpeters and violinists) when listening to sine tones or musical tones played on their instrument of training or on a different instrument. These investigators recorded larger dipole moments in both right and left temporal regions for tones played on a musical instrument than for sine tones, this effect being more salient for the instrument of training. Although this finding does not specify the electrophysiologic responses specific to musical timbre, it suggests that a certain degree of neural plasticity associated with musical timbre has taken place over the right and the left hemisphere in skilled musicians.

Two positron emission tomography studies were also carried out in this domain. The first one demonstrated bilateral parietotemporal activations with right greater than left frontotemporal asymmetries[52] using the timbre subtest of Seashore's battery.[22] More recently, Platel *et al.*[53] identified activations in the superior and middle frontal gyrus within the right hemisphere during the detection of timbral changes. However, these foci of activity are difficult to attribute purely to timbre processing, because variations in pitch and rhythm were also present, requiring selective attention to the timbre dimension. Nevertheless, converging evidence has been obtained with functional magnetic resonance imaging (fMRI). One of these studies was designed to examine auditory imagery of musical timbre. To verify if timbre imagery shows the right-sided bias generally reported in timbre perception, Halpern and her collaborators[54] studied the neural substrate underlying timbre perception and mental imagery of musical instruments. The subject's task consisted of making similarity judgments of musical instruments, either played or imagined. Consistent with lesion studies, the perception task activated several areas in the anterosuperior temporal gyrus with a right-sided asymmetry. When the judgments concerned not only heard timbres but also imagined timbres, posterior temporal areas were activated, relative to the visual imagery control condition, again with a slight right hemisphere predominance. The imagery task also activated left frontal areas that were attributed by the authors to the working memory component of maintaining the auditory image cued by the names of the musical instruments. This pattern of results showing that timbre similarity judgments of single tones appear to be mediated largely by the right temporal lobe structures confirms and extends previous results obtained in patients during a perceptual task.[42] The other fMRI study was designed to specify the characteristics of cerebral responses associated with differences in melodic stimuli based on timbre.[55] The two timbres used in this study were taken from extreme positions of a psychological space[56] and differed in terms of various acoustic cues (e.g., attack time, spectral centroid, and spectral flux). Throughout the experiment, listeners were exposed to blocks of different melodies played in either one timbre or the other. Because both experimental conditions involved listening to timbres, the neural network shared by the two conditions was cancelled out by the subtractive method, and the pattern of activation characterizing timbre perception in general could not be visualized. However, the investigators identified bilateral activations specific to one of the two timbres (for instance, timbre A), as compared to the other (timbre B), in the superior temporal gyrus (dorsomedial surface) and the superior temporal sulcus.

Conversely, when activations specific to timbre B compared to timbre A were examined, the authors found no significant results, suggesting that each area activated by timbre B was also activated by timbre A. This surprising pattern of activations was interpreted by the authors as resulting from the rapid temporal changes in spectral energy distribution that characterized timbre A. However, we also suggest that the melodic context in which timbre is presented may have influenced timbre perception, as we already pointed out,[42,44] by obscuring or enhancing the importance of certain parameters that may recruit the left temporal lobe areas in addition to those on the right.

CONCLUSIONS

The findings obtained in lesion and brain imaging studies generally support the involvement of the right auditory temporal areas, and more specifically of its anterior part, in processing spectral and temporal envelopes of musical timbre. However, controversies still exist about the possible contribution of left temporal lobe structures in timbre perception. Although we must remain cautious in interpreting these data, the results of lesion studies emphasize the necessity of comparing findings obtained with different perceptual paradigms and various types of stimuli. We demonstrated the relevance of evaluating perceptual ratings of similarity judgments to infer the physical distances between stimuli that are actually perceived by the patient, instead of assessing only the ability to differentiate same from different pairs of stimuli in a discrimination task. Similarly, we showed that similarity judgments of musical timbre can be differently affected in single tones compared to melodies, the relative weighting of the various acoustical features being different in these two conditions. Finally, cerebral activations obtained in neuroimaging and deficits resulting from brain damage do not provide exactly the same information. As reported in several brain imaging studies, left temporal lobe regions might well be involved in timbre processing, but it does not mean that these structures are essential for this cognitive ability. Unlike lesion studies that permit inferences about the indispensable structures for a specific cognitive processing, imaging studies in normal participants allow identification of cerebral structures normally involved in this function, which include not only anterior but also posterosuperior temporal lobe areas.

REFERENCES

1. AMERICAN STANDARDS ASSOCIATION: Acoustical Terminology. 1960. S1.1-1960. American Standards Association. New York.
2. GREY, J.M. 1977. Multidimensional perceptual scaling of musical timbres. J. Acoust. Soc. Am. **61:** 1270–1277.
3. GREY, J.M. & J.W. GORDON. 1978. Perceptual effects of spectral modifications on musical timbres. J. Acoust. Soc. Am. **63:** 1493–1500.
4. MILLER, J.R. & E.C., CARTERETTE. 1975. Perceptual space for musical structures. J. Acoust. Soc. Am. **58:** 711–720.
5. PLOMP, R. 1970. Timbre as a multidimensional attribute of complex tones. *In* Frequency Analysis and Periodicity Detection in Hearing. R. Plomp & G.F. Smoorenburg, Eds. :397–414. Sijthoff. Lieden.
6. WESSEL, D.L. 1994. Timbre space as a musical control structure. Comp. Mus. J. **3:** 45–52.

7. PITT, M.A. 1994. Perception of pitch and timbre by musically trained and untrained listeners. J. Exp. Psychol. Hum. Percept. Perform. **20:** 876–986.

8. CUSACK, R. & B. ROBERTS. 2000. Effects of differences in timbre on sequential grouping. Percept. Psychophysiol. **62:** 1112–1120.

9. SINGH, P. & A. BREGMAN. 1997. The influence of different timbre attributes on the perceptual segregation of complex-tone sequences. J. Acoust. Soc. Am. **102:** 1943–1953.

10. IVERSON, P. 1995. Auditory stream segregation by musical timbre: effects of static and dynamic acoustic attributes. J. Exp. Psychol. Hum. Percept. Perform. **21:** 751–763.

11. SINGH, P. 1987. Perceptual organization of complex-tone sequences: a trade off between pitch and timbre? J. Acoust. Soc. Am. **82:** 886–899.

12. BOUCHER, R. & M.P. BRYDEN. 1997. Laterality effects in the processing of melody and timbre. Neuropsychologia **35:** 1467–1473.

13. PRIOR, M. & G.A. TROUP. 1988. Processing of timbre and rhythms in musicians and non-musicians. Cortex **24:** 451–456.

14. RUSHFORD, K. & D. MURRAY. 1977. Left-right differences in the processing of instrument tone segments. Council Res. Music Ed. **52:** 1–6.

15. KALLMAN, H.J. 1978. Can expectancy explain reaction time ear asymmetries? Neuropsychologia **16:** 225–228.

16. KALLMAN, H.J. & M.C. CORBALLIS. 1975. Ear asymmetry in reaction time to musical sounds. Percept. Psychophysiol. **17:** 368–370.

17. PAQUETTE, C. & I. PERETZ. 1997. Role of familiarity in auditory discrimination of musical instrument: a laterality study. Cortex **33:** 689–696.

18. RASTATTER, M.P. & A.J. GALLAHER. 1982. Reaction-times of normal subjects to monaurally presented verbal and tonal stimuli. Neuropsychologia **20:** 465–473.

19. BRANCUCCI, A. & P. SAN MARTINI. 1999. Laterality in the perception of temporal cues of musical timbre. Neuropsychologia **37:** 1445–1451.

20. SAN MARTINI, P. *et al.* 1994. Prevalent direction of relective lateral eye movements and ear asymmetries in a dichotic test of musical chords. Neuropsychologia **32:** 1515–1522.

21. MILNER, B. 1962. Laterality effects in audition. *In* Interhemispheric Relations and Cerebral Dominance. V. Mountcastle, Ed. :173–201. John Hopkins University Press. Baltimore.

22. SEASHORE, C.E., D. LEWIS & J.L. SAETVEIT. 1940. Revision of the Seashore measures of musical talents. University of Iowa Student Aims Program Research. Iowa City University. Iowa Press.

23. CHASE, R. 1967. Report of research in discussion following B. Milner: Brain mechanism suggested by studies of temporal lobes. *In* Brain Mechanisms Underlying Speech and Language. C.H. Millikan & F.L. Darley, Eds. Grune & Stratton. New York.

24. TRAMO, M.J. & M.S. GAZZANIGGA. 1989. Discrimination and recognition of complex tonal spectra by the cerebral hemispheres: differential lateralization of acoustic-discriminative and semantic associative functions in auditory pattern perception. Soc. Neurosci. **15:** 421.2.

25. CHOBOR, K.L. & J.W. BROWN. 1987. Phoneme and timbre monitoring in left and right cerebrovascular accident patients. Brain Lang. **30:** 278–284.

26. MAZZUCHI, A. *et al.* 1982. A case of receptive amusia with predominant timbre perception defect. J. Neurol. Neurosurg. Psychiatry **45:** 644–647.

27. KOIKE, A. *et al.* 1996. Preserved musical abilities following right temporal lobectomy. J. Neurosurg. **85:** 1000–1004.

28. CARMON, A. & I. NACHSON. 1971. Effect of unilateral brain damage on perception of temporal order. Cortex **7:** 410–418.

29. EFRON, R. 1963. Temporal perception, aphasia and déjà vu. Brain **86:** 403–424.

30. EHRLÉ, N., S. SAMSON & M. BAULAC. 2001. Processing of rapid auditory information in epileptic patients with left temporal lobe damage without dysphasia. Neuropsychologia **39:** 525–531.

31. ROBIN, D.A., D. TRANEL & H. DAMASIO. 1990. Auditory perception of temporal and spectral events in patients with focal left and right cerebral lesions. Brain Lang. **39:** 539–555.

32. SAMSON S., N. EHRLÉ & M. BAULAC. 2001. Cerebral substrates for musical temporal processes. Ann. N.Y. Acad. Sci. **930:** 166–178.
33. SWISHER, L. & I.J. HIRSH. 1972. Brain damage and the ordering of two temporally successive stimuli. Neuropsychologia **10:** 137–152.
34. TALLAL, P. & F. NEWCOMBE. 1978. Impairment of auditory perception and language comprehension in dysphasia. Brain Lang. **5:** 13–24.
35. BALZANO, G.J. 1964. What are musical pitch and timbre? Music Percept. **3:** 297–314.
36. BERGER, K.W. 1964. Some factors in the recognition of timbre. J. Acoust. Soc. Am. **36:** 1888–1891.
37. SCHOUTEN, J.F. 1968. The perception of timbre. *In* Report of the 6th International Congress on Acoustics, Tokyo. GP6–2.
38. SAMSON, S. & R.J. ZATORRE. 1994. Contribution of the right temporal lobe to musical timbre discrimination. Neuropsychologia **32:** 231–240.
39. BELIN, P. *et al.* 1998. Lateralization of speech and auditory temporal processing. J. Cognit. Neurosci. **10:** 536–540.
40. LIÉGEOIS-CHAUVEL, C. *et al.* 1999. Specialization of left auditory cortex for speech perception in man depends on temporal coding. Cereb. Cortex **9:** 484–496.
41. TALLAL, P., S. MILLER & R.H. FITCH. 1993. Neurobiological basis of speech: a case for the preeminence of temporal processing. Ann. N.Y. Acad. Sci. **682:** 27–47.
42. SAMSON, S., R.J. ZATORRE. & J.O. RAMSAY. 2002. Deficits of musical timbre perception after unilateral temporal-lobe lesion revealed with multidimensional scaling. Brain **125:** 1–13.
43. WHITFIELD, I.C. 1985. The role of auditory cortex in behavior. *In* Cerebral Cortex. A. Peters & E.G. Jones, Eds. :329–349. Plenum. New York.
44. SAMSON, S., R.J. ZATORRE. & J.O. RAMSAY. 1997. Multidimensional scaling of synthetic musical timbre: perception of spectral and temporal characteristics. Can. J. Psychol. **51:** 307–315.
45. KRUSKALL, J.B. 1964. Multidimensional scaling by optimizing goodness of fit to a nonmetric hypothesis. Psychometrika **49:** 1–27.
46. SHEPARD, R.N. 1962. The analysis of proximities: multidimensional scaling with an unknown distance function. Psychometrika **27:** 125–140.
47. RAMSAY, J.O. 1982. Some statistical approaches to multidimensional scaling data. J. R. Statist. Soc. A. **145:** 285–312.
48. JONES, S.J., O. LONGE & M. VAZ PATO. 1998. Auditory evoked potentials to abrupt pitch and timbre change of complex tones: electrophysiological evidence of 'streaming'? Electroencephalogr. Clin. Neurophysiol. **108:** 131–142.
49. AUZOU, P. *et al.* 1995. Topographic EEG activations during timbre and pitch discrimination tasks using musical sounds. Neuropsychologia **33:** 25–37.
50. CRUMMER, G.C. *et al.* 1994. Neural processing of musical timbre by musicians, nonmusicians, and musicians possessing absolute pitch. J. Acoust. Soc. Am. **95:** 2720–2727.
51. PANTEV, C. *et al.* 2001. Timbre-specific enhancement of auditory cortical representations in musicians. NeuroReport **12:** 169–174.
52. MAZZIOTTA, J.C. *et al.* 1982. Tomographic mapping of cerebral metabolism: auditory stimulation. Neurology **32:** 921–937.
53. PLATEL, H. *et al.* 1997. The structural components of music perception. A functional anatomical study. Brain **120:** 229–243.
54. HALPERN, A. *et al.* 2002. An fMRI study of timbre perception and imagery. Proceedings of the 7th International Conference on Music Perception and Cognition, Sydney.
55. MENON, V. *et al.* 2002. Neural correlates of timbre change in harmonic sounds. NeuroImage **17:** 1742–1754.
56. McADAMS, S. *et al.* 1995. Perceptual scaling of synthesized musical timbres: common dimensions, specificities, and latent subject classes. Psychol. Res. **58:** 177–192.

Recognition of Interleaved Melodies

An fMRI Study

CAROLINE BEY AND ROBERT J. ZATORRE

Montreal Neurological Institute, McGill University, Montreal, Quebec H3A2B4, Canada

ABSTRACT: An fMRI study of interleaved melody recognition was conducted to examine the neural basis of the bottom-up and top-down mechanisms involved in auditory stream segregation. Hemodynamic activity generated by a mixed sequence was recorded in eight listeners who were asked to recognize a target melody interleaved with distractor tones when the target was presented either before or after the composite sequence. fMRI results suggest that similar cortical networks were involved in both conditions, including bilaterally the auditory cortices within the superior temporal gyrus as well as the thalamus and the inferior frontal gyrus. However, when listeners heard the melody before they had to extract it from the mixture, neural activation in the inferior frontal operculum was significantly enhanced bilaterally; no change in auditory cortical activity was detected.

KEYWORDS: auditory stream segregation; interleaved melody recognition; fMRI

INTRODUCTION

When listening to polyphonic music, listeners must separate voices played by different instruments or by the same instrument in different frequency registers. This ability to segregate sounds coming from different environmental sources has been called auditory scene analysis.[1] This analysis is preattentive in part. However, acquired knowledge about sounds and sound sequences, such as music and speech, can help us to extract them from a complex scene via top-down mechanisms. Results of electrophysiologic studies tend to confirm the existence of these two components.[2,3] However, the cerebral structures involved in these processes are still unknown. Therefore, we conducted an fMRI study on interleaved melody recognition to examine the neural basis of the bottom-up and top-down mechanisms involved in auditory stream segregation.

METHOD

Two unfamiliar six-tone melodies, one of which was interleaved with distractor tones, were presented successively to eight listeners who were required to decide

Address for correspondence: Caroline Bey, FPSE, UNI MAIL, 40 bd du Pont d'Arve, 1205 Geneva, Switzerland.
caroline_bey@yahoo.fr

Ann. N.Y. Acad. Sci. 999: 152–154 (2003). © 2003 New York Academy of Sciences.
doi: 10.1196/annals.1284.017

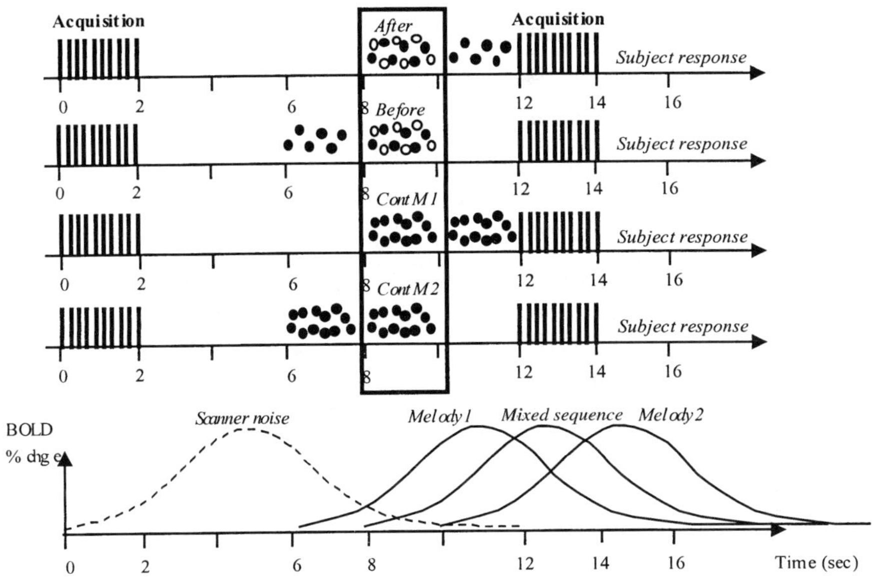

FIGURE 1. Visual illustration of the experimental design.

whether the melodies were identical or different.[4] Using a single-trial fMRI procedure with clustered volume acquisition,[5] we recorded hemodynamic activity generated by the mixed sequence when the target melody was presented either *before* or *after* the composite sequence and also in a control condition consisting of a simple melody discrimination task when participants were listening to the first or second 12-note melody (FIG. 1).

RESULTS AND DISCUSSION

The fMRI results suggest that similar cortical networks were involved in both conditions, including the auditory cortices within the superior temporal gyrus bilaterally as well as the thalamus and inferior frontal gyrus. However, when listeners heard the melody before they had to extract it from the mixture, neural activation in the inferior frontal operculum was significantly enhanced bilaterally; no change in auditory cortical activity was detected (FIG. 2). Results of the control conditions were similar, suggesting that this effect could be explained by the sequence position and is not unique to the top-down mechanism involved in stream segregation. Thus, the effect may be related to a matching process, that is, the recognition of a target stored in working memory involving frontal cortical mechanisms in both control and experimental tasks.[6] Alternatively, there may be an asymmetry in the hemodynamic response, leading to greater activity in the second scanned sequence, contrary to our

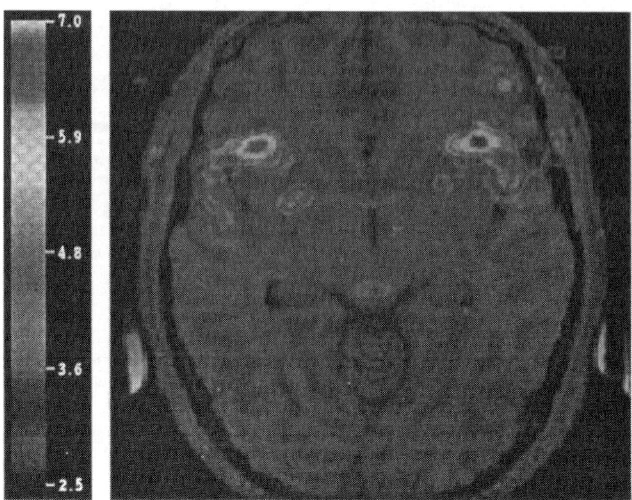

FIGURE 2. Horizontal section (z = −7) showing significant activation in the *before* versus *after* comparison. Scale indicates z score values.

assumption based on preliminary data and previous studies.[5] Further study is needed to examine this point.

ACKNOWLEDGMENTS

This research was supported by grants from the Canadian Institutes of Health Research and the Fyssen Foundation. We wish to thank Albert S. Bregman and Stephen McAdams for their helpful suggestions on this research.

REFERENCES

1. BREGMAN, A.S. 1990. Auditory Scene Analysis: The Perceptual Organization of Sound. MIT Press. Cambridge, MA.
2. ALAIN, C., S.R. ARNOTT & T.W. PICTON. 2001. Bottom-up and top-down influences on auditory scene analysis: evidence from event-related brain potentials. J. Exp. Psychol. Hum. Percept. Perform. **27:** 1072–1089.
3. SUSSMAN, E., W. RITTER & J.H.G. VAUGHAN. 1999. An investigation of auditory streaming effect using event-related brain potentials. Psychophysiology **36:** 22–34.
4. BEY, C. & S. MCADAMS. 2002. Schema-based processing in auditory scene analysis. Percept. Psychophysiol. **64:** 844–854.
5. BELIN, P. *et al.* 1999. Event-related fMRI of the auditory cortex. NeuroImage **10:** 417–429.
6. ZATORRE, R.J. & J.R. BINDER. 2000. Functional and structural imaging of the human auditory system. *In* Brain Mapping: The Systems. A.W. Toga & J.C. Mazziotta, Eds. :365-402. Academic Press. Los Angeles.

Electrical Brain Responses to Descriptive versus Evaluative Judgments of Music

ELVIRA BRATTICO,[a] THOMAS JACOBSEN,[b] WOUTER DE BAENE,[c] NOA NAKAI,[d] AND MARI TERVANIEMI[a]

[a]Cognitive Brain Research Unit, Department of Psychology, University of Helsinki FIN-00014, Finland

[b]Institut für Allgemeine Psychologie, Leipzig Universität, Leipzig D-04103, Germany

[c]Department of Experimental Psychology, University of Ghent, Ghent 9000, Belgium

[d]Sibelius Academy, Helsinki FIN-00251, Finland

ABSTRACT: The present study was aimed at finding neural correlates of aesthetic versus descriptive listening of the same musical cadences. Results showed that aesthetic listening generated greater right frontocentral negativities than did descriptive listening, indicating distinct cortical mechanisms for aesthetic versus descriptive processing of music.

KEYWORDS: aesthetics; event-related potentials; musical cadences; evaluative processing

The cognitive processes of subjective evaluation have mainly been considered so variable between individuals as not to be an appropriate object for neurophysiological investigation. However, recent studies have opened new frontiers in this respect. For instance, the subjective euphoric experience of "chills" that subjects feel when listening to their favorite music has been measured and localized in precise areas of the brain.[1] This euphoric emotion might be related to aesthetic appreciation, where the perceptual, representational, and evaluative activities, focused on the form and meaning of a piece of art, have an interconnected role.[2] However, the experiment of Blood et al.[1] did not address the question of whether evaluative versus descriptive listening tasks, besides the music material presented, might as well differentiate the evoked brain activation. Previously, by using the dichotic listening method, a left ear advantage (corresponding to right hemispheric dominance) for affective judgments of melodies was found, whereas no ear preference was observed for non-affective judgments concerning melody coherence.[3] Furthermore, in the visual modality, electrical brain potentials differentiated symmetry (descriptive) versus beauty (evaluative) judgments of abstract pictures[4]: an early frontocentral negative deflection was elicited by "not-beautiful" judgments, whereas larger posterior, positive deflections

Address for correspondence: E. Brattico, Cognitive Brain Research Unit, Department of Psychology, P.O. Box 9, University of Helsinki FIN-00014, Finland. Voice: +358-9-191-29457; fax: +358-9-191-29450.

elvira@cbru.helsinki.fi.

Ann. N.Y. Acad. Sci. 999: 155–157 (2003). © 2003 New York Academy of Sciences.
doi: 10.1196/annals.1284.018

reflected "symmetric" judgments. To our knowledge, no studies have so far been conducted on the neural substrates of different ways of listening to music. The aim of the present experiment was to compare electrical brain activity during descriptive versus evaluative judgments of the same musical cadences.

Fifteen nonmusicians were measured with a 25-electrode electroencephalogram (EEG) while they were listening to 180 cadences consisting of five chords (4 of 500 ms in length, the last of 1000 ms). The cadences were of 30 types transposed over 6 musical scales: 10 with correct, 10 with ambiguous, and 10 with incorrect ending chords. Prompted by a visual cue, subjects were asked to judge the cadences: when "C" appeared, subjects judged the correctness or incorrectness of the cadence; when "L" appeared, whether they liked or disliked it. Comparison of the waveforms averaged according to the task showed larger negativities at the central electrodes for the aesthetic as compared to the descriptive task at 500–700 ms from the cadence onset and at 2200–2400 ms (after the manipulated ending chord, occurring at 2000 ms). Since the cadence chords ended at 3000 ms, the larger negativities for the evaluative (aesthetic) task indicate that more neural resources are devoted to prepare an evaluative listening. Moreover, these negativities were larger over the left hemisphere for the descriptive but not for the aesthetic task, concordant with the hypothesis that this hemisphere is specialized for analytic processing.[5]

Comparison of the waveforms averaged according to the subjects' answers for the "like/dislike" task showed larger negative deflections over the right hemisphere (at F4) at 2300–3600 ms (not before the onset of the manipulated ending chord) for the "disliked" as compared to the "liked" answers. Furthermore, brain responses locked to the "liked" answers were greater over the left hemisphere (at F3) than the right, probably indicating a lateralization effect of sound pleasantness/unpleasantness.[6]

Comparisons of the "correct/incorrect" answers at 3300–3600 ms (1300 ms after the onset of the last chord) showed a larger negativity at F3 for the "correct" judgments, whereas "incorrect" answers did not show any lateralization. This result, discrepant from the literature where incorrect chords evoke earlier different brain responses from correct chords, can be explained in the present experiment by the use of ending chords purposely not clearly correct or incorrect in order to avoid the easy correlation with liking/disliking. To conclude, the present data show that brain mechanisms underlying descriptive and evaluative processing of the same musical cadences are dissociable in their response morphology and time dynamics by using event-related potentials. Further investigations for a more detailed localization of the brain structures involved are desirable.

REFERENCES

1. BLOOD, A.R. & R. ZATORRE. 2001. Intensely pleasurable responses to music correlate with activity in brain regions implicated in reward and emotion. Proc. Natl. Acad. Sci. USA **98:** 11818–11823.
2. ARGENTON, A. 1996. Arte e cognizione. Introduzione alla psicologia dell'arte. Raffaello Cortina. Milano.
3. GAGNON, L. & I. PERETZ. 2000. Laterality effects in processing tonal and atonal melodies with affective and nonaffective task instructions. Brain Cognit. **43:** 206–210.
4. JACOBSEN, T. & L. HOFEL. 2001. Aesthetics electrified: an analysis of descriptive symmetry and evaluative aesthetic judgment processes using event-related brain potentials. Emp. Stud. Arts **19:** 177–190.

5. BEVER, T.G. & R.J. CHIARELLO. 1974. Cerebral dominance in musicians and nonmusicians. Science **185:** 537–539.
6. DAVIDSON, R.J. 1993. The neuropsychology of emotion and affective style. *In* Handbook of Emotion. M. Lewis & J.M. Haviland, Eds. :143–154. Guilford Press. New York.

Cortical Correlates of Acquired Deafness to Dissonance

ELVIRA BRATTICO,[a] MARI TERVANIEMI,[a] VESA VALIMAKI,[c]
TITIA VAN ZUIJEN,[a] AND ISABELLE PERETZ[b]

[a]*Cognitive Brain Research Unit, Department of Psychology, University of Helsinki FIN-00014, Finland*

[b]*Department of Psychology, University of Montreal, Montreal QC, H3C 3J7, Canada*

[c]*Laboratory of Acoustics and Audio Signal Processing, Helsinki University of Technology, HUT FIN-02015, Finland*

ABSTRACT: Patient I.R., who had bilateral lesions in the auditory cortex but intact hearing, did not distinguish dissonant from consonant musical excerpts in behavioral testing. We additionally found that the electrical brain responses did not differentiate musical intervals in terms of their dissonance/consonance, consistent with the idea that this phenomenon depends on the integrity of cortical functions.

KEYWORDS: dissonance; brain lesions; mismatch negativity (MMN); event-related potentials (ERPs)

In all musical cultures, the human auditory system is suggested to be differentially sensitive to certain musical intervals.[1] For instance, listeners can classify musical intervals termed consonant by theorists as pleasant and harmonious, and dissonant intervals as unpleasant and disharmonious.[2] Several investigators[3] suggested that this phenomenon originates in the basilar membrane of the internal ear. However, recent data point to a more cortical locus for these effects. The mismatch negativity (MMN),[4] a cortical electrical brain response representing the automatic detection of the change in a repetitive auditory environment, was greater to a dissonant change in a consonant context than to a consonant change in a consonant context, even if this change was smaller in pitch distance, contradicting evidence that MMN increases its amplitude with larger pitch shifts.[5,6] Further evidence comes from a patient (I.R.) with bilateral lesions in the auditory cortex, found to be deaf to dissonance.[7] However, this patient was never studied with functional brain measures. The aim of this study was to further understand the cortical mechanisms of I.R.'s defect of distinguishing dissonant from consonant sounds.

Address for correspondence: E. Brattico, Cognitive Brain Research Unit, Department of Psychology, P.O. Box 9, University of Helsinki FIN-00014, Finland. Voice: +358-9-191-29457; fax: +358-9-191-29450.
elvira@cbru.helsinki.fi

Ann. N.Y. Acad. Sci. 999: 158–160 (2003). © 2003 New York Academy of Sciences.
doi: 10.1196/annals.1284.019

With 40-channel electroencephalography (EEG) we measured I.R., a right-handed 47-year-old patient with bilateral lesions in the auditory associative cortices resulting in irreversible auditory impairments in the musical domain exclusively. Her language comprehension and general intellectual functions were intact. Two of I.R.'s matched control subjects (M.M. and S.F.) were tested as well. Simultaneous musical intervals that were sufficiently far apart in pitch to be discriminable by I.R. were presented in two different contexts: (1) a "consonant" context where the standard (frequent) interval was a perfect fourth (with a 3:4 frequency ratio) and the three deviant (infrequent, $P = 0.2$ in total) intervals presented in three separate blocks were a dissonant major seventh (8:15 ratio), a consonant major sixth (3:5 ratio), and a consonant perfect octave (1:2 ratio); (2) a "dissonant" context in which the standard dissonant interval was a major seventh and the three deviants ($P = 0.2$ in total), presented in separate blocks, were a consonant perfect eleventh (3:8 ratio), a dissonant augmented eleventh (16:45 ratio), and a dissonant minor tenth (5:12 ratio). During the EEG recordings performed in three separate experimental sessions the subjects were asked to keep their attention away from the sounds and to concentrate on watching a movie. Following the last experimental session of the EEG recordings, two behavioral tasks were performed: in the first, subjects were asked to write on a piece of paper if two intervals in an interval pair were the same or different; in the second task, subjects were asked to judge on a scale from 1 to 7 the pleasantness or unpleasantness of each musical interval. In line with previous results,[5] control subjects were expected to respond with larger cortical responses to the discrepant deviant intervals (the dissonant in the consonant context and the consonant in the dissonant context) despite their being closer in pitch distance to the standards (6 semitones) than to another deviant (7 semitones). Patient I.R. would prefer to respond to the acoustic pitch distance and thus have the largest response to the largest pitch interval.

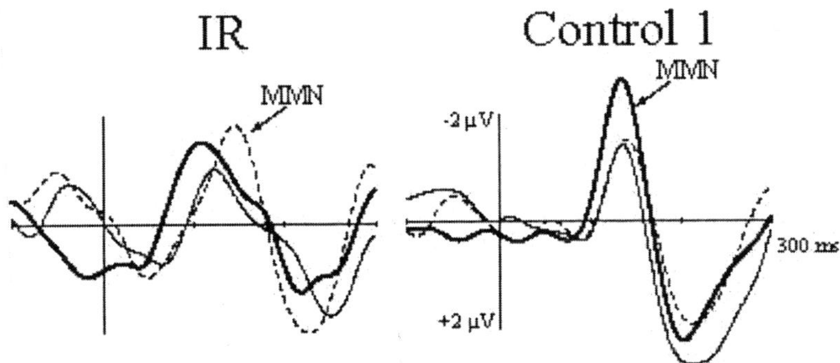

FIGURE 1. Difference waves where brain responses to deviants minus standards are grouped between the two consonant and dissonant context conditions. (*Left*) The mismatch negativity (MMN) responses of patient I.R. are shown. (*Right*) The MMN responses of one of the two controls are illustrated.

Behavioral results showed that I.R.'s hit rate in the discrimination task of an interval pair was 60%, whereas that of control subjects was 99%. Patient I.R.'s lower rate might be due to her faster habituation and fatigue, the behavioral task being performed after the ERP session. Moreover, both control subjects differentiated their pleasantness/unpleasantness evaluation according to the dissonance/consonance of musical intervals, whereas I.R. did not. To extract the MMN brain response, difference waves were calculated by subtracting the waveforms elicited by the standard intervals from the waveforms elicited by the deviants. The latency and morphology of the negative peaks in all subjects and for all conditions corresponded to the notion of the MMN. In particular, the control subjects showed a larger MMN to the dissonant interval in the consonant context and to the consonant interval in the dissonant context, indicating that the perceptual quality of the intervals mattered more than the pitch distance of their tone components (FIG. 1, right side). On the contrary, I.R.'s MMN reacted in an incoherent manner, being larger in both contexts to the deviant intervals that were closer in pitch distance to the standards (FIG. 1, left side).

In conclusion, our preliminary results suggest that patient I.R.'s brain lesions have altered the neural substrates underlying MMN generation, which reflects sensitivity to a change from consonance to dissonance.

REFERENCES

1. DOWLING, W.J. & D.L. HARWOOD. 1986. Music Cognition. Academic Press. San Diego.
2. KAMEOKA, A. & M. KURIYAGAWA. 1969. Consonance theory, Part I: Consonance of dyads. JASA **45:** 1451–1459.
3. PLOMP, R. & J.M. LEVELT. 1965. Tonal consonance and critical bandwidth. JASA **38:** 549–560.
4. NÄÄTÄNEN, R. 1992. Attention and Brain Function. Erlbaum. Hillsdale, NJ.
5. BRATTICO, E. *et al.* 2000. Processing of musical intervals in the central auditory system: an event-related potential (ERP) study on sensory consonance. *In* Proceedings of ICMPC6. C. Woods *et al.*, Eds. :1110–1119. Keele University. Keele, UK.
6. TIITINEN, H. *et al.* 1994. Attentive novelty detection in humans is governed by preattentive sensory memory. Nature **372:** 90–92.
7. PERETZ, I. *et al.* 2001. Cortical deafness to dissonance. Brain **124:** 928–940.

Estimating Internal Drift and Just Noticeable Difference in the Perception of Continuous Tempo Drift

A New Method

SOFIA DAHL AND SVANTE GRANQVIST

Department of Speech, Music and Hearing, Royal Institute of Technology, KTH, SE-10044 Stockholm, Sweden

ABSTRACT: Is there such a thing as an internal representation of a "steady tempo" and is this representation itself free from tempo drift? To investigate this question, we propose a new method for studying detection of continuous tempo drift.

KEYWORDS: tempo drift; tempo perception; just noticeable difference (JND); internal drift

INTRODUCTION

The aim for this exploratory study was to investigate the extent to which a continuous tempo drift is perceivable. Whether internal representation of a steady tempo exists or not is still an open question; however, it seems reasonable to presume that precision in the production of tempo is closely linked to perception.

Earlier studies indicate that the just noticeable difference (JND) for tempo detection is in the range of 2%. This JND is about 10 times the maximum tempo drift measured for professional drummers in a study by Dahl.[1] The tempo drifts in this study were typically about 0.05%/interval, or 0.25 ms/interval at 120 beats per minute (nominal beat separation 500 ms).

In contrast to other investigations of tempo detection, longer stimuli were used and smaller drift magnitudes were tested. While investigating perceptual threshold for tempo drift, we did not rule out the possibility that the internal "clock" can have an inherent tempo drift. In practice, this means that some listeners will perceive that tempo is increasing when, in fact, it is decreasing, and vice versa.

Address for correspondence: Sofia Dahl, Department of Speech, Music and Hearing, Royal Institute of Technology, KTH, SE-10044 Stockholm, Sweden. Voice: +46 8 7907857; fax: +46 8 7907854.

sofia.dahl@speech.kth.se; svante.granqvist@speech.kth.se

Ann. N.Y. Acad. Sci. 999: 161–165 (2003). © 2003 New York Academy of Sciences.
doi: 10.1196/annals.1284.020

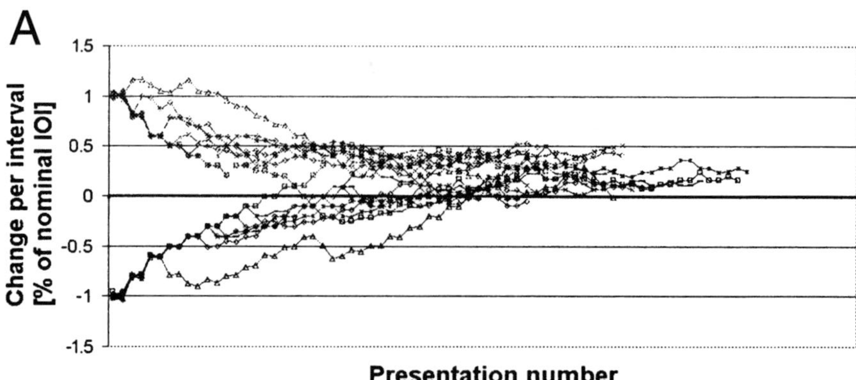

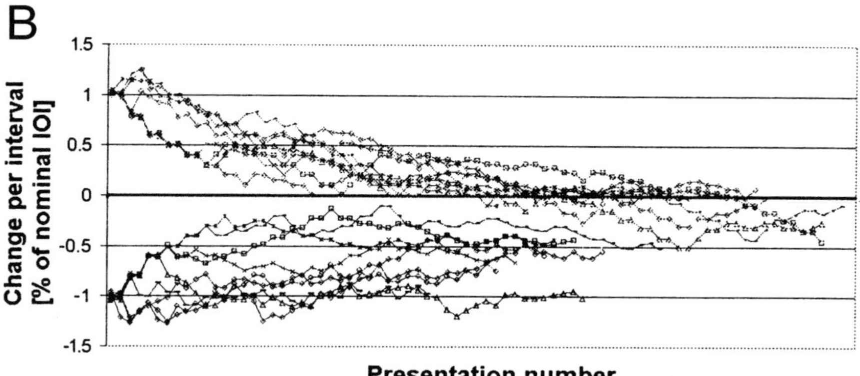

FIGURE 1. Response series for two subjects. Subject A.A. (**A**) has an internal drift (ID) corresponding to a decreasing tempo and a small just noticeable difference (JND). Subject G.S. (**B**) has an ID corresponding to a increasing tempo and a larger JND.

METHOD

The method uses a modification of the method for Parameter Estimation by Sequential Testing (PEST). Several click sequences are presented to the listener in each test, and depending on the listener's response (correct or incorrect) the magnitude of the tempo drift is modified for the next presentation. Two correct responses in a row are required to assume that the response is correct. Thus, the tempo drift converges towards the 75% correct response level.

Each session consists of several series running in parallel. At the start of each test session, each series is designated as either "increasing" or "decreasing." Because of the designation, the correct response to a series designated "decreasing" is always "decreasing," also when the physical drift is increasing. Thus, the physical drift of a series can shift sign and cross the zero line as the response curves converge around a perceived isochronous tempo. The average endpoints across the "increasing" and "decreasing" series, respectively, give a span of nondiscriminable drift. This span was interpreted as twice the just noticeable difference (JND), centered around an internal reference, the internal drift (ID). Examples of the response series for two subjects with different ID and JND can be seen in FIGURE 1.

EXPERIMENT

Seven subjects did three listenening sessions of 40 minutes each. Each session consisted of six parallel series, three designated as increasing and three designated as decreasing. The nominal start tempo had a beat separation of 500 ms (120 beats/min). The number of intervals per presentation was 10 (corresponding to 5 s). To minimize the recognition of an "absolute tempo," some randomization was applied to the initial nominal tempo for each stimulus. There was also an initial silence randomized between 0.5 and 1 second.

ANALYSIS AND RESULTS

FIGURE 2 shows the individual and mean internal drifts for the seven subjects. The internal drift differs significantly between subjects, but each subject is consistent. No significant differences were noted between sessions for any subject.

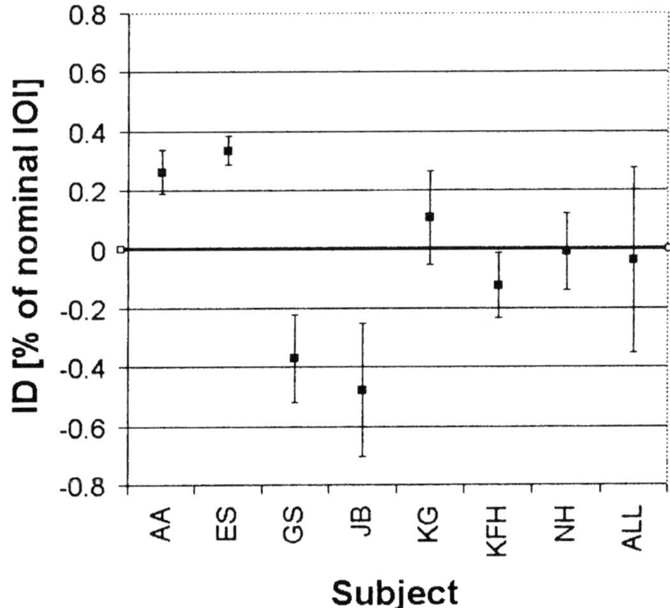

FIGURE 2. Individual and mean internal drift (ID) for the seven subjects. *Vertical error bars* indicate standard deviation.

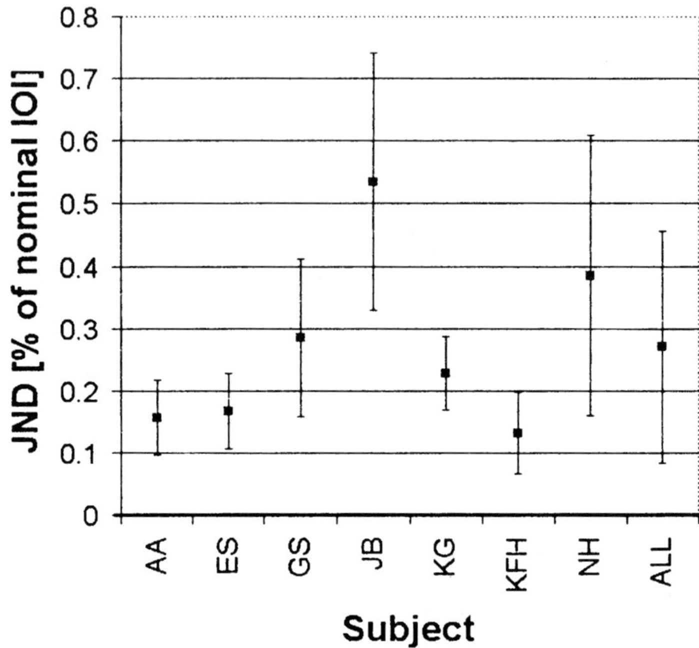

FIGURE 3. Individual and mean just noticeable difference (JND) for the seven subjects. *Vertical error bars* indicate standard deviation.

The JNDs for the subjects are shown in FIGURE 3. The mean JND across the seven subjects was 0.27% of nominal IOI, which corresponds to a change of 1.35 ms per interval.

CONCLUSION

The values for just noticeable difference (JND) were small compared to previously reported estimates (e.g., see Refs. 2 and 3). However, our results correspond to measured production data.[1,3] The method is also successful in estimating internal drift (ID). This ID was consistent within subjects, but differed between subjects. Further investigation could shed light on how IDs depend on the nominal tempo (cf. Ref. 4).

ACKNOWLEDGMENTS

We would like to thank the subjects who participated in the test and Joakim Westerlund for help with the statistical tests.

REFERENCES

1. DAHL, S. 2000. The playing of an accent. Preliminary observations from temporal and kinematic analysis of percussionists. J. New Music Res. **29:** 225–234.
2. DRAKE, C. & M.-C. BOTTE. 1993. Tempo sensitivity in auditory sequences: evidence for a multiple-look model. Percept. Psychophys. **54:** 277–283.
3. MADISON, G. 2001 Functional modelling of the human timing mechanism. Published doctoral dissertation, Uppsala University, Sweden.
4. VOS, P.G., M. VAN ASSEN & M. FRANEK. 1997. Perceived tempo change is dependent on base tempoand direction of change: evidence for a generalized version of Schulze's (1978) internal beat model. Psychol. Res. **59:** 240–247.

Congenital Amusia Interferes with the Ability to Synchronize with Music

SIMONE DALLA BELLA[a,b] AND ISABELLE PERETZ[a]

[a]Department of Psychology, University of Montreal, Montreal, Quebec, Canada

[b]Department of Psychology, Ohio State University, Columbus, Ohio 43210, USA

ABSTRACT: Eight adults with a music-specific learning disability (i.e., tone deafness, but we prefer the term "congenital amusia") were asked to tap along with music (e.g., Ravel's *Bolero*) and with nonmusical isochronous sequences (i.e., noise bursts). The amusic persons' tapping performance was poorly synchronized with music compared to that of nine matched control participants. By contrast, synchronization with the noise bursts was normal, suggesting that amusic persons' timing difficulty is limited to music.

KEYWORDS: congenital amusia; synchronization

INTRODUCTION

The ability to synchronize with music (e.g., tapping the foot along with music) is a basic skill that is shared by musicians and nonmusicians.[1–3] This ability develops spontaneously and precociously.[4] Yet, a few individuals may never acquire this ability. They suffer from congenital amusia, a learning disability that is specific to the musical domain.[5,6] This disorder has been diagnosed in adults who show above average intellectual, memory, and language abilities. As shown in a previous study,[6] these amusic individuals exhibit a deficit when performing basic musical tasks, such as discriminating music on the basis of pitch information and recognizing well-known tunes. The underlying cause for the disorder is ascribed to a defect in fine-grained pitch perception.[5,6] The objective of the present study is to examine whether persons with congenital amusia are also impaired in timing tasks, as suggested by Ayotte and collaborators.[6]

METHOD

Eight persons with congenital amusia (age range 51–62 years) who participated in a previous study[6] and a group of nine matched control participants (age range 50–67 years) performed two tapping tasks. Participants were asked to tap with their dominant hand in two different conditions: (1) in time along with a musical piece

Address for correspondence: Simone Dalla Bella, Department of Psychology, Ohio State University, 1885 Neil Ave., Columbus, OH 43210, USA. Fax: 614-292-5601.
dalla-bella.2@osu.edu

Ann. N.Y. Acad. Sci. 999: 166–169 (2003). © 2003 New York Academy of Sciences.
doi: 10.1196/annals.1284.021

TABLE 1. Performance obtained by amusic subjects and matched controls in the music condition

Tasks	A2	A3	A4	A7	A8	A9	A10	A11	Controls (SD)
Ravel's *Bolero*									
Taps (*n*)	85	81	170	102	90	76	125	78	101(39)
Mean ITI (ms)	903	838	395	574	855	882	467	882	749 (235)
Standard deviation of ITI (ms)	75	176[c]	59	60	41	26	63	36	46 (34)
Successful synchronizations (*n*)[a]	27	15	5[b]	0[b]	10[b]	0[b]	3[b]	73	74 (31)
La Bottine Souriante									
Taps (*n*)	100	106	171	90	95	101	97	101	113 (35)
Mean ITI (ms)	494	464	291	544	527	493	508	494	463 (88)
Standard deviation of ITI (ms)	25	39	33	41	32	22	40	21	29 (14)
Successful synchronizations (*n*)	83	4[c]	7[c]	9[c]	34[b]	90	40[b]	95	94 (27)
Stayin' Alive									
Taps (*n*)	110	111	152	106	98	112	108	113	124 (37)
Mean ITI (ms)	578	575	423	592	658	580	578	580	546 (97)
Standard deviation of ITI (ms)	26	39	66[c]	48[b]	44[b]	25	64[c]	29	26 (8)
Successful synchronizations (*n*)	100	27[b]	0[c]	26[b]	12[b]	93	32[b]	102	111 (34)

[a]Number of successful synchronizations: number of taps occurring ± 10% of the inter-beat interval around the expected beat

[b] > 2 SDs from the mean of the control group.

[c] > 3 SDs.

(music condition); and (2) in time with isochronous sequences of noise bursts (isochronous condition). In the music condition, participants heard the beginning (about 1 minute and 30 seconds) of the recorded performance of three well-known musical excerpts: Ravel's *Bolero*, a piece of instrumental folk music from *La Bottine Souriante*, and Bee Gees' *Stayin' Alive*. In the isochronous condition, participants heard isochronous sequences of short noise bursts at six different tempi (inter-onset interval (IOI): 150, 300, 450, 600, 750, and 1500 ms), each for a duration of 45 seconds. In both conditions, participants were asked to tap in time to the auditory stimuli (music or isochronous sequences), as regularly as possible. Each tapping condition was performed twice. Tapping responses were recorded onto an IBM think-pad computer and aligned with the presented sound with 1-ms accuracy.

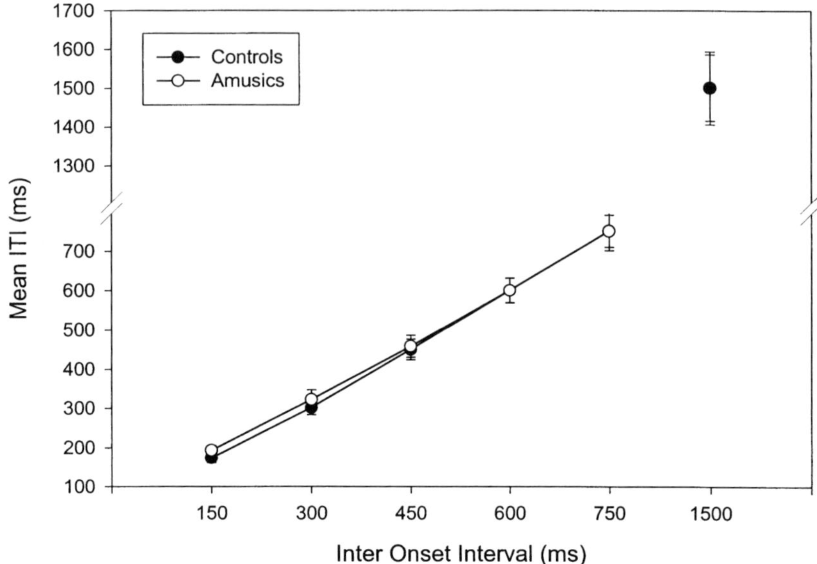

FIGURE 1. Performance obtained by amusic subjects and matched controls in the isochronous condition.

RESULTS AND DISCUSSION

For each excerpt in the music condition, the number of produced taps, the mean and standard deviation (SD) from the mean of inter-tap intervals (ITI), and the number of successful synchronizations (number of taps occurring ± 10% of the inter-beat interval around the expected beat) are reported in TABLE 1. The performance of each amusic participant was compared to matched controls' average performance. As indicated by the larger SDs, amusic participants tapped in an irregular fashion along with music, compared to controls. This is particularly apparent with the third excerpt (*Stayin' Alive*). Moreover, most of the amusic subjects have marked difficulty in synchronizing with music, as attested by the smaller number of successful synchronizations compared to controls. By contrast, the amusics' performance closely matched that of controls when tapping with the isochronous sequences, as seen in FIGURE 1.

The present study supports our previous suggestion that congenitally amusic adults have difficulties with tapping in time to music. This evidence rests on objective data in the present study, whereas it was based on judges' evaluation in our prior work. Tapping performance was clearly abnormal in 6 of the 8 amusic subjects. The fact that all amusics can synchronize with sequences of nonmusical sounds suggests that the impairment does not arise from a poor ability to tap along with sounds in general. Rather, the difficulty is most likely due to the amusic subjects' deficit in music perception, which would prevent them from extracting the musical beat.

REFERENCES

1. SNYDER, J. & C.L. KRUMHANSL. 2001. Tapping to ragtime: cues to pulse finding. Music Percept. **18:** 455–489.
2. JONES, M.R. & P. PFORDRESHER. 1997. Tracking musical patterns using joint accent structure. Canad. J. Exp. Psychol. **51:** 271–291.
3. LARGE, E.W. 2000. On synchronizing movements to music. Human Movement Sci. **19:** 527–566.
4. DRAKE, C., M.R. JONES & C. BARUCH. 2000. The development of rhythmic attending in auditory sequences: attunement, referent period, focal attending. Cognition **77:** 251–288.
5. PERETZ, I., J. AYOTTE, R.J. ZATORRE, *et al.* 2002. Congenital amusia: a disorder of fine-grained pitch discrimination. Neuron **33:** 185–191.
6. AYOTTE, J., I. PERETZ & K. HYDE. 2002. Congenital amusia: a group study of adults afflicted with a music-specific disorder. Brain **125:** 238–251.

Exploration of Roughness by Means of the Mismatch Negativity Paradigm

WOUTER DE BAENE,[a] ANDRÉ VANDIERENDONCK,[a] MARC LEMAN,[b]
ANDREAS WIDMANN,[c] AND MARI TERVANIEMI[d]

[a] Department of Experimental Psychology, Ghent University, Ghent, Belgium

[b] Institute for Psychoacoustics and Electronic Music, Department of Musicology,
Ghent University, Ghent, Belgium

[c] Institüt für Allgemeine Psychologie, Universität Leipzig, Leipzig, Germany

[d] Cognitive Brain Research Unit, Department of Psychology, University of Helsinki,
Helsinki, Finland

ABSTRACT: A mismatch negativity study was set up to find the neural correlates
of roughness perception. The results suggest that when the sounds are not
attended to, roughness is reflected by the mismatch positivity as evidenced at
the mastoid electrodes.

KEYWORDS: roughness; mismatch negativity; event-related potential (ERP)
correlates

The unpleasant character of sounds caused by the perception of amplitude fluctuation in the range of 20–200 Hz is called roughness (or sensory dissonance).[1] It characterizes the texture of a sound in terms of impure or unpleasant qualities. The focus of this study is on the main parameter influencing roughness of amplitude-modulated tones: degree of modulation (= modulation index m).[2] Roughness increases with increasing modulation index, reaching a maximum at an m value of 1.2. For larger values of m, roughness slowly decreases again.[3] To find the neural correlates of roughness perception, the auditory event-related brain potentials (ERPs) were recorded using the mismatch negativity (MMN) paradigm. The MMN is a frontally maximal negative ERP component, elicited when an infrequent deviant stimulus is presented in a flow of repetitive standard stimuli. [4] The MMN generally co-occurs with a polarity inversion at the mastoids (labeled mismatch positivity or MMP) and implies an incongruence between the sensory input of the deviant and the information of the standard stimuli that is encoded by the auditory system in an echoic memory trace.[5] Several studies suggested that the amplitude of the MMN reflects the size of the discrepancy between standard and deviant tones.[6] In this study, we used a 200-ms pure sinusoidal tone (f_c = 1000 Hz, f_m = 70 Hz, 60-dB SPL, ISI =

Address for correspondence: Wouter De Baene, Department of Experimental Psychology,
Ghent University, Henri Dunantlaan 2, B-9000 Ghent (Belgium). Voice: 32-9-2646435; fax: 32-
9-2646496.
Wouter.DeBaene@UGent.be

Ann. N.Y. Acad. Sci. 999: 170–172 (2003). © 2003 New York Academy of Sciences.
doi: 10.1196/annals.1284.022

300 ms) as the standard stimulus and varied the modulation index (m = 0.2, 0.5, 0.8, 1.0, 1.2, 1.4, 1.7, or 2.0) in deviant tones. In the inattentive condition, in which the subjects watched a silent movie, the MMN amplitude at Fz increased with increasing modulation index, whereas the MMP amplitude at the mastoids increased with increasing modulation index up to 1.2, followed by a decrease (FIG. 1). This suggests that in the inattentive condition, the MMN at Fz reflects a physical difference be-

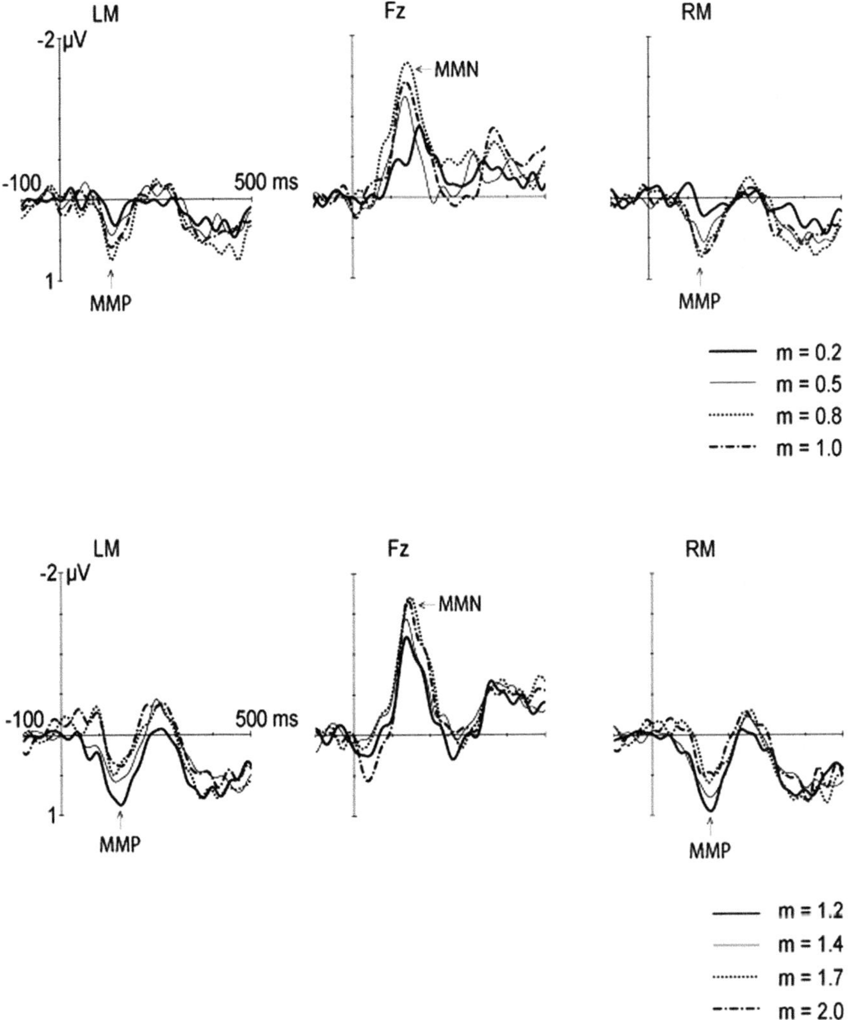

FIGURE 1. (*Top*) The grand average difference waves (ERP's deviant tones − ERP's standard tones) for the deviants with modulation indices of 0.2, 0.5, 0.8, and 1.0 at Fz and left and right mastoid (LM and RM) in the inattentive condition. (*Bottom*) The MMN difference waves for the deviants with the modulation indices 1.2, 1.4, 1.7, and 2.0 in the inattentive condition.

tween standard and deviant tones, whereas the MMP at the mastoids reflects a difference in roughness between standard and deviant tones. Due to a large overlap between the MMN and the N2b in the attentive condition (a detection task), no independent information could be extracted from these two components. We conclude that roughness is reflected by the MMP at the mastoids in the inattentive condition.

REFERENCES

1. VON HELMHOLTZ, H. 1863. Die Lehre von den Tonempfindungen als physiologische Grundlage für die Theorie der Musik. Georg Olms Verlagsbuchhandlung. Hildesheim.
2. ZWICKER, E. & H. FASTL. 1990. Psychoacoustics: Facts and Models. Springer-Verlag. Berlin.
3. MATHES, R.C. & R.L. MILLER. 1947. Phase effects in monaural perception. J. Acoust. Soc. Am. **19:** 780–797.
4. NÄÄTÄNEN R., A.W.K. GAILLARD & S. MÄNTYSALO. 1978. Early selective-attention effect on evoked potentials reinterpreted. Acta Psychol. **42:** 313–329.
5. NÄÄTÄNEN, R. 1992. Attention and Brain Function. Erlbaum. Hillsdale.
6. AMENEDO, E. & C. ESCERA. 2000. The accuracy of sound duration representation in the human brain determines the accuracy of behavioural perception. Eur. J. Neurosci. **12:** 2570–2574.

"Out-of-Pitch" but Still "In-Time"

An Auditory Psychophysical Study in Congenital Amusic Adults

KRISTA L. HYDE AND ISABELLE PERETZ

Department of Psychology, University of Montreal, C.P. 6128 succ. Centre-ville, Montreal (Qc), H3C 3J7, Canada

Research Centre of the Institut Universitaire de Gériatrie de Montréal, Montreal (Qc), H3W 1W5, Canada

ABSTRACT: Congenital amusia is a lifelong disability, commonly known as tone deafness, that prevents afflicted individuals from developing basic musical skills despite normal audiometry and above-average intellectual, memory, and language skills. Although it is estimated that 4% of the general population would be born with such a musical handicap, the underlying cause is presently unknown. Recently, we proposed that this disorder could be traced to a deficit in pitch perception on the basis of a single case.[4] Here we provide psychophysical evidence for the existence of a generalized defect that is both fine grained and specific to pitch because time is unaffected.

KEYWORDS: congenital amusia; pitch perception

The existence of congenital amusia has been conjectured for more than one century;[1–3] however, it is only recently that systematic explorations of this disorder have been undertaken.[4,5] In the present study, the amusic group consisted of 10 self-declared amusic volunteers who were carefully screened in order to rule out a general learning disorder. All amusic subjects had a high level of education and had a normal exposure to music during childhood, but had been unsuccessful in mastering basic musical skills. All failed in at least two of the six tests of a battery that covers discrimination of pitch contour, pitch intervals, key, rhythmic grouping, meter, and memory of novel but conventional music. As shown in TABLE 1, the amusic subjects are impaired on melodic tests requiring the discrimination of stimuli on the basis of pitch-related cues. By contrast, half the afflicted individuals display normal discrimination of the same stimuli on the basis of rhythmic cues. All experience auditory difficulties that are limited to music, as illustrated by the fact that they fail to recognize familiar tunes (in the absence of lyrics), whereas they recognize the corresponding lyrics as well as do the controls. The 10 control participants had no musical training (or deficiency) and were matched to the amusic subjects in age, gender, education, and handedness.

Address for correspondence: Prof. Isabelle Peretz, Department of Psychology, University of Montreal, C.P. 6128 succ. Centre-ville, Montreal (Qc), H3C 3J7, Canada. Voice: (514) 343 5840; fax: (514) 343 5787.

Isabelle.Peretz@umontreal.ca

Ann. N.Y. Acad. Sci. 999: 173–176 (2003). © 2003 New York Academy of Sciences.
doi: 10.1196/annals.1284.023

TABLE 1. Characteristics and mean (standard error) scores on musical tests

	Amusic ($n = 10$)	Control ($n = 10$)	t test
Subject characteristics			
Age (yr)	57 (1.6)	58 (1.7)	ns
Gender	3 M, 7 F	3 M, 7 F	
Education (yr)	15 (0.6)	17 (0.7)	ns
Musical discrimination			
Melodic (MBEA)	59.8 (2.3)	88.1 (2.9)	$P < 0.001$
Rhythmic (MBEA)	71.9 (4.9)	90.4 (3.6)	$P < 0.05$

Note: The melodic scores are averaged over 3 tests examining the use of contour, scale, and intervals, each respectively, in the discrimination of melodies. The rhythmic scores correspond to the discrimination of the same set of melodies in which interchanges of note durational values could be inserted. Each test consists of 30 trials. The tests are taken from the Montreal Battery of Evaluation of Amusia (MBEA) (http://www.fas.umontreal.ca/psy/grplabs/lnmcg/website/).

In the present study, the amusic and control subjects performed two psycho-acoustic tasks aimed at evaluating their ability to detect a pitch or a time change within a five-tone sequence. All tones were synthesized in a piano timbre, with tone duration of 100 ms and inter-tone onset interval (IOI) of 350 ms. The task consisted of responding "yes" whenever they detected a change in a constant pitch and iso-chronous sequence and of responding "no" when unable to detect a change. In the pitch-change condition, the pitch of the fourth tone was parametrically modified in half of the sequences. The pitch change occurred in equal proportions at one of five interval distances, ranging from 25 to 300 cents (where 100 cents = 1 semitone) in both ascending and descending directions. The presence and the size of the pitch changes could not be predicted because they were presented in a random fashion. Similarly, in the time condition, a time change was inserted before the fourth tone in half of the sequences, resulting in its advance or delay at one of five temporal distances. These ranged from 8 to 16% of the IOI. In each condition, subjects underwent an initial training session of 40 practice sequences followed by 360 experimental sequences.

As shown in TABLE 2, the amusic group can detect pitch changes of 200 cents and above, but they exhibit difficulties at smaller distances. They perform below normal controls at 100 cents [$t_{18} = 2.32$, $P < 0.05$] and can barely detect changes of 50 and 25 cents. The impairment was observed in each amusic individual. Lack of task understanding cannot account for the deficit, because the amusic subjects can detect the large pitch changes (over 200 cents). This performance also highlights the fine-grained nature of the pitch disorder. The analysis of variance (ANOVA) yielded a significant interaction between group (amusic subjects versus controls) and pitch distance, with $F(4,72) = 95.36$, $P < 0.001$. In the time-change detection task, the amusic subjects' performance matches that of normal controls even at subtle time differences, with data that conform to literature values[6] (TABLE 2). The ANOVA revealed no group effect ($F(1, 18) = 2.9$, n.s.) or significant interaction with temporal distance ($F < 1$). The overall pattern of results is supported by an ANOVA considering condition (pitch and time) and distance as within-subjects factors and group

TABLE 2. Pitch and time discrimination results

		Levels of pitch change (in cents)				
		25**	50**	100*	200	300
Pitch	Amusic	22.0 (5.0)	70.6 (6.3)	93.0 (2.3)	98.3 (0.4)	99.5 (0.3)
	Control	98.5 (0.5)	98.5 (0.9)	98.9 (0.9)	99.6 (0.4)	99.8 (0.2)
		Levels of time change (%IOI)				
		8	10	12	14	16
Time	Amusic	50.5 (5.2)	62.8 (4.8)	72.2 (3.9)	83.1 (3.5)	90.1 (2.5)
	Control	54.7 (4.9)	66.6 (4.7)	77.0 (4.1)	83.8 (3.2)	91.8 (1.9)

Mean percentage (and standard error) of hits (correct detection of a change) minus false alarms (incorrect detection of a change) for each group of subjects at increasing pitch and time distances. **P <0.001; *P <0.05.

(amusic subjects and controls) as between-subjects factor. A highly significant interaction was observed between the three factors, $F(9, 162) = 76.69$, P <0.001, reflecting the marked difference in performance between amusic subjects and controls in the pitch task only.

The results show that adults with musical impairments have an auditory perceptual deficit in discriminating pitch but not time changes between successive tones. Such a pitch deficit can account for the musical disorder. The defect is apparent at a distance of 1 semitone, which corresponds to the building block of Western musical scales. Considering that this difficulty emerged in a basic tone context, it is likely that the same deficit will be amplified in more complex conditions as in music. Together the results provide converging evidence that a fine-grained discrimination of pitch is at the root of congenital amusia.

Although fine temporal discrimination appears to be spared in this music-specific disorder, half of the afflicted individuals are impaired in rhythmic discrimination in a musical context (TABLE 1). We surmise that the variable rhythmic difficulty is a cascade effect of a faulty pitch-processing system. We construe that a poor pitch perceptual system might impinge upon the development of the entire musical system.[3,4] Accordingly, fine-grained pitch perception may be viewed as an essential component around which the musical system develops in a normal brain. In this respect, music diverges from speech. Music seems to be more dependent on fine resolution of pitch, whereas speech appears contingent upon fine resolution of time.[7]

Having established congenital amusia as arising from an aberrant pitch perceptual system narrows down the possible neural loci where an anomaly can be uncovered. Based on recent evidence from studies using neuroimaging techniques[8] and human lesion data,[9] neural deviations in pitch-based networks are expected to be found in the right auditory cortex of congenital amusic individuals.

ACKNOWLEDGMENTS

We thank our subjects for their continued cooperation and Robert Zatorre for his insightful comments. This work was supported by a postgraduate scholarship from the Canadian Fonds de la Recherche en Santé du Québec (to K.H.) and by a grant from Canadian Institutes of Health Research (to I.P.).

The present research was approved by the ethics committee of the Institut Universitaire de Gériatrie de Montréal and informed consent was obtained from all subjects.

REFERENCES

1. ALLEN, G. 1878. Note-deafness. Mind **10:** 157–167.
2. GESCHWIND, N. 1984. The brain of a learning-disabled individual. Ann. Dyslexia **34:** 319–327.
3. KALMUS, H. & D. FRY. 1980. On tune deafness (dysmelodia): frequency, development, genetics and musical background. Ann. Hum. Genet. **43:** 369–382.
4. AYOTTE, J., I. PERETZ & K. HYDE. 2002. Congenital amusia: a group study of adults afflicted with a music-specific disorder. Brain **125:** 238–251.
5. PERETZ, I. *et al.* 2002. Congenital amusia: a disorder of fine-grained pitch discrimination. Neuron **33:** 185–191.
6. EHRLÉ, N., S. SAMSON & M. BAULAC. 2001. Processing of rapid auditory information in epileptic patients with left temporal lobe damage. Neuropsychologia **39:** 525–531.
7. ZATORRE, R.J., P. BELIN & V.B. PENHUNE. 2002. Structure and function of auditory cortex: music and speech. Trends Cognit. Sci. **6:** 37–46.
8. PATTERSON, R.D. *et al.* 2002. The processing of temporal pitch and melody information in auditory cortex. Neuron **36:** 767–776.
9. TRAMO, M.J., G.D. SHAH & L.D. BRAIDA. 2002. Functional role of auditory cortex in frequency processing and pitch perception. J. Neurophysiol. **87:** 122–139.

Evoked Potentials of the Human Auditory Cortex: Sensitive to the Harmonic Series?

STEVE J. JONES

Department of Clinical Neurophysiology, The National Hospital for Neurology and Neurosurgery, London WC1N 3BG, UK

ABSTRACT: Cortically generated auditory evoked potentials (N1 and P2) were recorded to frequency changes of harmonic and inharmonic complex tones comprising four sinusoidal components. The responses obtained when the frequencies suddenly became stationary after a period of 16/s changes were significantly shorter in latency when the frequencies were harmonically related, possibly implying a process of periodicity detection.

KEYWORDS: auditory cortex; auditory evoked potentials; complex tones; harmonic series

INTRODUCTION

The origin of the harmonic and melodic intervals occurring in music probably consists in the mathematically related frequencies emitted when a resonant object vibrates (integer multiples of f_0). This "harmonic series" tends to group together perceptually, so that the resonating object (e.g., a musical instrument) is usually heard as a single entity.[1] It is not known to what extent harmonic grouping may be universal among mammals, but in the animal literature there is no evidence for neurons at any level that respond in a specific way to harmonic series. In human subjects, our previous work using synthesized musical instrument tones has distinguished two cortical generators of auditory evoked potentials, both giving rise to obligatory N1 and P2 potentials.[2–4] The "C potentials" (CN1, CP2) are produced at the onset of change in the distribution of spectral energy (pitch or timbre) after a period of constancy, while more anteriorly distributed "M potentials" (MN1, MP2) are produced at the offset of regular changes. The object of this study was to determine whether either of the underlying processes is influenced by the harmonicity of the tone, that is, are the responses to harmonic and inharmonic complex tones any different from one another?

Address for correspondence: Dr. S. J. Jones, Department of Clinical Neurophysiology, The National Hospital for Neurology and Neurosurgery, Queen Square, London WC1N 3BG, UK. Voice: +44 20 7837 3611, ext. 4109; fax: +44 20 7713 7743.
sjjones@ion.ucl.ac.uk

Ann. N.Y. Acad. Sci. 999: 177–179 (2003). © 2003 New York Academy of Sciences.
doi: 10.1196/annals.1284.024

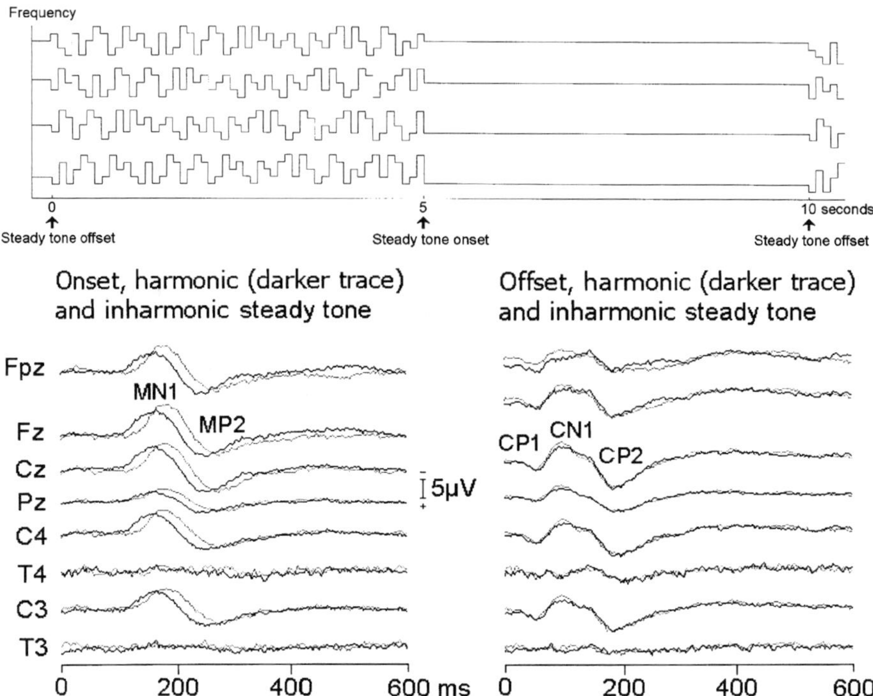

FIGURE 1. (*Upper*) Schema of Part B. (*Lower*) M potentials at the onset of a steady complex tone following a 5-second period of 16/s pseudorandom frequency changes, and C potentials at the offset of the tone after 5 seconds of steadiness (beginning of 16/s changes), mean of 8 subjects.

METHODS

Eight normally hearing subjects aged 20–50 years were tested while awake and reading a book. Electrodes were located at Fpz, Fz, Cz, Pz, C3, C4, T3, and T4, referred to the dorsal neck. The recording bandwidth was 1 to 200 Hz, digitized at 1 kHz. The recording epoch was 450 or 600 ms, groups of 30 or 50 responses were averaged, and three repetitions were made. The tones were continuous, approximately 45 dB above threshold, with smooth transition. In Part A, the stimulus rotated regularly around six complex tones, each made of four sinusoids, with changes occurring every 1 second. The musical pitches of the individual sinusoids were: C4, B4, F5, A5 (inharmonic); D4, C#5, G5, B5 (inharmonic); C#4, C5, F#5, A#5 (inharmonic); D#4, D5, G#5, C6 (inharmonic); B3, A#4, E5, G#5 (inharmonic); C4, C5, G5, C6 (harmonic). In Part B, four simultaneous sinusoids changed pseudorandomly at 16 changes/s for 5 seconds within a range of 5 semitones (A#3 to D4, A#4 to D5, F5 to A5, A#5 to D6), before resting for 5 seconds on C#4, C5, F#5, A#5 (inharmonic) or C4, C5, F5, C6 (harmonic).

RESULTS

In Part A, the latency of the CP2 was marginally shorter to the harmonic complex, but individual differences were nonsignificant in paired t tests. In Part B, there were no significant differences in the C potentials produced at the offset of the steady tones (FIG. 1). However, at the onset of the harmonic steady tone, the latencies of both MN1 and MP2 were significantly shorter as compared with the onset of the inharmonic steady tone (latencies after the last frequency change; MN1 188.8 ± 14.0 compared with 206.3 ± 0.2 ms, $P <0.01$; MP2 282.5 ± 20.1 as compared with 296.6 ± 16.8 ms, $P <0.02$).

DISCUSSION

In accordance with the animal literature on unit responses in the auditory cortex, the scalp C potentials showed little or no sensitivity to whether or not the frequencies of the complex tone were harmonically related. The M potentials, on the other hand, did show such sensitivity. The M potentials are believed to represent a process of spectrotemporal pattern analysis;[2,3] their elicitation at the sudden occurrence of "no change" may therefore involve a mechanism resembling autocorrelation to detect repetition of the sound waveform. The steady harmonic complex following a 5-second period of pseudorandom changes would have a straightforward periodicity identical to that of the fundamental (C4 in musical notation, periodicity 3.82 ms), such that within 5 ms or less autocorrelation should reveal that a change had not occurred at the expected time. The inharmonic complex had no straightforward periodicity; hence, the determination of "no change" by spectrotemporal analysis might involve a slower process of analyzing and comparing the periodicities of the individual sinusoids. The perceptual "grouping" phenomenon[1] may also be due to detection of the short overall periodicity of a harmonic complex tone rather than the mathematical relation between its constituent frequencies. If the universal human understanding of musical harmony and melody (in the sense that a tune is easily recognized, even when played at a variety of pitches) is also due to periodicity analysis, the generator of the M potentials in the vicinity of the auditory cortex may represent part of its neurophysiologic substrate.

REFERENCES

1. BREGMAN, A.S. 1990. Auditory Scene Analysis. MIT Press. Cambridge, MA.
2. VAZ PATO, M. & S.J. JONES. 1999 Cortical processing of complex tone stimuli: mismatch negativity at the end of a period of rapid pitch modulation. Cognit. Brain Res. **7:** 295–306.
3. JONES, S.J., M. VAZ PATO & L. SPRAGUE. 2000. Spectro-temporal analysis of complex tones: two cortical processes dependent on retention of sounds in the long auditory store. Clin. Neurophysiol. **111:** 1569–1576.
4. JONES, S.J. & N. PEREZ. 2001. The auditory "C-process": analysing the spectral envelope of complex sounds. Clin. Neurophysiol. **112:** 965–975.

Evoked Potentials to Test Rhythm Perception Theories

MARIJTJE L.A. JONGSMA,[a,b] PETER DESAIN,[a] HENKJAN HONING,[a,c] AND CLEMENTINA M. VAN RIJN[b]

[a]*Music, Mind, Machine Group and*

[b]*Biological Psychology, University of Nijmegen, NICI \ Nijmegen Institute of Cognition and Information, Nijmegen, The Netherlands*

[c]*Music Department \ ILLC, University of Amsterdam, Amsterdam, The Netherlands*

ABSTRACT: The general aim of this study was to investigate how rhythmic information is processed by the brain and how a mental representation of a rhythm leads to expectancies about events in the near future. We investigate this by means of EEG recordings from which evoked potentials (EPs), resulting from sensory and cognitive neural activity, are extracted.

KEYWORDS: rhythm; evoked potentials; expectancy; P_3; omissions; temporal information processing

INTRODUCTION

Evoked potential (EP) measurements have been proved to be well suited for studying aspects of music cognition.[1] It has long been known that expectancy modulates EPs. When expectancy is violated, auditory EPs (AEPs) typically show a large positive wave, the P_3.[2] A similar wave can be measured if a stimulus is expected yet omitted from a regular temporal pattern, the omission evoked potentials (OEPs).[3,4]

There are several theories concerned with rhythm perception.[5–9] These theories lead to different predictions about when a following event is maximally expected given a rhythmic sequence, thus predicting different AEP and OEP results.

METHODS

Musicians ($n = 14$) and nonmusicians ($n = 14$) participated in the experiment. All participants signed a written informed consent.

In experiment 1, we presented probe-beats on either the 1/3, 1/2, or 2/3 position within a test bar. Probe beats were preceded by two bars of either a duple- or triple-meter context (FIG. 1A). We hypothesized that sequential processing of rhythmic patterns[5] would lead to a maximal context effect on probe beats presented at the 1/3

Address for correspondence: Dr. Marijtje L.A. Jongsma, NICI\University of Nijmegen, P.O. Box 9104, 6500 HE Nijmegen, The Netherlands. Voice: +31-24-3616278; fax: +31-24-3616066. jongsma@nici.kun.nl

Ann. N.Y. Acad. Sci. 999: 180–183 (2003). © 2003 New York Academy of Sciences. doi: 10.1196/annals.1284.025

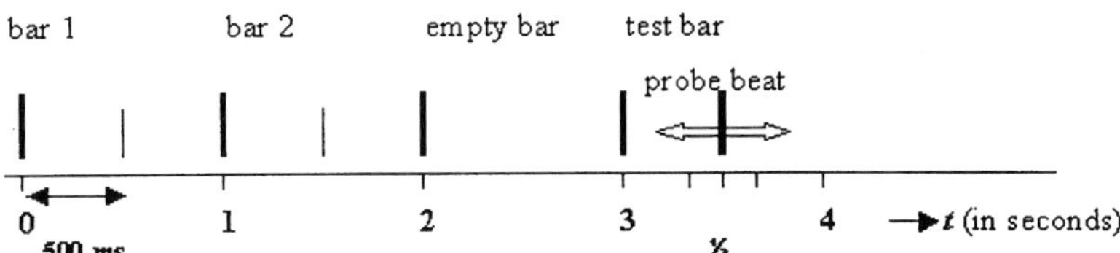

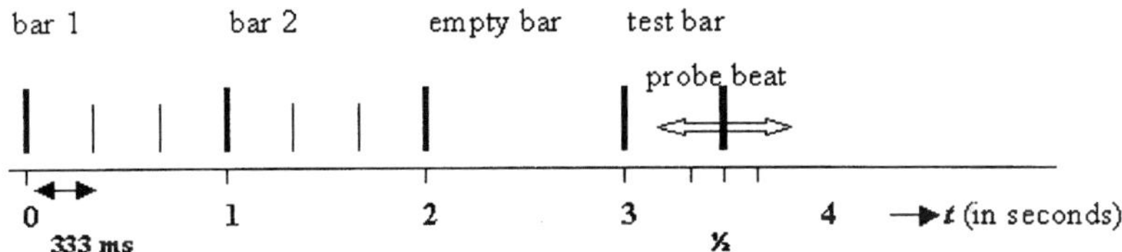

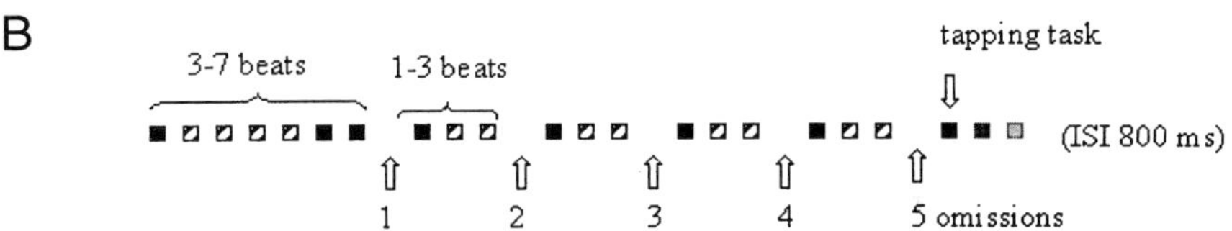

FIGURE 1. (**A**) Paradigm experiment 1; (**B**) paradigm experiment 2.

position, whereas hierarchical processing[5–9] would lead to a maximal context effect on probe beats presented at the ¼ position. Besides the AEP P_3 component, behavioral ratings, reflecting how well probe beats fit the preceding metric context (scale 1–7) were obtained.

In experiment 2, trials ($n = 90$) contained five stimulus omissions (ISI 800 ms). Three types of trials were presented, with one, two, or three beats interspersed among a total of 5 omissions (FIG. 1B). The task of the participant was to silently count the five omissions and to tap along with the first beat after the fifth omission.

RESULTS

In experiment 1, we found in nonmusicians a maximal difference between metric contexts on the 1/3 position probe beat. However, in musicians a maximal difference between metric contexts on the ½ position probe beat occurred, this for both the AEP P_3 amplitude (FIG. 2A and B) and the ratings (FIG. 3A and B).

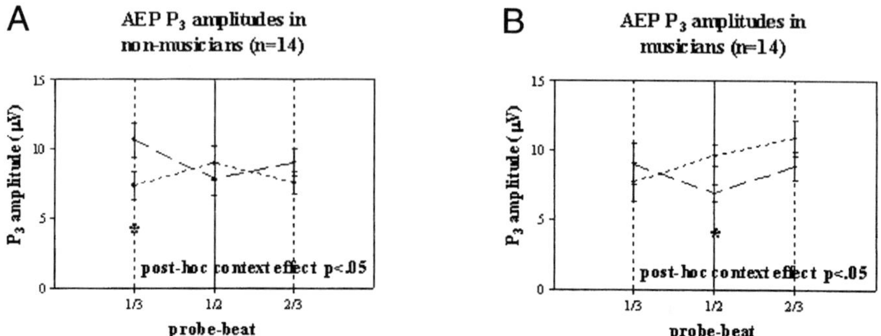

FIGURE 2. (A) AEP P_3 amplitudes in nonmusicians ($n = 14$); (B) AEP P_3 amplitudes in musicians ($n = 14$).

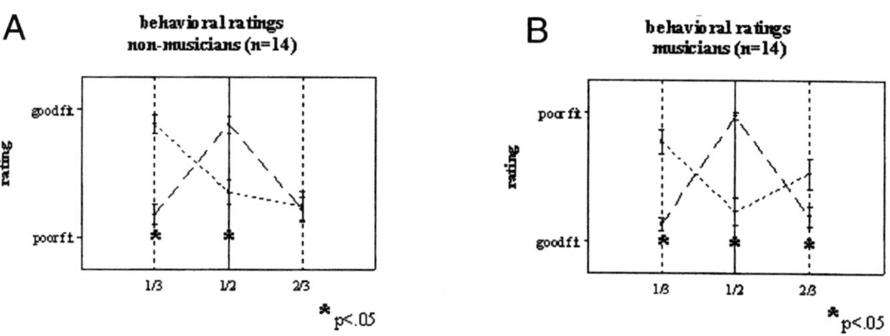

FIGURE 3. (A) Behavioral ratings in nonmusicians ($n = 14$); (B) behavioral ratings in musicians ($n = 14$).

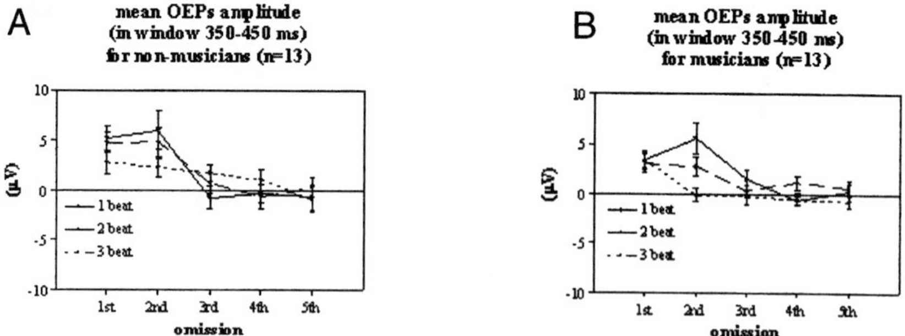

FIGURE 4. (A) Mean OEPs amplitude (in window 350–450 ms) for nonmusicians ($n = 13$); (B) mean OEPs amplitude (in window 350–450 ms) for musicians ($n = 13$).

In experiment 2, we found that OEPs did occur in response to unpredictable stimulus omissions, that is, the first two omissions in each trial type. OEPs were maximal in the window 350–450 ms after omission onset over Pz (FIG. 3A and B). However, musicians showed no OEPs in response to the second stimulus omissions in the three-beat interspersed trials. This result showed that musicians could predict this second omission (i.e., if after the first stimulus omission three beats have occurred, then the following event must be another omission). Thus, musicians used higher-order stimulus characteristics for making predictions.

DISCUSSION

Our results support the view that temporal patterns are processed sequentially in nonmusicians and hierarchically in musicians. Metric expectations thus influence both behaviors, the AEP P_3 and the OEPs. Therefore, we argue that the AEP P_3 and OEPs can be used to test theoretical predictions regarding rhythmically induced expectancies.

ACKNOWLEDGMENT

This project was supported by the Netherlands Organization for Scientific Research (NWO Project 451-02-026).

REFERENCES

1. BESSON, M. & A.D. FRIEDERICI. 1998. Language and music: a comparative view. Music Percept. **16:** 1-10.
2. CASTRO, A. & F. DIAZ. 2001. Effect of the relevance and position of the target stimuli on P300 and reaction time. Int. J. Psychophysiol. **41:** 43-52.
3. RUCHKIN, D.S., S. SUTTON, R. MUNSON, *et al.* 1981. P300 and feedback provided by absence of the stimulus. Psychophysiology **18:** 271–282.
4. JONGSMA, M.L.A., A.M.L. COENEN & C.M. VAN RIJN. 2002. Omission evoked potentials (OEPs) in rats and the effects of diazepam. Psychophysiology **39:** 229–235.
5. MARTIN, J.G. 1972. Rhythmic (hierarchic) versus serial structure in speech and other behaviour. Psychol. Rev. **79:** 487–509.
6. POVEL, D. 1981. Internal representation of simple temporal patterns. J. Exp. Psychol.: Human. Percept. Perform. **7:** 3–8.
7. PALMER, C. & C.L. KRUMHANSL. 1990. Mental representations for musical meter. J. Exp. Psychol. **16:** 728–741.
8. LARGE, E.W. & C. PALMER. 2002. Perceiving temporal regularity in music. Cognit. Sci. **26:** 1–37.
9. DESAIN, P. & H. HONING. 1994. Advanced issues in beat induction modelling: syncopation, tempo and timing. *In* Proceedings of the 1994 International Computer Music Conference. :92–94. International Computer Music Association. San Francisco.

Perceiving Musical Scale Structures

A Cross-Cultural Event-Related Brain Potentials Study

CHRISTIANE NEUHAUS

Institute of Musicology, University of Hamburg, Hamburg, Germany

ABSTRACT: In this study, event-related brain potentials (ERPs) are used to investigate the processing of musical scale structures from a cross-cultural perspective. ERP reactions reveal that universal listening strategies per se are modified by culture.

KEYWORDS: P300; processing negativity; musical scales; music perception; comparative musicology

INTRODUCTION

The present event-related brain potentials (ERPs) study was designed to study the perception and processing of musical scales from a cross-cultural point of view. The P300 component was used to indicate underlying cognitive processes.

METHODS

Five German, five Turkish, and five Indian musicians aged 20–54 years, all male with normal hearing, participated in the ERP experiment. The stimulus material consisted of four heptatonic scales: European major and harmonic minor scales, the Thai scale made of equal steps, and the makam Hicaz of Turkish art music. Stimuli were presented binaurally through earphones.

Each scale tone was generated synthetically; it was based on a pulse wave of the sound catalogue of a programmable synthesizer. Interval sizes were precisely adjusted with the "pitch fine" control. Stimulus duration was 200 ms; the interstimulus interval (stimulus onset to stimulus onset) was 540 ms. Each trial, consisting of eight scale tones, started after a 2-second rest.

Within one experimental block, two of the four scale types were presented according to the classic oddball paradigm, one scale type as the standard form (45 repetitions, 75% of the time) and the other one as the deviant form (15 repetitions, 25% of the time), resulting in 60 trials per block. Tested combinations were (a) major

Address for correspondence: Dr. Christiane Neuhaus, Max Planck Institute of Cognitive Neuroscience, Section "Magnetencephalography," Muldentalweg 9 D-04828 Bennewitz/Leipzig, Germany.

neuhaus@cns.mpg.de

Ann. N.Y. Acad. Sci. 999: 184–188 (2003). © 2003 New York Academy of Sciences.
doi: 10.1196/annals.1284.026

versus Thai; (b) major versus minor; (c) makam Hicaz versus Thai; (d) major versus makam Hicaz; and (e) Thai versus major. Subjects had to perform two tasks within a block: (1) the silent counting of the deviant scales as well as (2) the notation of the internal structure of the standard and deviant scales.

Scalp electrodes were attached to the midline with electrode placements at Fz, Cz, and Pz. A vertical electrooculogram was registered to detect eye movements. The time constant was 0.3 seconds.

Unexpectedly, many of the graphs revealed additional negative shifts with an onset latency of at least 430 ms. Thus, data input for statistical analysis consisted of maximum amplitude values within two time intervals (270–430 ms and 430–540 ms after stimulus onset).

A one-factor analysis of variance was computed (ONEWAY, dependent variable: M1T1P1 to M2T8P3 [M for "mode," i.e., standard or deviant status, T for "scale tone," and P for "electrode placement"]; independent variable: CULTURE [three levels, i.e., German, Turkish, and Indian groups]); for paired comparisons, the Scheffé test was applied; results were considered significant at $P < 0.05$ and $P < 0.01$.

RESULTS AND DISCUSSION (A SELECTION)

Block 1 (Major Standard Scale versus Thai Deviant Scale)

Scale Tone 2

For all subjects, Thai tone 2 evokes a negative deflection with a local amplitude maximum in the second time interval. The negative shift points to the effort for cog-

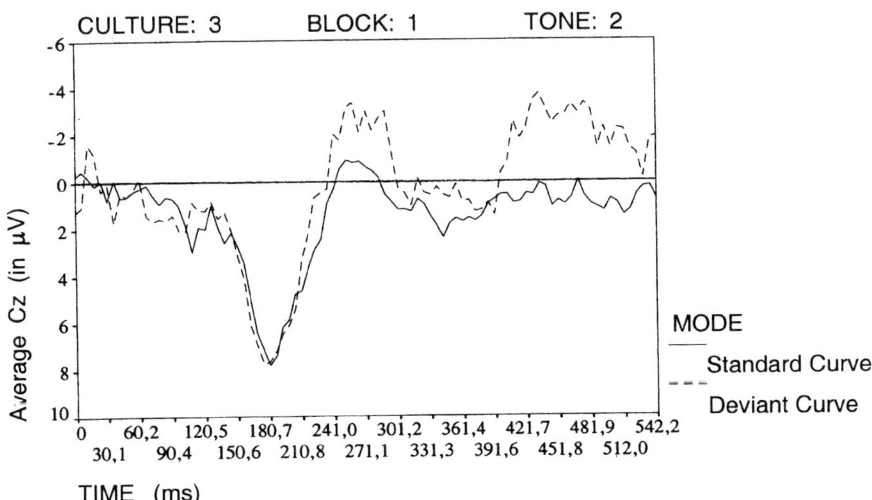

FIGURE 1. Grand average event-related brain potentials (ERPs) of Indian musicians (*n* = 5 subjects). Bioelectrical reactions to tone 2 of the major standard scale (*solid line*) and tone 2 of the Thai deviant scale (*broken line*), scalp site Cz. Analyzed time interval: 430– 540 ms.

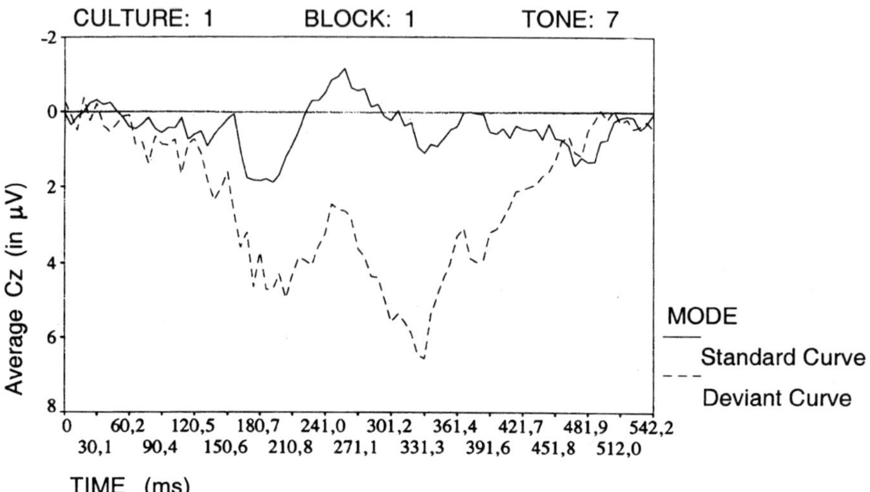

FIGURE 2. Grand average event-related brain potentials (ERPs) of German musicians ($n = 5$ subjects). Bioelectrical reactions to tone 7 of the major standard scale (*solid line*) and the Thai deviant scale (*broken line*), scalp site Fz. Analyzed time interval: 270–430 ms.

nitive processing after sensorial perception. Thus, it is labeled *processing negativity.* It is generally found after clear pitch identification, wherever differences in frequency between the deviant and the corresponding standard scale tone have a sufficient amount. Hence, its function is considered to be a bioelectrical indicator of interval judgment based on "categorical perception"[1] (FIG. 1).

Scale Tone 7

German musicians respond to the seventh deviant Thai tone with a clear P300 component at Fz, Cz, and Pz. For them, Thai tone 7 obviously causes a violation of the "leading tone expectancy." This violation elicits a context-updating process,[2] that is, a revision of the culture-intrinsic, overlearned template of the major scale. The updating process is indicated by a large P300 (FIG. 2).

Scale Tone 8

German musicians reveal a long-lasting positivity after presentation of the eighth deviant Thai tone; with this large P300, they seem to respond to the termination of the whole scale pattern ("Gestalt in time" or "closure of a cognitive epoch," wording of Verleger[3]). Indian musicians, however, do not develop a positive shift within this time window. Their missing reaction can be traced back to the traditional practice of playing scales and ragas in a combination of upward and downward movement, termed "arohana" and "avarohana"[4] (FIGS. 3 and 4).

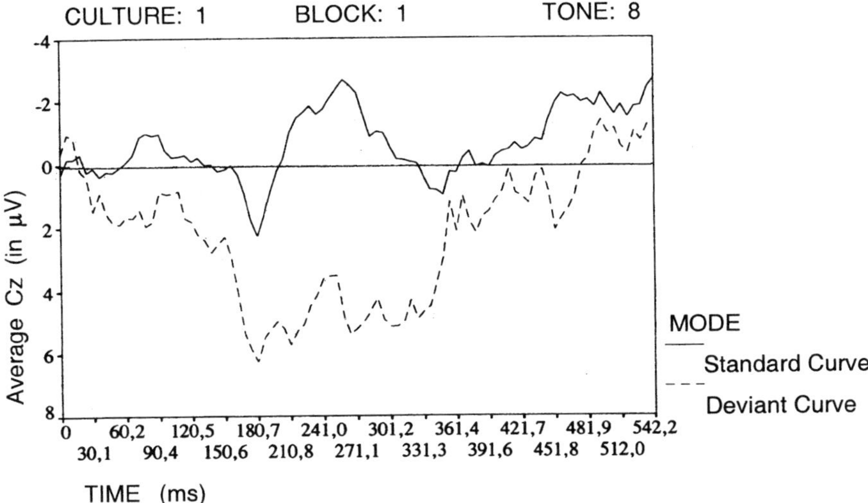

FIGURE 3. Grand average event-related brain potentials (ERPs) of German musicians. Electrophysiological reactions to tone 8 of the major standard scale (*solid line*) and tone 8 of the Thai deviant scale (*broken line*), scalp site Cz.

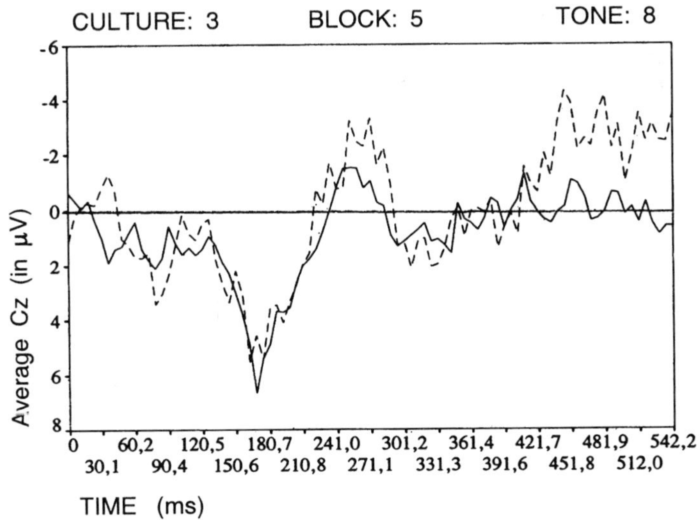

FIGURE 4. For comparison, the event-related brain potentials (ERPs) of Indian subjects, tone 8, block 5. Bioelectrical reactions to tone 8 of the Thai standard scale (*solid line*) and tone 8 of the major deviant scale (*broken line*), scalp site Cz.

CONCLUSION

Overlearned listening strategies applied to culture-specific scale material, that is, the interaction between universal mechanisms of perception and culturally "imprinted" musical contents,[5] have a significant influence on all investigated brain reactions.

REFERENCES

1. SCHNEIDER, A. 1994. Tone system, intonation, aesthetic experience: theoretical norms and empirical findings. Systematische Musikwissenschaft **2:** 221–254.
2. DONCHIN, E. 1981. Surprise! ... Surprise? Psychophysiology **18:** 493–513.
3. VERLEGER, R. 1986. Die P3-Komponente im EEG: Literaturübersicht, Diskussion von Hypothesen, Untersuchung ihres Zusammenhangs mit langsamen Potentialen. Profil. München.
4. JAIRAZBHOY, N.A. 1971. The rags of North Indian music: their structure and evolution. Faber & Faber. London.
5. HARWOOD, D.L. 1979. Contributions from psychology to musical universals. The World of Music **21:** 48–61.

Singing: A Selective Deficit in the Retrieval of Musical Intervals

DANIELE SCHÖN,[a,b] BOGDAN LORBER,[c] MARTIN SPACAL,[a] AND CARLO SEMENZA[a,d]

[a]*Department of Psychology, University of Trieste, Trieste, Italy*

[b]*INPC - CNRS, Marseille, France*

[c]*Department of Clinical Neurology, Division of Neurology, University Medical Centre, Ljubljana, Slovenia*

[d]*B.R.A.I.N. Neuroscience Centre, University of Trieste, Trieste, Italy*

ABSTRACT: A case of dissociation between discrimination and retrieval of musical information in a patient with a lesion of the right hemisphere is described. This patient has lost the ability to correctly retrieve a musical interval when required to sing. This occurs in the presence of unimpaired interval discrimination and correct retrieval of temporal patterns and melodic contour (direction).

KEYWORDS: singing; dissociation; discrimination; musical retrieval

While the dissociation of comprehension and production deficits in language has been reported extensively, it has rarely been the case in the music domain, and the descriptions are mostly anecdotal (Mann, 1898, and Jossmann, 1926; cited in Ref. 1). Indeed, most studies in the neuropsychology of music are concerned with the perceptual aspects of music cognition, mainly discrimination and recognition (for a recent review see Ref. 2). However, not only is the study of the neuropsychology of music performance important for musicians and neuropsychologists for rehabilitation issues, but also it can be of interest for the theory of music perception, in that both performance and perception are closely tied and probably share some abstract representations and neural networks.

Here we describe a patient (I.P.) with a premorbid high level of musical expertise, who, at the age of 55 years, had an ischemic lesion involving almost the entire vascular territory of the posterior branches of the right middle cerebral artery, including the inferior right frontal gyrus, a large portion of the posterior temporal lobe, and an inferior portion of the parietal lobe (FIG. 1). On recovery, he had no sequels and returned to work as top manager, although part time (75%). The patient showed good comprehension and production in both Slovene and English. Prosodic processing was also preserved. He showed no signs of dyscalculia, ideomotor

Address for correspondence: Dr. Daniele Schön, INPC, C.N.R.S, 31, Chemin Joseph-Aiguier, 13402 Marseilles Cedex 20, France. Voice: 0033 (0) 4 91 16 41 13; fax: 0033 (0) 4 91 77 49 69. danschon@lnf.cnrs-mrs.fr

Ann. N.Y. Acad. Sci. 999: 189–192 (2003). © 2003 New York Academy of Sciences.
doi: 10.1196/annals.1284.027

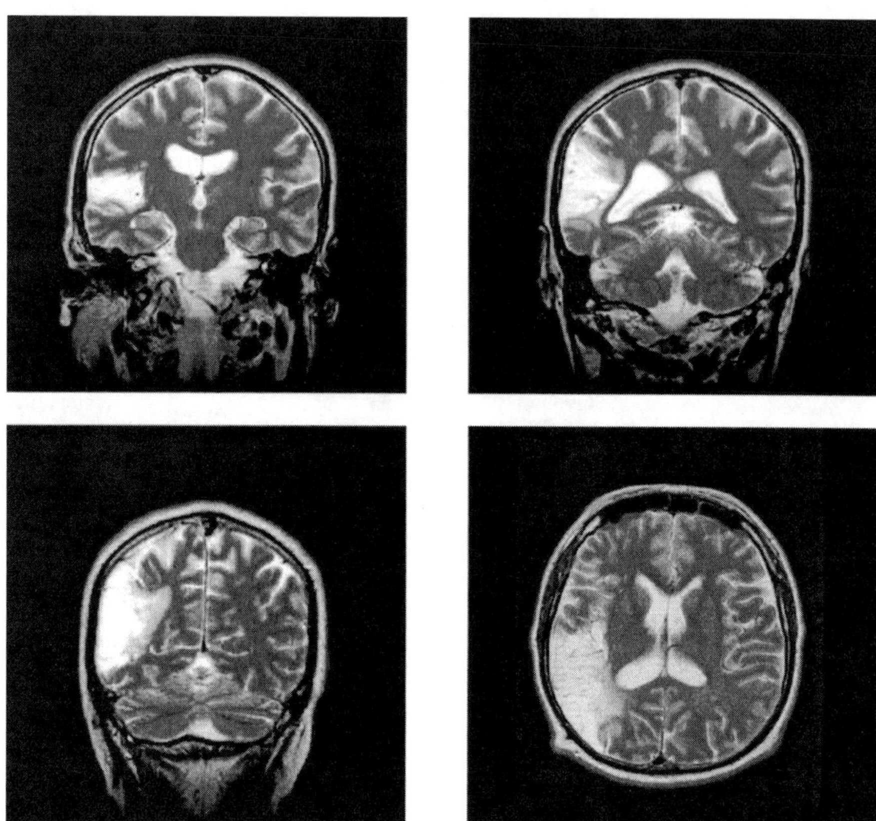

FIGURE 1. Three coronal and one horizontal T2-weighted images showing the lesion. The left side of the images corresponds to the right side of the brain.

apraxia or constructive apraxia, or severe attentional deficits. His short memory was mildly impaired (auditory digit span of 4). Premorbidly, he had been a high-level amateur tenor for at least 35 years. He sang regularly in public concerts as well as on private occasions. He could also improvise singing a second voice in order to harmonize traditional songs. He had had 8 years of music education. The following reported musical tests were the results of several testing sessions distributed across 14 months, beginning when patient I.P. was about 1 year postonset.

Initially, we tested I.P. with a series of musical tests adapted mainly from Schön et al.[3] Differently from other musical tests,[4,5] we used short sequences that could be used in both discrimination and reproduction tasks and that took into account the mild short-memory impairment of I.P. A matched group of 5 high-level amateur classical musicians was used as the control (mean age 60 years; university level education; one singer and four instrumentalists).

Results of general musical testing showed that I.P. could well identify familiar tunes and different musical meters. He was, however, at chance when asked to identify minor or major chords. He could perform well several discrimination tasks that had pitch or rhythm as the independent variable. He could also distinguish major and minor chords or triads and dyads (e.g., D-F-A and D-A). Finally, in the test of tonality, when required to judge the congruity of the final note in nonfamiliar melodies, I.P. performed at ceiling, thus showing that he was capable of expecting and perceiving the tonic of novel musical pieces.

This patient complained about his lost pleasure in listening to music and about his inability to sing in a vocal quartet and in choir. However, the results of this musical screening did not correspond to what would be expected in an amusic patient, since patient I.P., except for the chord identification task, performed all administered tests well.

Within production tasks patient I.P. showed severe impairment in tasks involving pitch, whereas he was good at tapping the tempo along with a metronome or with complex music, and he also performed well in a rhythm reproduction task.[6] The only pitch task that he could perform was to reproduce individual notes separated by a long silence (via MIDI). When he was asked in a preliminary test to sing tunes by heart (twice), he repeatedly missed several intervals and changed tonality throughout the piece. Using note-to-note scoring, in which each interval is judged independently of the original tonality, the patient committed 43 errors over 294 notes (3 contour direction and 40 interval size errors). When singing from the score, the performance was similar (1 contour direction and 33 interval size errors out of 240 notes). Finally, the patient in this task and in the following pitch production tasks administered during the testing sessions was aware of his errors, but he was reluctant to correct himself.

To investigate his deficit in more depth, we used two specific tasks. In the first, the patient heard two successive notes and was asked to immediately sing them. In the second task, there were four notes, the first two being the same as those in the previous task. Each task was comprised of 64 items. In both tasks, we manipulated the distance and the stability (consonance) of the intervals to be sung. In the second task, which was to test the possible facilitation effect that a "good" tonal context may have on pitch reproduction, two other notes followed the first two notes in order to obtain a "good" musical form (e.g., the descending major fourth became part of a cadence: V, II, III, I). Results were similar in the two tasks. Overall accuracy was 60%. About 32% of the trials contained a wrong interval, 7% a correct interval sung out of tune, and 1% an error in the contour direction. The control group was almost at ceiling in both tasks. Grouping the errors as a function of the criterion of stability/consonance showed that the dissonant intervals (38%) gave rise to more errors than the consonant intervals (26%, $P <0.001$, McNemar). On the contrary, grouping as a function of distance did not show any effect on performance.

This patient shows a dissociation between pitch discrimination and production. The functional models proposed to account for music processing disturbances are mainly limited to the recognition system.[2,7] Our case shows that, indeed, we can have an intact music recognition system (MRS) with deficits limited to the output system. We can go beyond this and begin to specify a functional model of music production. As patient I.P. does not seem to have a major problem in the production of rhythmic patterns, we think that melodic and temporal organization networks are dissociable not only at the recognition but also at the production level. Moreover, the fact that (1) the patient can reproduce single notes, (2) production errors are mainly wrong notes rather than notes out of tunes, and (3) there are more errors with dissonant than consonant intervals, all claim for a deficit in retrieving the correct musical representation of the interval.

Finally, according to previous anatomical models,[8] we expected the right hemisphere damage to induce a deficit in contour processing; however, this was not the case. Our patient, having a lesion of the right hemisphere, showed not only a contour production that was correct, but also an impaired pitch interval production. Previous studies mainly investigated nonmusicians. Thus, it may be that musicians, relative to music processing, have more adaptive brain substrates and more distributed neuronal network in both hemispheres.[9]

REFERENCES

1. BENTON, A.L. 1977. The amusias. *In* Music and the Brain. M. Critchley & R.A. Henson, Eds. :378–397. Heinemann. London.
2. PERETZ, I. 2001. Music perception and recognition. *In* The Handbook of Cognitive Neuropsychology. B. Rapp, Ed. :519–540. Psychology Press. Philadelphia.
3. SCHÖN, D., C. SEMENZA & G. DENES. 2001. Naming of musical notes: a selective deficit in one musical clef. Cortex **37:** 407–421.
4. STEINKE, W.R., L.L. CUDDY & L.S. JAKOBSON. 2001. Dissociations among functional subsystems governing melody recognition after right-hemisphere damage. Cognit. Neuropsychol. **18:** 411–437.
5. PERETZ, I. 1993. Auditory atonalia for melodies. Cognit. Neuropsychol. **10:** 21–56.
6. FRIES, W. & A. SWIHART. 1990. Disturbance of rhythm sense following right hemisphere damage. Neuropsychologia **28:** 1317–1323.
7. CARROL-PHELAN, B. & P.J. HAMPSON. 1996. Multiple components of the perception of musical sequences: a cognitive neuroscience analysis and some implications for auditory imagery. Music Percept. **13:** 517–561.
8. PERETZ, I. 1990. Processing of local and global musical information by unilateral brain-damaged patients. Brain **113:** 1185–1205.
9. ALTENMÜLLER, E.O. 2001. How many music centers are in the brain. *In* The Biological Foundations of Music. R. Zatorre & I. Peretz, Eds. Ann. N.Y. Acad. Sci. **930:** 232–259.

Audiovisual Interactions in Music Reading

A Reaction Times and Event-Related Potentials Study

DANIELE SCHÖN[a,b] AND MIREILLE BESSON[a]

[a]CRNC – CNRS, Marseilles, France

[b]Dipartimento di Psicologia, Università di Trieste, Trieste, Italy

ABSTRACT: The general aim of this experiment was to investigate the processes involved in reading musical notation and to study the relationship between written music and its auditory representation. Our main interest was to determine if musicians can develop expectancies for plausible or implausible auditory events on the sole basis of the visual score. Results showed that musicians can clearly expect auditory endings on the basis of visual information. These findings enliven the discussion on the question of whether music reading is actually music perception.

KEYWORDS: audiovisual interactions; music reading; reaction times; event-related brain potentials

An important aspect in the study of reading, be it music or language, is to understand the types of representation that are involved. As for language, music reading may involve phonologic and graphemic representations and also motor and auditory representations. Indeed, musicians most often read a score and play it at the same time. To do this, they have to identify the signs on the score, produce the appropriate motor action, check the tuning, and correctly interpret each note musically. Therefore, it would be difficult to claim that music reading is only a visuomotor coding task. Rather, it seems to belongs to the realm of music perception (for a review, see Ref. 1). Indeed, it has been demonstrated that in reading, musical memory is sensitive to musical structure.[2] Musicians may have, to a certain extent, an auditory-like representation of the written music before they actually play it.[3,4]

To test the hypothesis that musicians have an auditory-like representation of written music, we investigated the relationship between reading and listening when both abilities are coupled in the same experimental task. If written music can induce musical expectancies (i.e., not only motor expectancies), this may influence the way music is perceived. In the present study, we wanted to determine whether reading a music score can influence the way an auditory event is later perceived. In a given context, some notes are stable[5] and are highly plausible, whereas other notes are

Address for correspondence: Daniele Schön, INPC, C.N.R.S, 31, Chemin Joseph-Aiguier, 13402 Marseilles Cedex 20, France. Voice: 0033 (0) 4 91 16 41 13; fax: 0033 (0) 4 91 77 49 69. danschon@lnf.cnrs-mrs.fr

Ann. N.Y. Acad. Sci. 999: 193–198 (2003). © 2003 New York Academy of Sciences.
doi: 10.1196/annals.1284.028

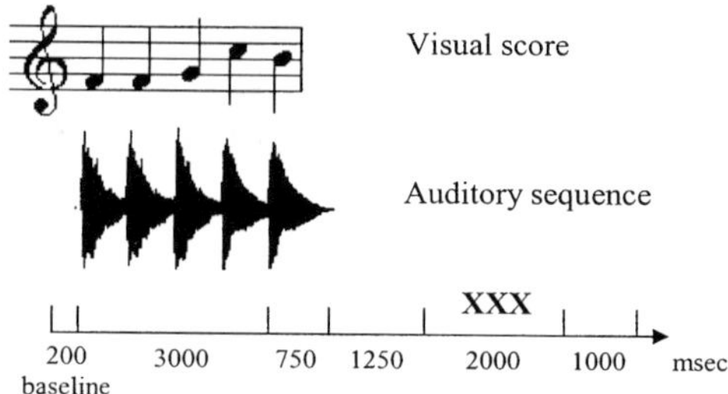

FIGURE 1A. Experimental design.

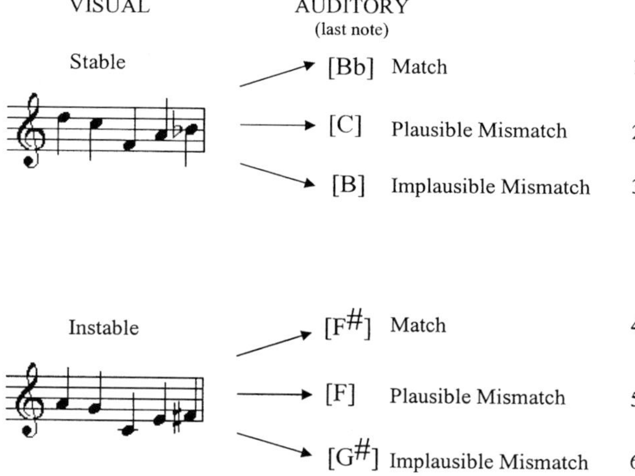

FIGURE 1B. Experimental conditions.

unstable and implausible, creating musical tension or surprise. Can the plausibility of an auditory event in a given musical context be modified on the basis of visual information?

Musicians were asked to judge whether the last note of a five-note auditory musical sequence matched or mismatched the information simultaneously provided on a score presented on a computer screen (FIG. 1A). There were six conditions (FIG. 1B): visual stable endings: (1) an auditory (stable) match, (2) an auditory plausible mismatch, and (3) an auditory implausible mismatch; visual unstable endings: (4) an auditory (instable) match, (5) an auditory plausible mismatch, and (6) an auditory implausible mismatch. While we predicted that musicians would be able to anticipate stable auditory events on the basis of the visual information provided by the score, the main question of interest was to find out whether they would also be able to do so for unstable auditory events? To answer these questions, we recorded reaction times (RTs) and the EEG from 28 electrodes.

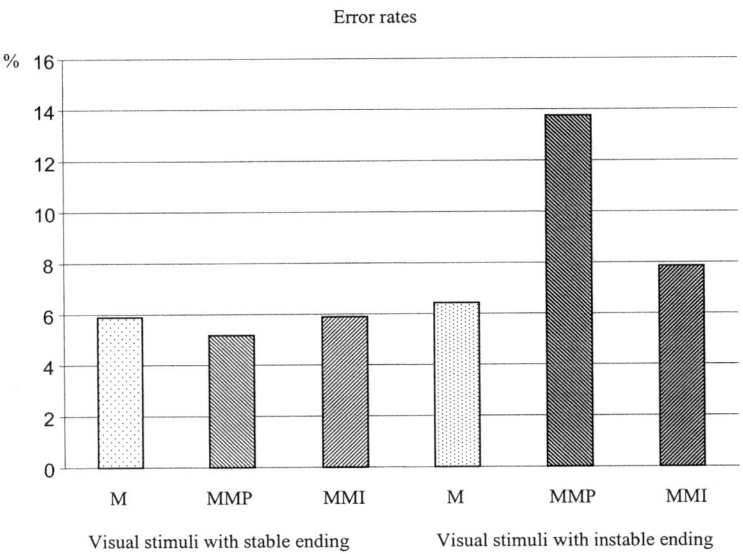

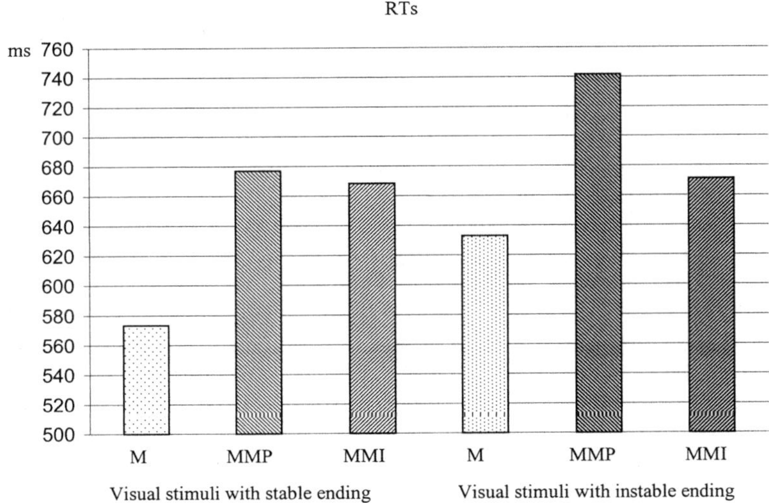

FIGURE 2. Error rates and reaction times. M, match; MMP, plausible mismatch; MMI, implausible mismatch.

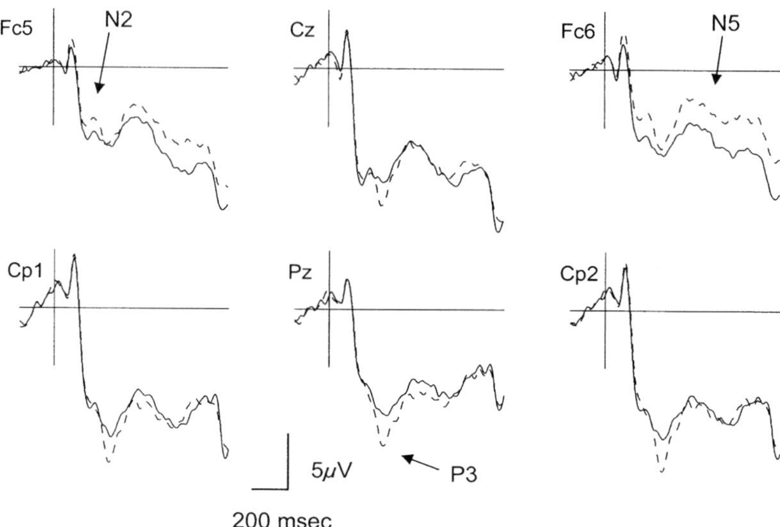

FIGURE 3. Variations in brain electrical activity time-locked to the stable matching (*solid line*, condition 1) and to the unstable matching conditions (*dashed line*, condition 4).

Behavioral results showed shorter RTs for stable than for unstable matching trials (FIG. 2). Two different reasons may account for this result. One is that the unstable matching condition may sound odd even though it matches the note expected on the basis of the visual stimulus. In other words, even if musicians can prepare themselves for unstable endings, the ending pitch remains somewhat unexpected and surprising, thus causing a delay in the response. The other reason involves the difficulty of anticipation. Indeed, it might be easier to anticipate a stable than an unstable final note. However, if this were true, a difficulty effect should also have been found when comparing the mismatches in the two different visual expectancy conditions, but this was not the case. Therefore, even though a difficulty effect may exist to a certain extent, our data seem to be better explained in terms of interference of the musical (tonal) expectancies in the expectancies created on the basis of the visual stimuli.

Overall, event-related potential (ERPs) results showed that the mismatching conditions differed from the matching conditions. Although this favors the capacity of musicians to anticipate or prepare themselves for an unstable ending, detailed analyses of our results show that such a direct and strong conclusion cannot be made without precautions. In the following discussion, we first consider the results in the auditory matching conditions and then in the auditory mismatching conditions.

When the auditory stimulus matches the note expected on the basis of the visual score (matching conditions, FIG. 3), the amplitude of early (ERAN[6]) and late (N5) negative components, as well as the amplitude of a positive component (P300), was larger for the unstable than for the stable ending condition. In other words and in line with the behavioral data, the ERP data show that the expectancy that musicians develop on the basis of the visual score is not as strong for unstable as for stable events. It seems then that musicians can anticipate a specific expected (i.e., stable) event, but cannot fully anticipate unexpected (i.e., unstable) events because they remain unexpected.

When the auditory stimulus mismatches the note expected on the basis of the visual score (mismatching conditions), several results need to be considered. First, independently of whether musicians are anticipating stable or unstable ending notes, implausible auditory mismatches are associated with larger N1, P3, and late positive components than are plausible auditory mismatches. Therefore, these results are in line with previous results in the ERPs and music literature[7] when purely auditory musical stimuli are used. However, in these studies, expected (matching) events were always compared with unexpected (mismatching) events. Therefore, it was not possible to determine whether the effects reported on the N1, P3, and late positivities were linked to a general mismatch process or to the violation of specific musical expectancies. By contrast, in the present study the comparison is between two mismatching events, one plausible and one implausible. Therefore, the fact that we found a differential effect, larger for implausible than for plausible auditory mismatches, clearly shows that these different ERP components are sensitive to specific musical expectancies.

Second, and very interestingly, the amplitude of the late positivity associated with implausible auditory mismatches is larger when musicians anticipate stable endings than when they anticipate unstable endings (FIG. 4). This result is clearly in line with our hypothesis that visual stable endings do create stronger musical expectancy than do visual unstable endings, thereby increasing the auditory mismatch effect. Moreover, this result is also in line with those just reported for the auditory matching stimuli, namely, that expectancy was not as strong for unstable as for stable matches.

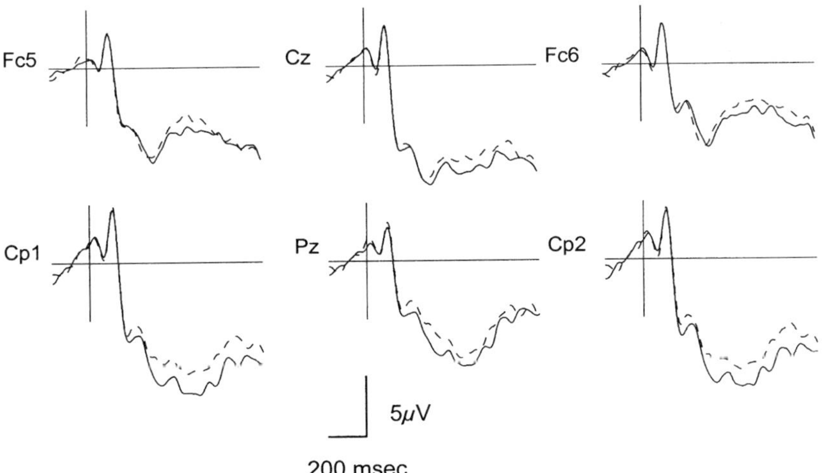

FIGURE 4. Variations in brain electrical activity time-locked to the implausible mismatch to the visual stable ending (*solid line*, condition 3) and to the implausible mismatch to the visual unstable ending (*dashed line*, condition 6).

In conclusion, musicians are clearly able to anticipate stable endings on the basis of visual information and seem able to do so for unstable endings, although to a lesser extent. Moreover, error rates, RTs, and electrophysiologic data showed that the auditory mismatch effect found in the present experiment is clearly modulated by the expectations built from the score. Strong interactions, therefore, seem to exist between visual and auditory musical codes, so that the representations built from the visual score interfere with the auditory perception of the musical sequences.

REFERENCES

1. SLOBODA, J.A. 1984. Experimental studies of musical reading: a review. Music Perception **2:** 222–236.
2. HALPERN, A. & G. BOWER. 1982. Musical expertise and melodic structure in memory for musical notation. Am. J. Psychol. **95:** 31–50.
3. SCHÖN, D., C. SEMENZA & G. DENES. 2001. Naming of musical notes: a selective deficit in one musical clef. Cortex **37:** 407–421.
4. SCHÖN, D. & M. BESSON. 2002. Processing pitch and duration in music reading: a RTs and ERPs study. Neuropsychologia **40:** 868–878.
5. KRUMHANSL, C. 1990. Cognitive Foundations of Musical Pitch. Oxford University Press. Oxford, UK.
6. KOELSCH, S., T. GUNTER, A.D. FRIEDERICI & E. SCHROGER. 2000. Brain indices of music processing: "nonmusicians" are musical. J. Cognit. Neurosci. **12:** 520–541.
7. BESSON, M. & MACAR. F 1987. An event-related potential analysis of incongruity in music and other non-linguistic contexts. Psychophysiology **24:** 14–25.

Preattentive Processing of Lexical Tone Perception by the Human Brain as Indexed by the Mismatch Negativity Paradigm

WICHIAN SITTIPRAPAPORN,[a] CHITTIN CHINDADUANGRATN,[a] MARI TERVANIEMI,[b] AND NAIPHINICH KHOTCHABHAKDI[a]

[a]*Clinical & Research Electroneurophysiology Laboratory, Neuro-Behavioural Biology Center, Institute of Science & Technology for Research and Development, Mahidol University at Salaya Campus, Nakhonpathom 73170, Thailand*

[b]*Cognitive Brain Research Unit, Department of Psychology, University of Helsinki, FIN-00014, Helsinki, Finland*

ABSTRACT: Mismatch negativity (MMN) was used to investigate the processing of the discrimination between native and non-native CV syllables in tonal languages. MMN elicited by the native word was greater than that elicited by the non-native word. Hearing a native-language deviant significantly altered the elicited MMN in both amplitude and scalp voltage field distribution, reflecting the presence of a long-term memory trace for spoken words in tonal languages.

KEYWORDS: brain; event-related potential (ERP); language; lexical tone; memory traces; mismatch negativity (MMN)

INTRODUCTION

Mismatch negativity (MMN), a component of the auditory event-related potential (ERP),[1] has been used to investigate the neural processing of speech and language[2] because it is considered to be a unique indicator of automatic cerebral processing of acoustic stimuli.[3] It was recently found that MMN in response to individual words is greater than that for comparable meaningless word-like stimuli[4] and pseudoword stimuli.[3] This enhancement reflects cortical memory traces for words formed during the subjects' previous language experience.[3,4] However, whether the human brain contains memory traces for lexical items in tonal languages that are uniquely engaged in speech perception remains an unresolved issue. This study investigates in detail whether the human brain has unique speech-specific mechanisms or treats lexical items in tonal languages equally.

Address for correspondence: Dr. Wichian Sittiprapaporn, Clinical & Research Electroneurophysiology Laboratory, Neuro-Behavioural Biology Center, Institute of Science & Technology for Research and Development, Mahidol University at Salaya Campus, Nakhonpathom 73170, Thailand. Voice: +66-0-2441-9321; fax: +66-0-2441-9743.

wichian_s@thailand.com

Ann. N.Y. Acad. Sci. 999: 199–203 (2003). © 2003 New York Academy of Sciences.
doi: 10.1196/annals.1284.029

MATERIAL AND METHODS

Subjects. Nine healthy right-handed volunteers (native Thai speakers aged 18–35 years) with normal hearing and no record of neurological disease were investigated in two separate experimental conditions.

Stimuli. Two stimuli were used, each consisting of a consonant-vowel (CV) syllable: a natural-speech word of Thai with falling tone /kʰâ/ and a Chinese morpheme with a similar tone /ta⁴/. Sequences of these two monosyllabic words were prepared and presented in two conditions: (1) in the native condition, /ta⁴/ was used as the standard and /kʰâ/ as the deviant stimulus; (2) in the non-native condition, the standard stimulus was /kʰâ/, whereas the deviant was /ta⁴/. All words were digitally edited to have an equal peak energy level in dB SPL, with the remaining data within each of the words scaled accordingly. Each condition was comprised of 500 syllables, arranged in randomized sequences of standard ($P = 90\%$) and deviant ($P = 10\%$) stimuli. The interstimulus interval (ISI) was 1.25 ms (offset-onset). Subjects were seated in an electrically and acoustically shielded chamber and instructed to ignore the auditory stimulation by reading a book of their choice.

Electroencephalographic Recording. Electrical activity of the subject's brain was continuously recorded (passband 0.01–100 Hz, sampling rate 128 Hz) with 20 active electrodes positioned according to the 10–20 International System (Electro-cap) and referred to linked earlobes, with an electrode between Fz and Fpz connected to ground.

EEG Data Processing. The recordings were later filtered offline. ERPs were obtained by averaging epochs, which started 100 ms before the stimulus onset and ended 400 ms thereafter; the 100-ms interval preceding stimulus onset was used as baseline. Epochs with a voltage variation exceeding ± 100 μV at any EEG channel or at either of the two electrooculographic (EOG) electrodes were discarded. MMN was obtained by subtracting the response to the standard from that to the deviant. For each experimental subject, the averaged MMN responses contained at least 125 accepted deviant trials in each condition. All responses were recalculated offline against average reference for further analysis.

Spatial Analysis. ERP voltages were transformed into reference-independent values by re-computing the voltages against average reference.[5] To quantify the amount of "spatial relief" or hilliness of the topographic fields, the scalp electric potential power or "global field power" (GFP) was calculated. The amplitude variability across the 20 electrodes was calculated by averaging the data, time point by time point, in the difference waveforms. The point of maximal GFP in the time segment between 144 and 204 ms after stimulus onset was detected. Low-resolution electromagnetic tomography (LORETA) was then applied to estimate the current source density distribution in the brain, which contributes to the electrical scalp field.[6] This resulted in 1153 current density values. Regions of interest (ROI) were then defined on the basis of local maxima of the LORETA distribution averaged over all nine subjects. For each subject, the distributions for the two different deviants were compared with paired *t* tests.

Statistical Analysis. The statistical significance of MMN was tested with one-sample *t* tests to verify the presence of the MMN component at the frontal (Fz) electrode site by comparing the mean amplitude of the 144–204-ms interval against a

TABLE 1. Overall mismatch negativity (MMN) amplitudes elicited by deviants

Condition	Trial type Standard Deviant	Mean MMN amplitude (μV)	P
Native	[ta^4] - [k^{h}â]	-2.41 ± 0.04	<0.0001
Non-native	[k^{h}â] - [ta^4]	-0.98 ± 0.06	<0.0001

Grand-average frontal (Fz) MMN amplitudes in μV ($\pm$ standard error of mean) elicited by deviants measured in the 144-204 ms (overall MMN) interval. The difference between conditions was tested by one-sample t-test.

hypothetical zero, separately in each condition. All results were expressed as mean $\pm$ SEM, and all were significant.

RESULTS

In both conditions, MMN was significant. The average latency value ($\pm$ SEM) for both conditions was 174 ± 30 ms. The amplitudes in a 60-ms interval around the peak (144–204 ms) were therefore calculated. The native (/khâ/ deviant) condition yielded larger MMN amplitudes than did the non-native (/ta^4/ deviant) condition (mean amplitude: -2.41 μV vs -0.98 μV; GFP = 1.00 vs 0.56, respectively; TABLE 1). The difference in the LORETA-ROI current densities distribution between conditions was defined. The LORETA images for the non-native condition demonstrated multiple sources: a predominant activity in the right temporal cortex (RT-ROI: *locx, locy, locz* = 0.033, -0.217, 0.215; 1.74 μA/mm^2) and in the medial prefrontal cortex (MF-ROI: *locx, locy, locz* = 0.033, -0.217, 0.430; 1.74 μA/mm^2). On the other hand, the LORETA images for the native condition demonstrated only one source: a predominant activity in the left temporal cortex (LT-ROI: *locx, locy, locz* = -0.020, 0.113, 0.215; 2.61 μA/mm^2). Thus, based on the local maxima of the grand average LORETA images, two ROIs and a single ROI were detected for the non-native and native conditions, respectively (FIG. 1). In the analysis of the scalp voltage field distribution, both conditions differed in amplitude and scalp voltage distribution of the response (FIG. 2).

CONCLUSION

The difference in cerebral lateralization between the auditory preattentive perception of native and non-native words in tonal languages was investigated. This study revealed that spoken words from the subject's native language elicited greater electric sources of the mismatch response, indicating a stronger contribution from the left than from the right auditory cortex. These responses can be activated in the absence of active attention to the auditory input and are probably available at early stages of cerebral speech processing. The findings illustrate how memory and language may interact in phonological processing at different stages of the prelexical perceptual process within the frontocentral areas. Thus, noninvasive methods for localizing electrical activity in the brain could make important contributions to the

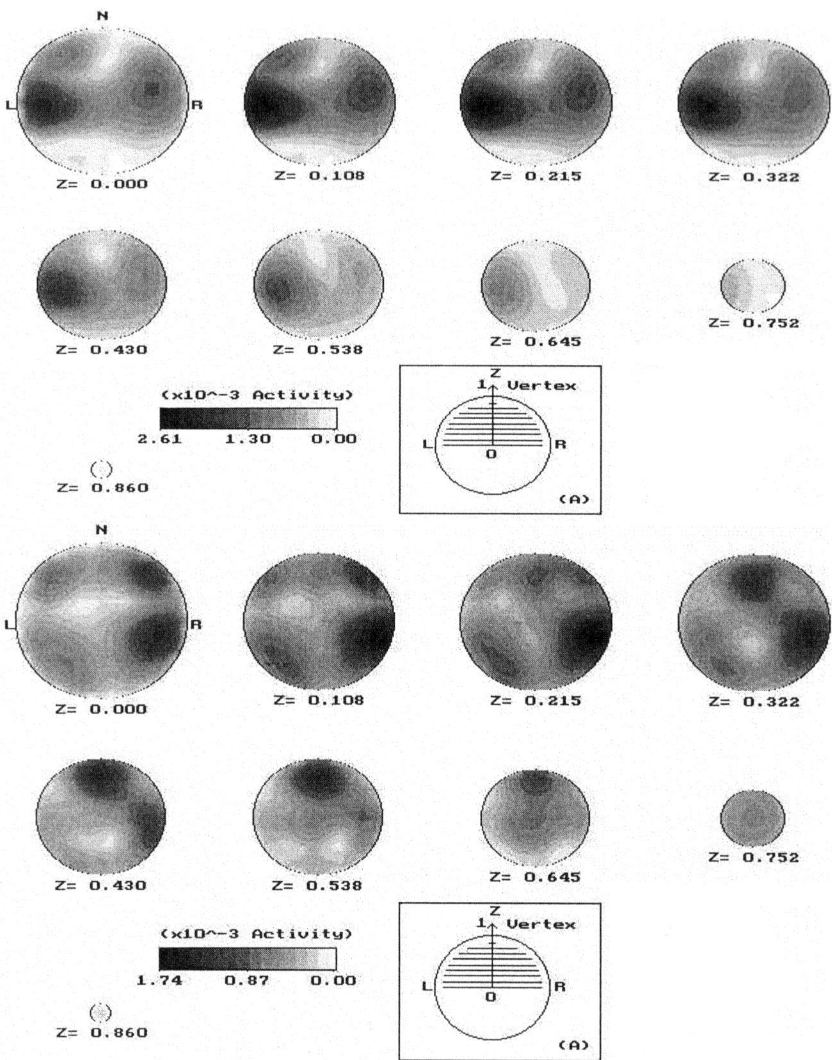

FIGURE 1. Schematic representation of mismatch negativity (MMN) regions of interest (ROI). The images are parallel horizontal brain slices viewed from the top of the head, with the indicated orientation (L, left; R, right; N, nasion) and elevations (Z) relative to the head radius of 7.8 cm. *Top views* illustrate x- and y-axes, the *posterior view* indicates the height of each brain slice (z-axis). (***Top panel***) Native condition, showing a left temporal locus (LT-ROI: *locx, locy, locz* = −0.020, 0.113, 0.215; 2.61 μA/mm^2). (***Bottom panel***) Non-native condition, showing right temporal (RT-ROI: *locx, locy, locz* = 0.033, −0.217, 0.215; 1.74 μA/mm^2) and medial frontal loci (MF-ROI: *locx, locy, locz* = 0.033, −0.217, 0.430; 1.74 μA/mm^2).

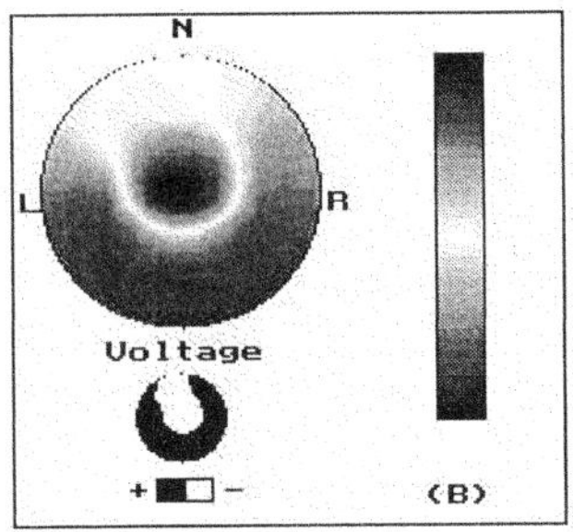
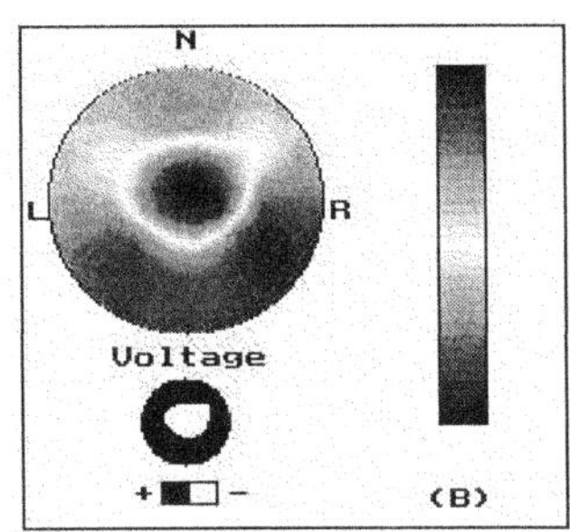

Native (Thai) condition   Non-native (Chinese) condition

FIGURE 2. Potential maps of electric mismatch negativity (MMN) responses (deviant-standard subtractions) evoked by deviants measured in the 144–204-ms (overall MMN) interval. Word-elicited MMNs differed from each other in amplitude and in scalp voltage distribution.

understanding of the relationships between changes in central nervous system activity and changes in pre-attentive functioning as indexed by the mismatch response.

ACKNOWLEDGMENTS

This study was supported by a UDC grant from Mahidol University, Thailand, and in part by a scholarship from the Mariani Foundation, Italy. We wish to thank Dr. Wolfgang Woerner and Dr. Colin Phillips for commenting on the manuscript. We also thank Dr. Jack Gandour for providing the stimuli for this study.

REFERENCES

1. NÄÄTÄNEN, R. & K. ALHO. 1995. Mismatch negativity: a unique measure of sensory processing in audition. Int. J. Neurosci. **80:** 317–337.
2. SHTYROV, Y. *et al.* 2000. Discrimination of speech and of complex nonspeech sounds of different temporal structures in the left and right cerebral hemispheres. NeuroImage **12:** 657–663.
3. SHTYROV, Y. & F. PULVERMÜLER. 2002. Neurophysiological evidence of memory traces for words in the human brain. NeuroReport **13:** 521–525.
4. PULVERMÜLER, F. 2001. Brain reflections of words and their meaning. Trends Cognit. Sci. **1:** 517–524.
5. LEHMANN, D. 1987. Principles of spatial analysis. *In* Handbook of Electroencephalography and Clinical Neurophysiology. A.S. Gevins & A. Remond, Eds. :309–354. Elsevier. Amsterdam.
6. PASCUAL, R.D. *et al.* 1994. Low-resolution electromagnetic tomography: a new method for localizing electrical activity in the brain. Int. J. Psychophysiol. **18:** 49–65.

Becoming a Pianist

An fMRI Study of Musical Literacy Acquisition

LAUREN STEWART,[a] RIK HENSON,[a,b] KNUT KAMPE,[a] VINCENT WALSH,[a] ROBERT TURNER,[b] AND UTA FRITH[a]

[a]*Institute of Cognitive Neuroscience, Alexandra House, London WC1N 3AR, UK*

[b]*Wellcome Department of Imaging Neuroscience, London WC1N 3BG, UK*

ABSTRACT: Musically naïve subjects were scanned using functional magnetic resonance imaging (fMRI) before and after they had been taught to read music and play keyboard. When subjects played melodies from musical notation after training, activation was seen in a cluster of voxels within the right superior parietal cortex consistent with the view that music reading involves spatial sensorimotor mapping.

KEYWORDS: pianist; fMRI; musical literacy

INTRODUCTION

When a child or adult starts to play the keyboard, a significant part of the initial musical training is devoted to learning to read musical notation. Musical pieces that, at first sight, appear meaningless in their written form will eventually be translated into a recognizable melody. Just as written language becomes meaningful and even compelling to read, so does musical notation. A key question is how the artificial process of sight-reading for keyboard performance becomes a natural process. How do brain areas become recruited for such a skill? The advent of functional magnetic resonance imaging (fMRI) has permitted longitudinal studies of the neural correlates of skill acquisition. Music reading is a skill that lends itself to such an approach, because only a small fraction of the population is musically literate and many are motivated to learn. Thus, a unique opportunity exists for investigating the acquisition of an artificial and culturally valued skill. The following represents a brief account of a study that was designed to look at music reading in a learning context.[1]

MATERIAL AND METHODS

Training. Subjects attended a 90-minute music lesson once a week for 15 weeks. Practical keyboard skills and music theory were taught to Grade 1 (Associated Board, UK) level.

Address for correspondence: Dr. Lauren Stewart, Institute of Cognitive Neuroscience, Alexandra House, 17 Queen Square, London WC1N 3AR, UK. Voice: (+44) (0)20 7679 1168; fax: (+44) (0)20 7813 2835.

l.stewart@ucl.ac.uk

Ann. N.Y. Acad. Sci. 999: 204–208 (2003). © 2003 New York Academy of Sciences.
doi: 10.1196/annals.1284.030

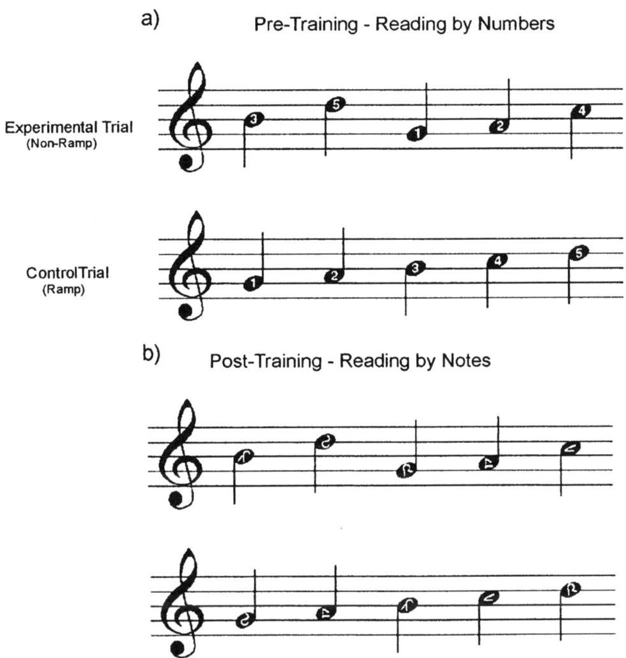

FIGURE 1. Explicit Music Reading Task. The explicit music reading task was different pre-training versus post-training.

FIGURE 2. Implicit Music Reading Task. The implicit music reading task was identical pre- and post-training.

Tasks Used during Scanning: (1) Explicit Music Reading Task (FIG. 1). This task required subjects to produce a series of keypresses in response to the appearance of a sequence of five musical notes. Before training, subjects used the superimposed numbers to play the sequences; after training, they relied solely on the musical notation. (2) Implicit Music Reading Task (FIG. 2). The task was identical before and after training. Subjects indicated whether the target (a single vertical line that extended above or below the five horizontal lines of the staff) was ascending or descending, using an arbitrary up/down mapping to the index and middle fingers.

Statistical Analysis. The explicit and implicit music reading tasks were analyzed separately for the pre-training and post-training sessions using statistical parametric mapping software (SPM99, Wellcome Department of Cognitive Neurology).

Statistical Parametric Mapping Software. Random effects analysis was used to isolate training-related activations (greater activation for experimental trials versus control trials, post-learning versus pre-learning).

RESULTS

Learning to Play a Melody: Explicit Music Reading Task. A training effect was seen in right superior parietal cortex (FIG. 3). Examination of the mean percentage signal change for the maxima of this region revealed a trial effect that, although significant before training, was even greater after training.

Effect of Exposure to Musical Notation: Implicit Music Reading Task. A training effect was seen in the left supramarginal gyrus, left inferior frontal sulcus, and right frontal pole (FIG. 4). Examination of the mean percentage signal change for the maxima of these regions revealed that all voxels exhibited the same relative pattern: a trial effect that was restricted to the post-training session.

Activations Common to Both Explicit and Implicit Music Reading. Inclusive masking revealed common training effects across the explicit and implicit music reading tasks in the bilateral superior parietal cortex, medial superior parietal cortex, and left postcentral gyrus.

DISCUSSION

Learning to Play a Melody: Explicit Music Reading Task. The dorsal visual processing stream, within which the superior parietal cortex resides, is known to be important for coding of spatial as opposed to featural aspects of visual stimuli (the "what"/"where" distinction).[2] A distinction has also been made between the visual perception of objects versus the control of action towards those objects (the "what"/ "how" dichotomy).[3] Whether the distinction made is one of "what versus where" or "what versus how," sight-reading for keyboard performance falls within the class of behaviors that the dorsal stream is known to subserve. First, the information relevant for performance is contained in the position of the note on the staff ("where"); second, musical performance relies on the use of this positional information to guide selection of the appropriate keypress ("how"). Note that our finding corresponds well to those of Sergent *et al.*[4] that relate to a PET study of sightreading in professional pianists, suggesting that they may be independent of skill level.

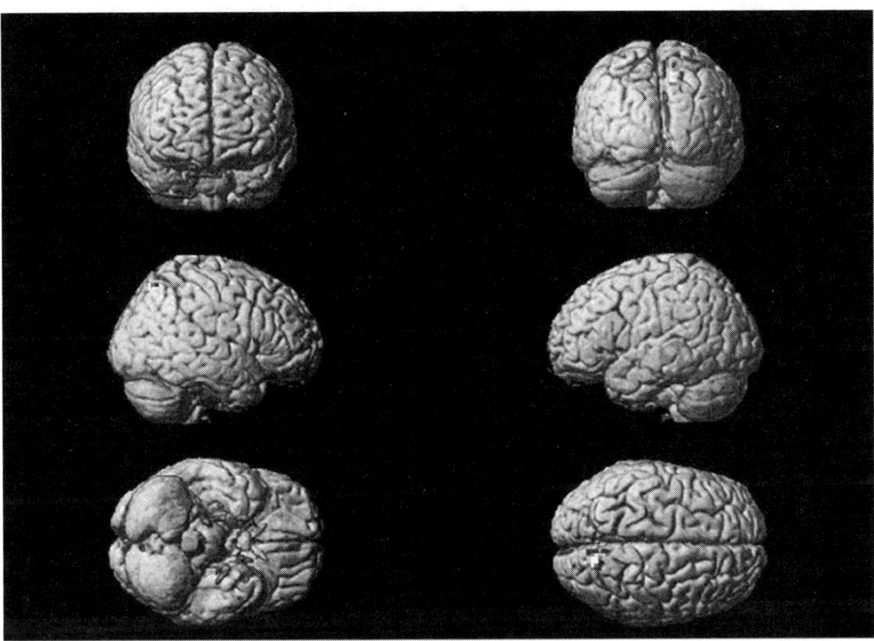

FIGURE 3. Explicit music reading. Statistical parametric map rendered onto a normalized structural image, showing activation that was greater for note to finger mapping (post-training) than for number to finger mapping (pre-training) for experimental trials minus control trials. Activation was seen in right superior parietal cortex.

Effect of Exposure to Musical Notation: Implicit Music Reading Task. During the course of training, subjects learned to make specific keypresses in response to particular musical notes. We suggest that the visual appearance of musical notes after training may be automatically and unconsciously interpreted as an instruction to act. For the purposes of performing the feature detection task, preparation of the learned musical response would be inappropriate and would be overridden by the preparation and execution of the task-relevant motor response. While the preparation and execution of the task-relevant response were common across both the pre- and the post-training sessions, the implicit preparation of a music-specific motor response would have only occurred after training. The supramarginal gyrus, especially on the left, is thought to be important in processes related to "motor intention."[5]

Commonalities in the Training Effect: Explicit and Implicit Tasks. We have suggested that the training effect shown by this area in the explicit task reflects a visuo-spatial sensorimotor translation between the notes on the staff and the appropriate keypresses. The fact that this region also showed a training effect in the implicit task leads us to conclude that the mere presence of musical notation may result in a similar translation even when such translation does not result in motor execution.

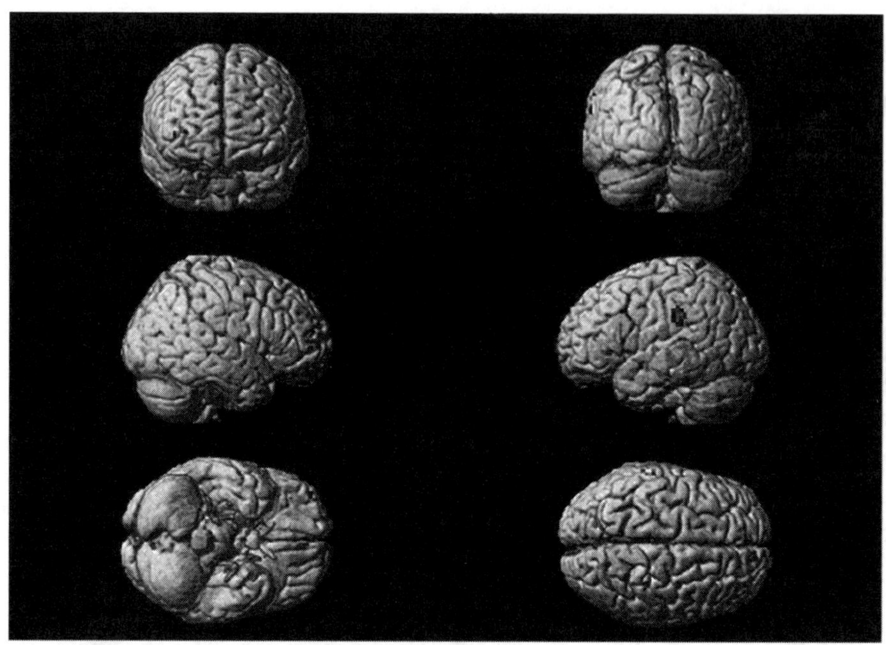

FIGURE 4. Implicit Music Reading. Statistical parametric map rendered onto a normalized structural image, showing activation that was greater post-training than pre-training for experimental trials minus control trials. Activation was seen in left supramarginal gyrus, post-central sulcus, medial parietal cortex, right cerebellum, and right frontal pole.

ACKNOWLEDGMENTS

This work was supported by the Medical Research Council. V.W. was supported by a Royal Society University Research Fellowship.

REFERENCES

1. STEWART, L., R. HENSON, K. KAMPE, *et al.* 2003. Brain changes after learning to read and play music. Neuroimage. In press.
2. MISHKIN, M. & L.G. UNGERLEIDER. 1982. Contribution of striate inputs to the visuospatial functions of parieto-preoccipital cortex in monkeys. Behav. Brain Res. **6:** 57–77.
3. GOODALE, M.A. & A.D. MILNER. 1992. Separate visual pathways for perception and action. Trends Neurosci. **15:** 20–25.
4. SERGENT, J., E. ZUCK, S. TERRIAH & B. MACDONALD. 1992. Distributed neural network underlying musical sight-reading and keyboard performance. Science **257:** 106–109.
5. RUSHWORTH, M.F., T. PAUS & P.K. SIPILA. 2001. Attention systems and the organization of the human parietal cortex. J. Neurosci. **21:** 5262–5271.

Activation of the Inferior Frontal Cortex in Musical Priming

BARBARA TILLMANN,[a,b] PETR JANATA,[a] AND JAMSHED J. BHARUCHA[a]

[a]Dartmouth College, Hanover, New Hampshire, USA

[b]CNRS-UMR 5020, Lyon, France

ABSTRACT: Musical contexts influence the processing of target events. Our study investigated the neural correlates of processing related and unrelated musical events presented as the last chord of eight-chord sequences.

KEYWORDS: music perception; musical expectancies; inferior frontal regions; prefrontal cortex; functional magnetic resonance imaging (fMRI)

In a musical context, listeners develop expectations for future events and these expectations influence perception. Behavioral studies investigated the influence of context on event processing with a musical priming paradigm.[1,2] A prime context (one chord or a chord sequence) is followed by a target chord that is more or less harmonically related. Harmonic relatedness means that in pieces from the Western tonal repertoire, the type of chord used as a target is frequently associated with events belonging to the prime context. In the musical priming paradigm, participants make speeded accuracy judgments on a perceptual feature of the target without explicitly judging the manipulated harmonic relations. Behavioral evidence shows that chord processing is facilitated (i.e., faster and more accurate responses) when the target is harmonically related to the prime than unrelated or less related. It has also been shown that related targets evoke a weaker P3b or a late positive component than unrelated ones.[3] Harmonic priming effects have been observed for musician and nonmusician listeners, an outcome pointing to the implicit nature of tonal harmonic knowledge and the robustness of the processes involved (cf. Bigand, this volume).

Our present study[4] investigated the neural correlates of musical priming effects using fMRI. In eight-chord sequences, the last chord (target) was either strongly related (a tonic chord) or unrelated (a chord belonging to a distant key). As in previous musical priming studies, half the targets were rendered acoustically dissonant for the experimental task, which involved performing speeded intonation judgments (consonant versus dissonant) about the targets. Behavioral data acquired during fMRI scanning replicated the facilitation effect of related over unrelated consonant

Address for correspondence: Dr. Barbara Tillmann, CNRS-UMR 5020 Neurosciences et Systemès Sensoriels, 50 Av. Tony Garnier, F-69366 Lyon Cedex 07, France. Voice: +33 (0) 4 37 28 74 93; fax: +33 (0) 4 37 28 76 01.

barbara.tillmann@olfac.univ-lyonl.fr

Ann. N.Y. Acad. Sci. 999: 209–211 (2003). © 2003 New York Academy of Sciences.

doi: 10.1196/annals.1284.031

targets. The overall activation pattern associated with target processing showed commonalities with networks previously described for target detection and novelty processing.[5,6] This activation network included frontal areas (inferior, middle, and superior frontal gyri, insula, anterior cingulate) and posterior areas (inferior parietal gyri, posterior cingulate) as well as thalamic nuclei and cerebellum. The characteristics of the targets, notably in how far the chord fits or violates the expectations built up by the prime context, influenced the activation levels of some network components. Increased activation was observed for targets that violated expectations based on either sensory-acoustic or harmonic relations. For example, activation in bilateral inferior frontal regions (inferior frontal gyrus, frontal operculum, insula) was stronger for unrelated than for related (consonant) targets. The strength of activation in these areas was also greater for dissonant targets than consonant targets. The sensitivity of the frontal operculum to target type and prime context is convergent with the outcome of MEG source localization of responses to musical expectancy violations[7] (cf. Koelsch and Friederici, this volume).

The data on musical context effects can be integrated with other data showing that Broca's area and its right homologue participate in nonlinguistic processes[8,9] and pointing to the role of these regions in temporal integration of information.[10] The activation patterns observed with musical material and semantic material[11] suggest an influence of association strength between events (stored in the listener's long-term memory knowledge): weakly associated events evoke stronger activation than do frequently associated events. Thus, the processing of low probability events requires more neural resources than does the processing of more familiar or prototypical events.

ACKNOWLEDGMENTS

This research was supported by National Institutes of Health Grant P50 NS17778-18.

REFERENCES

1. BHARUCHA, J.J. & K. STOECKIG, 1986. Reaction time and musical expectancy: priming of chords. J. Exp. Psychol. Hum. Percept. Perform. **12:** 403–410.
2. BIGAND, E. *et al.* 1999. Effect of global structure and temporal organization on chord processing. J. Exp. Psychol. Hum. Percept. Perform. **25:** 184–197.
3. REGNAULT, P., E. BIGAND & M. BESSON. 2001. Event-related brain potentials show top-down and bottom-up modulations of musical expectations. J. Cognit. Neurosci. **13:** 241–255.
4. TILLMANN, B., P. JANATA & J.J. BHARUCHA. 2003. Activation of the inferior frontal cortex in musical priming. Cognit. Brain Res. **16:** 145–161.
5. LINDEN, D.E.J. *et al.* 1999. The functional neuroanatomy of target detection: an fMRI study of visual and auditory oddball tasks. Cereb. Cortex **9:** 815–823.
6. KIEHL, K.A. *et al.* 2001. Neural sources involved in auditory target detection and novelty processing: an event-related fMRI study. Psychophysiology **38:** 133–142.
7. MAESS, B. *et al.* 2001. 'Musical syntax' is processed in the Broca's area: an MEG-study. Nature Neurosci. **4:** 540–545.
8. MÜLLER, R.-A., N. KLEINHANS & E. COURCHESNE. 2001. Broca's area and the discrimination of frequency transitions. Brain & Language **76:** 70–76.

9. ADAMS, R.B. & P. JANATA. 2002. A comparison of neural circuits underlying auditory and visual object categorization. NeuroImage **16:** 361–377.

10. FUSTER, J.M. 2001. The prefrontal cortex: time is of the essence. Neuron **30:** 319–333.

11. WAGNER, A.D. *et al.* 2001. Recovering meaning: left prefrontal cortex guides controlled semantic retrieval. Neuron **31:** 329–338.

Analyzing Pitch Chroma and Pitch Height in the Human Brain

JASON D. WARREN,[a,b] STEFAN UPPENKAMP,[c] ROY D. PATTERSON,[c]
AND TIMOTHY D. GRIFFITHS[a-c]

[a]*Wellcome Department of Imaging Neuroscience, Institute of Neurology,
Queen Square, London, United Kingdom*

[b]*Auditory Group, University of Newcastle Medical School, Newcastle-upon-Tyne,
United Kingdom*

[c]*Centre for the Neural Basis of Hearing, Department of Physiology,
University of Cambridge, Cambridge, United Kingdom*

ABSTRACT: The perceptual pitch dimensions of chroma and height have distinct representations in the human brain: chroma is represented in cortical areas anterior to primary auditory cortex, whereas height is represented posterior to primary auditory cortex.

KEYWORDS: pitch; pitch helix; pitch height; pitch chroma; human auditory cortex; planum temporale

HYPOTHESIS

There is a longstanding debate concerning the dimensionality of pitch perception[1,2]: in this functional magnetic resonance imaging (fMRI) experiment, we tested the hypothesis that the two dimensions of the musical pitch helix,[3] chroma and height, have distinct mappings in the human brain.

METHODS

Stimuli were sequences consisting of harmonic complexes in which pitch chroma and pitch height were varied between successive complexes while the total energy and spectral region remained fixed.[4] Pitch chroma was varied randomly across one octave (60–120 Hz) of the chromatic musical scale by varying f_0 from note to note, and pitch height was also varied randomly across about one octave by attenuating the amplitude of the odd harmonics in 2-dB steps. The brain blood-oxygen-level–dependent response was measured in a sparse fMRI protocol in which 192 brain volumes were acquired from each of 10 normal subjects. Subjects listened passively

Address for correspondence: Prof. T.D. Griffiths, Auditory Group, University of Newcastle Medical School, Framlington Place, Newcastle-upon-Tyne NE2 4HH, United Kingdom. Voice: +44 (0)191 222 3442; fax: +44 (0)191 222 6706.
t.d.griffiths@ncl.ac.uk

Ann. N.Y. Acad. Sci. 999: 212–214 (2003). © 2003 New York Academy of Sciences.
doi: 10.1196/annals.1284.032

to sound sequences with: (1) fixed pitch chroma, fixed pitch height; (2) changing pitch chroma, fixed pitch height; (3) changing pitch height, fixed pitch chroma; and (4) changing pitch chroma, changing pitch height. Images were preprocessed and analyzed using statistical parametric mapping implemented in SPM99 software (http//:www.fil.ion.ucl.ac.uk/spm).

RESULTS

In a fixed effects analysis of the group data, significant activation was demonstrated in each of the key contrasts at the $P <0.05$ voxel level of significance after correction for multiple comparisons over the entire brain volume. Changing pitch chroma minus fixed chroma and changing pitch height minus fixed height both activated lateral Heschl's gyrus (HG) and anterolateral planum temporale (PT) (FIG. 1A). After masking to exclude voxels activated by both contrasts, chroma change produced specific activation extending forward from HG into planum polare (PP) bilaterally (FIG. 1B), whereas height change produced specific activation extending posteriorly into PT bilaterally (FIG. 1C). Individual subject analyses consistently revealed a similar pattern of activation to that of the group.

CONCLUSIONS

Our results demonstrate that human pitch processing is hierarchical: common processing of pitch dimensions occurs in lateral HG and PT; specific processing of pitch chroma information occurs anterior to HG in PP, consistent with auditory information analysis that is not source dependent;[5,6] and specific processing of pitch

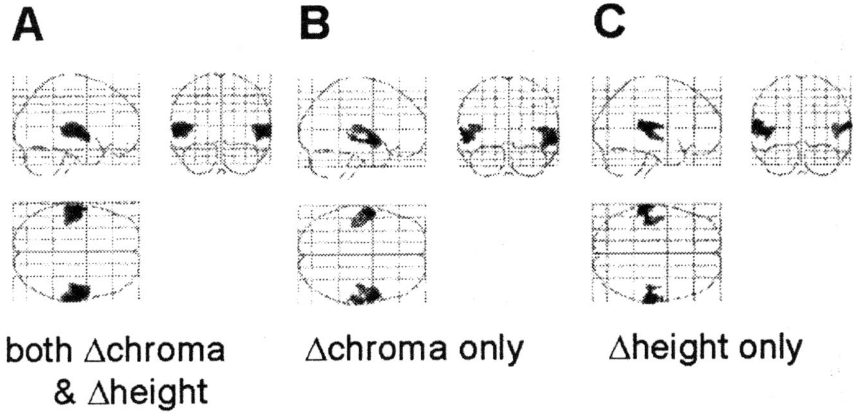

FIGURE 1. Group statistical parametric maps rendered as "glass brain" projections with voxel significance threshold $P <0.05$ (corrected for multiple comparisons over the whole brain volume). In the contrasts of changing pitch chroma minus fixed chroma and changing pitch height minus fixed height, voxels shown were activated: **(A)** *both* by chroma change *and* by height change; **(B)** *only* by chroma change; and **(C)** *only* by height change.

height information occurs in posterior PT, supporting a role for PT in sound object identification.[7]

Both monkey[8] and human[9] data suggest distinct "what" and "where" cortical auditory processing streams; however, this is disputed.[10] We suggest that the human homologues of macaque anterior belt areas process acoustic information streams, whereas the posterior human areas segregate sound objects. Furthermore, we propose a dual organization of the human auditory cortices: anterior areas process information streams (such as musical melody and speech prosody), whereas posterior areas initiate auditory scene analysis, crucial for the identification of sound objects (such as voices and musical instruments).

ACKNOWLEDGMENTS

This work was supported by the Wellcome Trust (J.D.W. and T.D.G.) and by UK Medical Research Council Grant G9901257 (S.U. and R.D.P.). We thank R. Frackowiak for helpful discussions.

REFERENCES

1. HELMHOLTZ, H.L.F. 1875. On the Sensations of Tone as a Physiological Basis for the Theory of Music, transl. A.J. Ellis, Ed. Longman's. London.
2. PANTEV, C. *et al.* 1989. Tonotopic organization of the auditory cortex: pitch versus frequency representation. Science **246:** 486–488.
3. SHEPARD, R.N. 1982. *In* The Psychology of Music. D. Deutsch, Ed. :344–390. Academic Press. New York.
4. PATTERSON, R.D. 1990. The tone height of multiharmonic sounds. Mus. Percept. **8:** 203–214.
5. ZATORRE, R.J., A.C. EVANS & E. MEYER. 1994. Neural mechanisms underlying melodic perception and memory for pitch. J. Neurosci. **14:** 1908–1919.
6. SCOTT, S.K., C.C. BLANK & R.J.S. WISE. 2000. Identification of a pathway for intelligible speech in the left temporal lobe. Brain **123:** 2400–2406.
7. GRIFFITHS, T.D. & J.D. WARREN. 2002. The planum temporale as a computational hub. Trends Neurosci. **25:** 348–353.
8. TIAN, B. *et al.* 2001. Functional specialization in rhesus monkey auditory cortex. Science **292:** 290–293.
9. MAEDER, P.P. *et al.* 2001. Distinct pathways involved in sound recognition and localization: a human fMRI study. NeuroImage **14:** 802–816.
10. BELIN, P. & R.J. ZATORRE. 2000. 'What', 'where' and 'how' in auditory cortex. Nature Neurosci. **3:** 965–966.

Part II: Brain Sciences versus Music

Introduction

DIEGO MINCIACCHI

Department of Neurological and Psychiatric Sciences, University of Florence, Florence, Italy

KEYWORDS: **brain; science; music**

As testified from the publication of *Music and the Brain*[1-3] to the more recent titles, *The Origin of Music* and *The Biological Foundations of Music*, the bridge between neurosciences and music has been consolidated. Brain sciences now propose detailed anatomical and physiological working models of the auditory system, and music offers eloquent results in perception research and psychoacoustics and also produces sound objects grounded on neuroscientific facts.

It is the first time, however, in the present book and especially in this section, that real brain research and real music come to a direct confrontation by bringing together researchers and musicians with different experiences and knowledge to stimulate possible cooperative development.

The opening contribution by Jones on the anatomical organization and functional properties of ascending auditory pathways traces the complex architecture and plan for the parallel processing of sound information in the central nervous system. After a thorough excursion into the brain circuits from which sound information flows, Jones comes to the thought-provoking conclusion that experimental studies of simple sound parameters, such as frequency analyses, are now of limited use and calls for the investigation of complex auditory behaviors using sounds, including music, that are more biologically relevant.

The contribution by Bentivoglio, on the other hand, reviews in detail the history of the correlation between musical skill and brain function. She clearly delineates the major key achievements and emphasizes the interpretational difficulties derived from the different theoretical approaches and the technologies available at different times. Remarkably, this sharp rendering of sparse and distant notions comes with a warning of the risk of the new "localizationist" approach to brain function offered by current functional neuroimaging studies.

Address for correspondence: Dr. Diego Minciacchi, Department of Neurological and Psychiatric Sciences, University of Florence, Viale Pieraccini 6, I-50134, Florence, Italy. Voice: +39-055-4279788; fax: +39-055-290662.

diego@neuro.unifi.it

Ann. N.Y. Acad. Sci. 1: 215–217 (2003). © 2003 New York Academy of Sciences.
doi: 10.1196/annals.1284.068

Additional challenges to the view of music as simply a composite result of elementary components (basically those derived from traditional assumptions such as, for the temporal domain, the opposition time/rhythm) are evident in the contribution by the Altenmüller group. A comparison of cortical activation patterns during the estimation of global and local musical time structures demonstrates the complexity of auditory cortical processing, which is far more prominent than the local versus global dichotomies in vision that have been described.

One of the leading organic systems to realize interactive music on the World Wide Web was described by Duckworth. The portrait of actual sonic possibilities, future implementation plans, and implications of perceptual and cognitive perspectives allows, even at this early stage, a basic description of a new form of sound communication, called virtual music. The natural inclusion of properties such as universal access, interactive participation, individual and unpredictable fruition, all-inclusive content, and multimedial character discloses a host of possibilities for the use and study of virtual music.

With its focus on musical forms, the contribution by Rosenboom explores the realm of self-organizing musical structures as they originate, coupling the brain to external musical systems. The paper presents a survey of several projects in which the author explores self-organizing musical forms in feedback paradigms using EEG information. The implications for music learning and active imaginative listening are discussed, and propositional music, music composed with a view towards including cognitive models as an intrinsic part of the creative activity, is introduced.

Roads approaches the issue of musical forms from the opposite side, exploring the sphere of sound particles or microsounds. Experimental music can now be shaped almost at the "atomic" level, and sounds used can be near the threshold of auditory perception. The paper describes the correlation of microsound properties with human perception of the temporal, intensity, and pitch domains, including pre-attentive and subliminal perception, and discusses the musical significance of microsounds.

Finally, I present a novel approach to the exploration of the universalities in the musical message. Information on the structure and function of the brain is used directly to build up sounds and musical forms. Relevant results are described, and major implications of the construction of sound objects are discussed.

The neurosciences and music are in a positive confrontation through the on-field experiences of researchers and musicians. The mass of knowledge brought together in this section gives this topic a unique perspective. Whereas from certain aspects the present scene would appear attractively seminal, the potential actually exists for a profound reassessment of methods to study the musical operations of the brain and to characterize the progressive compositional parameters that are applicable in music.

The dream expressed by Macrobius in the *Commentarium in somnium Scipionis* by Cicero becomes a little closer: "...omnis habitus animae cantibus gubernatur [...] ...dat somnos adimitque, nec non curas et immittit et retrahit, iram suggerit, clementiam suadet, corporum quoque morbis medetur, nam hinc est quod aegris remedia praestantes praecinere dicuntur." ("...Every soul condition is governed by the song [...] (it) induces sleep and awakes, similarly it gives and removes troubles, it excites anger, advises mercy, it is also remedy for body illnesses, hence who offers relief to sick people is named singer.")[4]

REFERENCES

1. CRITCHLEY, M. & R.A. HENSON, Eds. 1977. Music and the Brain. William Heinemann Med. B. Ltd. London.
2. WALLIN, N.L., B. MERKER & S. BROWN, Eds. 2000. The Origin of Music. MIT Press. Cambridge, MA.
3. ZATORRE, R.J. & I. PERETZ, Eds. The Biological Foundations of Music. 2001. Ann. N.Y. Acad. Sci. **930:** 1–462.
4. THEODOSIUS, A.M. 1963. Commentarium in Somnium Scipionis (2.3, 9). J. Willis, Ed. :105. B.G. Teubner. Leipzig.

Chemically Defined Parallel Pathways in the Monkey Auditory System

EDWARD G. JONES

Center for Neuroscience, University of California, Davis, Davis, California 95616, USA

ABSTRACT: The pathways ascending through the brain stem to the medial geniculate complex of the thalamus can be distinguished by immunostaining for the calcium binding proteins parvalbumin and calbindin and by the properties of the neurons in the subdivisions of the medial geniculate complex in which they terminate. The parvalbumin pathway, ascending from the central nucleus of the inferior colliculus, is the more direct and terminates in the ventral nucleus. The calbindin pathway is more diffuse in its origins and terminates in the dorsal and medial nuclei. Ventral nucleus neurons are sharply tuned, tonotopically organized and consistent in their responses. They project to core areas of the auditory cortex characterized by high parvalbumin immunoreactivity and by similar neuronal properties. Neurons in the dorsal and medial nuclei are not frequency specific or tonotopic and are labile in their responses. They project more diffusely to belt areas of the auditory cortex in which parvalbumin immunoreactivity is reduced and in which neuronal responses are less specific than in the core. The belt areas are the origins of streams of corticocortical connections leading into the temporal, parietal, and frontal lobes. These routes can be differentially engaged in functional imaging studies of monkeys responding to biologically significant sounds.

KEYWORDS: calbindin; parvalbumin; tonotopicity; pitch perception; complex sounds; medial geniculate complex; thalamocortical connections; auditory cortex; corticocortical connections; PET imaging

INTRODUCTION

Electrophysiologic studies of the upper levels of the auditory system have been dominated by the characterization of neurons in terms of their characteristic (or "best") frequencies and by their organization in tonotopically organized ensembles. Although this type of analysis may reveal those neurons and pathways that will be important in pitch perception, the understanding of the neuronal systems that underlie the ability to appreciate complex sound patterns such as music will require analysis at a greater level of complexity. In what follows, I review the organization of the medial geniculate complex of the monkey thalamus and its projection areas in the cerebral cortex from the perspective of recent immunocytochemical studies that permit the delineation of parallel pathways running from brain stem to cortex and in

Address for correspondence: Dr Edward G. Jones, Center for Neuroscience,1544 Newton Court, Davis, CA 95616. Voice: 530-757-8747; fax: 530-754-9136.
ejones@ucdavis.edu

Ann. N.Y. Acad. Sci. 999: 218–233 (2003). © 2003 New York Academy of Sciences.
doi: 10.1196/annals.1284.033

relation to correlative studies of the properties and connectivity of neurons in thalamus and cortex that may make them candidates for engaging in the perception of complex sounds.

STRUCTURE OF THE MEDIAL GENICULATE COMPLEX

The medial geniculate body of the thalamus is the gateway to the centers for auditory perception in the cerebral cortex. In all mammals, including primates, it is a complex of three major nuclei (FIGS. 1 and 2): ventral or principal, dorsal or posterior, and medial or magnocellular.[1] The ventral nucleus consists of medium-sized, tightly packed, and darkly staining cells that have a tendency to form dorsoventral or obliquely oriented rows. The dorsal nucleus caps the ventral nucleus and expands posteriorly to fill the whole cross-sectional area of the medial geniculate body. It consists of small- to medium-sized cells that are paler staining and less densely packed than those in the ventral medial geniculate nucleus. The medial or magnocellular nucleus contains relatively few cells, the most obvious of which are the large, deeply staining cells that give the nucleus its name, but it contains populations of smaller cells as well.

The ventral nucleus and commonly an anterior division of the dorsal nucleus are characterized by the fact that most of their cells are immunoreactive for the calcium-binding protein parvalbumin and both possess a densely parvalbumin immunoreactive neuropil[2] (FIG. 1).

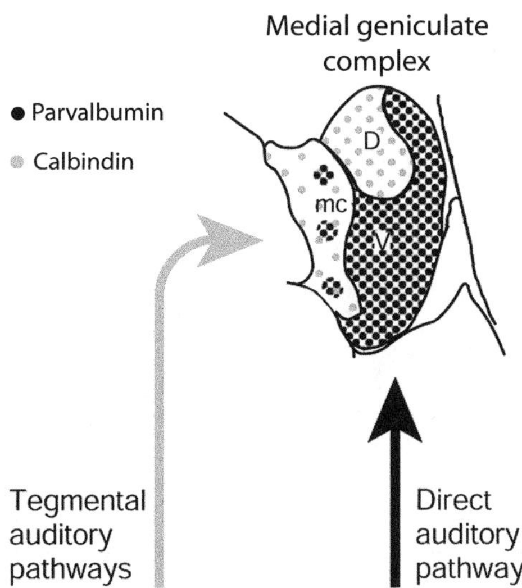

FIGURE 1. A frontal section through the middle of the medial geniculate complex of a macaque monkey, showing its ventral, dorsal, and medial or magnocellular divisions, each characterized by different patterns of cellular expression of the calcium binding proteins, parvalbumin and calbindin, and each innervated by brain-stem pathways in which the fibers also express the same proteins. Based on Molinari *et al.*[2] and unpublished work.

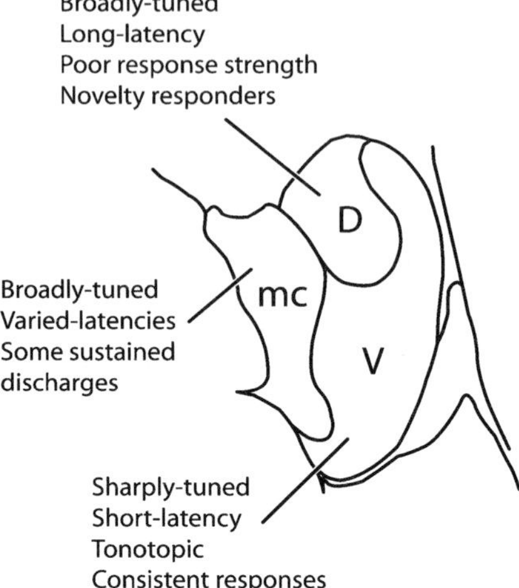

FIGURE 2. The typical responses to auditory stimuli of neurons in the different divisions of the medial geniculate complex. Based on work by Calford[10] in the cat.

The remainder of the dorsal nucleus contains more widely dispersed cells, only a few of which are parvalbumin immunoreactive. It also lacks a parvalbumin-dense neuropil, but it contains large numbers of neurons stained for another calcium-binding protein, 28-kDa calbindin[2] (FIG. 1).

The magnocellular nucleus contains large cells that are calbindin positive separated by islands of smaller parvalbumin-positive cells (FIG. 1).

PARALLEL PATHWAYS IN THE AUDITORY BRAIN STEM AND THALAMUS

The ascending input to the ventral nucleus is the most direct (oligosynaptic) route by which ascending auditory influences reach the medial geniculate complex and cerebral cortex. It arises in the central nucleus of the inferior colliculus of the ipsilateral side and ascends in the brachium of the inferior colliculus to the ventral nucleus. Its fibers are all parvalbumin immunoreactive (FIG. 1). A weaker input arises from the contralateral central nucleus (reviewed in Rouiller[3] and de Ribaupierre[4]). The ventral cochlear nuclei, the trapezoid body and lateral lemniscus and their associated nuclei, the central nucleus of the inferior colliculus, and the brachium of the inferior colliculus all stand out in the monkey brain stem by their heavy parvalbumin immunostaining for cells and/or fibers.

The parvalbumin-immunoreactive fibers of the brachium enter the ventral and anterodorsal medial geniculate nuclei and form the dense parvalbumin-positive neuropil of these nuclei. The remainder of the dorsal nucleus, by contrast, contains very weak neuropil staining, implying that few parvalbumin-positive fibers arising in the ventral nucleus of the inferior colliculus terminate there. Instead, the posterodorsal nucleus contains a weakly calbindin-immunoreactive neuropil, made up of calbindin-positive fibers entering from the lateral midbrain tegmentum and probably arising in large part from the calbindin cells that predominate in the pericentral and external nuclei of the inferior colliculus. The magnocellular nucleus receives local concentrations of parvalbumin- or calbindin-rich fibers deriving from both ascending pathways.

Fibers enter the ventral nucleus in serial order from the brachium of the inferior colliculus, giving the ventral nucleus an organizational pattern that can be viewed as a series of lamellae.[5] These lamellae are the basis for the tonotopic representation found in the nucleus. Fibers conveying impulses related to low-pitched sounds terminate laterally and those conveying impulses related to higher-pitched sounds terminate progressively more medially, thus setting up the tonotopic organization demonstrable in the nucleus by physiological recording techniques.[6–8] Most neurons of the central nucleus of the inferior colliculus have sharp frequency tuning, and the nucleus is tonotopically organized.[3] This organization is thus projected onto the ventral nucleus of the medial geniculate complex.

The other nuclei of the medial geniculate complex are the end stations for ascending fibers that come from more diverse sources and that ascend somewhat diffusely in the lateral midbrain tegmentum. These pathways are dominated by fibers that are calbindin immunoreactive (FIG. 1). Many of the inputs conveyed in these fibers are polysynaptic. The pericentral and external nuclei of the inferior colliculus and deep brain-stem regions adjacent to the inferior colliculus are the major sources of fibers to the dorsal nuclei, but other fibers may come directly from lower brain-stem sources as well. These nuclei of the inferior colliculus and adjacent regions that project to the dorsal nuclei are not tonotopically organized, and their neurons display labile properties and broad frequency tuning, comparable to neurons in the dorsal medial geniculate nucleus.[3]

The magnocellular medial geniculate nucleus, although primarily auditory in its connectional relationships, is also to some extent a region in which ascending auditory, medial lemniscal, spinothalamic, and vestibular fibers converge, although this may pertain only to cells in anterior parts of the nucleus.[9,10] The inferior colliculus and deep layers of the superior colliculus are the major sources of input to the magnocellular nucleus. The fibers are a mixture of parvalbumin- and calbindin-immunoreactive types (FIG. 1). Widely scattered cells in the external and parts of the central nuclei of the inferior colliculus may be the source of collicular afferents.[11] Because the external nucleus appears to be a region of convergence of auditory and dorsal column-lemniscal fibers, this projection may account for the somatosensory properties of some cells in the magnocellular medial geniculate nucleus and adjacent regions.[12]

PROPERTIES OF NEURONS IN DIVISIONS OF THE MEDIAL GENICULATE COMPLEX

Neurons in the ventral nucleus of the medial geniculate complex that receive their inputs from the direct brain-stem pathway are tightly coupled to their sources of cochlear input and are tonotopically organized. Neurons in the other nuclei that receive their inputs from the diffuse pathways display far more labile or complex properties and are less well organized tonotopically, if at all[10,11] (FIG. 2).

A cell in the ventral medial geniculate nucleus will respond to wide ranges of pure tone stimuli delivered to one or both ears; however, at intensities of sound pressure close to threshold, the cell responds only to a certain "best frequency" that is characteristic for that cell.[13] In the ventral division, cells with similar best frequencies tend to be arranged in lamellae stretching through most of the anteroposterior extent of the nucleus. The lamellae are arranged sequentially in tonotopic order so that a microelectrode advanced from lateral to medial across the ventral nucleus in cats and monkeys encounters cells whose best frequencies change systematically from low to high, except in the medial part of the cat ventral nucleus where the coiling of the lamellae makes tonotopicity difficult to demonstrate and where the higher frequency lamellae may be interrupted.[8] The representation of high frequencies medially is expanded in comparison with that of low frequencies, which are packed into the lateral few hundred microns.

Neurons in the ventral medial geniculate nucleus have extremely uniform properties that provide a basis for the relay of acoustic information to the cerebral cortex with little decrement in temporal or frequency-specific properties. They have high levels of spontaneous activity, sharp frequency tuning, and tonotopic organization and the shortest latency responses to acoustic stimuli.[10]

The sharply frequency-tuned cells of the ventral medial geniculate nucleus are influenced by stimuli applied to either ear: the effects mediated by the two ears can both be excitatory (EE cells) or stimulation of one ear may cause excitation and that of the other, inhibition (EI cells). EE cells, however, predominate,[10] and the effect mediated by the contralateral ear always leads that mediated by the ipsilateral ear, reflecting the fact that the most direct pathway to the ventral nucleus is the crossed one from the contralateral cochlear nuclei,[3] the predominance of contralateral inputs being already established in the central nucleus of the inferior colliculus.

In monkeys, the stimulus-response properties of ventral nucleus cells appear very similar to those of the cat.[7] In the awake squirrel monkey,[14] most neurons respond to a range of species-specific vocalizations, but they display complex, time-locked, and neuron-specific responses to the calls. These may include inhibitory and excitatory patterns that follow sequences in the stimulus and selective responses to modulations in amplitude of certain frequencies or to the direction of frequency modulations in a call.

Tonotopicity has not been described in the dorsal and magnocellular nuclei of the medial geniculate complex. Many cells encountered in the dorsal nucleus of the cat do not respond to auditory stimuli at all, and those that do have very labile responses[10,11] (Fig. 2). They respond at long latency (up to 100 ms), usually to a broad range of frequencies, often irregularly, and display little or no frequency tuning. Commonly, they habituate to repetitive stimuli, but they can be novelty sensitive and restore their firing patterns in response to small variations in a stimulus. The weakness of the response may reflect the multiple relays along the auditory pathways that provide input to these nuclei as well as an unusual sensitivity to anesthesia.

The medial or magnocellular nucleus in most species contains cells whose properties seem to reflect the presence of the populations of cells of different sizes, different chemistry, and with different cortical projections (Figs. 1, 2, and 6). Many cells of the magnocellular nucleus are sharply tuned with best frequency curves resembling those of cells in the ventral nucleus. They have short latency responses to pure tone stimuli and show evidence of binaural interaction, with the leading effect being mediated by the contralateral ear.[10,11] Others, however, if they respond at all, respond to wide ranges of frequency and have labile, long latency responses (FIG. 2). Some cells in the magnocellular nucleus that respond to auditory stimuli may also respond to tactile and vibratory stimuli.[9,12]

CHEMICAL ARCHITECTURE OF THE AUDITORY CORTEX IN MONKEYS

The parcellation of the auditory cortex of macaque monkeys, originally made by Merzenich and Brugge[15] on the basis of multiunit mapping, has been modified in other multiunit mapping studies,[16–21] which have more recently been allied with immunocytochemical[22] or histochemical[23] investigations, which have led to adjustments in the extent of areas and in the placement of areal borders, along with a series of new names. Pandya and Sanides[24] originally proposed that the auditory areas on the supratemporal plane of the primate cortex could be visualized as consisting of a primary auditory core area of distinctive cytoarchitecture and a surrounding belt of secondary auditory areas. The core is distinguished by dense immunoreactivity for parvalbumin, reflecting the dense input to this region from the ventral nucleus of the medial geniculate complex (FIG. 3). The surrounding belt areas show diminishing intensity of parvalbumin immunostaining as they become further and further removed from the core (FIG. 3). Mapping studies have shown that the core consists of an AI area, with a tonotopic representation in which high frequencies are represented posteriorly and low frequencies anteriorly. At its anterior border and still forming part of the core is the field originally called RL by Merzenich and Brugge[15] and now usually called R.[20–22] There is a reversal in the progression of best frequencies at the border, implying that the R field contains a tonotopic organization that is reversed with respect to that in AI (FIG. 4). Both components of the core (AI and R) are distinguished not only by a highly granular cytoarchitecture but also by dense immunostaining for the calcium binding protein parvalbumin[22] and by dense histochemical staining for acetylcholinesterase.[20,23] The fields of the surrounding belt have not been mapped in their entirety, but reversals in tonotopic sequences, connection tracing, and differences in the pattern of immunostaining for parvalbumin suggest the presence of multiple fields (FIGS. 3 and 4). There is no complete agreement on their exact extents, the names they should be called, or their connectivity, but a consensus is gradually emerging.

Around the two core areas, Morel and Kaas[19] and Hackett et al.[23,25] recognized a rostromedial field lying medial to the primary core, a field called rostrotemporal lying anterior to the core, two fields (anterolateral and posterolateral) lying lateral to the core, and both caudal and caudomedial fields lying caudal to the core. Rauschecker et al.[26] showed reversals in tonotopic representations along the lateral aspect of the supratemporal plane that suggested the presence of three auditory fields, which they called anterolateral, middle lateral, and caudolateral. They probably correspond to the anterolateral, posterolateral, and caudal fields of Morel et al.[20]

Jones et al.[22] and Molinari et al.,[2] in dividing the areas of the supratemporal plane of the macaque monkey cortex on the basis of different immunocytochemical staining for parvalbumin, found that the number of cortical neurons stained for parvalbumin does not vary greatly from area to area and that the major differences in the

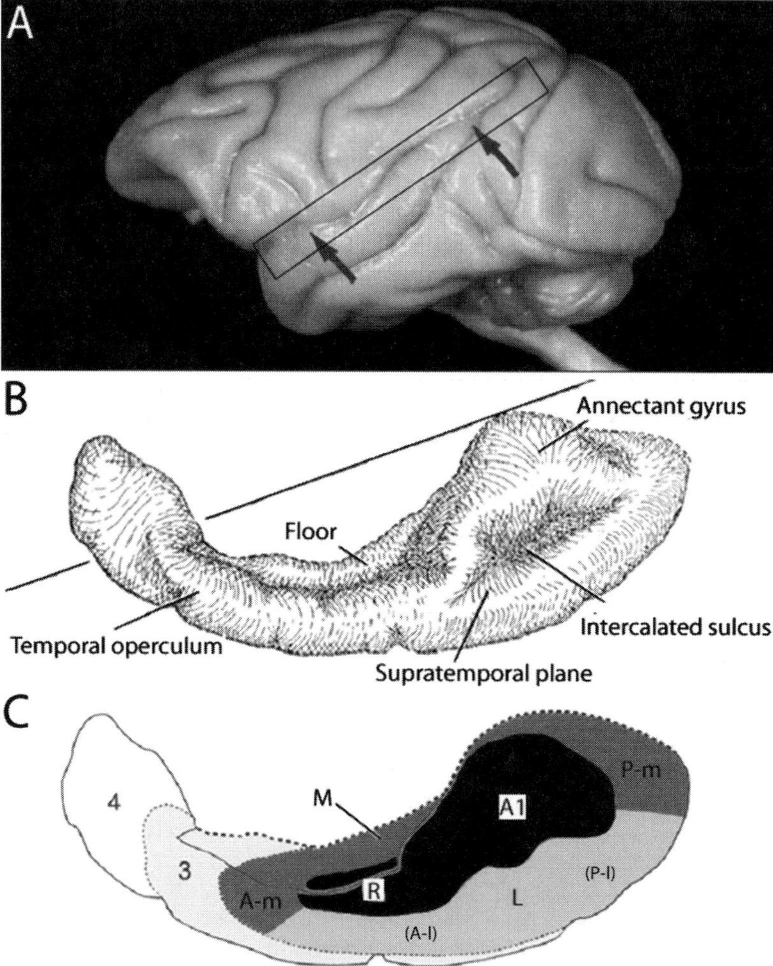

FIGURE 3. (**A**) Lateral view of the brain of a macaque monkey, showing the lateral sulcus (*boxed area*) and the location of auditory cortical areas (*arrows*). (**B**) Drawing of the surface of the lower bank of the lateral sulcus, showing the supratemporal plane and the annectant gyrus, an incipient Heschl's gyrus, on which the primary auditory areas are located. (**C**) Reconstruction of the same region as **C**, showing the locations of the core region of heaviest parvalbumin immunostaining and the surrounding belt areas in which parvalbumin immunostaining progressively declines. Based on Jones *et al.*[22]

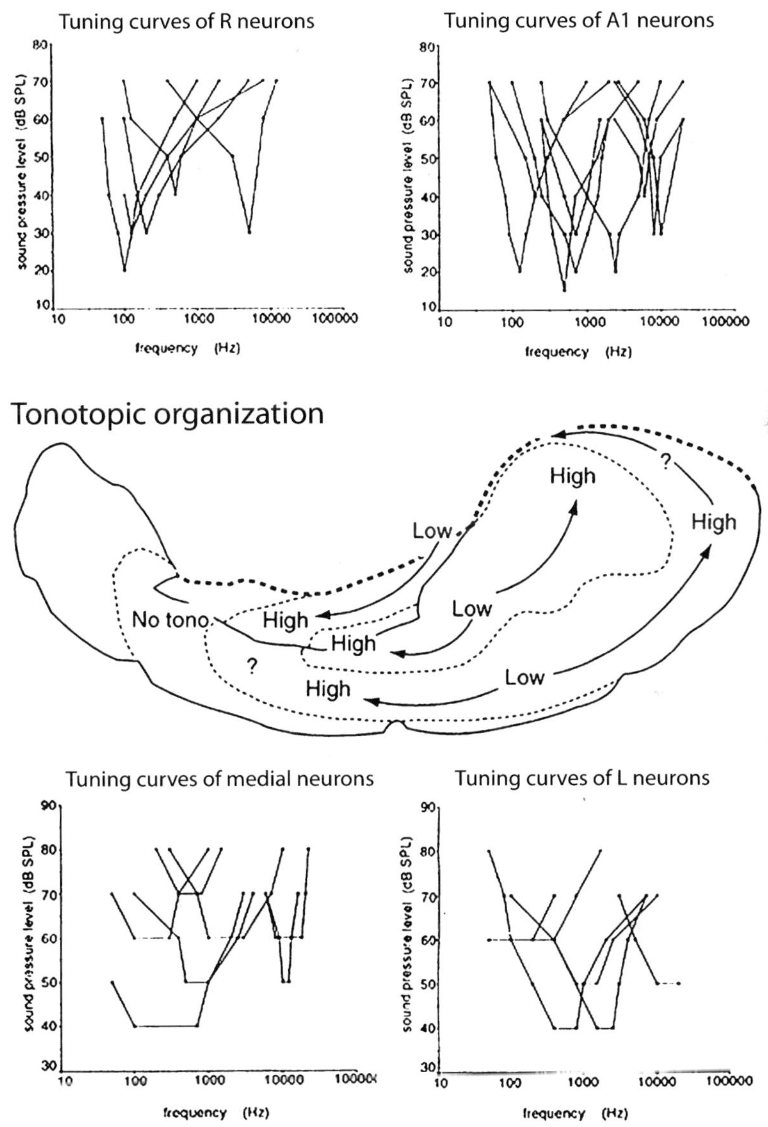

FIGURE 4. Tonotopic organization in the core ad belt areas, based on the results of multiunit mapping of responses to pure tone stimuli (*middle figure*), and the tuning curves of individual neurons in the two core areas (R and A1, above) and in two belt areas ((below). Based on Kosaki *et al.*[21]

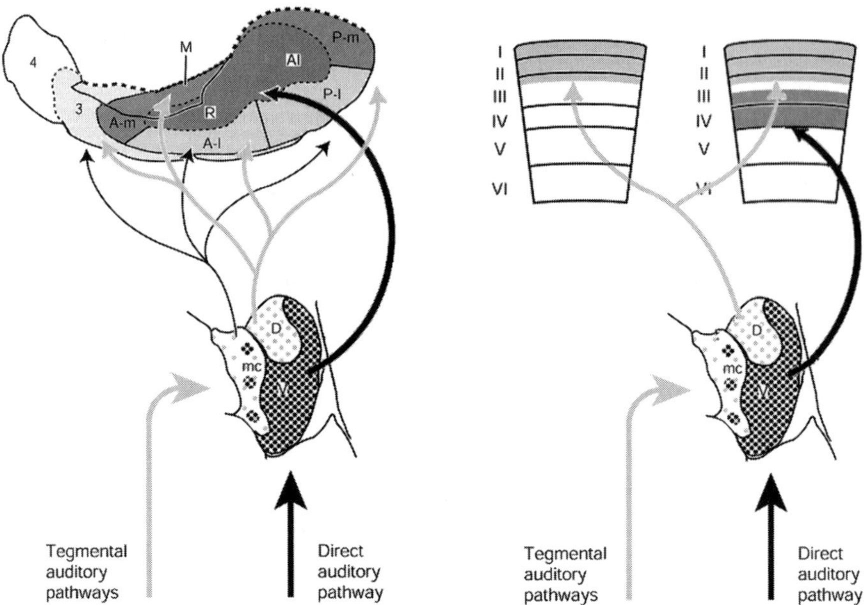

FIGURE 5. (*Left*) Projections of the different divisions of the medial geniculate complex of the macaque monkey to the areas of the auditory cortex. (*Right*) Laminar distributions of parvalbumin- and calbindin-immunoreactive cells in the divisions of the medial geniculate complex. Nuclei dominated by calbindin cells project diffusely and their axons terminate in superficial layers of the cortex; nuclei dominated by parvalbumin cells project in topographic fashion to selected areas and their axons end in middle layers. Based on Molinari *et al.*[2] and Hashikawa *et al.*[28,29]

intensity of immunostaining depend on the degree of staining of fiber plexuses in layers III, IV, and VI. The stained plexuses represent mainly thalamocortical fibers, and their different intensity reflects the relative proportions of parvalbumin immunoreactive cells in the medial geniculate complex projecting to each area (FIG. 5). Four principal zones can be distinguished on the supratemporal plane: a central region of densest immunostaining, more or less coextensive with the cortex of the small annectant gyrus commonly found on the posteromedial aspect of the supratemporal plane in the larger macaques, coincides with the auditory koniocortex. Surrounding this core zone is a zone of moderately dense immunostaining, which is heavier anteromedial and posteromedial to the core and somewhat less dense lateral to the core. Embracing the second zone is a third zone of much weaker parvalbumin immunostaining, extending out onto the surface of the superior temporal gyrus and around the anterior aspect of the second zone. The third zone is embraced, in turn, by an outermost zone in which parvalbumin immunostaining of fiber plexuses is essentially absent. It occupies the anterior end of the temporal operculum and extends to the temporal pole and over the surface of the middle and inferior temporal gyri.

In mapping the parvalbumin-immunostained regions on the basis of best frequency responses to pure tone stimuli, Kosaki *et al.*[21] demonstrated that the central core zone of densest parvalbumin immunoreactivity contained two representations of the auditory frequency range. High frequencies (>20 kHz) are represented posteromedially. On moving anteriorly, there is a gradual reduction in the frequencies represented, then a reversal and a progressive increase until, at the anterior border of the core, neurons respond again to frequencies as high as 20 kHz. This reversal in the representation enables fields AI and R to be delineated within the primary core region.

Reversals of frequency representation in the lateral division of the zone of somewhat reduced parvalbumin immunostaining surrounding the central core permits the identification of two fields probably corresponding to the anterolateral and posterolateral fields of Morel et al.[20] and to the anterior and middle fields of Rauschecker et al.[26] A posteromedial region, which probably contains the posterior field of Rauschecker et al., could not be divided into fields corresponding to the caudal and caudomedial fields of Morel et al., although there were suggestions of a tonotopic reversal in the middle of the posteromedial field.

A medial (M) field containing a tonotopic representation, with low frequencies represented posteriorly and high frequencies represented anteriorly, is confined to the floor and adjacent lateral bank of the inferior limiting sulcus of the insula and seems to be equivalent to the medial field of Merzenich and Brugge,[15] but it is much thinner than the rostromedial field of Morel et al.,[20] which is drawn as though extending for a considerable distance on to the insula. The anterior, high frequency representation in the M field is separated from the high frequency representation at the anterior end of the anterolateral field by a part of the surround zone in which tonotopic order is indistinct. This is the anteromedial (A-M) field, which may correspond to the rostrotemporal field of Morel et al.

Neurons in the fields lying lateral to the core fields are relatively unresponsive to pure tone stimuli and prefer more complex sounds, including species-specific calls.[21,26] Neurons here have broader tuning curves than do neurons in the AI and R core regions (FIG. 4), and those in the medial and anteromedial fields are essentially not tuned at all.[21] Outside the part of the belt immediately surrounding the core, in the outermost zones of weak or absent parvalbumin immunoreactivity, neurons cannot be driven by tonal stimuli, although they may respond to white noise.

THALAMOCORTICAL RELATIONSHIPS IN THE MACAQUE AUDITORY SYSTEM

The predominant input to fields AI and R is by far that from the ventral nucleus of the medial geniculate complex[2,17–20,25] (FIG. 5). The predominant inputs to the fields of the belt region are from the dorsal nucleus, and there appears to be a suborganization in which the major inputs to different areas of the belt come from different divisions of the dorsal nucleus.[2]

CHEMICALLY-DEFINED PARALLEL PATHWAYS IN THE MONKEY AUDITORY THALAMOCORTICAL SYSTEM

The parcellation of the monkey auditory cortex on the basis of parvalbumin-immunoreactive staining depends on the relative concentrations of parvalbumin-immunoreactive fibers in layer IV and deep layer III of each area.[22] The intensity of parvalbumin fiber staining is directly correlated with the relative concentrations of parvalbumin-immunoreactive cells in the ventral and dorsal nuclei of the medial geniculate complex and inversely correlated with the relative concentrations of cells immunoreactive for calbindin in the same nuclei[2] (FIG. 1). These differences among the nuclei are also correlated with the relative proportions of cells projecting to middle and superficial layers of the auditory cortex.

The vast majority of relay cells in the monkey ventral medial geniculate nucleus are parvalbumin immunoreactive (FIG. 1). The nucleus contains only a few, very widely dispersed calbindin cells. The anterior division of the dorsal nucleus contains a majority of parvalbumin cells, but the number of calbindin cells is increased. The posterior division of the dorsal nucleus contains approximately equal numbers of parvalbumin and calbindin cells anteriorly, but the posterior part of the nucleus, which caps the whole medial geniculate complex, contains almost exclusively calbindin-immunoreactive cells (FIG. 1). Apart from the ventral nucleus, the density of calbindin cells does not change greatly throughout the ventral and two dorsal nuclei, and they have been described as forming a diffuse matrix of cells akin to that found in the ventral posterior and ventral lateral nuclei.[27] The magnocellular nucleus contains islands of cells in which parvalbumin- or calbindin-positive types predominate.

Experiments involving injections of one retrogradely transported fluorescent dye into middle layers of the auditory cortical fields of monkeys and placement of a second dye on the surface to affect layers I and II show that parvalbumin cells project only to the middle layers and calbindin cells to the superficial layers[2,28] (FIG. 5).

Injections of tracers into middle layers of the dense parvalbumin-immunoreactive core fields AI and/or R invariably label a focus of parvalbumin-positive cells in the ventral nucleus. Surface applications of tracer to these fields label a few scattered calbindin cells in the ventral and dorsal nuclei and a few cells of both types in the magnocellular nucleus. Middle layer injections in the fields of the moderate-to-dense parvalbumin-immunoreactive belt around AI and R label a focus of parvalbumin cells in anterior or posterior divisions of the dorsal nucleus, while surface applications label scattered calbindin cells throughout the dorsal nucleus and in the magnocellular nucleus. Injections in the parvalbumin-weak fields outside these regions invariably label parvalbumin and calbindin cells in the magnocellular nucleus. Hence, parvalbumin cells form the basis of a topographically organized projection from the ventral nucleus to middle layers of the two core fields and from the dorsal nuclei to one or more of the immediately surrounding fields. The calbindin cells form the basis of a diffusely projecting system that terminates in superficial layers and is unconstrained by the borders of cortical fields. The magnocellular nucleus, although projecting to more than one field, also possesses this organization because its parvalbumin cells project to layer IV of single fields, whereas its calbindin cells project to layer I of multiple fields.[29] The monkey auditory thalamocortical system thus reflects the division of the subcortical auditory pathways into two chemically distinct parallel streams. One characterized by parvalbumin leads through the tonotopically organized nuclei to the primary core areas of auditory cortex and to the surrounding fields with which they are most closely connected. The other, characterized by the presence of calbindin, is more diffusely projected onto these and wider areas of cortex.

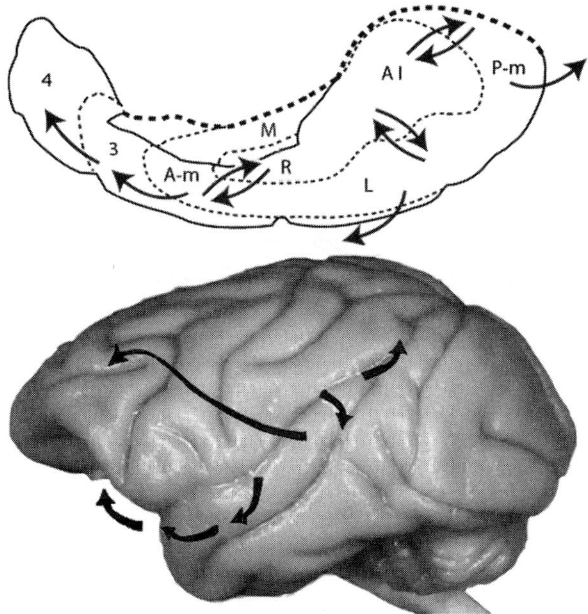

FIGURE 6. Corticocortical connections between the auditory areas of the supratemporal plane and the outflow pathways leading to temporal, parietal, and frontal lobes. Based on Jones *et al.*[22] and other published sources.

CORTICOCORTICAL PATHWAYS EMANATING FROM THE AUDITORY CORTEX

Various patterns of corticocortical connections link the central core auditory fields to those of the inner surround, and further projections connect the inner surround to outer regions and to more distant areas of cortex.[17–21,24,25,30] From the immediate surround, three streams of corticocortical connections can be traced into areas of association and limbic cortex[31,32] (FIG. 6). These include a stream arising from more anterolateral areas and passing towards orbital regions of prefrontal cortex and anterior entorhinal areas; a stream arising from lateral areas and passing towards lateral prefrontal and posterior entorhinal areas; and a stream arising from posterolateral areas and passing towards lateral prefrontal, posterior parietal, and posterior cingulate cortex. It is likely that these different streams are engaged in processing different aspects of auditory perception and with different aspects of auditory behavior.

IMAGING STUDIES OF AUDITORY PROCESSING IN MONKEY CEREBRAL CORTEX

In studies carried out in collaboration with Dr. T. Hashikawa and J. He, we used positron emission tomography (PET) to identify the areas of cortex in the Japanese

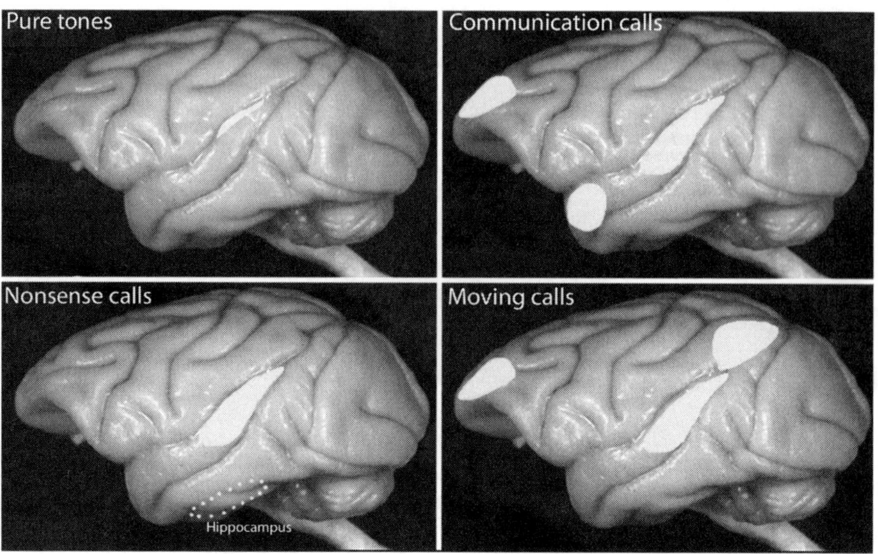

FIGURE 7. Regions of highest $H_2^{15}O$ uptake (*pale areas*) observed by positron emission tomography in sedated monkeys exposed to various auditory stimuli. From unpublished data of Hashikawa *et al.*[33]

macaque (*Macaca fuscata*) that showed highest activity in response to pure tone stimuli or to species-specific vocalizations[33] (FIG. 7).

Monkeys were anesthetized with nitrous oxide. Binaural stimuli were presented at the external auditory meatus. Brains were scanned in horizontal slices, and regional cortical blood flow was measured by uptake of $H_2^{15}O$.

As expected, presentation of pure tone stimuli resulted in enhanced regional blood flow in the primary core region of the auditory cortex. Subtraction of the signals elicited by stimuli of different frequencies (10 vs 200 Hz) showed that low tones were represented posteriorly and high tones anteriorly. No other areas of cortex showed significantly enhanced activity in response to pure tones. By contrast, the presentation of recordings of communication calls ("coo" sounds) led to activation of a large area of the superior temporal gyrus thought to represent both the lateral belt areas of the supratemporal plane and the more laterally located cortex that is the main target of corticocortical connections emanating from these areas. There was also activation of an area of prefrontal cortex located anteromedial to the principal sulcus, of an area at the anterior tip of the temporal lobe, as well as of the amygdala and striatum. Presentation of a nonsense sound (reversed "coo"), although activating the upper superior temporal cortex, did not lead to activation of the other cortical areas, although it revealed increased activity in the hippocampus. A coo sound, presented as though moving in space, resulted in enhanced activity in the same areas as the stationary coo sound, but there was also activation of the cortex of the inferior parietal lobule, an area of parietal cortex not active under the other stimulus conditions (FIG. 7). It is likely that these different patterns of cortical activation reflect differential involvement of the three or four corticocortical pathways leading outwards from the auditory areas of the supratemporal plane. The biologically significant sounds are clearly more effective in engaging these pathways than pure tonal stimuli.

CONCLUSIONS

Analysis of the auditory pathways in the upper levels of the neuraxis has lagged behind that of the other sensory systems. There has nevertheless been a slow advance of anatomical knowledge about the organization of the medial geniculate complex and about the division of the auditory cortex, especially that of monkeys, into a primary core area and surrounding belt areas containing multiple, tonotopically organized representations of the auditory periphery. Studies of chemically defined brain stem and thalamocortical pathways in monkeys have revealed the presence of two parallel pathways running through the medial geniculate complex that are probably functionally very distinct, although most data upon which this probability rests come from studies in cats rather than monkeys. The pathways running most directly from the contralateral ventral cochlear nucleus through the ventral medial geniculate nucleus to the core region of the auditory cortex are highly ordered tonotopically, and their neurons are highly reliable in their capacity to transfer short latency information relevant to pitch perception to the cerebral cortex. The parallel pathway ascending from multiple brain-stem sources, relaying in dorsal and medial nuclei of the medial geniculate complex and projected mainly to cortical regions surrounding the core area, is less highly ordered from the perspective of best frequency analysis and appears to convey information that may be of more subtle import than pitch. The seemingly more diffuse spread of thalamocortical fibers emanating from the medial and dorsal divisions of the medial geniculate complex and the corticocortical connections within the cortex ensure that information ascending in the less direct pathway can be brought together with that ascending in the more direct ventral nucleus pathway at an early state in cortical processing. In the core area itself where the terminations of thalamocortical axons from the medial/dorsal nuclei end in superficial layers and those from the ventral nucleus in the middle layers, the neurons of the core are likely to be detectors of coincident inputs arriving in the two pathways. The characteristics of neurons in the belt areas around the core areas and the projections of these belt areas into parietal, temporal, and frontal regions concerned with auditory behaviors far more complex than mere pitch discrimination make the homologous areas likely candidates for being involved in the perception of music in humans. Studies of the medial geniculate nuclei and auditory cortical areas on the basis of best frequency analysis and tonotopic organization may have reached a point of limited usefulness. The challenge is now to examine these brain regions from a perspective that will lead to a better understanding of their roles in the analysis of sounds and patterns of sounds including music that are of greater biological relevance.

REFERENCES

1. Jones, E.G. 1985. The Thalamus. Plenum. New York.
2. Molinari, M., M.E. Dell'Anna, E. Rausell, *et al.* 1995. Auditory thalamocortical pathways defined in monkeys by calcium binding protein immunoreactivity. J. Comp. Neurol. **362:** 171–194.

3. ROUILLER, E.M. 1997. Functional organization of the auditory pathways. *In* The Central Auditory System. G. Ehret & R. Raymond, Eds. :3–96. Oxford University Press. New York.

4. DE RIBAUPIERRE, F. 1997. Acoustical information processing in the auditory thalamus and cerebral cortex. *In* The Central Auditory System. G. Ehret & R. Raymond, Eds. :317–398. Oxford. New York.

5. MOREST, D.K. 1965. The laminar structure of the medial geniculate body of the cat. J. Anat. **99:** 143–160.

6. AITKIN, L.M. & W.R. WEBSTER. 1972. Medial geniculate body of the cat: organizations and responses to tonal stimuli of neurons in ventral division. J. Neurophysiol. **35:** 365–380.

7. ALLON, N., Y. YESHURUN & Z. WOLLBERG. 1981. Responses of single cells in the medial geniculate body of awake squirrel monkeys. Exp. Brain Res. **41:** 222–232.

8. IMIG, T.J. & A. MOREL. 1985. Tonotopic organization in ventral nucleus of medial geniculate body in the cat. J. Neurophysiol. **53:** 309–340.

9. POGGIO, G.F. & V.B. MOUNTCASTLE. 1960. A study of the functional contributions of the lemniscal and spinothalamic systems to somatic sensibility. Bull. Johns Hopkins Hosp. **106:** 266–316.

10. CALFORD, M.B. 1983. The parcellation of the medial geniculate body of the cat defined by the auditory response of single units. J. Neurosci. **3:** 2350–2364.

11. CALFORD, M.B. & L.M. AITKIN. 1983. Ascending projections to the medial geniculate body of the cat: evidence for multiple, parallel auditory pathways through thalamus. J. Neurosci. **3:** 2365–2380.

12. LEDOUX, J.E., D.A. RUGGIERO, R. FOREST, *et al.* 1987. Topographic organization of convergent projections to the thalamus from the inferior colliculus and spinal cord in the rat. J. Comp. Neurol. **264:** 123–146.

13. ROSE, J.E., N.B. GROSS, C.D. GEISLER & J.E. HIND. 1966. Some neural mechanisms in the inferior colliculus of the cat which may be relevant to localization of a sound source. J. Neurophysiol. **29:** 288–314.

14. ALLON, N. & Y. YESHURUN. 1985. Functional organization of the medial geniculate body's subdivisions of the awake squirrel monkey. Brain Res. **360:** 75–82.

15. MERZENICH, M.M. & J.F. BRUGGE. 1973. Representation of the cochlear partition on the superior temporal plane of the macaque monkey. Brain Res. **50:** 275–296.

16. IMIG, T.J., M.A. RUGGIERO, L. . KITZES, *et al.* 1977. Organization of auditory cortex in the owl monkey (*Aotus trigvirgatus*). J. Comp. Neurol. **171:** 111–128.

17. AITKIN, L.M., M. KUDO & D.R.F. IRVINE. 1988. Connections of the primary auditory cortex in the common marmoset, *Callithrix jacchus jacchus.* J. Comp. Neurol. **269:** 235–248.

18. LUETHKE, L.E., L.A. KRUBITZER & J.H. KAAS. 1989. Connections of primary auditory cortex in the New World monkey, *Saguinus.* J. Comp. Neurol. **285:** 487–513.

19. MOREL, A. & J.H. KAAS. 1992. Subdivisions and connections of auditory cortex in owl monkeys. J. Comp. Neurol. **318:** 27–63.

20. MOREL, A., P.E. GARRAGHTY & J.H. KAAS. 1993. Tonotopic organization, architectonic fields, and connections of auditory cortex in macaque monkeys. J. Comp. Neurol. **335:** 437–459.

21. KOSAKI, H., T. HASHIKAWA, J. HE & E.G. JONES. 1997. Tonotopic organization of auditory cortical fields delineated by parvalbumin immunoreactivity in macaque monkeys. J. Comp. Neurol. **386:** 304–316.

22. JONES, E.G., M.E. DELL'ANNA, M. MOLINARI, *et al.* 1995. Subdivisions of macaque monkey auditory cortex revealed by calcium binding protein immunoreactivity. J. Comp. Neurol. **362:** 153–170.

23. HACKETT, T.A., T.M. PREUSS & J.H. KAAS. 2001. Architectonic identification of the core region in auditory cortex of macaques, chimpanzees, and humans. J. Comp. Neurol. **441:** 197–222.

24. PANDYA, D.N. & F. SANIDES. 1973. Architectonic parcellation of the temporal operculum in rhesus monkey and its projection pattern. Z. Anat. Entwickl.–Gesch. **139:** 127-161.

25. HACKETT, T.A., I. STEPNIEWSKA & J.H. KAAS. 1998. Subdivisions of auditory cortex and ipsilateral cortical connections of the parabelt auditory cortex in macaque monkeys. J. Comp. Neurol. **394:** 475–495.
26. RAUSCHECKER, J.P., B. TIAN & M. HAUSER. 1995. Processing of complex sounds in the macaque nonprimary auditory cortex. Science **268:** 111–114.
27. JONES, E.G. 2001. The thalamic matrix and thalamocortical synchrony. Trends Neurosci. **222:** 500–520.
28. HASHIKAWA, T., E. RAUSELL, M. MOLINARI & E.G. JONES. 1991. Parvalbumin- and calbindin-containing neurons in the monkey medial geniculate complex: differential distribution and cortical layer specific projections. Brain Res. **544:** 335–341.
29. HASHIKAWA, T., M. MOLINARI, E. RAUSELL & E.G. JONES. 1995. Patchy and laminar terminations of medial geniculate axons in monkey auditory cortex. J. Comp. Neurol. **362:** 195–208.
30. PANDYA, D.N., M. HALLETT & S.K. MUKHERJEE. 1969. Intra- and interhemispheric connections of the neocortical auditory system in the rhesus monkey. Brain Res. **14:** 49–65.
31. PANDYA, D.N. & E.H. YETERIAN. 1990. Prefrontal cortex in relation to other cortical areas in rhesus monkey: architecture and connections. Prog. Brain Res. **85:** 63–94.
32. ROMANSKI, L.M., B. TIAN, J. FRITZ, *et al.* 1999. Dual streams of auditory afferents target multiple domains in the primate prefrontal cortex. Nat. Neurosci. **2:** 1136.
33. HASHIKAWA, T., J. HE, T. KAKIUCHI, *et al.* 1995. Functional activation of auditory cortex: a PET study in Japanese macaque. Neurosci. Abstr. **21:** 1178.

Musical Skills and Neural Functions

The Legacy of the Brains of Musicians

MARINA BENTIVOGLIO

*Department of Morphological and Biomedical Sciences,
University of Verona, Verona, Italy*

ABSTRACT: An overview of the history of debates on the correlation of musical skills with neurological functions in health and disease is presented. Selected biographical sketches of composers (Hildegard von Bingen, Mozart, Donizetti, Mussorgsky, and Ravel), whose neurological disease may have influenced musical creativity, are discussed. The search for information on the localization of skills in the brains of musicians is reviewed. The relation of mental ability to brain structure is a prominent theme in the history of neuroscience, and the effort to localize musical skills dates back to the excesses of phrenology in the early nineteenth century. The phrenological tables included an "organ of music" among the sites subserving intellectual capabilities, mapped on the basis of palpation of the head bumps. Since the second half of the nineteenth century, when the study of brain physiology, anatomy, and pathology had a remarkable development and impact on the neurosciences, structural features of the brain, particularly the cerebral cortex, of individuals with peculiar talents, including musical skills, have been examined to search for clues on the localization of mental phenomena. These studies, which continued in the twentieth century, are currently difficult to validate in view of the rigor imposed by current scientific standards. However, the issue of localization of functions in the musical brain is still debated and is now at the forefront of the neurosciences, exploiting especially functional neuroimaging. The historical overview of these problems is certainly exemplary of progress of knowledge, but also warns against excessive "localizationist" efforts.

KEYWORDS: history of neuroscience; music; phrenology; cerebral cortex; famous brains; localization of function

INTRODUCTION

The neural correlates of mental skills have been debated since the birth of the neurosciences and are still in the forefront of this field of research. Historically, this problem has been related to that of localization of functions in the brain, particularly the cerebral cortex, since the end of the nineteenth century, when the cortical mantle was accepted as the site of higher neural functions, including motor control and

Address for correspondence: Dr. M. Bentivoglio, Department of Morphological and Biomedical Sciences, Medical Faculty, Strada Le Grazie 8, 37134 Verona, Italy. Voice: +39-045-8027158; Fax: +39-045-8027163.
marina.bentivoglio@univr.it

Ann. N.Y. Acad. Sci. 999: 234–243 (2003). © 2003 New York Academy of Sciences.
doi: 10.1196/annals.1284.034

mental activity. Musical skills have attracted interest since the beginning of this debate, to which they gave the special flavor of a search for a structural substrate of creativity and unique artistic skills.

This overview attempts to provide a synthetic historical account of the debates on musical skills and neural functions. Neurological dysfunction and its relation to musical creativity are dealt with first, presenting selected medical biographical sketches of music composers. The localization of skills in the cerebral cortex of musicians is discussed from a historical point of view, starting from the phrenological approach in the early nineteenth century.

NEUROLOGICAL DYSFUNCTION AND MUSICAL CREATIVITY

Can music be the product of a diseased brain and/or be influenced by neurological disease? Can neurological dysfunction enlighten us about the neural substrate of musical skills? It is hard to provide definite answers to these questions in scientific terms, and there is the risk of falling into the rhetoric of genius and madness.[1] However, inasmuch as mental activity is the product of the brain, it should not be surprising that musical skills can be influenced by brain disease, but it may be surprising that such influence can be both positive and negative, as shown by the biographical stories of some musicians.

Hildegard von Bingen (1098-1179), a remarkable woman and a German prioress, produced major theological works, visionary writings, and music. She was the first medieval woman composer of sacred music, and most of her musical work has survived. Hildegard, known as the "Sybil of the Rhine," had visionary experiences and suffered from a nervous disorder. In her writings *Scivias* (*Scire* or *vias Dominis* or *vias lucis*), she described her visions, summing up the Christian doctrine and the history of salvation. Hildegard also authored medical writings, including a medical encyclopedia (*Liber simplicis medicinae*) and notes for a medical handbook (*Liber compositae medicinae*). Her visions, preceded by aura and followed by loss of consciouness and sickness, also inspired her music (handed down as *Carmina Hildegardis*) and the texts of her songs (for example, the hymns of *Symphonia harmonie celestium revelationum*). Hildegard may have suffered from severe migraine attacks, but most probably had epileptic fits or "hysteroepilepsy."[2]

Jumping ahead several centuries, the premature death of Wolfgang Amadeus Mozart (1756-1791), traditionally ascribed to infectious causes, was recently rediscussed and neurological disease was hypothesized.[3,4] This could have played a role in Mozart's well known premonitions of death and thus in the composition of the *Requiem in D minor* and other works. A skull believed to be Mozart's, saved by the successor of the gravedigger, exhibits a left temporal fracture, raising the question of chronic subdural hematoma. This would be consistent with several falls and could have caused the weakness, headaches, and fainting Mozart experienced in 1790 and 1791. Aggressive bloodletting to treat suspected rheumatic fever could have decompensated such a lesion to produce Mozart's death in December 1791.

The mental insanity of the Italian composer Gaetano Donizetti (1797-1848) has also been discussed in relation to his creativity.[5] Donizetti was affected by neurosyphilis (as were other musicians, including Franz Schubert[6]), which progressed over several years. In the last years of Donizetti's life, his mind had so deteriorated that he had to be admitted to an asylum, where he died in a state of mental derangement and physical paralysis. It has been suggested that the composer's disease influenced his ability to create powerful scenes of psychosis in his operas.[5] In particular, Donizetti portrayed in *Anna Bolena* the mental disorder of Anne Boleyn imprisoned in the tower of London, and in *Lucia di Lammermoor* he portrayed the psychosis of a girl with hallucinations. Donizetti died 13 years after *Lucia*'s premiere.

The Russian composer Modest Petrovic Mussorgsky (1839-1881) combined exceptional creative ability with psychiatric and neurological disorders (alcoholism, Wernicke's encephalopathy). Mussorgsky's skills have also been discussed in relation to his psychopathology.[7] His first important orchestral work, *Night on the Bare Mountain,* and his operas *Salambô* and *Boris Godunov* were written in between psychiatric and neurological episodes, until heavy drinking left him incapable of sustained creative effort.

With the French composer Maurice Ravel (1875-1937), one of the most extensively studied cases of a brain-damaged musician, neurological disease may have influenced his writing of music from the beginning, leading then to his inability to compose.[8–10] Ravel suffered from an undiagnosed progressive neurodegenerative disease (the brain was never examined, because permission for autopsy was not granted) that probably started at the end of the 1920s, before the creation of the *Bolero* and the two piano concertos (*Concerto for the Left Hand* and the *Piano Concerto in G*). It has been suggested that the fascinating musical perseveration of the *Bolero* may have been due to early symptoms of the composer's illness.[11] Ravel's neurological disease caused then Wernicke's aphasia, agrafia, alexia, and apraxia, indicating involvement of the left hemisphere. In the last years of his life, Ravel became incapable of writing music, although he had preserved intact perceptual auditory abilities and musical memory, indicating that these functions are subserved by different neural circuits.

PHRENOLOGY AND MUSIC

Localization of functions in the brain is a prominent theme in the history of science, neuroscience, and neurology, dating from ancient speculations, including Hippocratic writers, and later Galen, searching for the seat of the soul within the brain.[12] Modern speculation led René Descartes initially to select the pineal body as the seat of the soul, drawing attention, however, away from the brain surface, the cerebral cortex.[12–14] At the beginning of the nineteenth century, the work of Franz Joseph Gall (1758-1828), who had obtained a medical degree at the University of Vienna, brought back attention to the localization of functions at the cerebral surface.[12–16] Joined by Johann Caspar Spurzheim (1776-1832), Gall was the inventor of cranioscopy (or "organology," that developed into phrenology, a term introduced by Spurzheim). Gall established the idea that the brain was the organ of the mind, made up of multiple "organs," so that the different functions of the brain were situated in specific sites, each representing a mental or moral activity. As they were on the surface of the brain and were undergoing hypertrophy or atrophy with use, they could be detected through the skull, and the palpation of head bumps could thus reveal intellectual abilities and personality traits of an individual. Gall believed that there were 26 or 27 such organs, but his followers increased the numbers. The

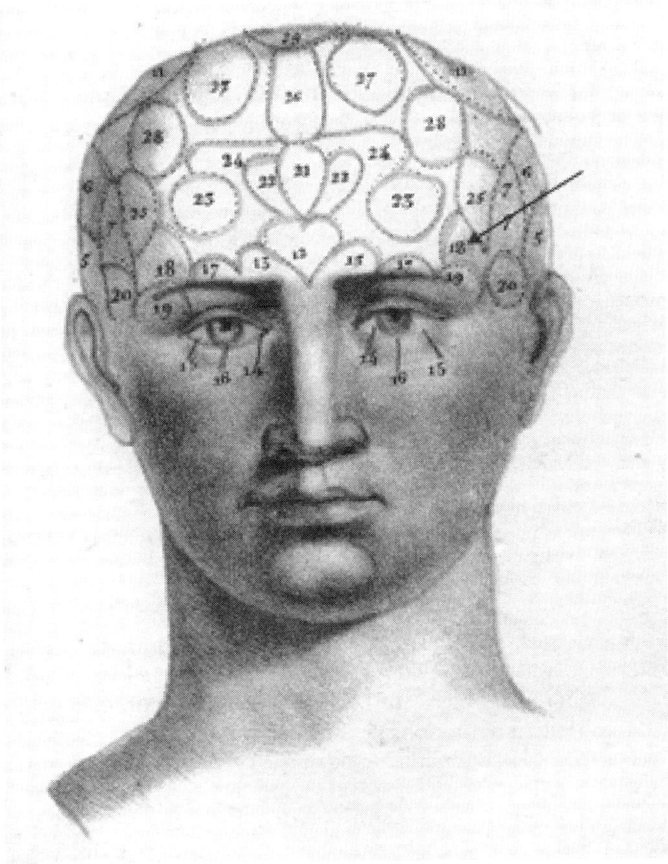

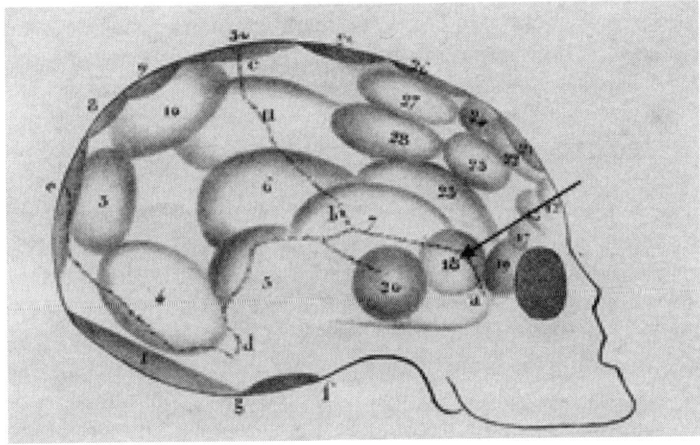

FIGURE 1. The system of Gall's organs on a front view of the head (***top***) and on a lateral view of the skull (***bottom***), as depicted in Ottin's Manual of Phrenology[16] in which N° 18 is the organ of music.

assumption that skull and brain morphology were similar was soon recognized as erroneous, and phrenology did not survive as a scientific approach to brain physiology and in general to the development of brain sciences. However, Gall's primary concept of the localization of mental function and his theory of the relation between the structure of the brain surface and psychological faculties had a deep and lasting influence.

In Gall's classification, the organ of music (the organ of the "relationships between sounds, musical memory, sentiment of melody and harmony"[16]) was N° 18[16] (or N° 17[14,15]), located laterally in the supraorbitary region, at the border between the inferior frontal and temporal regions (FIG. 1), and corresponded to N° 32 ("melody") in Spurzheim's phrenological classification. Gall had identified this organ by palpation of the head of several musically talented individuals and had first recognized it in Mozart's head.[17] Gall's organ of music is described as follows in a French treaty of phrenology[16]:

> *Situ.* This organ is located immediately above the external angle of the eye and, when it is very developed, results in square foreheads, inflated in the lateral part of the head. *Natural history.* Music cannot be explained by hearing, as colors cannot be explained by sight. An innate talent for both these intellectual manifestations should thus be acknowledged, and it has to be conceived that the ear hears sounds, that the voice sings them, as the eye sees the colors and the hand poses them in a painting; in one word, it is necessary to admit the existence of an intellectual faculty that conceives tones, a memory that stores them and an instinct that stimulates us to produce it.

Among the numerous followers of the phrenological doctrine, the American brothers Lorenzo Niles Fowler (1811-1896) and Orson Squire Fowler (1809-1887) set up an enterprise, the L.N. Fowler & Co., which marketed very popular phrenological heads and porcelain busts. In Fowler's busts, the phrenological numbering was changed into written definitions (FIG. 2). The most famous of these phrenological heads was presented and marketed again in 1916 as the new Fowler's phrenological bust, which depicted more than 100 regions or organs. The organ of "tune and modulation" was placed close to the original Gall's organ of music (FIG. 2). It is worth noting that the organ of music was surrounded by those of mathematical skills ("order and calculation;" "time and measure") both in the original Gall's organology and in Fowler's bust; luckily, in Fowler's bust the organ of music was also located very close to that of "humor and mirthfulness" (FIG. 2).

CORTICAL STRUCTURE AND MENTAL SKILLS

As soon as brain anatomy, physiology, and pathology underwent a remarkable development in the second half of the nineteenth century, the study of the structure of the brain of individuals with peculiar skills and intellectual capabilities was examined to search for clues on mental phenomena.[12,13,18] This line of research started in the mid-1850s in Germany, where Rudolph Wagner (1805-1864) studied "élite brains," including the brain of the physicist and mathematician, Carl Friedrich Gauss (1777-1855). Wagner described a remarkably convoluted appearance of Gauss' brain, but he could not confirm peculiar features in the brain of other intellectuals. Research along this line was pursued by other pioneers in the neurosciences, including the Swedish anatomist, Gustaf Magnus Retzius (1842-1919). Retzius examined the brain of the mathematician Sonja Kovaleski (1850-1891)[19]

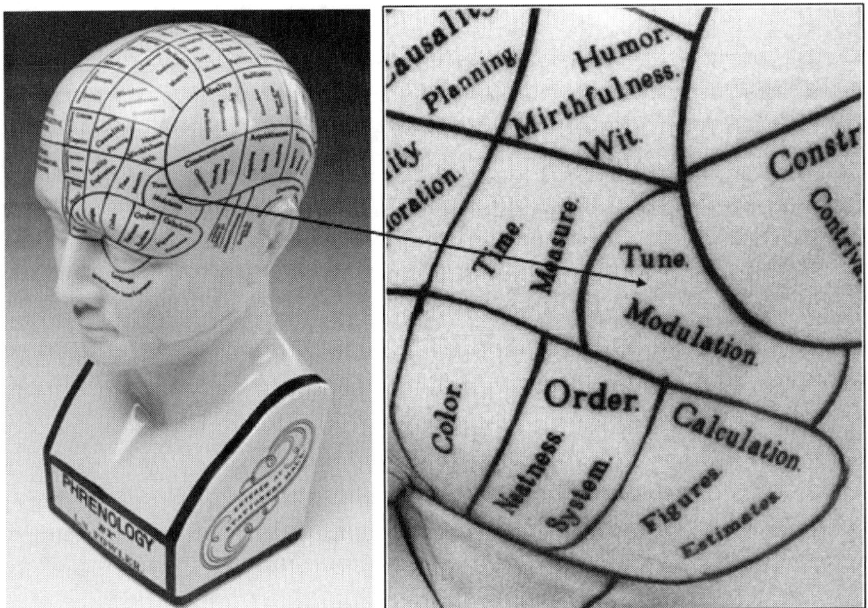

FIGURE 2. Localization of musical skill in Fowler's phrenological bust (courtesy of Jochen Richter).

and that of other intellectuals, including individuals with musical talents, such as the astronomer Hugo Gyldén, who had been musically gifted. In his studies (reviewed by Meyer[20]), Retzius emphasized the development of frontal and parietal cortical areas in these brains.

At the beginning of the twentieth century, Edward Anthony Spitzka (1876-1922) conducted a most extensive survey of the brains of individuals with superior intellectual skills.[13,20,21] This approach was also pursued by Oskar Vogt (1870-1959) and his wife Cécile (1875-1962).[12,13,18,20,22] The Vogts provided outstanding contributions in the field of cytoarchitectonics and collected a number of human brains, including "élite brains." Oskar Vogt was also requested to examine the brain of Vladimir Ilyich Lenin (1870-1924), the intellectual leader of the Russian October Revolution.[18,22] The Moscow Brain Research Institute founded for the study of Lenin's brain had also collected brains of musicians, such as those of the Armenian composer and conductor, Alexander Spendiaroff (1871-1928), the tenor, Leonid Sobisnow (1872-1934), and the Armenian composer, Aram Khatshaturjan (1903-1978),[22] but the results of these investigations do not seem to have been published. In the enduring efforts for a correlation of cortical structure and mental skills, features of the brain of physicist Albert Einstein (1879-1955) have been debated also in recent studies.[23]

CORTICAL STRUCTURE AND MUSICAL SKILLS

The anatomical investigations of the brain of musicians have been exhaustively reviewed by Meyer.[20] It will only be briefly recalled here that in a series of works published between 1906 and 1913, Auerbach[24] provided remarkable contributions on the brains of eminent musicians, including conductors Felix Mottl (1856-1911) and Hans von Bülow (1830-1894), teacher of music Naret Koning, singer Julius Stockhausen (1826-1906), and cellist Bernhard Cossmann (1822-1910). Auerbach concluded that the middle and posterior thirds of the superior temporal gyrus were highly developed in these brains and that the supramarginal gyrus was equally well developed. In the brain of the singer Stockhausen, Auerbach observed striking development of the left second frontal convolution.

Moreover, Spitzka demonstrated interest in the brains of musicians. In particular, Spitzka[21] mentioned, among other data on the skulls and the brains of musicians, that the brain of Robert Schumann (1810-1856) exhibited numerous and finely fashioned acoustic striae on the floor of the fourth ventricle and noted that the temporal fossae of Schumann's skull were more capacious than the frontal fossae. Spitzka[21] also reported that the brain of the Bohemian composer, Bedrich Smetana (1824-1884), who had died of paralytic dementia, weighed 1250 g and presented atrophy of the convolutions, dilatation of the ventricles, and atrophy of the auditory nerves (Smetana had become deaf). In addition, Spitzka[21] noted that the brain of the Hun-

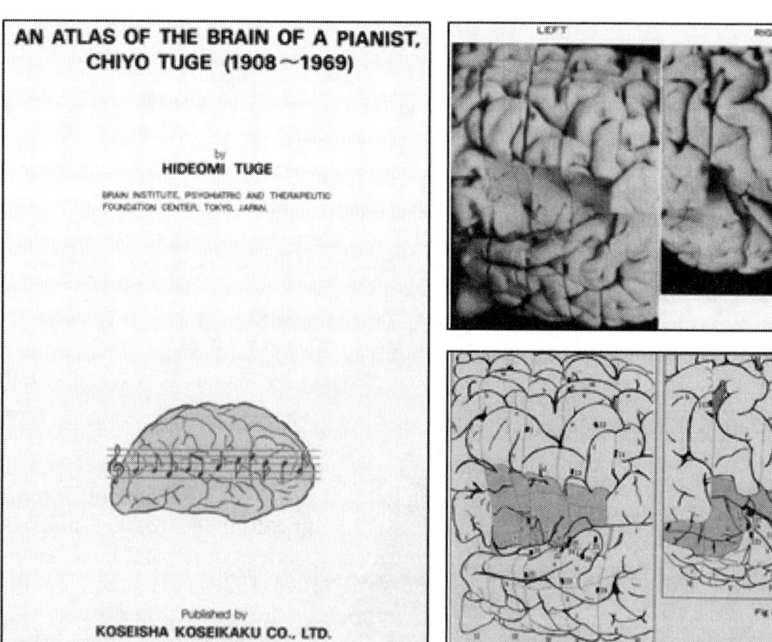

FIGURE 3. Cover page of Tuge's Atlas[27] and plate XIX of the Atlas showing Heschl's gyrus of the hemispheres.

garian violinist, Rudolf Lenz, weighed 1636 g (the brain had been found to be "somewhat softened" after its removal) and exhibited great expansion of the subparietal regions, especially on the right side.

The paper by I. Somogyi on the morphological correlate of musical abilities, published in 1930,[25] recalled attention not only on the analysis of the cortical surface, but also on microscopic cytoarchitectonic analysis. This paper described the brains of two musicians, the Magyar opera tenor, Georg Pogány, and the Hungarian violinist, Johann Nagy.[25]

In addition to the studies reviewed by Meyer,[20] two other investigations are worth commenting upon. Dorothée Beheim-Schwarzbach published in 1974 a study on human brains of the Vogts' collection,[26] focusing on the relations between musical skills and cortical structure and including a microscopic study. She compared the brains of a craftsman, a politician, an industrialist, and a philologist with musical talents, a violin player and composer, and a woman from New Guinea. In the brain of the musically talented individuals, and especially in that of the violinist, she found marked development of the first transverse gyrus.

In 1975, a peculiar atlas (FIG. 3) was published by the Japanese neuroanatomist, Hideomi Tuge (1905-1983), who had examined the brain of his wife, pianist Chiyo Tuge, who died from liver cancer.[27] The main findings of this extensive study (70 tables) can be summarized as follows: the central sulcus was considered to be relatively longer than normal, the supramarginal gyrus and temporal lobe were well developed, and Heschl's gyrus on the left side was remarkably larger than that on the right.

CONCLUDING REMARKS

The modern progress of knowledge on the neural correlates of musical skills[28,29] is based nowadays on several different approaches, especially on imaging technology, as also presented in several contributions to this volume. The overview just presented emphasizes that such progress follows at least two centuries of speculations, based on the theoretical approaches and technologies available at different times. The anatomical studies on the brain of musicians are currently difficult to interpret in view of the obvious interindividual variability of the described features and of the rigor imposed by current scientific standards. The progress made over the last two centuries on the localization of functions in the brain is certainly striking and reassuring. However, the history of the debates on the neural correlates of musical skills also provides a warning against a too precise "localizationist" approach: the exciting and novel mapping studies on functional neuroimaging of the "new generation of cartographers"[30] should certainly avoid its becoming a "cyberphrenology."[31]

ACKNOWLEDGMENTS

The author is deeply grateful to Dr. Jochen Richter for his advice and help. The help of Gerhard Martin and Michael Hagner in the collection of the material is also gratefully acknowledged.

REFERENCES

1. LOMBROSO, C. 1910. The Man of Genius. Walter Scott Publ. Co. London.
2. BROWN, A. & J. BARNES. 1994. Hildegard von Bingen (1098-1179): mystic, musician and medic. Adler Mus. Bull. **20:** 5–10.
3. BARONI, C.D. 1997. The pathobiography and death of Wolfgang Amadeus Mozart: from legend to reality. Hum. Pathol. **28:** 519–521.
4. DRAKE, M.E., JR. 1994. Mozart's chronic subdural hematoma. Neurology **44:** 2417–2418.
5. PESCHEL, E. & R. PESCHEL. 1992. Donizetti and the music of mental derangement: Anna Bolena, Lucia di Lammermoor, and the composer's neurobiological illness. Yale J. Biol. Med. **65:** 189–200.
6. ROLD, R.L. 1995. Schubert and syphilis. J. Med. Biogr. **3:** 232–235.
7. LERNER, V. 1998. The pathography of composers: Modest Mussorgsky. J. Med. Biogr. **6:** 175–181.
8. ALAJOUANINE, T. 1948. Aphasia and artistic realization. Brain **71:** 229–224.
9. SERGENT, J. 1993. Music, the brain and Ravel. Trends Neurosci. **16:** 168–172.
10. AMADUCCI, L., E. GRASSI & F. BOLLER. 2002. Maurice Ravel and right-hemisphere musical creativity: influence of disease on his last musical works? Eur. J. Neurol. **9:** 75–82.
11. CYBULSKA, E. 1997. Bolero unraveled. A case of musical perseveration. Psychiat. Bull. **21:** 576–577.
12. CLARKE, E. & C.D. O'MALLEY. 1996. The Human Brain and Spinal Cord. Norman Pbl. San Francisco.
13. FINGER, S. 1994. Origins of Neuroscience. Oxford University Press. New York & Oxford.
14. CLARKE, E. & K. DEWHURST. 1984. Histoire Illustrée de la Fonction Cérébrale. Editions R. Dacosta. Paris.
15. ZOLA-MORGAN, S. 1995. Localization of brain function: the legacy of Franz Joseph Gall (1758–1828). Annu. Rev. Neurosci. **18:** 359–383.
16. OTTIN, N.J. 1835. Précis Analytique et Raisonné du Système du Docteur Gall, sur les Facultés de l'Homme et les Functions du Cerveau. 6th Edition. H. Dumont. Bruxelles.
17. NATALUCCI, G. 1888. Musica e Sistema Nervoso. D. Natalucci. Civitanova-Marche, Italy.
18. BENTIVOGLIO, M. 1998. Cortical structure and mental skills: Oskar Vogt and the legacy of Lenin's brain. Brain Res. Bull. **47:** 291–296.
19. RETZIUS, G. 1900. Das Gehirn des Mathematikers Sonja Kovaleski. Biologische Untersuchungen neue Folge **9:** 1–16. Aftonbladets Druckeri. Stockholm.
20. MEYER, A. 1977. The search for the morphological substrate in the brains of eminent including musicians: a historical review. *In* Music and the Brain. C. MacDonald & R.A. Henson, Eds. :255–281. Heinemann Medical Books. London.
21. SPITZKA, A. 1907. Article IV. A study of the brains of six eminent scientists and scholars belonging to the American Anthropometric Society, together with a description of the skull of Professor E.D. Cope. Trans. Am. Philos. Soc. **21:** 175–308.
22. RICHTER, J. 2000. Rasse, Elite, Pathos. Centaurus Verlag. Herbolzheim, Germany.
23. WITELSON, S.F. 1999. The exceptional brain of Albert Einstein. Lancet **353:** 2149–2153.
24. AUERBACH, S. 1906–1913. Zur Lokalisation des musikalischen Talentes im Gehirn unad am Schädel. Arch. Anat. Physiol. (Anat. Abtllg.) 1906: 197–230; 1908: 31–38; 1911: 1–10; 1913 (Suppl.): 89–96.
25. SOMOGYI, I. 1930. Über das morphologische Korrelat der musikalischen Fähigkeiten. Monat. Psychiat. Neurol. **75:** 138–169.
26. BEHEIM-SCHWARZBACH, D. 1974. Cytoarchitektonik der Dorsalfläche der 1. Temporalwindung links (T 1) bei sechs menschlichen Gehirnen (darunter vier Elitegehirne) der Sammlung von C. und O. Vogt. Z. mikrosk.-anat. Forsch. **88:** 325–363.
27. TUGE, H. 1974. An Atlas of the Brain of a Pianist, Chiyo Tuge (1908–1969). Koseisha Koseikaku Co, Ltd. Tokyo, Japan.

28. MÜNTE, T.F., E. ALTENMÜLLER & L. JÄNCKE. 2002. The musician's brain as a model of neuroplasticity. Nat. Rev. Neurosci. **3:** 473–478.
29. ZATORRE, R.J. & C.L. KRUMHANSL. 2002. Mental models and musical minds. Science **298:** 2138–2139.
30. KOSIK, K.S. 2003. Beyond phrenology, at last. Nature Rev. Neurosci. **4:** 234–239.
31. HAGNER, M. 2001. Das Genie und sein Gehirn. Jahrbuch 2001 des Collegium Helveticum der ETH Zurich. :187–211.

Brain Processing of Meter and Rhythm in Music

Electrophysiological Evidence of a Common Network

HELEN KUCK, MICHAEL GROSSBACH, MARC BANGERT, AND
ECKART ALTENMÜLLER

*Institute for Music Physiology and Musicians' Medicine,
Hannover University of Music and Drama, 30161 Hannover, Germany*

ABSTRACT: To determine cortical structures involved in "global" meter and
"local" rhythm processing, slow brain potentials (DC potentials) were recorded from the scalp of 18 musically trained subjects while listening to pairs of
monophonic sequences with both metric structure and rhythmic variations.
The second sequence could be either identical to or different from the first one.
Differences were either of a metric or a rhythmic nature. The subjects' task
was to judge whether the sequences were identical or not. During processing of
the auditory tasks, brain activation patterns along with the subjects' performance were assessed using 32-channel DC electroencephalography. Data were
statistically analyzed using MANOVA. Processing of both meter and rhythm
produced sustained cortical activation over bilateral frontal and temporal
brain regions. A shift towards right hemispheric activation was pronounced
during presentation of the second stimulus. Processing of rhythmic differences
yielded a more centroparietal activation compared to metric processing. These
results do not support Lerdhal and Jackendoff's two-component model,
predicting a dissociation of left hemispheric rhythm and right hemispheric
meter processing. We suggest that the uniform right temporofrontal predominance reflects auditory working memory and a pattern recognition module,
which participates in both rhythm and meter processing. More pronounced
parietal activation during rhythm processing may be related to switching of
task-solving strategies towards mental imagination of the score.

KEYWORDS: rhythm processing; meter processing; musicians; brain activation;
local and global processing

INTRODUCTION

In contrast to pitch or melody processing, relatively few neurobiological investigations have been concerned with processing of musical time structures. These studies have yielded contradictory results. Besides the fact that various methods were

Address for correspondence: Prof. Dr. med. Eckart Altenmüller, M.D., M.A., Institute for
Music Physiology and Musicians' Medicine, Hanover University of Music and Drama, Hohen-
zollernstr. 47, 30161 Hannover, Germany. Voice: 0049 (0) 511 3100 552; fax: 0049 (0) 511 3100
557.
altenmueller@hmt-hannover.de

Ann. N.Y. Acad. Sci. 999: 244–253 (2003). © 2003 New York Academy of Sciences.
doi: 10.1196/annals.1284.035

applied and different populations with respect to musical expertise were investigated, inconsistent definitions of "musical time structures" may have contributed to these discrepancies. With respect to temporal structures, for example, three levels of organization may be distinguished: meter, pulse (grouping), and rhythm. Rhythm is defined as the serial relation of durations between different acoustical events in a train of sounds, that is, rhythm represents a serial durational pattern, whereas pulse or grouping is based on gestalt principles and depends, among other physical characteristics, on the relative proximity in time of sound events.[1] Meter, in contrast, involves a temporal invariance in terms of the regular recurrence of pulses marking off equal durational units, which can be organized as measures. Meter, therefore, represents a more complex acoustical gestalt, because its perception and production require information on sound intensity (accented and unaccented events) and on periodicity of rhythmic events, the latter based on integration of information over longer time periods. Perception and creation of meter are a prerequisite of the musician's ability to make music "swing."

The cerebral mechanisms underlying the processing of rhythm and meter are largely unknown. In a group of patients with unilateral right- or left-hemispheric brain damage, Peretz[2] found spared metric judgment in the presence of disrupted rhythmic discrimination, irrespective of whether the right or left hemisphere was lesioned. In a more recent investigation of 65 patients who had undergone unilateral temporal cortectomy for the relief of intractable epilepsy, this dissociation between rhythmic and metric judgment was confirmed.[3] For meter processing, a critical involvement of the anterior part of the superior temporal gyrus was found, whereas rhythm processing seemed to rely more on the posterior parts of the right superior temporal gyrus. In a very elegant multimodal auditory and visual paradigm, Penhune *et al.*[4] demonstrated in chronic epileptic patients a modality-specific dissociation with isolated involvement of the anterior secondary auditory areas in the right temporal lobe during processing of acoustically presented time structures. In an earlier study,[5] we investigated 20 patients who had had small unilateral cerebrovascular cortical lesions. Using a discrimination paradigm similar to the test published by Peretz,[2] processing of acoustically presented rhythms and meters was assessed. Detailed analysis of the individual patterns of neuropsychological deficits revealed a hierarchical organization, with an initial right-hemisphere recognition of meter followed by identification of rhythm via left-hemisphere subsystems. In addition, individual aspects of musicality and musical behavior as well as musical knowledge contributed to the formation of neuronal subsystems underlying the perception of musical time structures.

With respect to the functional anatomy of rhythm processing in normal subjects, new studies exist. Patel *et al.*[6] investigated brain activation with the PET technique during regularly and irregularly presented series of tones. They found a pronounced involvement of the left frontal Broca region and concluded that language processing and rhythm processing might be closely related to each other. Contradictory results emerged from a PET study by Penhune *et al.*,[7] testing the perception and reproduction of regular isochronous or complex novel time structures in both the auditory and the visual modality. Auditory perception of rhythm produced an activation of the right planum temporale. Because the design of the study additionally required production of rhythms, activation of the somatosensory cortices and of the cerebellar hemispheres was demonstrated. In an fMRI work, processing of simple (1:2, 1:3,

1:4) and complex (1:2.5, 1:3.5) time relationships was compared.[8] Simple rhythmic relationships yielded activation of left prefrontal and parietal brain areas, whereas complex relationships were processed in right prefrontal, premotor, and parietal regions. Although not discussed by the investigators, the results can be interpreted in the light of Lerdhal and Jackendoff's two-component model.[9] According to their model, rhythm and meter sense rely on two different cognitive operations, which may be processed in different hemispheres: processing of rhythm requires a left-hemispheric "local"-level, serial cognitive operation; processing of meter a right-hemispheric "global"-level, holistic strategy linked to grouping or chunking mechanisms. Applied to the fMRI experiment, simple rhythmic relationships can be understood as "local tasks," because they are accessible for analytic and sequential "counting" strategies. By contrast, this processing mode is not accessible for perception of the complex stimuli, which therefore had to be analyzed in a holistic way as auditory gestalt.

Summarizing the results of lesion and brain imaging studies, a puzzling and in many instances contradictory variety of findings emerged. The present study attempts to contribute to the clarification of the neuronal substrates of the processing of musical time structures. To assess cortical activation patterns during processing of musical time structures, the topographic distribution of sustained surface negative DC-potential shifts was recorded using scalp electrodes. Because these DC potentials reflect activation of the underlying cortex,[10] their local distribution reveals task-specific patterns that correspond to the brain structures specifically involved in the processing of the respective task.

METHODS

Subjects and Stimuli

With respect to findings indicating that cortical activation patterns during music processing may be influenced by musical expertise,[11] only trained musicians were included in the present study: 18 experienced right-handed musicians (10 males, 8 females) with at least 5 years of formal musical training were included. They either played a melodic instrument (violin, viola, violoncello, and clarinet), piano, or sang in a semiprofessional choir. None of the participants played percussion. Subjects were between 21 and 39 years of age (mean 25.5).

In a same–different paradigm, subjects had to rate whether two subsequent acoustic time sequences of 4 seconds' duration each, were same or different. Each stimulus pair consisted of two monophonic sequences, sampled as MIDI piano sounds and played at the pitch of one-line b flat (for an example, see FIG. 1). Stimuli were structured in time and in intensity of the beats. They consisted of sequences of notes with "regular" temporal intervals. The temporal sequences of subsequent notes were composed of 1:2, 2:1, 1:3, 3:1, 1:4, 4:1 ratios, or, expressed in musical terms, of classical time values of notes such as crochets, quavers, triplets, and semiquavers. Intensity variations were interspersed at regular time intervals to produce the perception of an underlying regular (metric) pulsation. The second stimulus was either identical to the first or differed with respect to its rhythmic or metric stucture. In the meter condition, the occurrence of the marked notes was shifted to a different meter,

Stimuli

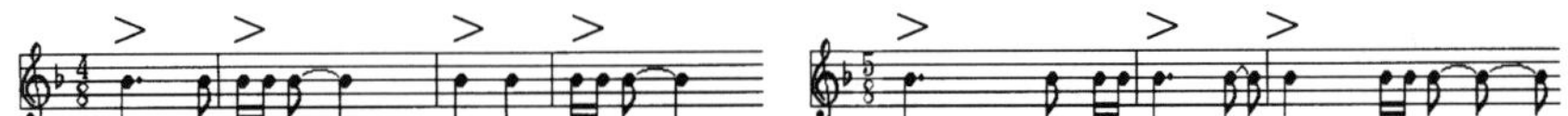

Meter   **changed**

Rhythm   **changed**

FIGURE 1. Examples of pairs of stimuli used for the "meter" (*upper row*) and the "rhythm" (*lower row*) condition.

changing for example a 4/8 meter into a 5/8 meter (see example in FIG. 1, upper row). In the rhythm condition, single notes were changed in value without affecting the underlying metric structure (see example in FIG. 1, lower row).

Tasks and Procedures

Subjects were seated comfortably in a dimly lit and acoustically shielded room. Stimuli were delivered via stereo speakers 1.5 m in front of the subjects, set at a comfortable sound level of 72 dBA measured at the listener's position. To minimize artifacts in the EEG recordings, subjects were requested to fix their gaze on a fixation point displayed on a 14-inch monitor in front of them and to avoid body movements or vocalizations while performing the tasks.

Subjects began each trial by pressing a button on a keyboard as soon as they felt ready. After a randomized time interval of 1,000–1,500 ms, the first stimulus was delivered, followed by a 2-second break and the subsequent second stimulus. Two seconds after presentation of both stimuli, subjects had to press one of two keys on the keyboard, indicating either a "same" or a "different" judgment of the stimuli delivered previously. Button presses were randomly balanced executed with the right/left hand. A total of 160 pairs of stimuli were presented. In the rhythm condition, 40 stimulus pairs were identical and 40 were different, and in the meter condition accordingly 40 pairs were identical and 40 were different. After the experiment (total duration about 2 hours), subjects filled in a questionnaire concerning eventual task-solving strategies (verbal, e.g., "counting" vs holistic, nonverbal strategies).

Data Acquisition and Data Analysis

DC potentials were recorded from 32 electrodes positioned according to the modified 10/20 system over left and right frontal, central, temporal, parietal, and occipital brain regions. Linked mastoid electrodes served as a reference. Impedance was reduced to less than 1 kOhm. To control artifacts arising from eye movements, the electrooculogram (EOG) was obtained. The frequency band of amplification ranged from DC to 100 Hz. Sampling rate was 200 Hz (SynAmps–Amplifier).

Each data-sampling epoch lasted 12 seconds. Baseline correction was performed offline, and trials contaminated by artifacts were excluded from further analysis. For each of the two conditions, about 30 artifact-free trials were averaged per subject. Based on the individual averages, grand averages across all subjects were derived. For selected time intervals, mean amplitudes of DC shifts referenced to baseline were calculated. Analyses of brain activation during processing of the first and the second stimulus were performed on the mean amplitude of DC-EEG evoked between the 2,000 and 4,000 ms after beginning of the stimulus. The first second of either stimulus presentation was excluded from analysis, because unspecific EEG components related to arousal and orientation occur in this time interval.

Data were subjected to MANOVA using the software packet "Statistica." A within-subjects factor Task (2 levels: rhythm vs meter) was tested. For correction of violations of the sphericity assumption, the Huynh-Feldt epsilon was used. Only those variables that after correction showed significant ($P < 0.01$) main effects or significant interaction terms were further analyzed using the appropriate T test. Because this procedure applies multiple T tests, the P values were Bonferroni adjusted. To further analyze the spatial origin of significance between means of the two conditions, electrodes were grouped according to regions of interests and contrasted against each other.

RESULTS

Behavioral Results and General Course of DC Potentials

On average, 69% of the tasks were answered correctly. There was no significant difference in the responses to metric (71% correct) or rhythmic (68% correct) variations. Differences between both types of tasks are therefore not attributable to different levels of difficulty.

The general course of the DC potential shifts is shown in FIGURE 2. A characteristic biphasic negative-going plateau-like shift reflects the presentation of the two stimuli. Maximal amplitudes can be detected over anterior frontotemporal and central electrode positions.

In FIGURE 3, the values of DC potential amplitudes are visualized topographically. In both stimulus conditions, maximal activation occurred over frontotemporal brain regions. The mean amplitudes during the second period are displayed in FIGURE 4. The second stimulus yielded an additional increase in amplitude in all electrode positions and a more pronounced lateralization towards the right frontotemporal brain areas.

DC-EEG

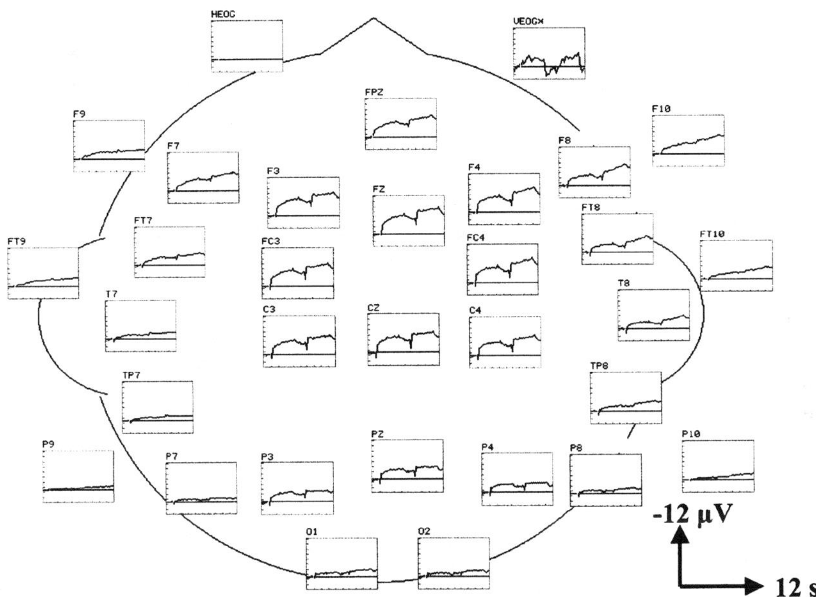

FIGURE 2. Grand average of all subjects during the "rhythm" condition displayed topographically. Each of the small rectangular insets represents DC potential shifts in one electrode position. Shown are slow potential shifts during 12 seconds, including 1-second baseline, 4-second presentation of the first stimulus, 2-second pause, and subsequent 4-second presentation of the second stimulus and 1-second postperiod. Averages obtained from electrodes over the left hemispheres are on the left side, those from the right hemisphere on the right side. Stimulation produces a widespread biphasic increase in surface-negative DC potential shift, with maximal amplitudes over bilateral frontocentral brain regions.

Results of Statistical Testing

Since the main factor Task was highly significant ($P < 0.01$), testing was continued using contrast analysis. Contrast analysis revealed that both rhythm and meter processing produced increased activation over the right frontotemporal regions (electrodes F4, F8, Fc4, F10, FT10, FT8, T8, $P = 0.001$). When contrasting rhythm versus meter processing, the first stimulus presentation did not reveal any differences, because stimuli were identical in nature. However, during the second presentation of paired stimuli, rhythmic alterations produced a more pronounced activation over centroparietal brain regions (Cz, Pz, C4, C3, P4, P3, TP7, TP8, P7, P8, $P < 0.001$), as shown in the difference plots displayed in FIGURE 5.

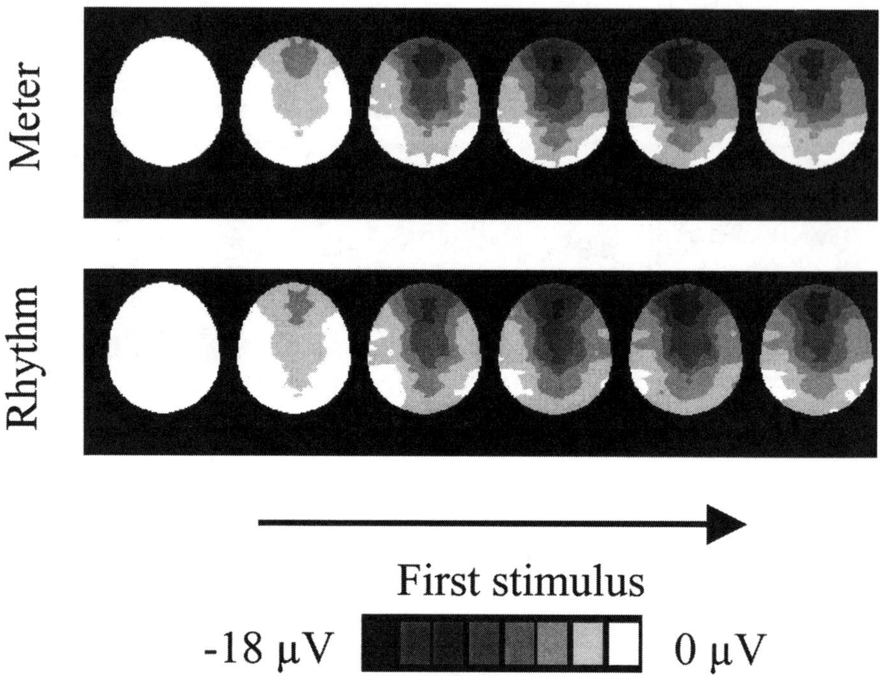

FIGURE 3. Brain maps representing meter (*upper row*) and rhythm (*lower row*) processing during the first stimulus presentation. Grand averages over all subjects are displayed topographically. Each diagram represents mean activation during 1 second, beginning with the baseline period. The subsequent four diagrams represent the 4 seconds of stimulus presentation. Activation is dark, inhibition is white (see microvolt scale). Brain diagrams are displayed as *top views*, frontal regions *up*, left hemisphere on the *left*, and right hemisphere on the *right*. As can be recognized, both meter and rhythm processing produce a prefrontal, frontal, and bilateral temporal activation pattern with only subtle lateralisation towards the right hemisphere.

DISCUSSION

To our knowledge, this is the first systematic study directly comparing cortical activation patterns during processing of local and global acoustically presented time structures. As local processing, we consider the auditory tracing of a single event in time in contrast to global processing, which in the present paradigm relies on the integration and detection of accents over a longer time span. The main results can be summarized as follows:

Processing of musical time structures produces sustained cortical activation, especially over frontotemporal brain regions with right hemispheric preponderance.

Processing of rhythmic differences yields more centroparietal activation compared to metric processing.

During processing of both metric and rhythmic stimuli, predominant right frontotemporal activation occurred. This lateralization effect was less pronounced during

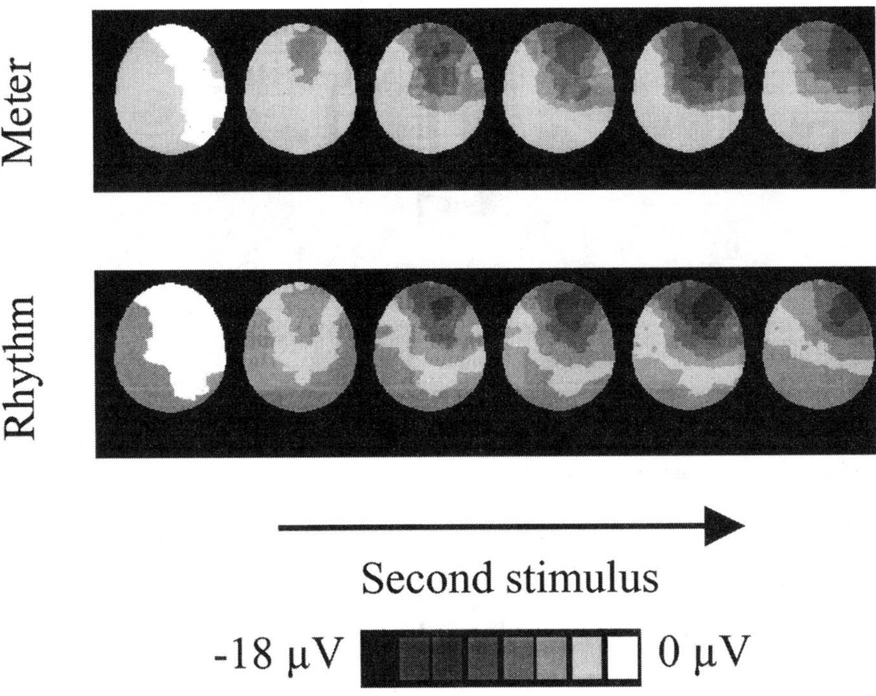

FIGURE 4. Brain maps representing meter (*upper row*) and rhythm (*lower row*) processing during the second stimulus presentation. Same conventions as in FIGURE 3. Compared to FIGURE 3, demonstrating brain activity during the encoding of the first stimulus, retrieval and comparison of the second stimulus produce even more electronegativity, with right-hemispheric lateralization. Rhythm processing produces a more pronounced activation over posterior parietal regions.

the encoding phase, when subjects listened to the first stimulus, and more pronounced during recall and comparison when subjects were exposed to the second stimulus. This result does not fit into Lerdhal and Jackendoff's two-component model,[1] predicting a dissociation of left-hemispheric rhythm and right-hemispheric meter processing. It should be mentioned in this context that up to now lesion studies investigating separately rhythm or meter processing equally failed to clearly demonstrate such a dichotomy.[3,4,6]

In contrast to the pronounced local/global dichotomies observed in the visual modality (for a review see Ref. 12), auditory perception seems to be organized in a more complex manner. According to Zatorre and Belin,[13] the left hemisphere auditory cortex is specialized for rapid temporal processing, whereas the right temporal lobe is specialized for spectral processing. These differences may be related to anatomic asymmetries in myelination and spacing of cortical columns in both hemispheres. In the present study, processing of time structures was more characterized by integration over time and pattern recognition than by rapid temporal processing (e.g., as it is required in language comprehension). We therefore suggest that the

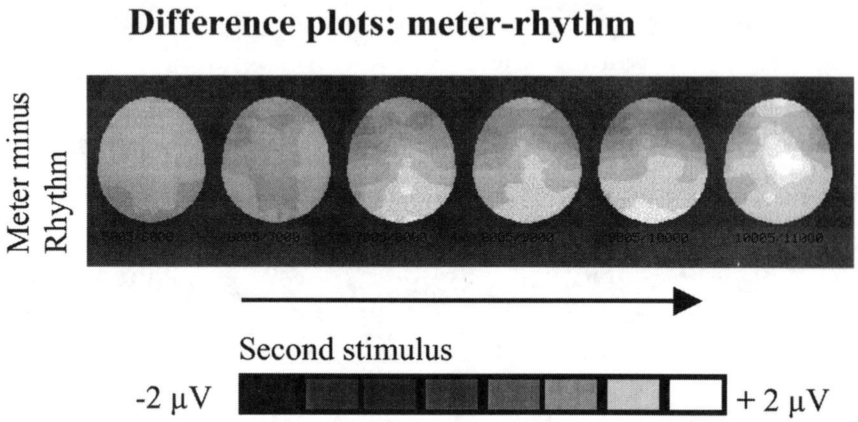

FIGURE 5. Difference plots, obtained after subtracting activation during rhythm processing from activation during meter processing. Only minor differences can be recognized. The *bright area* over posterior regions corresponds to relatively higher activation during rhythm processing in parietal brain areas.

right temporofrontal predominance reflects auditory working memory and a pattern recognition module, which is participating in both rhythm and meter processing.

A surprising finding was the more pronounced parietal activation during rhythm processing. Although several brain imaging studies have demonstrated activity in the parietal lobe during temporal processing of music,[9,14] this finding was never commented in detail. From studies investigating musical auditory imagery,[15,16] it is known that mental imagery of a musical score produces a similar parietal activity. We could speculate that during the rhythm condition subjects switched their task-solving strategy and relied to a greater extent on their inner eye and their mental imagination of the score. However, as a posthoc speculation, not reflected in the questionnaires filled in by our subjects, this finding remains to be verified in a prospective study designed to control this variable.

ACKNOWLEDGMENTS

This research was supported by Grants Al 269/4-2 and Al 269/4-3 from the Deutsche Forschungsgemeinschaft.

REFERENCES

1. DRAKE, C. 1998. Psychological processes involved in the temporal organization of complex auditory sequences: universal and acquired processes. Music Percept. **16:** 11–26.
2. PERETZ, I. 1990. Processing of local and global musical information by unilateral brain-damaged patients. Brain **113:** 1185–1205.

3. LIÉGEOIS-CHAUVEL, C., I. PERETZ, M. BABAÏ, *et al.* 1998. Contribution of different cortical areas in the temporal lobes to music processing. Brain **121**: 1853–1867.
4. PENHUNE, V.B., R.J. ZATORRE & W.H. FEINDEL. 1999. The role of auditory cortex in retention of rhythmic patterns as studied in patients with temporal lobe removals including Heschl's gyrus. Neuropsychologia **137**: 315–331.
5. SCHUPPERT, M., T.F. MÜNTE, B.M. WIERINGA, *et al.* 2000. Receptive amusia: evidence for cross-hemispheric neural networks underlying music processing strategies. Brain **123**: 546–559.
6. PATEL, H., C. PRICE, J.C. BARON, *et al.* 1997. The structural components of music perception. A functional anatomical study. Brain **120**: 229–243.
7. PENHUNE, V.B., R.J. ZATORRE & A.C. EVANS. 1998. Cerebellar contributions to motor timing: a PET study of auditory and visual rhythm reproduction. J. Cognit. Neurosci. **10**: 752–765.
8. SAKAI, K., O. HIKOSAKA, S. MIYAUCHI, *et al.* 1999. Neural representation of rhythm depends on its interval ratio. J. Neurosci. **19**: 10074–10081.
9. LERDAHL, F. & R. JACKENDOFF. 1983. A Generative Theory of Tonal Music. MIT Press. Cambridge.
10. ALTENMÜLLER, E. & C. GERLOFF. 1998. Psychophysiology and EEG. *In* Electroencephalography. E. Niedermeyer & F. Lopes da Silva, Eds. :637–655. Williams & Wilkins. Baltimore.
11. SCHLAUG, G. 2001. The brain of musicians: a model for functional and structural adaptation. *In* The Biological Foundations of Music. R.J. Zatorre & I. Peretz, Eds. Ann. N.Y. Acad. Sci. **930**: 281–299.
12. IVRY, R.B. & L.C. ROBERTSON. 1998. The Two Sides of Perception. MIT Press. Cambridge, MA.
13. ZATORRE, R.J. & P. BELIN. 2001. Spectral and temporal processing in human auditory cortex. Cereb. Cortex **11**: 946–953.
14. BROCHARD, R., A. DUFOUR, C. DRAKE, *et al.* 2000. Functional brain imaging of rhythm perception. *In* ICMPC6 Proceedings. C. Woods, G. Luck, R. Brochard, *et al.*, Eds. Keele University. UK.
15. BEISTEINER, R., E. ALTENMÜLLER, W. LANG, *et al.* 1994. Musicians processing music: measurement of brain potentials with EEG. Eur. J. Cognit. Psychol. **6**: 311–327.
16. NAKADA, T., Y. FUJII, K. SUZUKI, *et al.* 1998. "Musical brain," revealed by high-field (3 Tesla) functional MRI. Neuroreport **9**: 3853–3856.

Perceptual and Structural Implications of "Virtual" Music on the Web

WILLIAM DUCKWORTH

Department of Music, Bucknell University, Lewisburg, Pennsylvania 17837, USA

ABSTRACT: The development of Cathedral, one of the first interactive works of music and art on the World Wide Web; the development of new virtual instruments, such as the PitchWeb, that allow listeners to participate actively and creatively; and the possibility that a new form of music is emerging are discussed. This new music, Virtual Music, is a new art form, within a new medium, producing a new type of interactive artistic experience, with the ability to draw in and engage its listeners in the creative process.

KEYWORDS: virtual music; cathedral; pitchweb; cyberspace; creativity on the Web

INTRODUCTION

I want to begin by thanking Diego Minciacchi for inviting me to participate. I consider it a great honor. Even though his invitation has been the source of some concern over what I should say, it has also given me the opportunity to organize my thoughts about this work and reflect back over what has been accomplished during the last 6 years, and for that I am extremely grateful.

I want to talk to you more as an artist than as an educator or researcher. My intention is to give you an insight into what composers working in some of the newer fields of musical experimentation are interested in and to offer some thoughts about newly emerging concepts of musical creativity and what this may someday mean for you.

The reason that I think it is important to speak as an artist is that I want to speculate about the future and what it might be like musically. I also want to raise the possibility that some forms of music on the World Wide Web will be continuous and regenerative and constantly in a state of global interaction and flux. And I want to offer the potential for community music being created and recreated online by people like you and me, acting together and working in creative partnerships, some in ways never before thought possible.

Adress for correspondence: Dr. William Duckworth, Department of Music, Bucknell University, Lewisburg, Pennsylvania 17837, USA. Voice: 201-869-9589; fax: 201-869-9841.
duckworth@monroestreet.com

Ann. N.Y. Acad. Sci. 999: 254–262 (2003). © 2003 New York Academy of Sciences.
doi: 10.1196/annals.1284.036

Ultimately, the point I want to make is that the Web is changing music, not only by offering a new set of tools, but also by shifting the focus of what it means to be musical, and that as a result of this shift, a new artistic paradigm is emerging with its own unique set of characteristics. Just as surely as listeners in the nineteenth century heard their music live and twentieth-century audiences experienced the majority of their music in broadcast or recorded form, listeners of the twenty-first century will obtain their music primarily from cyberspace. Even now, many people can see the promise and the potential for a universal jukebox with immediate Internet access to any music ever recorded, and some of us can envision more interactive experiences and the possibility of a worldwide cyberband connecting a global artistic community online.

Certainly, as a composer, I know that my own work during the last decade has undergone a profound change. My concepts of who my listeners are, what they hear, and what the circumstances are in which they hear it have all undergone major transformations. Until the mid-1990s, for example, most of my music was written for traditional forces. It wasn't until 1996 that media artist Nora Farrell and I began creating a work for the Web.

So it is important to keep in mind that what I am going to describe as fact was almost unimaginable, and certainly physically impossible, a mere decade ago. In 1992, for example, there were an estimated 50 sites on the World Wide Web. What traffic there was moved at a rate of 9600 baud, or about 50 times slower than the speed of the average 56K connection today and 500 times slower than broadband. And it is equally important to remember both how recently this paradigm shift began taking place and how far and how fast this new world of the Web has propelled the new art form of Virtual Music.

Now over a long period of time, how this new paradigm may affect how people listen to music, what they might hear, or the questions this may raise neurologically is, at this point, anybody's guess. But what I can speak about is where we are now, how we got here, and what the next step is, at least for Nora and me. Along the way, I also want to characterize and define this new art form.

Where I would like to begin is with Cathedral,[1] first, by talking about the new tools now at our disposal and then by discussing the change in artistic focus they are helping to bring about.

CATHEDRAL

Nora Farrell and I started work on a piece for the Web in Amsterdam in March 1996. We were there for a week to explore a new MIDI device being developed at STEIM, which is a government-sponsored center offering artists technological expertise. But during that week, we actually spent more time in the cafes talking about the Internet, how fast it was growing, and how we might begin to use it artistically. We decided to create a piece for the Web, not only because we recognized the vast artistic potential of the Internet, but also because we could see the possibility of creating an entirely new musical and social experience online.

Even in these earliest discussions, we knew that we wanted Cathedral to be an interactive website with web-based musical instruments that anyone could play, and we wanted to make a place on the Web for acoustic music through a series of live webcasts and performances online. Remember that most of this was impossible when we first discussed it. But fortunately, the technology stayed current with our vision and we were able to realize everything that we had created that week in our minds.

Cathedral went online in June 1997; at the time, there were fewer than a million sites on the World Wide Web, and fewer than 2% of those made any sounds at all. This first version of the website included streaming audio, streaming video, animation, images, and texts. Our goals at that point were:

- to create an imaginative, ongoing artistic experience that builds community,

- to blur the distinctions separating composers, performers, and audiences, a process begun by John Cage in the 1950s, if not by Erik Satie 50 years before that, and, ultimately,

- to offer each individual listener the ability to create his or her own unique musical experience online.

In its most fundamental form, Cathedral today consists of three primary components:

- a website featuring a variety of interactive musical, artistic, and text-based experiences,

- a group of virtual instruments that allow listeners to participate actively and creatively, and

- an Internet band that gives periodic live performances and offers listeners focused moments in which to come together and play music in community online.

THE WEB SITE

Conceptually and thematically, Cathedral is the story of five mystical moments in time. These moments, which we call the Building, the Bomb, the Pyramid, the Web, and the Dance, represent for us significant points of human creativity and collective actions. Specifically, these moments are the building of Chartres Cathedral, the first detonation of the atomic bomb, the building of the Great Pyramid at Giza, the founding of the World Wide Web, and the inception of the Native American Ghost Dance religion. For Nora and me, these five moments signify events of continuing importance for the future, because they raise issues of religion and spirituality, power and self-destruction, lost civilizations, the future of individuals and societies, and the natural and supernatural, and, by extension, the fate of the indigenous people of the world. On the Cathedral website, there are 32 areas designed not as answers, but as points of reflection and contemplation for these five moments. For our listeners, these moments provide points around which to attach meaning and extend narrative, creating a multilayered artistic form and encouraging a return to mythic thinking and the reintegration of sonic materials.

In addition, the website contains a Codex that houses the 32 basic texts that are the inspiration for Cathedral; a Chronicle that is a reflection on Cathedral and its five moments in the form of a hypernovel; a Stage where live webcasts by the Cathedral

Band occur and where previous webcasts are archived; and a set of three Virtual Instruments called Sound Pool, Chaos, and PitchWeb.

The music on the site consists of four types and includes *computer music* written specifically for the Web and existing as either sound, MIDI, or MP3 files that are streamed to the site and heard in real time; *acoustic music* written for traditional ensembles and incorporated into the site through the live webcasts and the archives; *improvised music* performed live and online by the Cathedral Band, and *interactive music,* which is played live by the listeners on the virtual instruments, sometimes with the band.

THE VIRTUAL INSTRUMENTS

When we decided to create the virtual instruments, our goal was to give our listeners an active role in the creative process by encouraging everyone who visits the Cathedral site to be both a listener and a performer. Additionally, we thought it was important to design instruments that have built-in levels of musicality, so they can be played by people of any musical ability.

At present, the most interactive of these instruments is the PitchWeb,[2] which is a web-based multiuser instrument with text chat capabilities. It is played by selecting and manipulating shapes that are mapped to sound samples contained in multiple banks of 64 sounds each. Users can select individual sounds from the sound palette, move them to the playing field, and play them in any order and combination with movements of the computer mouse. These sound/shapes can also be resized and overlapped to create polyphonic passages, and in the CD-ROM version the resulting sound patterns can be saved and reopened at another time.

On a simpler level, PitchWeb users may also create music by typing in words or phrases in any language and having the instrument automatically convert them into musical sounds. Although in its most complex form, the PitchWeb may be played directly on the computer keyboard, like a synthesizer. And with the online multiuser version of the instrument, performers can affect the composite sound individually by having their changes updated instantly across the network and heard by everyone at the same time.

The earliest version of the PitchWeb was debuted online from The Franklin Institute Science Museum in Philadelphia in October 1998, when it was played with the Relâche Ensemble by me as well as by audience members on their laptops and on computer terminals stationed around the hall. The multiuser version of the instrument debuted in December 2001, when PitchWeb bands in Boston, New York, and Atlanta performed during a 48-hour Cathedral webcast. And it was used again in the summer of 2002 during a week of Cathedral Band performances and online activities from the Brisbane Powerhouse in Australia.

I would like to add that in those Powerhouse concerts, both Nora and I had the opportunity to mix our online PitchWeb players into the performance directly from the stage where we were performing. It was something that we both found added a new dimension to the experience, because we could converse with them through text chat even as their sounds were appearing live. It gave us the impression of simultaneously being on stage and online.

THE CATHEDRAL BAND

The aspect of Cathedral that most incorporates live performance into the website is the Cathedral Band.[3] We think of it as an Internet band that plays both in concert and online. The instrumentation, which can vary, always includes a DJ who plays drum'n'bass and ambient samples, plus the PitchWeb and the PitchWeb MUD, or multiuser domain, which allows us to bring in players worldwide. The rest of the band varies, depending on the location of the concert and their availability. So far, we have performed with electric guitar, trombone and toys, mixed percussion, Tibetan folk, Chinese pipa, Hindustani vocals, didjeridu, and AJ Sabatini, who, as the Chronicler, is the voice of Cathedral. In performance, the Chronicler tells the story of one of the five moments, expanding, though not explaining, its references and associations.

Our live Cathedral webcasts began in 1998 from the Spoleto Festival USA and have continued from Philadelphia, Seattle, New York, Phoenix, and Brisbane. In December 2001, we produced a 48-hour webcast streaming 34 concerts live from 5 continents and, in the process, creating the largest festival of Internet music ever held. During the weekend, our listeners experienced Tanzanian folk songs sung by Jonathan Hart Makwaia; the pipa playing of Wu Man; an audio tour of Krakow and Chicago; keyboard performances from Tokyo and Buenos Aires; Australian birdcalls and poetry from New Zealand; as well as the collective contributions of Pitch-Web bands playing day and night.

Incidentally, one thing we learned pretty quickly about putting music online is that webcasting a concert is not the same thing as creating a concert for the Web. It's a different auditory experience altogether. It's much more personal and more intimate, and there is far less of a sense of being in a communal space in cyberspace as there is, say, when sitting with other people inside a concert hall. On the Web, it is just you and your computer, everywhere and yet nowhere, all at the same time. And in that environment, music streamed in real time, we've found, is an intimate, up close, and in-your-head experience.

NEW ARTISTIC FOCUS

But artistically, what does this all add up to—a website, a PitchWeb, and an Internet band? And what's different, formally and perceptually, from musical experiences of the past? Certainly, as a work of art, Cathedral is unlike any previous concert or theatrical model. To mention the most obvious reason why, the form of Cathedral is not expressed linearly in time. In fact, it's highly nonlinear. And time is no longer a factor in a piece of music that is always available; that has no beginning, middle, or end; and that no two people experience in the same order or for the same length of time. So, if nothing else, Cathedral is a basic reconfiguration of the relation between musical time and space or, in this case, cyberspace.

Now on first glance this may seem to create artistic chaos, because Cathedral can only be experienced in fragments and, furthermore, appears to have "centers everywhere and boundaries nowhere," to borrow a phrase from Marshall McLuhan's last book, "The Global Village." But what we've found in the 6 years that we have been doing this work is that there are new unifying artistic forms unique to cyberspace that can be intuited and that the true identity of a work like Cathedral is *only* comprehended over time, through its many personalities. In cyberspace, these musical experiences appear like a series of worlds unfolding in real time. There is obvious order, but no real map, and no "correct" way to go forward, because development is unfolding in all directions at once, creating an interactive matrix of possibilities and causing a plurality of forms to occur, all of which are individual, none of which are, in any sense, alike. Or, to look at it in yet another way, as musical space transforms in cyberspace, the artistic perceptive abilities of people will be expected to expand in order to comprehend three-dimensional thematic pathways that are to be taken with different pacings and different understandings of the flowing of time.

The result is a structure that is, at once, "simultaneous, discontinuous, and dynamic," to borrow another McLuhan phrase. McLuhan, who is distinguishing between different modes of viewing the world with this phrase, goes on to say that a simultaneous, discontinuous, and dynamic structuring of information can plunge man "into a new form of knowing, far from his customary experience tied to the printed page," and he suggests that knowing itself is being recast and retrieved in this newer, nonlinear, 360-degree, acoustic form.[4]

But is there any evidence to support our own artistic observations about this transformation of form and time on the Web? Yes, some, and I would like to mention the work of two researchers. As early as 1992, Marcos Novak, a transarchitect, artist, and theorist currently at UCLA, suggested in an essay titled "Liquid Architectures in Cyberspace," that on the Web "time itself may pulse, now passing faster, now slower." According to him, it is the infinite possibilities of relating forms that drives Internet time, and he thinks that we should abandon the metaphor of a flowing river of time in favor of the concept of metamorphic liquidity, architectures that, in his words, "breathe, pulsate, and transform from one form to another."[5]

And Dante Tanzi, who is a member of the technical and musical staff in the Laboratory of Musical Information at the University of Milan and who has been considering the questions of musical form and time on the Web for almost a decade, believes that "the mobile and immaterial architecture of cyberspace leads communication away from a sequential and linear style, compelling subjective time to flow again and again around decisions already made."[6]

He says that on the Web, "the differentiation between [the various] forms of time, [the] accumulation of performances, and [the] presence of decentralized configurations of subjectivity, upset not only [our] ideas of [musical] object and musical process but also [alter] the subjective point of view from which those ideas have been drawn."[7]

In a recent essay he goes even further in saying that the impermanence of contexts and the pluralities of time forms on the Web establish new conditions for musical awareness, leading to diverse types of imaginative and symbolic operations. And he observes that "we are asked to change our attitude towards the identification of musical events when music becomes a space inhabited and manipulated by many people, although the result is still to be perceived as a single process."[7]

Now the common theme running throughout all of these scenarios, it seems to me, is the necessity of acknowledging and accepting a new type of information exchange, one that encourages nonflowing musical time. The Web has created a new communications model that is more liquid that linear and has multiple points of access and a seemingly infinite number of communication paths. Certainly, this model is more directly related to the functioning of a neural network than the music-as-sonic-artifact concept of the past. Furthermore, the nonlinear, time-curving nature inherent in these new Web technologies, when distributed in large-scale throughout the network, offers, at least in principle, a new metaphor for consciousness.

VIRTUAL MUSIC

For Nora and me, it is this concept of a musical space created by many people, but perceived as a single unified process that is, we think, the most interesting and the one that holds the most artistic promise for the future, because in this type of environment, creativity is not isolated in one or two individuals, but rather is distributed throughout the network. And in this scenario—this resonant and interpenetrating process—not only are the nonlinear forms and content of musical experiences free to evolve and merge, but there is also a continuous invitation to listeners to synthesize and recreate. In this sense, Virtual Music is a new art form, within a new medium, producing an entirely new type of interactive artistic experience. And with this degree of decentralized authority, not only is there a redefinition of musical object and musical space, as well as a restructuring of the traditional composer, performer, and listener roles, but eventually it also becomes necessary to redefine music as well.

But how does this actually happen? How does music, or any other art form, get redefined? It happens through a paradigm shift. A paradigm is a mutually agreed upon set of beliefs that filters, and in some ways controls, the way we see and experience the world, assigning, as it does, validity to some concepts and withholding it from others. A good example is Einstein's theory of relativity, which replaced an earlier set of assumptions about time and space. Another is Darwin's theory of evolution. Mindsets such as these determine the kinds of questions that are asked and this, in turn, informs and influences our definition of the world. Remember that people once thought the earth was flat and acted accordingly. A paradigm shift occurred when they decided that it was round. The result of a paradigm shift always produces a new perspective within the collective consciousness. And in the last decade, I would suggest, there has begun a fundamental paradigm shift in the arts, as well as in the rest of our lives, from a media culture to a cyber culture, with a global audience and global communications and experiences, and worldwide creative interactivity. To restate this in more musical terms, we are moving from the ancient world of hearing, through the modern world of listening, to the cyber world of creative interactivity.

Now the proof of this paradigm shift, of course, is that the artists already get it and are employing the Web in all kinds of new creative ways that go far beyond commerce and information exchange. And with Virtual Music, we now have a set of characteristics that challenge some of our most basic musical assumptions. For example, Virtual Music is not an artifact. The music is as it is, when it is; if you return to it later, it will be different. Cathedral exists only in real time. Our live webcasts are interactive events that people at home not only hear but also participate in, and even affect the course of, in some small but not insignificant way. So for the first time, we have a universal form of music that has the ability to draw in and engage its listeners in the creative process and to decentralize the act of creating music in favor of community-driven decisions and actions. The problem for the composer, of course, is how to encourage the interactive features of cyberspace while still preserving some form of overall creative control that defines the work as his or her own. It is within this transforming and interactive context that the cyber artists of the future will need to reconsider and redefine the artistic parameters of what it means to be a composer.

But we can, even at this early stage, identify some of the basic characteristics of Virtual Music:

- It's **universal**; anyone with a computer and Internet access can be involved.

- It's **interactive**; people can participate in it and alter it in ways that they, as well as everyone else, can hear. And because it's interactive, it becomes a form of music that begins to participate in its own development, a process easier to characterize with neurologic or biologic metaphors, rather than with mechanical ones.

- It's an **individual experience,** different for each listener, not only because it's nonlinear and without beginning, middle, or end, but also because it's listened to differently, because it occurs in individual home environments.

- Its content is **inclusive, not exclusionary**. There is no one predominant or acceptable method, or sound, or style, or set of instruments, or performing environment for the World Wide Web. And there probably never will be. In cyberspace, there's room for everything; the possibilities appear endless; the future of the Web is filled with sound.

- At its core, it's **multimedia**, with video, even today, sometimes playing an almost equal role with audio. And we can expect this re-conception of video to increase substantially as broadband capabilities become more universally available. Furthermore, through some application of virtual reality or tele-immersion, Virtual Music will eventually engage all of our senses and literally transport us artistically to other worlds.

Now, while this last characteristic is still a dream, there are some people currently working to make it a reality. Jaron Lanier, the inventor of "virtual reality" and a musician himself, is the lead scientist on the National Tele-Immersion Initiative, a consortium of universities studying next-generation Internet technologies. He recently told me that he can envision a day when tele-immersion will be able to draw in the real world, rather than create a virtual one, and give the illusion that "people in different cities are actually in the same room." That, he said, "would change the nature of the game of on line music."[8]

And I think that as neuroscientists studying music, one of the key elements to keep an eye on in the future is the degree to which the distinction between art and gaming becomes increasingly blurred. Think of a time, for instance, when the artistic experience has the potential of being as entertaining, interactive, and engaging as the gaming experience, plus as meaningful and thought provoking as all good art. Or consider the possibility that Virtual Music has the potential to become a live musical organism, living in cyberspace, growing and changing course because of the collective actions of its users, and you begin to get a glimpse of where we may be going.

And what about the role of the artist in this scenario, a scenario in which future musical instruments may well be connected to XBOXES and look like Gameboys? According to McLuhan, again from "The Global Village," it is the role of the artist, as a cultural mediator, "to keep the community in conscious relation to the changing and hidden ground of its preferred objectives." To put this in more practical terms, as artists and as scientists, we must stay open to the possibility that the way we make art, perceive music, and interact with our audience will most likely be changing in the coming years. Hopefully, Nora and I are helping to facilitate this change with Cathedral. Hopefully, I have been able to open a door onto this work for you here.

REFERENCES

1. Cathedral website may be found at http://www.monroestreet.com/Cathedral
2. PitchWeb may be found at http://www.monroestreet.com/Cathedral/pitchweb/
3. Cathedral Band archive may be found at http://www.monroestreet.com/Cathedral/stage/
4. McLuhan, M. & B.R. Powers. 1989. The Global Village: Transformations in World Life and Media in the 21st Century. Oxford University Press. Cambridge.
5. Novak, M. 1992. Liquid architectures in cyberspace. *In* Cyberspace: First Steps. M. Benedikt, Ed. The MIT Press. Cambridge.
6. Tanzi, D. n.d. Time, Proximity And Meaning On The Net. CTheory.net, 2000. Found at http://ctheory.net/text_file.asp?pick+124
7. Tanzi, D. 2001 Observations about Music and Decentralized Environments. Leonardo **34:** 431–436.
8. Lanier, J. 2002. The interview may be found in the October 2002 edition of the American Music Center's on-line magazine NewMusicBox, found at http://www.newmusicbox.org (located in the archives).

Propositional Music from Extended Musical Interface with the Human Nervous System

DAVID ROSENBOOM

*California Institute of the Arts, School of Music,
Santa Clarita, California 91355-2397, USA*

ABSTRACT: Results obtained from projects in which self-organizing musical structures spontaneously arise through electrical interface between the brain and generative musical systems are surveyed. This provides a springboard for examining important paradigm shifts taking place in our thinking about what musical forms can be and how this might influence efforts to increase our understanding of the underlying neural dynamics. Implications of this work for the design of music curricula are considered, emphasizing the importance of active imaginative listening. A view of composing, termed "propositional music," is introduced in which the proposition of cognitive models of music is an ongoing part of creative musical activity.

KEYWORDS: propositional music; self-organizing musical forms; music learning; music and brain; music and electroencephalogram; brain music interface; brainwave music; neuromusic

THE BRAIN AS A MICROCOSM FOR THEORETICAL SYSTEMS EXPLORATIONS IN MUSIC

Whenever in history there has emerged an era of enriched inquiry into the nature of things and the natural phenomena of human beings, intense interactivity among the arts and the dominant systems of inquiry at the time has fueled the essential creative impulses of both. Since the development of what we now refer to as *scientific* inquiry, this interdisciplinary interactivity has been particularly vigorous. It has helped us glimpse possible unity in the multiplicity of ways we view the universe and to begin comprehending what a tiny portion of the brain's immense activity is actually comprised of what we call the conscious mind.

In this fascinating and newly developing, interdisciplinary field of neuroscience and music, it is of paramount importance—lest we accrue only a library of relatively inert facts—that we maintain keen awareness of the great breadth of human experiences that may be called *music* and that neuroscientists avoid the fatal, past mistakes

Address for correspondence: David Rosenboom, Dean, School of Music, California Institute of the Arts, 24700 McBean Parkway, Santa Clarita, CA 91355-2397, USA. Voice: 661-253-7816; fax: 661-255-0938.

david@music.calarts.edu

Ann. N.Y. Acad. Sci. 999: 263–271 (2003). © 2003 New York Academy of Sciences.
doi: 10.1196/annals.1284.037

of anthropologists when they approached studying new cultures while carrying hidden presumptions about what cultural models can be and therefore both missed the fundamental paradigms on which these unfamiliar cultures were based and, worse, sometimes destroyed them in the process. Similarly, musical artists must be aware of the differences and similarities in the two methods of inquiry. Both are full of creative investigation, hypothetical models, experimental methods, hypotheses, precise expressions, and esthetic musings. The most serious difference lies in the scientists' focus on experimental verification of purported facts, which can be shared reliability within a community, and the musicians' license to invent whole worlds of potential reality and then create and explore them as if they were indeed naturally and substantially manifest.

Since the beginning of our investigation of the brain as a substrate of human experience, artists have engaged in parallel explorations. The analysis of vision in both art and the science of visual perception is a prime example.[1] Music has had a similar relation with the study of auditory perception. Being largely a shared activity among multiple participants, music has also explored interactive processes. Cybernetics and systems theory played an important role, and now, explorations with neural networks, genetic algorithms, and complex adaptive systems are widespread in experimental music. Cognitive modeling—as part of building musical models—has become a component, whether overtly or subtly inherent, in compositional practice.[2]

SELF-ORGANIZING MUSICAL FORMS IN A FEEDBACK PARADIGM WITH EEG FEATURES

In other publications, I have surveyed musical and artistic projects that originated, in part, through applying results from brain science, emphasizing those employing direct monitoring of biological phenomena, such as EEG, EMG, EKG, GSR, and respiration in the creation and performance of original works of music, visual art, kinetic art, and dance.[3,4] The field has grown very wide since then with projects ranging from music related to DNA codes, genetic mapping, and artificial life, to genetic engineering and prosthetic body extensions as art forms. My own work began using statistical and spectral analyses of EEG and other measures in the creation of musical forms, installations, and performances in the late 1960s.

My 1972 work, *Portable Gold and Philosophers' Stones*,[4,5] used measures of the stability and coherence of prominent EEG frequency components—employing spectral decomposition and autocorrelation statistics—to direct the evolution of an electronic, musical landscape. An increase in the coherence of these components resulted in expanding both the range and the detail of control over electronic sounds that was given to features of the performers' EEGs. This piece was particularly effective in creating easily identifiable and differing results when the performers engaged in different styles of meditation, performed by focusing on active versus passive listening, and they used different techniques to control the distribution of attention.

Soon, this work evolved in the direction of using auditory event-related potentials (AEPs) to investigate the perception of *musical forms*. With this term I refer to the structures and shapes contained in the way certain musical parameters, such as pitch, loudness, timbre, degrees of variation in rhythm sequences, relatedness to a pitch center or harmonic fundamental, the succession of pattern groupings in a hierarchical array, movement of sound in space, and the apparent complexity of sound aggregates change during the course of a performance.

By the mid-1970s, I had developed an interactive, feedback tool with which to carry this out[3] and which I used in my 1976–77 composition, *On Being Invisible*.[5] In brief, the system worked as follows. Computer algorithms were developed that could automatically generate an array of musical forms realized with electronic sounds. Depending upon experimental or musical purpose, the process could begin with either pre-composed forms and procedures or sounds generated by means of stochastic techniques using probabilities. In either case, the sound output was analyzed with a partial model of musical perception, a key component of which involved a *unidirectional, rate-sensitive, difference detector* applied to the contours—changes in time—of several musical parameters. The primary output from this analysis was a proposed parsing of the sound stream into presumably, perceptually meaningful groupings. The boundaries separating these groupings were further hypothesized to be places where one would be highly likely to find shifts in or arousal of attention like that often associated with the production of significant features in the AEP. The most useful of these AEP features was the P300 wave, which is associated with the consolidation of percepts into memory. In other words, it was predicted by the model that a listener's brain would process these as having significance in the emerging musical form. The next step was testing the prediction. At first, this involved looking for changes in coherent-wave statistics towards desynchronization as an indicator of attention shift. Later, AEPs were recorded from an active listener—also considered to be a performer—and analyzed by means of signal averaging and adaptive template matching. This template-matching procedure was intended to speed up the process of obtaining results on which the remainder of the feedback process could be based with reasonable confidence, without requiring the accrual of a large number of waveforms contributing to an average. The amplitudes of P300 and sometimes N100, known as the *attention wave*, were used to decide if the prediction was confirmed by large or growing amplitudes or denied by small amplitudes or missing AEPs. If the results yielded confirmation, then in the generative compositional model, the likelihood would be increased that the kinds of changes in sounds or sound patterns associated with this confirmation would occur again. If the predictions were not confirmed, then some kind of mutation process would be invoked, usually through stochastic procedures, to move the sounds into new, uncharted territory. As this process continued, the computer system would store the sound sequences for which parsing with reliably confirmed results were obtained. These were called *clangs*, after composer/theorist James Tenney's terminology.[6]

Eventually, after enough clangs were stored, a hierarchical structure builder would be activated, and sequences containing several clangs would be assembled and output. In this process, the model of perception would analyze the recent history of the ordering of these groups and make predictions based on calculating an expectancy value for each one at the current point in the sequence. If an event with high expectancy was output, it was considered unlikely to be associated with a large-component AEP. If one with low expectancy occurred, a large-component AEP was predicted. Again, confirmation or disconfirmation of the predictions with AEP data was used to either consolidate or mutate the output stream and build hierarchical groupings of clangs into *sequences* and *collections*. In addition to expectancy, the process involved other calculations, such as the strength of each prediction, the weighting of changes in each parameter, the recent context of variance in each parameter, and masking in time of small changes by recent, large changes. *On Being Invisible* has been described as an *attention-dependent sonic environment*.

Some of the most musically interesting performances were characterized by waves of convergence towards patterned musical behavior and divergence away to more unpredictable sounds. These waves would move back and forth in a dynamic fashion, which for the listener, was imbued with its own kind of drama. The metaphor of invisibility in the title was invoked for many reasons, one of which refers to the role of an individual interacting within a system larger than himself in deciding when to initiate action, to initiate change by volitional means, and to simply be a part of a larger process.

It is very important to emphasize that when analyzing all this in retrospect, it became clear that the best results were obtained when the performer/participants were engaged in what I will call *active imaginative listening*. It is also important to note that the musical forms in this work are emergent, not knowable in advance, yet imbued with rich structure. Indeed, active imaginative listening to emerging forms may be the most important factor driving the evolution of music in our culture.

It is clear that no cause and effect model has so far succeeded in illuminating the spontaneous emergence of particular musical forms and styles over the course of cultural history. It is more likely that such forms emerge from interactions among the underlying, component parts in a complex adaptive, sociobiological aggregate and the overarching cultural dynamics. Such a view is termed a *holarchic model of musical morphogenesis*. Similarly, although stimulus and response models have been effective in illuminating fundamental components in the brain's neural circuitry, they have not been as effective in illuminating the more global dynamics of perception and cognitive organization that operate in elaborate, participatory experiences, such as those associated with high-level musical forms. A more global view of emergent processes may be required.[7]

A fruitful musical spin-off from this work has been the utilization of the partial model of musical perception in building an intelligent musical instrument to play in improvisations and use in composition. I wrote a computer program, called HFG, Hierarchical Form Generator, that has the capability of parsing musical inputs, the nature of which is entirely unknown in advance.[8] HFG uses similar principles for tracking parametric changes, parsing the musical input into chunks, and storing the results in memory. These parsed chunks are then made available for recall by mapping them onto the available stimuli of the instrument (for example, the keys on a keyboard). A library of transformation procedures is also provided with which the chunks can also be altered along with the means to output them singly or in hierarchical groupings. All this takes place in real time, under the control of the improvising musician. The result is a new kind of instrument, which has been worthy of extensive practice in order to learn to *play* the models and the algorithms. This kind of playing includes the added component of interacting with something that is modeled on aspects of musical feature perception presumed to imitate those taking place in the brain. Some of the musical results have been recorded and released.[9,10]

In recent years, I have returned to this system, most notably in a new composition, which I describe as a self-organizing, interactive, multimedia, chamber opera, called *On Being Invisible II (Hypatia Speaks to Jefferson in a Dream)*.[11] The sequencing of this work's nonlinear narrative content, the ordering of projected, digital video images, and the structure of accompanying electronic sounds all result from a similar analysis of the brainwaves of two performers. Consequently, the form of the opera cannot be known in advance. In this case, new techniques resulting from advances in dynamic systems analysis are also employed.

Recently, our work has been informed by studies relating musical perception and the dimensional complexity of brain activity. Some of these experiments explore what is known as the nonlinear resonance hypothesis of music perception in which relationships are drawn between the complexity of music stimuli, the complexity of observed EEG waveforms, levels of musical training and experience, and the presumed activation of neural assemblies necessary to process rich associations. In these studies, complexity is given a precise mathematical definition—commonly used in studying chaotic systems—involving calculating the correlation dimension of a time series. It has been shown that subjects' ratings of the subjective complexity of the music stimuli followed the mathematical complexity measures.[12] The complexity of EEG waveforms and music stimuli are strongly related, but also strongly affected by the level of subjects' musical training and experience. Mayer-Kress[13] has also presented an interesting and related sonification experiment with the EEG. In this exercise, rapid short-time synchronization events in the EEG, presumed to accompany cognitive events and perception, were mapped onto a musical texture of independent instrument sounds that were drawn into transient synchronization as concomitant events were detected in the EEG.

I have used these ideas in further composition and feedback paradigms. In a work for instrumentalists, *Two Lines*, I used correlation dimension ideas in the analysis and construction of a long melodic line.[14] In this case, the investigation was into the imperfections and instabilities that appear while a musician attempts to perform a very difficult, proprioceptive, musical, and motor task requiring great stability, playing a precise, smooth, musical drone for a very long time. In this piece, the drone was eventually removed, while its instabilities were retained, highly magnified, orchestrated, and turned into music. Although on initial exposure it may sound chaotic, phase portrait plots of pitch transitions in the remaining melody reveal interesting tendencies towards certain pitch attractors.[15] Perhaps, these expose some of the dynamics of motor output programs and compensating corrective actions in response to perception. Analysis of its dimensional complexity is also interesting in how it relates to the overall musical gestalt. One could also think of this as using active listening as an investigative tool for developing insight into a complex phenomenon.

In *On Being Invisible II...*, this new tool of measuring dimensional complexity can also be used in the feedback paradigm. First, electronic sound is produced in the same way as in the earlier version of *On Being Invisible*, and AEP testing provides feedback driving the hierarchical organization of sounds. In addition, the dimensional complexity of the EEG time series data is correlated with that for musical parameters, and feedback is given to the performers for concurrence or nonconcurrence in the two dimensionality measures. Much more work with this and related techniques is being planned.

In a more recent piece, *Four Lines*,[16] dimensional complexity is used in another way. This time, an AEP time line is constructed, showing the occurrence of AEPs recorded simultaneously from two brainwave performers during a particular performance of *On Being Invisible II*.... This is analyzed and correlated with the time line of parsed musical changes in two corresponding lines of electronic music. The juxtaposition of continuity with extremes in these lines is compared with their changing complexity measures and used in composing music that attempts to capture and explore the apparent organization of perceived musical segments and aspects of the psychological time experienced in these performances. Again, if one listens openly without perceptual bias, this sonic landscape offers an opportunity to use the enormous integrative powers of auditory apprehension to explore how we hear in such an abstract landscape and how our musical minds assemble experiences for us combining perception with imagination. Listening while observing how we internally synthesize sound apperceptions and learning from that is part of the objective. It takes great discipline and practice to accomplish this, but it is well worth the effort.

IMPLICATIONS FOR MUSIC LEARNING

In studying the interaction of the brain with sounds made to be somehow responsive to electrochemical indicators of brain activity, especially when working with sounds in their raw form, that is, not already imbedded in a presumed model of what music is, many interesting things about musical attention emerge. Expectancy and surprise, redundancy and novelty, are, of course, important factors. Most individuals, no doubt, seek after optimal, mental activation levels, summing external and internal experience. Things falling outside what is comfortable, outside their associated *Wundt curves* mapping positive affective responses to stimuli are the most likely to desynchronize the EEG.[17] These are also context sensitive, determined by degrees of variance among sound parameters during their recent histories. When viewed over larger time-space scales, they also become cultural objects full of referents. The referents gain significance as, over time, meanings are invented by each individual and attached to raw sound forms. Some of these become adopted by cultures and then taught to young populations as a way of speeding up the learning process. However, this is accomplished at the great cost of excluding huge potential vocabularies of sound as having musical potential. Sound is made musical most significantly through active imaginative listening, which then may be applied through the proactive actions of attempting to imbue sound forms with aspects of one's musical intelligence. That creates *musical forms*. However, musical forms have been misapprehended as ruling agents rather than emerging properties of interacting components. All such forms are impermanent.

I have described elsewhere a collection of results regarding our ability to record neuroelectrical concomitants for a variety of musical parsing actions, including novel things like segmentation and fusion of musical lines being shifted in phase, correct and incorrect identification of musical stimulus targets, and even AEPs associated with overlaid, imagined events, such as emphasizing rhythmic groupings of repetitive stimuli where no acoustic correlate for such grouping exists, and long-latency AEP components for imagined sound events that are highly expected but nonoccurring.[3] Many of these carry import for various skilled musical practices.

ACTIVE IMAGINATIVE LISTENING AND CREATIVE MUSIC MAKING

In curriculum design it is important to understand how active imaginative listening and composition share significant commonalties. Both involve construction—whether overtly conscious or not—of symbolic mental models, making discoveries through differentiation, parsing, and mental reorganization, distilling efficient representations for synthesizing memory images by means of hierarchical grouping, integrating the resulting engrams into the ongoing neural dynamics of the organism, and subsequently, resynthesizing from memory during recall or association through resonant correlation. In active listening, this *compositional act* manifests in the construction of internal, mental images, whereas the isolated, *pre-compositional acts* of composers focus on creating rich, associative, sonic environments within which rewarding, active listening may be encouraged. Both composition and active listening are essential components for completing meaningful musical experiences.

In studies measuring affective reactions to sequences of tones with various amounts of stimulus variation and complexity, the usual, inverted U-shaped, *Wundt curve* is obtained for randomly selected subjects.[18] However, when the subjects are selected for high levels of musical experience and training, the curve shows generally greater positive measures. It has even been suggested that for subjects who may be considered super-musicians, the curve tends to flatten out. In other words, positive—not neutral—affect is reported for all stimulus sequences. The best musicians, those who listen most actively and imaginatively, can find interesting content or invent interesting meaning for nearly any complex of sounds. Far from being non-discriminating, such artists are furthermore able to take sound in its raw entirety, listen analytically without requiring a presumed syntax, and perform the next steps of choice making as they practice their art and create rich musical experiences. Such musicians, when participating in one of my feedback-based, self-organizing, musical experiences, will produce results that may not fall inside expected paradigms for perception of standard musical forms. The sonic results may be quite extraordinary. This also reminds us of the natural, perceptual world of the child.

It is well known that certain innate potentialities in children become replaced by other competencies as specialized learning progresses. The trick in education is to preserve these earlier potentialities, while helping the child acquire the practical competencies needed to realize ideas creatively, as does the super-musician. In both experimental work and music pedagogy, however, we are often driven to obtain the biggest effect for the least investment. This requires predefining the effect desired, which carries with it the danger of limiting the definition of music competency so that it encompasses only that which can be obtained by this minimal investment. This can be catastrophic for both music and science.

The essential goals for the future of our music curricula must include the reintegration of performance, composition, and improvisation as basic to music learning, all informed by active imaginative listening. Current research suggests that when one is fully engaged in this kind of listening, accompanied by free form analysis and imaginative synthesis of memory engrams, nearly all the brain's available resources are called into play.[19] The total musician must be educated early and not too late. Improvisation should be an integral part of musicianship training. Musical invention should always be encouraged in parallel with acquiring musical discipline. Improvisation should be taught at the beginning of training. Improvisation is a deep discipline, requiring intensive practice and the creative construction of one's individual practice methods. Improvisation and mature performance demand spontaneous interpretive plasticity, requiring rapid modification of motor output patterns, which cannot be known in advance. During the development of such high levels of musicianship the cerebellum may be loaded with a very large array of potential output patterns that can be quickly activated. The composition of this array requires deeply informed training and practice, for it will determine, in part, aspects of a musician's technical and stylistic capacities. Collaboration encouraging specialists such as performers and composers to exchange roles helps in learning. Musicianship should always be taught with students' primary sound-making vehicles, with instruments as well as the voice, to fully integrate the results. What is learned in theory exercises should always be played. Different kinds of notation and visual representations for music should be explored, even at an early age. Teachers should play with their students to encourage aural learning alongside technical, physical, and notational learning. The key is <u>listening</u> — deep listening and parsing of newly presented material is essential for bringing coherence to subsequent, real-time performance through re-structuring or re-forming what has been perceived. Improvisation—or spontaneous music making—is the vehicle that transports the full spectrum of musical realization from the realm of abstraction to that of actualization—with the full engagement of intellect, intuition, imagination, proprioception, and physical and psychological being. The total human becomes the total musician. That's how we learn.[20]

PROPOSITIONAL MUSIC

Elsewhere, I have written extensively about what I call "propositional music."[21] This term refers to a particular style of musical thinking in which the act of composing includes proposing complete musical realities, emphasizing the dynamic emergence of forms through evolution and transformation. It presupposes no extant model of music and no predefinition of a proper critical stance about music. It views composers as having the license and ability to propose complete cognitive models of music. I propose that if neuroscience and music are to collaborate effectively in an endeavor that purports to illuminate both, something like this way of thinking about music will be essential.

REFERENCES

1. VITZ, P.C. & A.B. GLIMCHER. 1984. Modern Art and Modern Science, The Parallel Analysis of Vision. Praeger. New York, NY.
2. ROSENBOOM, D. 1987. Cognitive modeling and musical composition in the Twentieth Century: a prolegomenon. Perspect. New Music **25:** 439–446.
3. ROSENBOOM, D. 1997. Extended Musical Interface with the Human Nervous System: Assessment and Prospectus. [Revised version, The MIT Press, electronic documents: http://mitpress2.mit.edu/e-journals/Leonardo/isast/monograph1/rosenboom.html] (Original work published 1990, San Francisco, CA: Leonardo Monograph Series, **1.**)
4. ROSENBOOM, D., Ed. 1976. Biofeedback and the Arts, Results of Early Experiments. Aesthetic Research Centre of Canada Publications. Toronto, Ontario.

5. ROSENBOOM, D. 2000. Invisible Gold, Classics of Live Electronic Music Involving Extended Musical Interface with the Human Nervous System. Audio CD. Pogus Productions. Chester, NY. **21022–2.**

6. TENNEY, J. 1986. Meta + Hodos and META Meta + Hodos. Frog Peak Music. Hanover, NH.

7. NUNEZ, P.L. 2000. Toward a quantitative description of large-scale neocortical dynamic function and EEG. Behav. Brain Sci. **23:** 371–398.

8. ROSENBOOM, D. 1992. Parsing real-time musical inputs and spontaneously generating musical forms (Hierarchical Form Generator, HFG). *In* Proceedings of the 1992 International Computer Music Conference. Computer Music Association. San Francisco, CA.

9. ROSENBOOM, D. & A. BRAXTON. 1995. Two Lines. Audio CD. Lovely Music Ltd. New York, NY. LCD3071.

10. ROSENBOOM, D. with T. SANKARAN & C. HADEN. 1993. Extended trio: sampler (1992), *On* Hallways, Eleven Musicians and HMSL. Audio CD: Frog Peak Music. Hanover, NH. FP002.

11. ROSENBOOM, D. 2000. On Being Invisible II (Hypatia Speaks to Jefferson in a Dream). On Transmigration Music. Audio CD. Centaur Records. Baton Rouge, LA. CRC2490.

12. BIRBAUMER, N.W. *et. al.* 1996. Perception of music and dimensional complexity of brain activity. Int. J. Bifurc. & Chaos **6:** 267–278.

13. MAYER-KRESS, G. 1994. Sonification of multiple electrode human scalp electroencephalogram (EEG). Poster presentation at ICAD'94, Santa Fe Institute. Santa Fe, NM.

14. ROSENBOOM, D. 1989. Two Lines. Music score. Frog Peak Music. Hanover, NH.

15. ROSENBOOM, D. 1996. *B.C.—A.D.* and *Two Lines*, two ways of making music while exploring instability. Perspect. New Music **34:** 208–226.

16. ROSENBOOM, D. 2001. Four Lines. Music score. Frog Peak Music. Hanover, NH.

17. VITZ, P. 1974. Experiments on affective reactions to 'music' stimuli. Scientific Aesthetics **IX:** 3–14.

18. VITZ, P. 1966. Affect as a function of stimulus variation. J. Exp. Psychol. **71:** 74–79.

19. HODGES, D.A. 1996. Neuromusical research: a review of the literature. *In* Handbook of Music Psychology. D.A. Hodges, Ed. :197–284. MMB Music. St. Louis, MO.

20. ROSENBOOM, D. 1996. Improvisation and composition—synthesis and integration into the music curriculum. *In* Proceedings, the 71st Annual Meeting. :19–31. National Association of Schools of Music. Reston, VA.

21. ROSENBOOM, D. 2000. Propositional music: on emergent properties in morphogenesis and the evolution of music, essays, propositions, commentaries, imponderable forms and compositional method. *In* Arcana, Musicians on Music. J. Zorn, Ed. :203–232. Granary Books/Hips Road. New York, NY.

The Perception of Microsound and its Musical Implications

CURTIS ROADS

*Media Arts and Technology Program and Center for Research in
Electronic Art Technology (CREATE), University of California,
Santa Barbara, California 93106, USA*

ABSTRACT: Sound particles or microsounds last only a few milliseconds, near
the threshold of auditory perception. We can easily analyze the physical properties of sound particles either individually or in masses. However, correlating
these properties with human perception remains complicated. One cannot
speak of a single time frame, or a "time constant" for the auditory system. The
hearing mechanism involves many different agents, each of which operates on
its own timescale. The signals being sent by diverse hearing agents are integrated by the brain into a coherent auditory picture. The pioneer of "sound
quanta," Dennis Gabor (1900–1979), suggested that at least two mechanisms
are at work in microevent detection: one that isolates events, and another that
ascertains their pitch. Human hearing imposes a certain minimum duration in
order to establish a firm sense of pitch, amplitude, and timbre. This paper traces disparate strands of literature on the topic and summarizes their meaning.
Specifically, we examine the perception of intensity and pitch of microsounds,
the phenomena of tone fusion and fission, temporal auditory acuity, and preattentive perception. The final section examines the musical implications of microsonic analysis, synthesis, and transformation.

KEYWORDS: music perception; temporal acuity; microsound; music composition; sound particles; grains; pulsars; pointillism

THE TIMESCALE OF SOUND PARTICLES

Music transpires on many timescales. For a given musical composition, the top-
most timescale is the macroform comprising the major sections of the work. Beneath
this layer are the myriad phrases or mesostructures within each section. Each phrase
can be partitioned into a collection of individual sound objects (or notes) that form
the sonic surface of the work. In 1946, the Nobel prize–winning physicist and inventor, Dennis Gabor, postulated the existence of another multilayered stratum beneath
the sonic surface: the microsonic hierarchy. In Gabor's view, all sonic phenomena
can be decomposed into collections of sound particles on a timescale of 1 to 100 ms.
He called these particles *sound quanta*. As Gabor observed, the timescale of sound
quanta is of special significance to auditory perception because it delimits the thresh-

Address for correspondence: Curtis Roads, Media Arts and Technology Program and
CREATE, 3431 South Hall, University of California, Santa Barbara, CA 93106. Fax: (805) 893-
2930.

clang@create.ucsb.edu

Ann. N.Y. Acad. Sci. 999: 272–281 (2003). © 2003 New York Academy of Sciences.
doi: 10.1196/annals.1284.038

old of hearing for such properties as pitch, timbre, amplitude, spatial position, and microevent order.[1–3]

For events that last less than about 1 ms, many mechanisms of sound recognition break down, relegating these events to subsymbolic status. For example, the individual samples of a linear pulse-code–modulated audio signal, which last 22 μs or less (depending on the sampling rate), can hardly be distinguished from one another. It takes a series of hundreds of individual samples to form an acoustic particle that the ear can distinguish as being unique.

Computer-based tools let us view and manipulate the microsonic layers, from which all acoustic phenomena emerge. Electronic musicians can synthesize new tones made up of streams or clouds of sonic particles. This is the domain of *particle sound synthesis,* which takes different forms depending on the properties of the particles and the strategy for particle generation. *Granular synthesis* and *pulsar synthesis* are two of the best-known techniques.[4–6]

In the field of psychoacoustics, much study has gone into the perception of microevents. The source material for these studies consists primarily of individual particles (impulses, tone and noise bursts, tone pips). Understandably, these studies do not address the musical potential of direct composition with sound particles. Since particle synthesis techniques are now widespread in electronic music circles, it is time to reconsider these studies in this light.

With modern sound analysis software, it is easy to analyze the acoustical properties of sound particles from myriad angles. However, correlating them with human perception remains complicated. We cannot paint a complete portrait of the relations between microsound and perception. The perception of microsound in musical contexts is intertwined with cognitive functions, both rational and emotional, that resist understanding from a scientific point of view. The background and mood of the individual listener are like a narrow filter imposed on the sonic sensation. No two persons hear with the same brain. We could say that the communication between the composer and the listener is "imperfect." This is not a flaw; rather, it is part of the fascination of music.

The remainder of this paper harvests results from scientific studies in the perception of microsound. I compare scientific studies with my own experiments in the studio and analyze the musical implications from the perspective of the composer of electronic music. Some of this material appeared in my book entitled "Microsound,"[6] but this paper is a new formulation based on additional research.

THE LIMITS OF TEMPORAL PERCEPTION

Sound has a powerful effect on human beings. As it penetrates the human auditory system, acoustic energy is converted from variations in air pressure (outer ear) into mechanical vibrations (middle ear), liquid waves (inner ear), and ultimately patterns of electrical impulses transmitted to the brain via the auditory nerve. In turn, these electrical patterns trigger biochemical changes in the brain and the body. Since the auditory system processes different media (mechanical, liquid, electrical, chemical) and employs multiple agents operating on their own timescales, one cannot speak of a single time frame or a "time constant" for the auditory system.[7] The brain integrates signals being sent by diverse hearing agents into a coherent auditory picture.

The earliest research on the latency of the human auditory system can be traced to the great pioneer of nineteenth-century acoustics, H. Helmholtz. He was trained as a surgeon and professor of anatomy. One of his first discoveries was the speed at which nerves conduct electrical impulses.[8] This and later research have determined that the nerve fibers conduct impulses at rates between 1 and 100 meters per second, with faster rates corresponding to larger nerve bundles.[9] This was the first physical measurement of the "thickness of the present,"[10] corresponding to the accumulation of sources of inherent latency in the human auditory system.

Since that time, ascertaining the temporal limits of sound perception has intrigued scientists.[10–14] Measurement is made complicated by the fact that it is hard to separate temporal effects from spectral effects. As is well known in signal processing, to change the duration of a sound simultaneously changes its spectrum. In particular, the shorter a sound, the broader its spectrum. To cite one example, subjects can discriminate a single click from a pair of clicks separated by as little as a few dozen microseconds. The single click and the double click have spectral differences that appear in the highest octave of perception. When noise is added to mask frequencies in this octave, the threshold value of the gap between the particles increases dramatically, indicating that the initial result was not a direct measure of temporal resolution.

Gabor[1–3] suggested that at least two mechanisms are at work in the detection of microsonic events: one that isolates events in time and space, and another that ascertains their pitch. In line with this theory, Edward G. Jones pointed out a bifurcation of function in his description of two parallel auditory pathways, the tegmental and the direct.[15] Zatorre et al.[16] point to a similar bifurcation of auditory pathways for "temporal" versus "spectral" information.

In their important book "Audition," Buser and Imbert[12] summarize a number of experiments with transitory audio-phenomena. The general result of these experiments is that for stimuli lasting less than 200 ms, many aspects of auditory perception change character. Different modes of hearing come into play in the perception of intensity, fusion/fission, and pitch. The next sections address these issues.

MICROTEMPORAL INTENSITY PERCEPTION

In general, subjective loudness diminishes with shrinking duration below 200 ms. Specifically, at low amplitudes, short sounds must be much greater in intensity than longer sounds in order to be perceptible. The difference in the detectability threshold is on the order of +20 dB SPL (approximately 1000%) for tone pips of 1 ms as opposed to those of 100 ms in duration.[12]

MICROTEMPORAL FUSION AND FISSION

Machine circuits can easily track pulse patterns in range of gigahertz, as any computer demonstrates. By contrast, the human auditory system is constrained by severe limits in its temporal resolution. Human beings lose track of pulse rhythms at tempi greater than about 20 Hz. If one tone pip follows less than 200 ms after another, the onset of the second tone pip will tend to be masked by the former, a time-lag phe-

TABLE 1. Average pitch recognition time as a function of frequency

	100	500	1000	5000
Frequency (Hz)	100	500	1000	5000
Minimum duration for pitch recognition (ms)	45	26	14	18

nomenon known as *forward masking,* which contributes to the illusion of continuous tone. When the onset time between pips is less than about 50 ms (corresponding to a particle emission rate of 20 Hz) human perception reaches "attentional limits" and groups the successive events into the illusion known as a continuous pitched tone.

Similarly, at fast tempi, a rapid sequence of tones at different pitches blends into a continuous "ripple." Here the auditory system is unable to successfully segment the incoming information into a temporal structure (i.e., to track its rhythm and melody) and simplifies the situation by interpreting it as a rapidly varying texture. Bregman describes these phenomena in his theory of *auditory streams.*[17–19] Conversely, a stream of particles that is already perceived as a fused tone can be made to segregate or disintegrate when the pitches, timbres, spatial positions, or amplitudes of the particles differ beyond certain thresholds, or more obviously when the onset time between events is sufficiently great.

Using sound analysis techniques such as the tracking phase vocoder[20] or matching pursuit analysis with wavelets,[21,22] we can make a sound such as continuous speech disintegrate (or coalesce) by deleting (or adding) individual particles on the Gabor matrix or time-frequency plane (Ref. 6, track 65 of the audio compact disc).

MICROTEMPORAL PITCH PERCEPTION

The acoustician Werner Meyer-Eppler was one of the most important figures in the early development of electronic music. (Karlheinz Stockhausen considers him to be his most important teacher.[24]) In the 1950s, Meyer-Eppler's experiments showed that the time needed to recognize the pitch of a tone is dependent on the tone's frequency, with the greatest pitch sensitivity in the middle frequency range between 1000 and 2000 Hz. TABLE 1, cited in Ref. 25, confirms this view.

As TABLE 1 shows, it takes only 14 ms to recognize a pitch of 1000 Hz, while tones both higher and lower than 1 kHz take longer to recognize. Doughty and Garner[11] divided the mechanism of pitch perception into two regions. They estimated that above about 1 kHz a tone must last at least 10 ms (ten cycles) to be heard as possessing the characteristic of pitch. Below 1 kHz, at least two to three cycles of the tone are needed. As Helmholtz[25] observed, the phenomenon of recognizable pitch evaporates below approximately 40 Hz.

MICROTEMPORAL AUDITORY ACUITY

Green[26] suggested that temporal auditory acuity (the ability of the ear to detect discrete events and discern their order) extends down to durations as short as 1 ms. Microevents that are less than about 2 ms in duration are heard as a click, but we can still change the waveform and frequency of the events to vary the timbre of the click.

Such spectral cues also help us detect whether a brief asymmetric particle is played forward or backward.

Detection of gaps in white noise extends down to the range of 2 ms.[27] For narrowband noise signals, the detection of gaps ranges from 22 ms for a band centered at 250 Hz to 3 ms for a gap centered at 8 kHz. Many studies show increased temporal acuity for sound events above 2 kHz. Detection of gaps in pure sinusoids goes down to 5 ms for "preserved phase" sines where the endpoint and beginpoint of an interrupted sine tone match.[27] Shorter events (in the range of microseconds) can be distinguished on the basis of amplitude, spectrum, and spatial position.

MICROTEMPORAL LOCALIZATION

Human beings sense delays in signal arrival times between the two ears with an accuracy of a few microseconds.[28] This interaural sensitivity is exploited by the auditory system in order to localize microsounds in space. The perception of the spatial position of a sound particle is distorted by a *localization blur* introduced by the human auditory system. Localization blur means that a point source sound produces an auditory image that spreads out in space. For brief Gaussian tonebursts, the horizontal localization blur is in the range of 0.8° to 3.3°, depending on the frequency of the signals.[29] The localization blur in the median plane (starting in front, then going up above the head and down to behind the head) is greater, on the order of 4° for white noise and becoming much greater (less accurate) for purer tones.

MICROTEMPORAL PREATTENTIVE AND SUBLIMINAL PERCEPTION

Certain sounds are too extreme in frequency or amplitude to be perceived; some transients are too short in duration. Yet the nervous system may react to these varieties of acoustic radiation, regardless of whether they penetrate the consciousness of the subject.

One of the most important measurements in engineering is the response of a system to a unit impulse.[30] Auditory neuroscientists have sought a similar type of measurement. The impulse response equivalent in the auditory system are the *auditory evoked potentials,* measured on the scalp, which follow stimulation by tone pips and clicks. The first response in the auditory nerve occurs about 1.5 ms after the initial stimulus of a click, in the realm of *preattentive perception.*[31] Preattentive perception performs a rapid analysis by an array of neurons, and combines this with past experience into a wave packet in its physical form, or percept in its behavioral form. The neural activities sustaining these acts of preattentive perception take place in the cerebral cortex. Sensory stimuli are preanalyzed in both the pulse and wave modes in intermediate stations of the brain. As Freeman[31] notes, in the visual system, complex operations in the retina and lower brain such as adaptation, range compression, contrast enhancement, and motion detection take place. Sensory stimuli activate *feature extractor* neurons that recognize specific characteristics. Comparable operations have been described for the auditory cortex. The last responses to a click occur some 300 ms afterward, in the medial geniculate body of the thalamus in the brain.[12] As Buser and Imbert indicate, the latent response is relevant to attention and expectation.

I should also mention *subliminal perception* or perception without awareness. It has been shown that the brain takes notice of very high frequency sounds above 22 kHz, whether or not the mind is consciously aware of them.[32,33] Another example of perception without awareness occurs when rhythmic sounds are modulated in ranges below the threshold of perception. Recent studies have shown that these modulations nonetheless enter in unconscious synchronized motor responses.[34–37]

A class of psychological studies has tested the influence of brief auditory stimuli on various cognitive tasks. In most studies, these stimuli are verbal hints to some task asked of the listener. Some types of influences have apparently been shown, but the results are not clear-cut. Part of the problem is theoretical: how does subliminal perception work? According to a cognitive theory of Reder and Gordon,[38] for a concept to be in conscious awareness, its activation must be above a certain perceptual threshold. Magnitude of activation is partly a function of the exposure duration of the stimulus. A subliminal microevent raises the activation of the corresponding element, but not enough to reach threshold. The brain's "production rules" governing awareness cannot fire without the elements passing threshold, but a subliminal microevent can raise the current activation level of an element enough to make it easier to fire a production rule later.

This is potentially significant for music. If the subliminal hints are brief musical cues (to pitch, timbre, spatial position, or intensity), then we can embed such events at pivotal instants, knowing that they will contribute to a percept without the listener necessarily being aware of their presence. Subtlety of effect is an important aspect of the aesthetic of composition with microsound. Barely perceptible variations in the properties of individual microevents—their onset time, duration, frequency, waveform, envelope, spatial position, and amplitude—lead to different aesthetic percepts on a higher timescale. For example, a shift of phase in one sound particle is imperceptible, but successive shifts in a series of particles produce a strong impression.

My colleague Stephen Pope has suggested that the domain of subliminal perception is really much broader than it is often assumed to be. For example, how many listeners in a concert hall are truly aware of the unfolding mechanisms at work in Beethoven's late string quartets, much less those of Bartok? They may hear every note, but they are not fully aware of the structural connections that the notes form. Only a detailed analysis reveals these structures. In this sense, the effect of art is largely subliminal.

CONCLUSION: THE MUSICAL IMPLICATIONS OF MICROSOUND

Music differs radically from language and speech.... [In music] we are dealing with a system of communication in which each piece of music consists of a few nonreferential items that are combined according to the prevailing stylistic "syntactic" rules of harmony, melody, timbres, rhythm, or musical form. Music is thus a game of combinatory acoustical constructs that the brain of the composer can conceive and that the listener should be able to learn or discover.... Contemporary music, by making use of new combinatory rules and the great novelty of its acoustic units and clusters, obliges the listener to expend greater effort in perceptual preparation and learning.

—O.S.M. Marin[39]

Musical culture is constantly evolving. In considering the application of neuro-science to music, we must be careful to define music in the broadest possible sense. Otherwise the discourse may be distorted by outmoded models. To cite an example, at a past conference "The Brain and Music," the composer Michael Tippett[40] portrayed microtonality (the use of non-equal-tempered melodic scales) as "a deviation from an unconscious or innate notation of tempered whole tones and semitones." Yet equal temperament is an artifact of European music history, not an "innate" characteristic of the human brain. The invention of equal temperament is well documented.[41] Microtonal scales are in widespread use in nonwestern cultures and by avant-garde musicians.

An individual listener can appreciate radically diverse forms of musical expression. This alone indicates that the foundations of music cognition extend far deeper than any traditional music theory dogma. Ultimately, music is rooted in the senso-motoric dynamics of the human nervous system. From the rhythms of the brain, to the nervous system and the gestures of the muscles that it controls: "Music is primarily a matter of biology."[42] This being so, many issues in music perception can be analyzed without reference to musical stimuli, per se, such as the perception of temporal intervals on multiple timescales.[43]

Human auditory perception is exquisitely attuned to the musical properties of microevents. Since the development of computer-based techniques in the early 1970s, it has been possible to compose music on this scale.[4] Graphical audio tools make it easy to zoom in and operate on microsounds.[44] As FIGURE 1 shows, a new generation of sophisticated sound analysis tools is emerging, based on classifying sounds according to dictionaries of particles.[21,22,45] In certain ways, these non-Fourier tech-

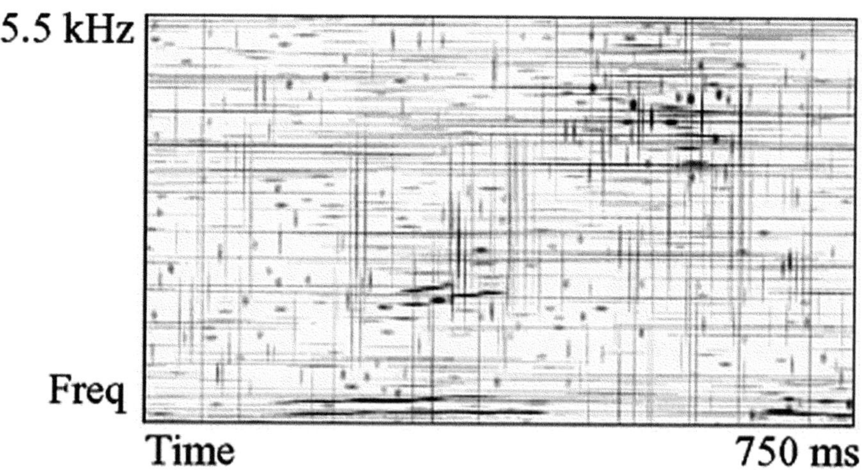

FIGURE 1. Time-frequency energy of a 750-ms speech signal (the word "wavelets") with white noise added, as viewed with the technique of matching pursuit wavelet analysis. This type of analysis is excellent at visualizing the highly localized particulate nature of sound energy. Both speech (*dark spots*) and white noise (*grey spots*) are seen as agglomerations of particles scattered on the time-frequency plane. From Davis *et al.*[22]

niques can mitigate the "acoustic uncertainty principle" that has vexed acoustic analysis in the past.[16] Moreover, the ability to go beyond mere visualization, to transform sounds in this domain, is fostering new and evocative forms of musical expression.

Techniques for the synthesis of microsonic sound particles are now in widespread use by composers of electronic music. My compositions *Clang-tint, Half-life, Tenth vortex, Eleventh vortex, Sculptor,* and *Volt air* are examples of multiscale compositions, where the organization of sound extends from the architecture of the macroform down to the tiniest of perceptible microevents. An important aspect of this work involves adjusting the silent spaces around sounds in order to obtain a precise microrhythm. In this fine editing mode, the ear is quite sensitive to adjustments on the timescale of a few hundred microseconds.

Just as it has become possible to sculpt habitats from fiberglass foam, the flowing structures that we create with microsound do not necessarily resemble the usual angular forms of musical architecture. To the contrary, they tend toward liquid-like or cloud-like structures. In the context of particle physics, Einstein[46] stated, "Atomism compels us to give up the idea of sharply and statically defining bounding surfaces of solid bodies." Likewise, particles of sound dissolve the solid notes into more supple materials that cannot always be measured in terms of definite intervals. As a result, sound objects may have "fuzzy edges," that is, ambiguous pitch and indefinite duration (due to particle evaporation, coalescence, mutation, and interaction with other layers of musical structure). We see close links to pointillist art theories involving thousands of brush strokes.[47]

Microvariations melt the frozen abstractions of traditional music theory such as continuous tone, pitch, instrument timbre, dynamic marking, and even event duration, reducing them to a constantly evolving stream of particle morphologies. Intervals may emerge, but they are not an indispensable grid. There is rather an interplay between intervallic and nonintervallic material.

Within these flowing structures, the quality of particle density—which determines the transparency of the material—takes on prime importance. An increase in particle density induces tone fusion. It lifts a cloud into the foreground, while a decrease in density causes evaporation, dissolving a continuous sound band into a pulsating rhythm or vaporous background texture. At a constant density, a change in the characteristics of the particles themselves induces mutation, an open-ended transformation. Such processes—coalescence, disintegration, evaporation, and mutation—are central to this new approach to musical organization.

Upon reading a draft of this paper, the composer and psychologist Dr. Gerard Pape made the following suggestion: Why rule out so-called non-scientific research that would ask individual listeners to describe what they hear when they are listening to microsounds, how grains or pulsars make them feel...?

I might point out that I run such experiments as often as I can, in the form of concerts. Although I have not yet handed out questionnaires to be filled out by the subjects, I have definitely noted a variety of reactions!

ACKNOWLEDGMENTS

I would like to credit Dr. Gerard Pape (Centre de Composition Musicale "Iannis Xenakis," Paris) and my colleague Stephen Pope, University of California at Santa

Barbara, for insightful comments on a draft of this manuscript, which led to many improvements and clarifications. I would also like to thank the Mariani Foundation and my colleagues at the conference on the neurosciences and music for the opportunity to learn more about this fascinating area of scientific inquiry.

REFERENCES

1. GABOR, D. 1946. Theory of communication. J. Inst. Elect. Eng. Pt. 3, **93:** 429–457.
2. GABOR, D. 1947. Acoustical quanta and the theory of hearing. Nature **159:** 591–594.
3. GABOR, D. 1952. Lectures on communication theory. Technical Report 238. Research Laboratory of Electronics. Massachusetts Institute of Technology. Cambridge, MA.
4. ROADS, C. 1978. Automated granular synthesis of sound. Comput. Music J. **2:** 61–62. Granular synthesis of sound. 1985. *In* Foundations of Computer Music. C. Roads & J. Strawn, Eds. :145–159. The MIT Press. Cambridge, MA.
5. ROADS, C. 2001. Sound composition with pulsars. J. Audio Eng. Soc. **49:** 134–147.
6. ROADS, C. 2002. Microsound. Book and audio compact disc. MIT Press. Cambridge, MA.
7. GORDON, J. 1996. Psychoacoustics in computer music. *In* The Computer Music Tutorial. C. Roads, Ed. :1053–1069. MIT Press. Cambridge, MA.
8. HELMHOLTZ, H. 1850. On the velocity of propagation of nerve impulses. Comptes Rendus 30, 33. Cited in H. Margenau's introduction to H. Helmholtz. 1885. On the Sensations of Tone. Translated by A. Ellis. New edit. 1954. Dover Publications. New York. The cited page is not numbered.
9. GALAMBOS, R. 1962. Nerves and Muscles. Anchor Doubleday. New York.
10. WINCKEL, F. 1967. Music, Sound, and Sensation. Dover Publications. New York.
11. DOUGHTY, J. & W. GARNER. 1947. Pitch characteristics of short tones I: two kinds of pitch thresholds. J. Exp. Psychol. **37:** 351–365.
12. BUSER, P. & M. IMBERT. 1992. Audition. MIT Press. Cambridge, MA.
13. MEYER-EPPLER, W. 1959. Grundlagen und Aufwendungen der Informationstheorie. Springer-Verlag. Berlin.
14. WHITFIELD, J. 1978. The neural code. *In* Handbook of Perception, vol. IV, Hearing. E. Carterette & M. Friedman, Eds. Academic Press. Orlando, FL.
15. JONES, E.G. 2002. Anatomical and physiological pathways to sound perception. Presented at the conference on the Neurosciences and Music. Venice, Italy. Ann. N.Y. Acad. Sci., this volume.
16. ZATORRE, R.J., P. BELIN & V. PENHUNE. 2002. Structure and function of auditory cortex: music and speech. Trends Cognit. Sci. **6:** 37–46.
17. BREGMAN, A. 1990. Auditory Scene Analysis. MIT Press. Cambridge, MA.
18. BREGMAN, A. 1997. When will we hear separate events in a sequence of sounds? Preprint 4582. Presented at the 103rd Conference of the Audio Engineering Society. Audio Engineering Society. New York.
19. BREGMAN, A. & J. CAMPBELL. 1971. Primary auditory stream segregation and perception of order in rapid sequences of tones. J. Exp. Psychol. **89:** 244–249.
20. MCAULAY, R. & T. QUATIERI. 1986. Speech analysis/synthesis based on a sinusoidal representation. IEEE Trans. on Acoustics, Speech, and Signal Processing **34:** 744–754.
21. MALLAT, S. & Z. ZHENG. 1993. Matching pursuit with time-frequency dictionaries. IEEE Trans. Signal Processing **41:** 3397–3415.
22. DAVIS, G., S. MALLAT & Z. ZHENG. 1994. Adaptive time-frequency decompositions with matching pursuit. Optical Eng. **33:** 2183–2191.
23. KURTZ, M. 1992. Stockhausen: A Biography. Faber and Faber. London.
24. BUTLER, D. 1992. The Musician's Guide to Perception and Cognition. Schirmer Books. New York.
25. HELMHOLTZ, H. 1885. On the Sensations of Tone. Translated by A. Ellis. New edit. 1954. Dover Publications. New York.
26. GREEN, D. 1971. Temporal auditory acuity. Psychol. Rev. **78:** 540–551.

27. MOORE, B. 1989. An Introduction to the Psychology of Hearing. Academic Press. London.
28. BLAUERT, J. 1997. Spatial Hearing. Revised edition. MIT Press. Cambridge, MA.
29. BOERGER, G. 1965. Die Lokalisation von Gausstönen. Ph.D. thesis. Technische Universität, Berlin.
30. RABINER, L. & B. GOLD. 1975. Theory and Application of Digital Signal Processing. Prentice-Hall. Englewood Cliffs, NJ.
31. FREEMAN, W. 1995. Chaos in the central nervous system. *In* Neural Modeling and Neural Networks. F. Ventriglia, Ed. :185–216. Pergamon Press. New York.
32. OOHASHI, T., E. NISHINA, N. KAWAI, *et al.* 1991. High frequency sound above the audible range affects brain electric activity and sound perception. Preprint 3207(W-1). Presented at the 91st convention of the Audio Engineering Society. Audio Engineering Society. New York.
33. OOHASHI, T., E. NISHINA, Y. FUWAMOTO & N. KAWAI. 1993. On the mechanism of hypersonic effect. *In* Proceedings of the 1993 International Computer Music Conference. S. Ohteru, Ed. :432–434. International Computer Music Association. San Francisco.
34. THAUT, M.H. 2002. Neurophysiology of neural timing networks in musical rhythm and rhythmic synchronization: basic science and clinical data. Presented at the conference on the Neurosciences and Music. Venice, Italy.
35. THAUT, M., T. BIN & M. AZIMI-SADJADI. 1998. Rhythmic finger-tapping sequences to cosine-wave modulated metronome sequences. Hum. Mov. Sci. **17**: 839–863.
36. MOLINARI, M. 2002. The neurobiology of rhythmic motor entrainment: a neurorehabilitation perspective. Presented at the Conference on the Neurosciences and Music. Venice, Italy.
37. MOLINARI, M., M. THAUT, C. GIOIA, *et al.* 2001. Motor entrainment to auditory rhythms is not affected by cerebellar pathology. Proceedings of the Society for Neuroscience 950.2. Society for Neuroscience. Washington, DC.
38. REDER, L. & J.S. GORDON. 1995. Subliminal perception: nothing special, cognitively speaking. *In* Cognitive and Neuropsychological Approaches to the Study of Consciousness. J. Cohen & J. Schooler, Eds. Lawrence Erlbaum Associates. Hillsdale, NJ.
39. MARIN, O.S.M. 1982. Neurological aspects of music perception and performance. *In* The Psychology of Music. D. Deutsch, Ed. :453–477. Academic Press. New York.
40. TIPPETT, M. 1978. Foreword. *In* Music and the Brain. M. Critchley & R. Henson, Eds. W. Heinemann Medical Books. London.
41. APEL, K. 1972. Harvard Dictionary of Music. Harvard University Press. Cambridge, MA.
42. WALLIN, N. 1991. Biomusicology. Pendragon Press. Stuyvesant, NY.
43. TOOP. R. 1982. A note—new music and neurobiologic research: can they meet? *In* Music, Mind, and Brain. M. Clynes, Ed. :387–398. Plenum. New York.
44. WENGER, E. & E. SPIEGEL. 1999. MetaSynth Manual. U&I Software. San Francisco.
45. GOODWIN, M. & M. VETTERLI. 1997. Atomic decompositions of audio signals. Proceedings of the IEEE ASSP Workshop on Audio Signal Processing. :4–7. IEEE. New York.
46. EINSTEIN, A. 1952. Relativity: The Special and the General Theory. 15th edit. of the original published in 1916. Three Rivers Press. New York.
47. HOMER, W.I. 1964. Seurat and the Science of Painting. MIT Press. Cambridge, MA.

Translation from Neurobiological Data to Music Parameters

DIEGO MINCIACCHI

Department of Neurological and Psychiatric Sciences, Degree Course in Sciences of Movement, University of Florence, Florence, Italy

ABSTRACT: Composers have explored different ways to use biological information for the realization of music. Throughout the decades, biological findings have been repeatedly indicated as a source of inspiration or a reservoir of extramusical material for musical composition. More radical and fertile are attempts to produce music systematically using biological data in processes called data sonification or biofeedback techniques. Presented here is a novel strategy of translation where populations of neurobiological data are converted into relational structures from which sound objects are generated by flexible and homogeneous control of the sound parameters. All brain data originate from experiments performed with standard anatomical and physiological techniques, and results of studies based on these experimental materials have already been published. During the translation processes, the information for every sound parameter (such as pitch, duration, envelope, and dynamics) is never derived from fixed transcriptions of data properties. Rather, the space and/or the time interrelations of data populations are used to obtain indexes for sound construction. In this way, equivalent sets of information are exploited to model, or sculpt, the different parameters of sound objects. Three examples from the last decade's personal productions are given. The first refers to the microformal aspects of sound aggregation and is based on data from a microstimulation experiment in the motor cortex. The second describes the earliest translation process developed for live performance with conventional instruments and is based on experiments using a conventional tract tracing technique to compare selected spinal-projecting cell populations in two differently organized brains. The third outlines a recent music production for three pianos based on data from experiments using the multiple fluorescent tract-tracing technique to simultaneously label different populations of thalamocortical neurons. The approach here described can potentially contribute to a unitary view of the different sound parameters and of the micro- and macroformal aspects of the compositional process.

INTRODUCTION

Since the early proposal by Köhler[1] on the common ground properties of forms and the landmark work by Lerdahl and Jackendoff[2] on the generative theory of tonal

Address for correspondence: Dr. Diego Minciacchi, Department of Neurological and Psychiatric Sciences, University of Florence, Viale Pieraccini 6, I-50134, Florence, Italy. Voice: +39-055-4279788; fax: +39-055-290662.

diego@unifi.it

Ann. N.Y. Acad. Sci. 999: 282–301 (2003). © 2003 New York Academy of Sciences.
doi: 10.1196/annals.1284.039

music, several studies have been devoted and hypotheses have been developed for outlining the universals in music. Many investigators have focused on the possible links between universals in music and innateness, competencies, and ethnomusicological questions.[3–9]

In the present article, evidence from a completely novel approach to the exploration of universalities in the musical message is presented. The issue is addressed by exploring the possibility of constructing sound objects aimed at independence from historical references but possessing, at the same time, a clear musical nature. Assuming that forms, and therefore musical forms, are subject to the general principles of isomorphism, the different properties and parameters of sound objects must also conform to these principles. By contrast, the ability to decode sound information and give it formalization and musical meaning resides in the brain and must be represented as networks of neurons and connections and as functional processing properties. An explicit short circuit is then formulated to couple musical form and the brain. Data on the structure and functions of the brain can be used directly as information to build up sounds and musical forms, rejecting compositional rules.

Scattered anticipation of this line of thinking can be found as different lines of research developed over the last 30 years. Biological information, applying conventional rules of music composition, has been employed to produce musical scores starting from gene coding sequences.[10,11] Results of these procedures of data sonification were not musically interesting, but they pointed out clearly the potentialities of biological information.

Organic and fruitful is the line of research by Lucier[12] who in 1965 composed the work, *Music for solo performer*, in which electrical brain activities are amplified through electroencephalogram scalp electrodes and are used, in live performance, to drive percussion instruments and tape recorders. Since the 1970s, Rosenboom has explored the theoretical foundations of biofeedback and realized music based on biofeedback techniques, making physical connections between externally generated sounds and the internal experience as represented by electrical brain activities.[13,14] Further attempts in this direction were made in the 1990s by Knapp and Lusked[15,16] who developed an interface/biocontroller, monitoring electrical activities of the body (electroencephalogram, electrooculogram, electrocardiogram, and electromyogram) and translating them into MIDI (Musical Instruments Digital Interface) control messages. Also in this case, some music was realized using the interface/biocontroller, but it remained an isolated experience.[17]

In the present approach, musical construction is entirely derived from information on the structure and functional properties of the brain, and brain data are used to produce all relevant parameters of the sound objects. We must first suppose that, as with the musical forms, brain structure and functions are also subjected to general principles of isomorphism and that at least some of the properties and parameters described in brain research are producing information adherent to these principles. The enormous mass of literature dealing with these general principles of the operative brain will not be discussed; however, I will only point out here that the structural or functional information coming from musically related structures of the brain is as relevant as information from any other brain region. In other words, to construct a sound object, information from the auditory cortex is as pertinent as information from, for example, somatosensory or visual cortical areas. This should not be surprising, as substrates of music elaboration are widely distributed in the brain and increase with the complexity of auditory processing to possibly involve the entire cortical mantle and cerebellum.[18]

Furthermore, the process of translating this theoretical frame into an effective ontogeny of sound objects requires profound rethinking of what exactly are the relevant compositional parameters for this music. I had to face the fact that existing compositional concepts of the treatment of sound parameters were only partially effective, especially considering conventional composition and conventional instruments. Definitely more intriguing, but also potentially practicable, were notions originating from linguistic approaches to the representation of music.[19,20] (See also the pioneering review by Roads.[21])

I am not yet at the point of describing or even proposing a comprehensive and stable set of compositional parameters developed to translate brain information into music. I will mention only two general notions that, at the state of the art, appear a consolidated ground and reliable frame, in my experience, for developing sound objects.

First, the process of modeling a sound object should conform to the same principles in the treatment of all different parameters of the musical form. This concept of uniformity in the rules of treatment should inform the musical construct down to the smallest sound event, that is, microsounds,[22] up to the largest macroformal parameters such as those determining the durations and the kinetic of dynamics.

Second, within any single parameter, the architecture of sound treatment must adhere to the basic relational nature of sound information processing and musical perception. This concept is well exemplified, for the frequency parameter, by the long-standing notion that, to elaborate tones, the numeric values of frequency ratios are more relevant than absolute frequency values.[7,23–25]

In the next sections are presented examples of the processes developed to realize some of the musical works composed during the last decade. For each example, information on the starting set of neurobiological data and the resulting musical outcome is given. The meaning, or significant idea, leading the process of translation into sounds is outlined. Then, the key attributes of the procedures used for translation from brain data to musical parameters are illustrated. Finally, selected examples of the scores composed using the described operations are shown.

"WESTBAU" FROM *DER DOM*, 1992

"Westbau"[a] has been realized as commission of the International Music Institute of Darmstadt. The electronic part of this work was produced using, for the first time, extensive translations of data from experimental brain studies into sound parameters.

Neurobiological data were obtained from an experiment of microstimulation of the motor cortex in monkeys.[26,27] Briefly, cortical microstimulations induced move-

---

[a]"Westbau" from *Der Dom*, for bass voice, five trombones, four double basses, and electronics. World premiere: International Society for Contemporary Music Festival: World Music Days, Open Doors Concert at the Teotihuacan Pyramids, Mexico City, Mexico, 1993. CD: Diego Minciacchi: *The Aforesaid*. Experimental Studio of the SWR Heinrich Strobel Foundation, Freiburg, and SWR Studios, Baden-Baden, Germany. Production: col legno, Produktion GmbH, Germany. WWE 20077, 2001. Nicholas Isherwood (bass voice), Barrie Webb (trombone), and Stefano Scodanibbio (double bass).

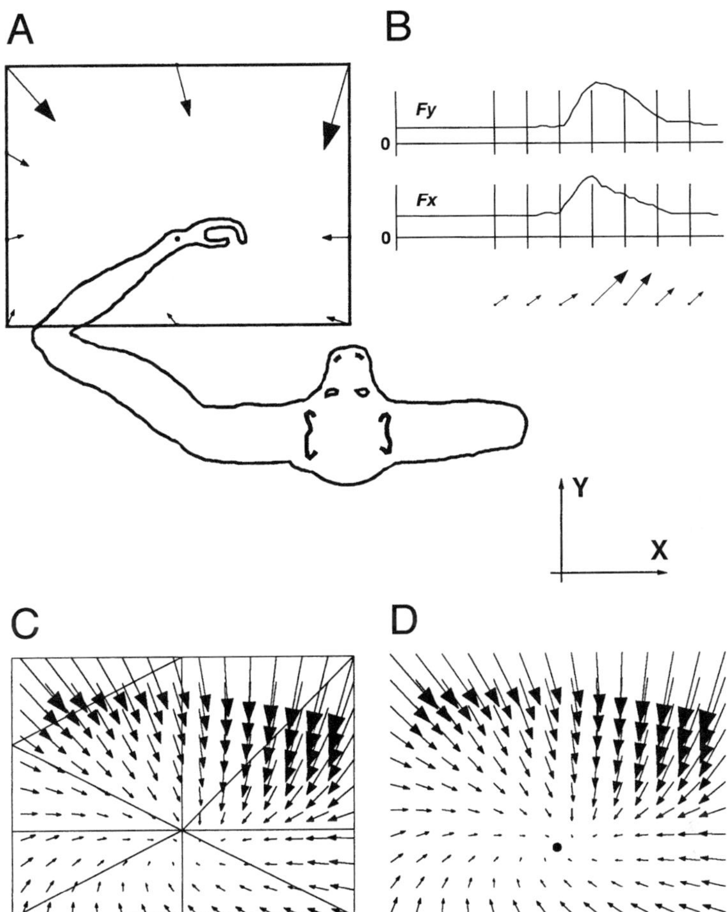

FIGURE 1. (A) Experimental design of microstimulation of the motor cortex in monkeys. For each cortical microstimulation site the limb was displaced, as starting configuration, in different parts of the working space. For each configuration of the limb, a microstimulation was delivered and the X-Y static forces were recorded at the wrist by a force transducer. (B) Each recording consisted of one pair of Y (Fy) and X (Fx) force components whose combination resulted in a set of force vectors (*bottom*). (C) Time-dependent force fields were then derived for the entire working space, and a location having X-Y forces 0-0 (D) was calculated and defined as the "equilibrium point."

ments of the upper limbs, which were recorded as X-Y forces at the wrist. During the experiment, for a single cortical microstimulation site, the limb was displaced, as starting configuration, in different parts of the working space. For each configuration of the limb, a microstimulation was delivered and X-Y forces were recorded. A force field was then reconstructed, and a location having X-Y forces 0-0 was calculated and defined as "equilibrium point" (FIG. 1). The original data were represented by sets of forces on the X-Y plane. These forces were time dependent, and the datasets, organized in stacks of force fields, described displacements of the limb endpoint as displacements of the equilibrium point (i.e., the trajectory).[28–31]

The strategy developed for the translation of this information into a system of sounds was basically a reexamination of the space in terms of sound parameters. It is known that brain constructs multiple representations of the working space, and "near," "far," "medial," and "lateral" regions can be distinguished.[32–34] Sound parameters were elaborated to reflect this uneven nature of the X-Y working space. In the Y dimension, the "equilibrium point" was considered the minimum possible pitch dispersion, this is, unison, or one single pitch. When moving up or down from the equilibrium point, the pitches differentiate. Going towards the "near space of the

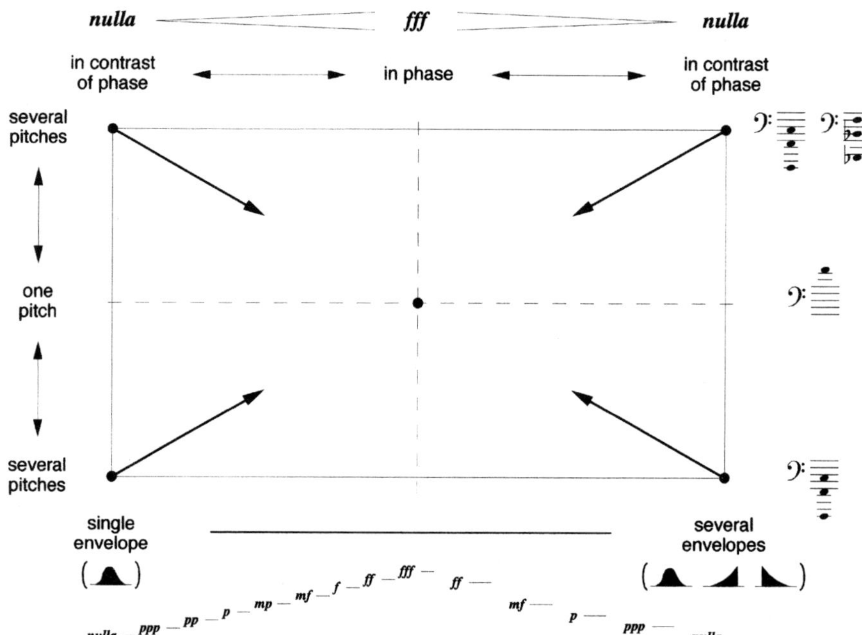

FIGURE 2. Drawing of the strategy developed for translation of force fields and their parent "equilibrium points" into the sound parameters, pitch and dynamics. In the center is one representative simplified force field having only four converging force vectors and the resulting equilibrium point. The equilibrium point, projected on the Y axis, is considered the minimum possible pitch dispersion (unison); moving up or down along the Y axis, the pitches differentiate (*left*). For smaller Y values, the "near space of the limb," pitches pertain to one series of harmonics; for larger Y values, the "far space of the limb," pitches pertain to more than one harmonic series (*right*). Following a similar process for dynamics, the equilibrium point, projected on the X axis, is considered the maximum possible phase congruence, that is, sound intensities are in phase (fff, *upper part of drawing*). When moving right or left along the X axis, the phase congruence decreases to reach minimum congruence, that is, sound intensities are in contrast of phase (nulla). For smaller X values, the "medial space of the limb," sounds have only one shape or envelope and intensities are controlled more finely; for larger Y values, the "lateral space of the limb," sounds have more envelopes and intensities are controlled more roughly (*lower part of drawing*).

limb," the pitches pertain to only one congruent series of harmonics; going the opposite direction, towards the "far space of the limb," the pitches pertain to more than one harmonic series. Similarly, in the X dimension, at the equilibrium point, sounds were considered as in phase, that is, maximum intensity, whereas at the borders of the space, sounds were considered in contrast of phase, that is, minimum intensity. In the "medial space of the limb," sounds have only one shape or envelope, and dynamics are controlled more finely. In the "lateral space of the limb," sounds have more envelopes, and dynamics are controlled more roughly (FIG. 2).

To realize the project, catalogues of sounds having different pitch, shape, and performing technique were recorded in the studio, mostly from the bass voice of Nicholas Isherwood. Single sounds (sound units, durations ranging 400–1200 ms) were then sculpted to obtain the configuration desired and were placed in time and space through a massive work of postproduction using a Studer Dyaxis-based editing system.

The result of this process was somewhat surprising. The sound aggregates obtained from translation of single stacks of forces fields had individual pitch contour, recognizable preferences for unequal-interval landmarks, and distinct architecture of dynamics. Also, the intermingling of sound units of different shapes, intensities, and harmonic properties resulted in a perceivable temporal patterning. Even without discussing esthetics, the major characteristics of a potential musical message were thus present.

The foregoing short description outlines the first attempt to shape a translation process from biological information into music to retain at least part of the key features of the original data and to use compositional processes lacking immediate historical references.

PROCEDURE PER SIMULAZIONI INDENTATE, 1995–1996

Procedure per simulazioni indentate[b] has been realized as commission of the Akanthos Ensemble. This work represents a good example of a translation process that had to be adapted to live performance with conventional instruments. Under these conditions of live performance and conventional instruments, keeping together instrumental feasibility and coherence to the original significance of data was a crucial task.

For *Procedure,* material from studies in which populations of spinal-projecting neurons were identified in the brain of a mutant animal was used and compared with homologous cell populations of normal animals.[35,36] The spinal-projecting neurons of mutant and normal animals were visualized by a conventional tract-tracing technique, acquired for analysis as digital images, and converted into X-Y charts. The basic set of brain information, in this case, was a comparison between pairs of cell populations having the same functional significance but pertaining to two differently organized brains. The original purpose of these studies was to identify strategies of reorganization consequent to the new functional condition induced by the mutation.

---

[b]*Procedure per simulazioni indentate,* for viola and ensemble (violin, cello, flute, clarinet, horn, piano, and percussion). World premiere: Tage für Neue Musik, Rottenburg, Germany,1996. Akanthos Ensemble, conductor: Aldo Brizzi.

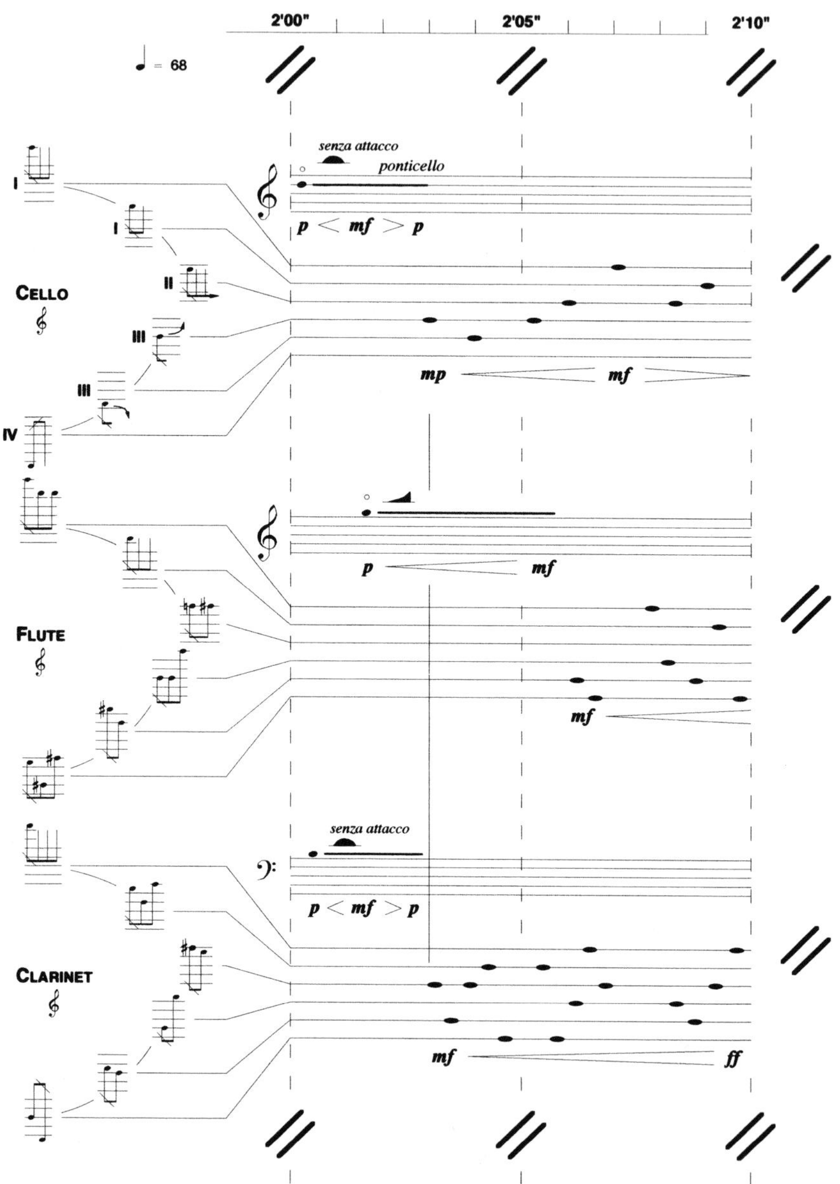

FIGURE 3. In this cut of the score for *Procedure*, the double system of notation adopted to represent sound units and conventional attacks for the cello, flute, and clarinet parts is illustrated. For each instrument the conventional attacks are written on the upper staff. Sound units are grouped and written as distinct "reservoirs" on the left part of the score and are individually related to horizontal lines. When the sound unit has to be performed, a mark (*black elongated dot*) is given.

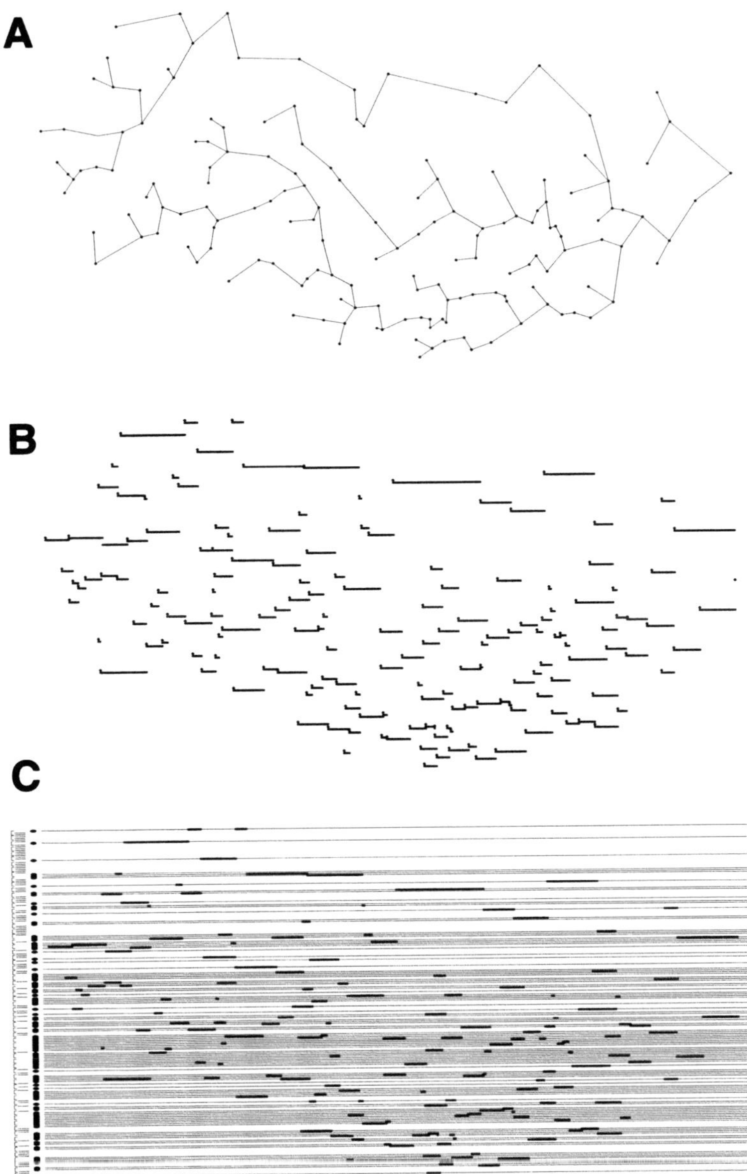

FIGURE 4. The process developed for a region of *Procedure* is illustrated. (**A**) One cell population is analyzed using an algorithm for finding the nearest neighbors in euclidean space. (**B**) Segments are projected on a horizontal plane and related to their data of origin. (**C**) Segments and data are then placed on the mixed pitches/instrumental space. Instrumental domains and pitches are derived as from the left heading. (Only segments are shown in this drawing, and instruments and pitches are not represented to ease reading.)

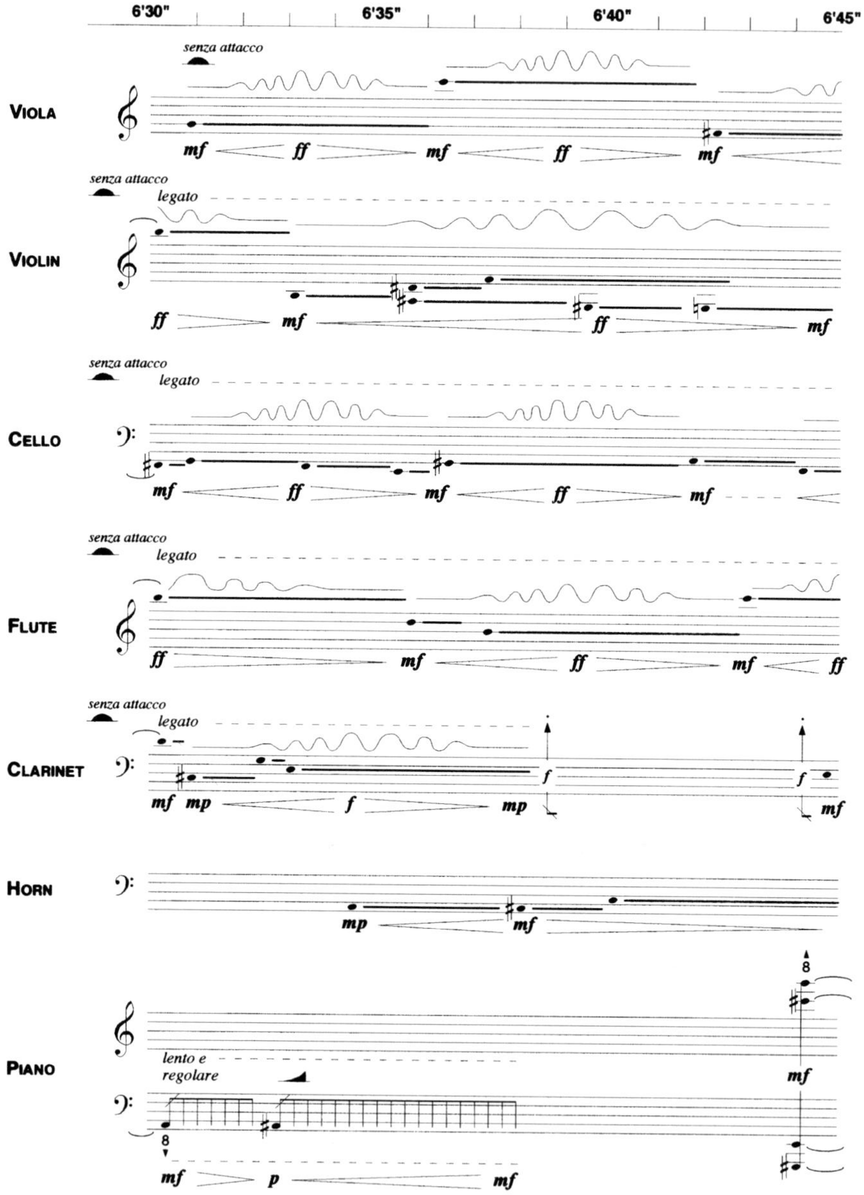

FIGURE 5. Cut of the score for *Procedure,* illustrating part of the region composed, as described in FIGURE 4.

The musical counterpart was to let the "normal" and the "reorganized" sets of data confront each other in the instrumental and time domains.

The data, original cells of the populations, were regarded as microformal units. Sound units were either single attacks or short groups of attacks (FIG. 3). The groups of attacks were small enough to still be considered simple sound events, but they included a distinctive musical character in addition to the pitch's contour (group envelope or shape). Remarkably, the process adopted to realize *Procedure*, although not going down to the domain of the Gabor's "sound quanta," appears somewhat similar to some techniques of electronic sound manipulation such as granular synthesis and pulsar synthesis.[22,37–41]

Streams of sound units were constructed by distributing the compositional particles in time and through the instrumental space. The original data locations were positioned on the new musical time/instruments planes, using different relative orientations of the two systems of reference. The reason to use this mobile shift of orientation was to distribute differentially, on time and through the instrumental space, the information contained in the planar original format of biological data. Using this general strategy of data translation, different local operations were performed through the score.

One of them is briefly described. For the realization of a region of *Procedure*, one population of neurons was analyzed using an algorithm for finding the nearest neighbors in euclidean space (using the software XYZ GeoBench for Macintosh developed by Peter Schorn at ETH Zurich). Segments were projected on horizontal plane, related to their data of origin, and placed on a grid containing selected instruments and pitches (FIG. 4). Results of this process are illustrated in FIGURE 5, where a cut of the final score is given. The time-related attacks and pitches are representations of the original data in the instrumental space and the durations reflect the lengths of the associated segments.

QUINTUM DESERTUM, 2000–2001

Quintum Desertum, as "pars I," was first realized in short version for three pianos and then in complete version as commission of the Ensemble SurPlus - "...antasten... 5. Internationales Pianoforum."[c] *Quintum Desertum* is a large musical form having a duration of more than 1 hour. The work was produced recently from fairly old data due to the complexity of the score and the demanding format of the live performance. In the following description, concerning the live piano parts, only the main characteristics of the translation procedure is outlined.

The original information was derived from three different populations of neurons contained in the same nuclear complex, the anterior intralaminar nuclei of the thalamus. Cells were visualized using the multiple fluorescent tract-tracing technique, acquired as digital images, and converted into X-Y charts. These cell populations are

---

[c]*Quintum Desertum pars I*, for three pianos, World premiere: Darmstadt Ferienkurse, Darmstadt, Germany, 2000. Pianos: Kaya Han, Maki Namekawa, Ueli Wiget. *Quintum Desertum*, for three pianos and electronics, World premiere: "...antasten... 5. Internationales Pianoforum," Heilbronn, Germany, 2001. Ensemble SurPlus, pianos: Serhiy Lunhu,Olga Krasotowa, Rei Nakamura; conductor: James Avery; electronics: Studio GRM, Paris.

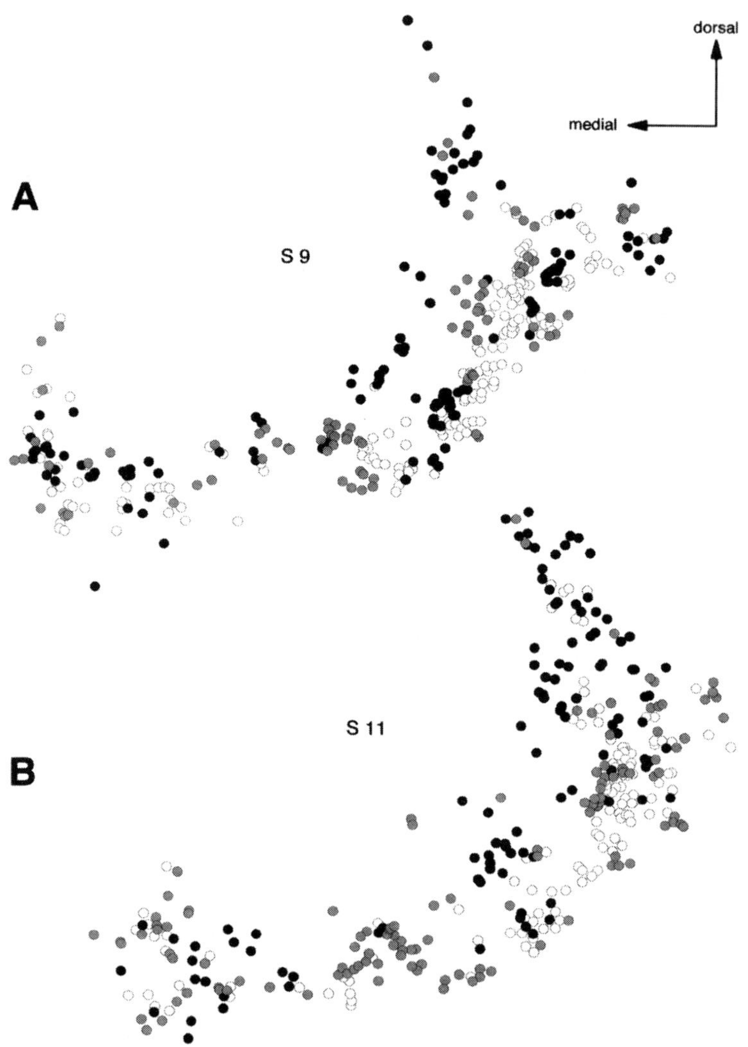

FIGURE 6. Two representative sets of data (in coronal sections) showing the three cell populations of the anterior intralaminar nuclei of the thalamus utilized to derive sound parameters for *Quintum Desertum*. Cells were visualized using the multiple fluorescent tract-tracing technique. These projecting thalamocortical neurons send information to the somatosensory cortical area S1 in the regions of representation of the face and upper and lower limbs, respectively. **A** is rostral, **B** is caudal. *White circles,* cells projecting to the S1 region of face representation; *gray circles*, cells projecting to the S1 region of upper limb representation; *black circles,* cells projecting to the S1 region of lower limb representation.

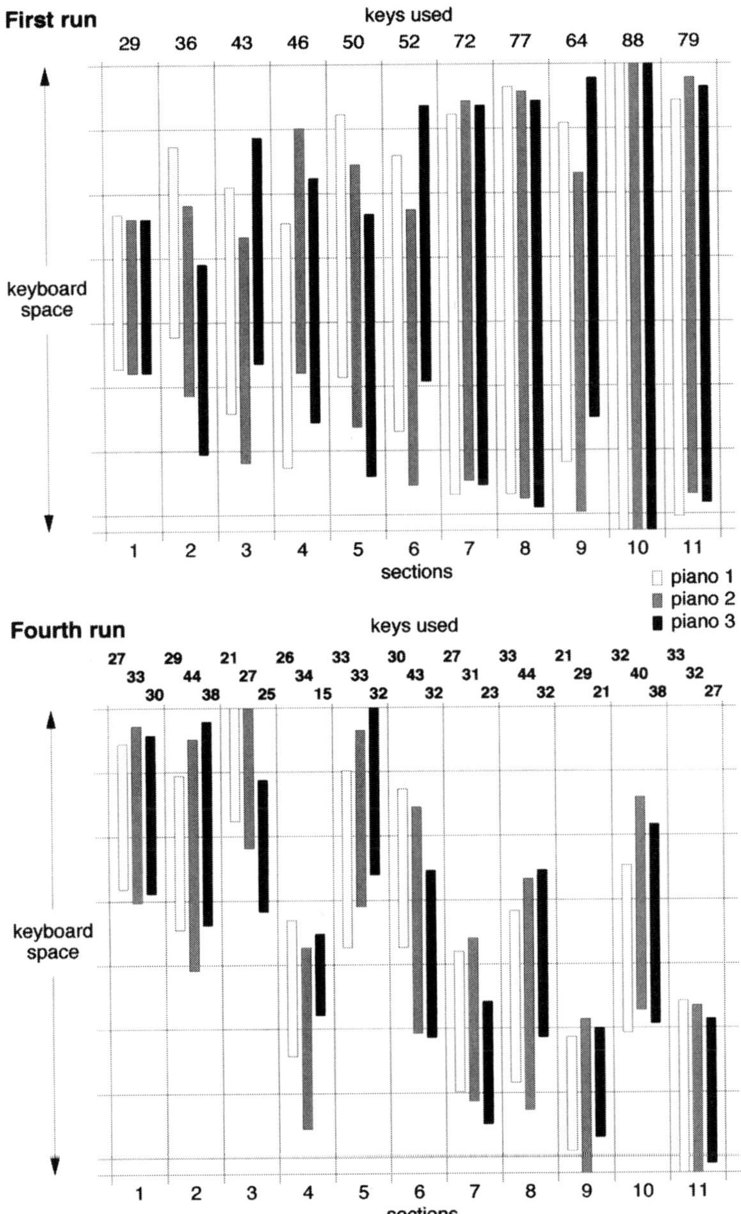

FIGURE 7. The process developed to relate the piano parts to the space of the keyboard and to determine the proportional associations among pianists. The *upper drawing* represents the procedure adopted for the first run (exposition) of data; the *lower drawing* represents the procedure for the fourth run. In *white* (piano 1) is indicated the dataset from the population projecting to the S1 region of face representation; in *gray* (piano 2) and *black* (piano 3) are those projecting to the S1 regions of the upper and lower limb representations, respectively.

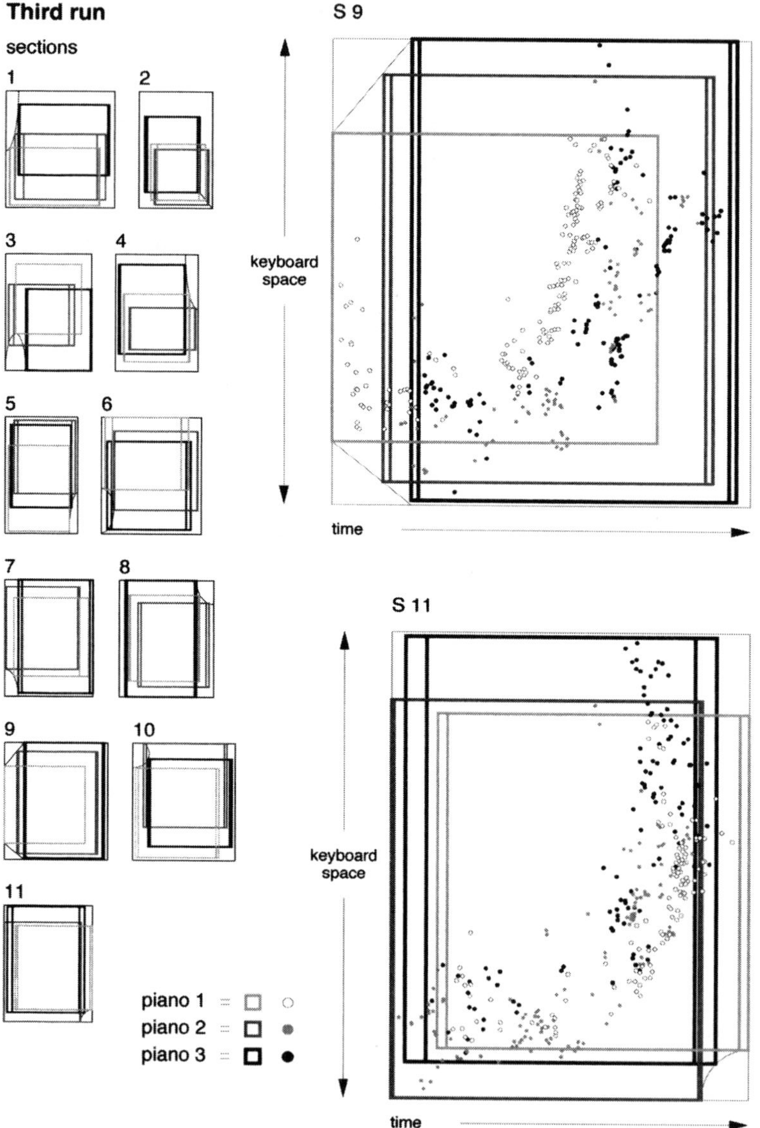

FIGURE 8. The datasets are differently related to each other every time a chart is presented. The drawing illustrates the procedure developed to change the ratios linking the original charts to the time and to the keyboard space for the third run of data.

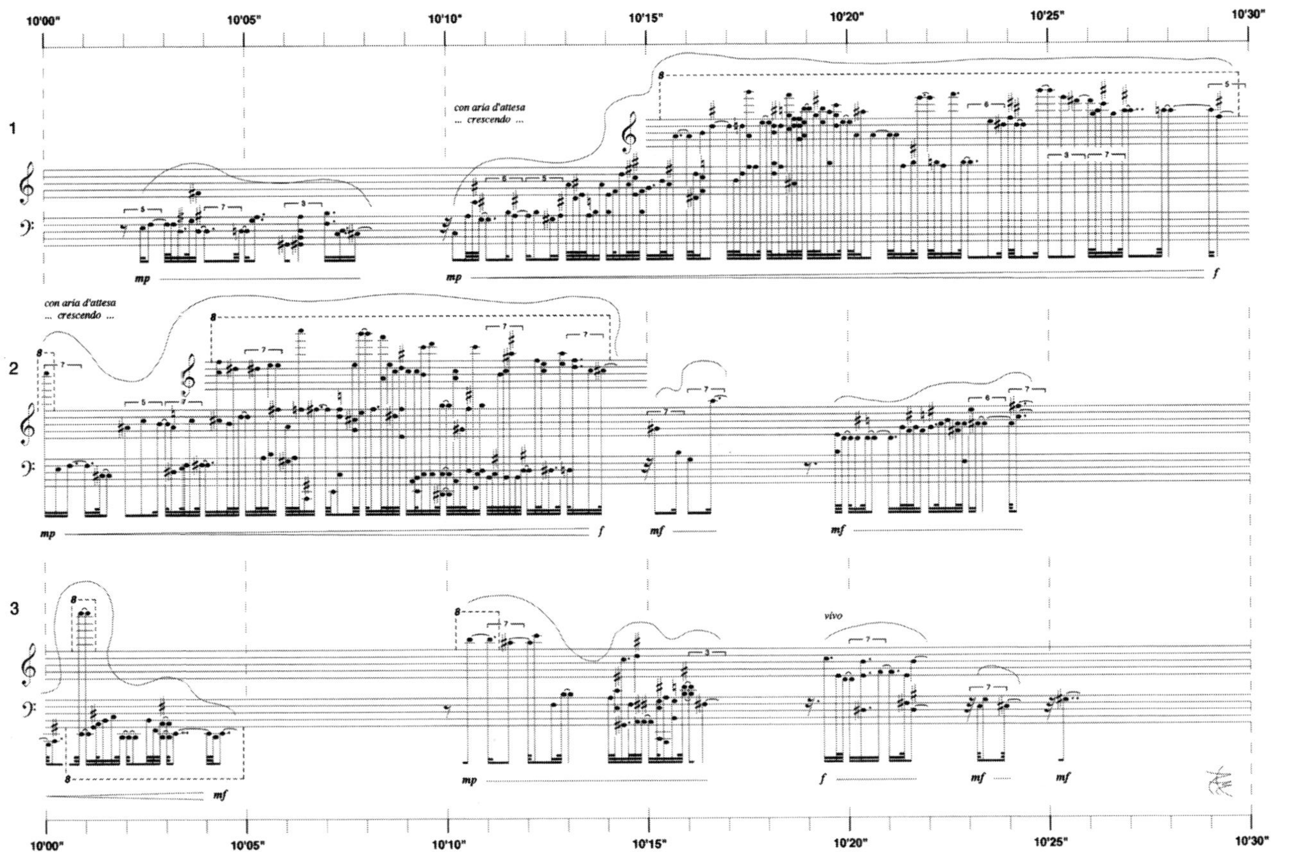

FIGURE 9. Cut of the score of *Quintum Desertum* for the three pianists (1, 2, and 3), illustrating that all graphs are from conventional notational systems.

located very close each other, often partially intermingled (FIG. 6). They receive mixed information from various brain regions and send results of their elaboration to three different subregions of a cortical area well organized in space to represent body parts: the somatosensory area S1. One population projects to the S1 region of representation of the face; the second and third populations project to the S1 regions of representation of the upper and lower limbs, respectively. Data were selected from studies of intralaminocortical projections to primary sensory areas.[42,43] The principal aim of the research was to understand how nonspecific information coming from the intralaminar thalamic nuclei was mapped onto cortical areas characterized by precise spatial distribution of the modality-specific primary input.

The work *Quintum Desertum* is organized to represent and observe, in a frame of always changing viewpoints, this set of three cell populations that are strictly inter-related among them but possessing individual functional significance. Here also the original cells were considered as sound units and were essentially represented as single attacks or as groups of attacks in the form of chords, when data appeared exceedingly concentrated. The X-Y charts of cells were translated into the space of the three piano keyboards by changing, for each presentation of the original data, the ratios relating the actual piano parts to the entire space of the keyboards and the pro-portional associations among pianists (FIG. 7).

Continuous changes of the ratios linking charts to time and to the keyboard space were also introduced, within a single presentation, to further elaborate changes of the mutual relation between the populations/pianos and thus of the perspective of viewing/listening them (FIG. 8). As result, the three populations of data—the three pianos—were differently related to each other every time a set of data (chart) is presented, and the process was designed to operate both within a single set and through sets.

In addition, since each pianist performs information derived from a single popu-lation and consequently is only concerned with information pertaining to a single region of the cortical somatosensory space, the three piano parts were approached differently to express distinctive sound and technical characteristics. Among the others, in the three piano parts the techniques of sound attack, the schemes of dura-tions, and the use of the sustaining pedals were handled differently. All sound parameters of this type, but using individual procedures, were managed with degrees of accuracy that were higher for the population related to the face representation, less precise for the population related to the upper limb, and lower for the lower limb-related population.

Finally, for *Quintum Desertum*, the results of the translation processes were con-verted, for the three performers, into conventional notation (FIG. 9).

GOING THE OPPOSITE DIRECTION

As an example of the potential importance of interchanges between brain sci-ences and creative activities, a recent preliminary result is mentioned here to signify how knowledge on brain organization can be stimulated by a musical aim. Searching for clues on local microorganization of data arrays to be used in large-scale musical forms, the possibility of analyzing the spatial behavior and microorganization of cell

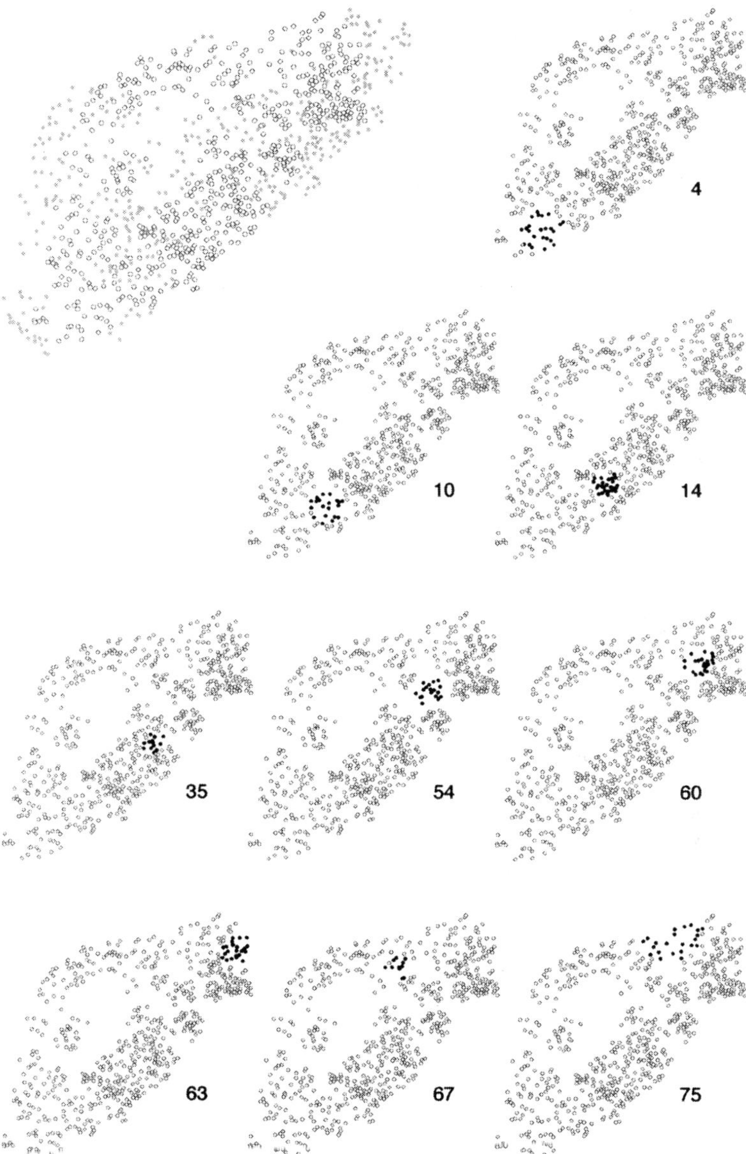

FIGURE 10. Parvalbumin-positive cells of one coronal section of the parabrachial area (*upper left*). The microstructural spatial organization of the cell population is analyzed by evaluating the X and Y coordinates and the areas of cell profiles. The nearest cells were correlated in terms of adjacency and areas of cell profiles. In the case of no correlation the cell was excluded (*gray dots*). In the case of correlation, the next closest cell was considered (*white empty circles*). Examples of the tuft-like groups of correlated cells as revealed by our analysis are represented (*black dots*) in the other smaller figures.

populations in the central nervous system was explored. A strategy to study the spatial organization of selected cell populations was thus elaborated.[d]

We tested this new strategy of spatial analysis on immunocytochemically visualized parvalbumin and calretinin positive cells within the parabrachial area of the mouse brain, a region that is not considered anatomically and functionally homogeneous. From digital images of the parabrachial area, the X and Y coordinates and areas of cell profiles of parvalbumin and calretinin positive neurons were collected and organized into vectors.

The microstructural organization of the two cell populations was then analyzed. The nearest cells were correlated in terms of adjacency and areas of cell profiles. When no correlation between areas of cell profiles and cell distance occurred, the cell was excluded. In the case of correlation the next closest cell was considered. Tuft-like groups were evident for both populations, suggesting that parvalbumin and calretinin cells in the parabrachial region possess individual microstructural organization (FIG. 10).[44]

This preliminary analysis, prompted originally by musical needs, appears potentially useful as a tool for investigation of brain spatial microorganization and for study of altered cell populations by revealing changes of cell sociology.

CONCLUDING REMARKS

The major implication of this approach to the construction of sound objects is to challenge the classical quest for universals in music. Even considering universals and their cognitive processing in a very broad sense (i.e., the advantages of unequal-interval scales over equal-interval scales, the pitch contour advantages based on small-integer ratios, and relevance of Gestalt principles of grouping), evidence emerges of the implicit—historical—distinction still assigned to the different characteristics of sound. The present approach, instead, is based on the uniform management of sound parameters, is respectful of individual extramusical data properties, and can operate in complex environments. Furthermore, the strategies of brain data translation outlined here have been actualized through several works that have been produced, released, and performed in a large series of live (festivals, concerts) and recorded (CD productions, radio broadcastings) conditions.

To process a given set of elements of musical material using extramusical information—series of numbers—has long been called serialism. Original ideas were developed by Arnold Schoenberg, Olivier Messiaen, and especially Anton von Webern and then inherited in the 1950s essentially by Karlheinz Stockhausen, who frequently spoke of "serial principles."[45] In his own words, "serialism means... having everything based on a limited number of different values for any (sound) parameter...."[46]

The present strategy, based on translation of neurobiological data into musical information, is not substantially different; it can, in fact, be considered a natural evolution of the same line of thinking. The only difference is in the choice of the values

[d]The work was developed, in collaboration with M.-R. Voegelin, D. Carretta, S. Ciabatti, F. Pinto, and C. Catini, at the Laboratory of Experimental Neuroanatomy and Neurophysiology, Department of Neurological and Psychiatric Sciences, University of Florence, Florence, Italy.[44]

used to manage sound parameters. While serialism takes into consideration a theoretic architecture of sound based on systems of numbers, the present strategy constructs sounds founded on numeric systems of biological significance. It surpasses the boundaries of the spiritual and democratic attitude of serialism,[46] to approach a more natural and biological concept of sound design.

REFERENCES

1. KÖHLER, W. 1929. Gestalt Psychology. Liveright. New York.
2. LERDAHL, F. & R. JACKENDOFF 1983. A Generative Theory of Tonal Music. MIT Press. Cambridge, MA.
3. TRAINOR, L.J., K.L. MCDONALD & C. ALAIN. 2002. Automatic and controlled processing of melodic contour and interval information measured by electrical brain activity. J. Cognit. Neurosci. **3:** 430–442.
4. CROSS, I. 2001. Music, cognition, culture, and evolution. *In* The Biological Foundations of Music. R.J. Zatorre & I. Peretz, Eds. Ann. N.Y. Acad. Sci. **930:** 28–42.
5. DRAKE, C. & D. BERTRAND. 2001. The quest for universals in temporal processing in music. *In* The Biological Foundations of Music. R.J. Zatorre & I. Peretz, Eds. Ann. N.Y. Acad. Sci. **930:** 17–27.
6. TREHUB, S. 2000. Human processing predisposition and musical universals. *In* The Origin of Music. N.L. Wallin, B. Merker & S. Brown, Eds. :427–448. MIT Press. Cambridge, MA.
7. IMBERTY, M. 2000. The question of innate competencies in musical communication. *In* The Origin of Music. N.L. Wallin, B. Merker & S. Brown, Eds. :449–462. MIT Press. Cambridge, MA.
8. NETTL, B. 2000. An ethnomusicologist contemplates universals in musical sound and musical culture. *In* The Origin of Music. N.L. Wallin, B. Merker & S. Brown, Eds. :463–472. MIT Press. Cambridge, MA.
9. SMITH, L.D. & R.N. WILLIAMS. 1999. Children's artistic responses to musical intervals. Am. J. Psychol. **112:** 383–410.
10. OHNO, S. & M. OHNO. 1986. The all pervasive principle of repetitious recurrence governs not only coding sequence construction but also human endeavor in musical composition. Immunogenetics **24:** 71–78.
11. OHNO, S. & M. JABARA. 1986. Repeats of base oligomers (N = 3n ± 1 or 2) as immortal coding sequences of the primeval world: construction of coding sequences is based upon the principle of music composition. Chem. Scripta **26B:** 43–49.
12. LUCIER, A. 1995. Reflexionen. :22–42, 300. MusikTexte. Köln, Germany.
13. ROSENBOOM, D. 1976. Biofeedback and the Arts: results of Early Experiments. A.R.C. Publications. Vancouver, Canada.
14. ROSENBOOM, D. 2003. Propositional music from extended musical interface with the human nervous system. *In* The Neurosciences and Music. G. Avanzini, C. Faienza, D. Minciacchi, L. Lopez & M. Majno, Eds. Ann. N.Y. Acad. Sci. **999:** this volume.
15. KNAPP, B.R. & H.S. LUSTED. 1990. A bioelectric controller for computer music applications. Comput. Mus. J. **14:** 42.
16. LUSTED H.S. & R.B. KNAPP. 1996. Controlling computers with neural signals. Scientif. Am. **275:** 82–87.
17. TANAKA, A. 1993. Musical technical issues in using interactive instrument technology. Proc. Int. Comput. Mus. Conf., Tokyo, 1993.
18. ALTENMÜLLER, E.O. 2001. How many music centers are in the brain? *In* The Biological Foundations of Music. R.J. Zatorre & I. Peretz, Eds. Ann. N.Y. Acad. Sci. **930:** 273–280.
19. MCADAMS, S. & D. MATZKIN. 2001. Similarity, invariance, and musical variation. *In* The Biological Foundations of Music. R.J. Zatorre & I. Peretz, Eds. Ann. N.Y. Acad. Sci. **930:** 62–76.
20. BESSON, M. & D. SCHON. 2001. Comparison between language and music. *In* The Biological Foundations of Music. R.J. Zatorre & I. Peretz, Eds. Ann. N.Y. Acad. Sci. **930:** 232–258.

21. ROADS, C. 1985. Grammars as representation for music. *In* Foundations of Computer Music. C. Roads & J. Strawn, Eds. :403–442. MIT Press. Cambridge, MA.

22. ROADS, C. 2003. The perception of microsound and its musical implications. *In* The Neurosciences and Music. G. Avanzini, C. Faienza, D. Minciacchi, L. Lopez & M. Majno, Eds. Ann. N.Y. Acad. Sci. **999:** this volume.

23. PLOMP, R. & W.J.M. LEVELT. 1965. Tonal consonance and critical bandwidth. J. Acoust. Soc. Am. **38:** 548–560.

24. SACHS, C. 1943. The Rise of Music in the Ancient World: East and West. Norton. New York.

25. RADULESCU, H. 2003. Brain and sound resonance, the world of self-generative functions, as a basis of the spectral language of music. *In* The Neurosciences and Music. G. Avanzini, C. Faienza, D. Minciacchi, L. Lopez & M. Majno, Eds. Ann. N.Y. Acad. Sci. **999:** this volume.

26. MINCIACCHI, D., F.A. MUSSA-IVALDI, S.E. GISTZER & E. BIZZI. 1990. Microstimulation of motor cortex in monkeys: a force field analysis. Soc. Neurosci. Abstr. **16:** 423, 178.9.

27. MINCIACCHI, D., E.G. JONES, P.L. STRICK, *et al.* 1991. Workshop: functional and structural properties of cortical areas related to movement. Proceedings of the Third IBRO Congress of Neuroscience, Montreal, August 4–9. :14, W47.

28. BIZZI, E., N. ACCORNERO, W. CHAPPLE & N. HOGAN. 1984 Posture control and trajectory formation during arm movement. J. Neurosci. **11:** 2738–2744.

29. MUSSA-IVALDI, F.A., N. HOGAN & E. BIZZI. 1985. Neural, mechanical, and geometric factors subserving arm posture in humans. J. Neurosci. **10:** 2732–2743.

30. SHADMEHR, R., F.A. MUSSA-IVALDI & E. BIZZI. 1993. Postural force fields of the human arm and their role in generating multijoint movements. J. Neurosci. **13:** 45–62.

31. HATSOPOULOS, N.G. 1994. Is a virtual trajectory necessary in reaching movements? Biol. Cybern. **70:** 541–551.

32. LEMAY, M. & L. PROTEAU. 2001. A distance effect in a manual aiming task to remembered targets: a test of three hypotheses. Exp. Brain Res. **140:** 357–368.

33. WEISS, P.H., J.C. MARSHALL, G. WUNDERLICH, *et al.* 2000. Neural consequences of acting in near versus far space: a physiological basis for clinical dissociations. Brain **123:** 2531–2541.

34. WOODIN, M.E. & A. ALLPORT. 1998. Independent reference frames in human spatial memory: body-centered and environment-centered coding in near and far space. Mem. Cognit. **26:** 1109–1116.

35. CARRETTA, D., M. SANTARELLI, D. VANNI, *et al.* 2001. The organisation of spinal projecting brainstem neurons in an animal model of muscular dystrophy. A retrograde tracing study on mdx mutant mice. Brain Res. **895:** 213–222.

36. SBRICCOLI, A., M. SANTARELLI, D. CARRETTA, *et al.* 1995. Architectural changes of the cortico-spinal system in the dystrophin defective mdx mouse. Neurosci. Lett. **200:** 53–56.

37. ROADS, C. 1978. Automated granular synthesis of sound. Comp. Music J. **2:** 61–62.

38. ROADS, C. 1985. Granular synthesis of sound. *In* Foundations of Computer Music. C. Roads & J. Strawn, Eds. :145–159. MIT Press. Cambridge, MA.

39. ROADS, C. 1991. Asynchronous granular synthesis. *In* Representations of Musical Signals. G. De Poli, A. Picciali & C. Roads, Eds. MIT Press. Cambridge, MA.

40. ROADS, C. 2001. Sound composition with pulsars. J. Audio Eng. Soc. **49:** 134–147.

41. ROADS, C. 2002. Microsound. Book and Audio Compact Disc. MIT Press. Cambridge, MA.

42. MINCIACCHI, D., A. GRANATO, A. ANTONINI, *et al.* 1995. Mapping subcortical extrarelay afferents onto primary somatosensory and visual areas in cats. J. Comp. Neurol. **362:** 46–70.

43. MINCIACCHI, D. & A. GRANATO. 1997. How relevant are subcortical maps for the cortical machinery? An hypothesis based on parametric study of extra-relay afferents to primary sensory areas. *In* Self-Organization, Computational Maps and Motor Control. P. Morasso & V. Sanguineti, Eds. :149-168. Elsevier/North-Holland. Amsterdam.

44. MINCIACCHI, D., M.-R. VOEGELIN, D. CARRETTA, *et al.* 2002. A tensor-based strategy for spatial analysis reveals the tuft-like micro-organization of parvalbumin- and calretinin-positive cell populations in the parabrachial region of the mouse. Soc. Neurosci. Abstr. 610.18.
45. STOCKHAUSEN, K. 1957. "...how time passes...," Die Reihe **3**, p10. English trans.: Cornelius Cardew 1959.
46. COTT, J. 1973. Stockhausen; Conversations with the Composer. Simon & Schuster. New York.

Roundtable II: A Common High-Level Ground for Scientists and Musicians

Introduction

CURTIS ROADS

University of California, Santa Barbara, California 93106, USA

KEYWORDS: scientists; musicians

Music perception is an extremely complicated cognitive activity. To fully explain the psychophysics by which music transforms the thoughts and emotions of a listener remains a distant goal. Nonetheless, scientists have harnessed high-technology tools in a search to unlock the mysteries surrounding the relation between consciousness and sound. But esthetic experience cannot be distilled to a process that is triggered by a specific stimulus. Specifically, there is no accounting for taste (not yet) from the neuroscience perspective.

This roundtable discussion presents a palette of intellectually colorful oppositions: overviews of the scientific research of Bigand, Molinari, and Thaut opposite the musical perspectives of contemporary composers Minciacchi and Radulescu. At issue was the nature of music cognition.

Dr. Bigand began with the provocative assertion that nonmusicians perform as well as highly trained musicians in high-level music cognition tasks. This is not as shocking as it appears at first glance, when we realize that traditional musical training is mainly concerned with specialized competencies associated with an instrument (e.g., finger and mouth positions in playing a wind instrument), sight reading a score in common music notation, and solfeggio. These mechanical coordination skills, which are the product of thousands of hours of repetitious practice, have little to do with the sensual experience of music listening, much less the analytical understanding of the inner workings of musical process and structure. Happily, the joy of dancing to a beat, reacting to expressive performance gestures, and recognizing global form are available to all listeners. As Dr. Bigand points out, many listeners are functionally experts even if they cannot verbally articulate their auditory experiences.

Psychoacoustics assumes that the human perceptual system is a measuring instrument, yielding results (responses and judgments) that can be systematically analyzed. It examines the relations between observed stimuli and responses and the reasons for those relations. By contrast, composition theory is the collection of ideas that a composer uses to organize his or her thoughts to realize a piece. As the

Address for correspondence: Curtis Roads, Media Arts and Technology and CREATE, 3431 South Hall, University of California, Santa Barbara, CA 93106.

Ann. N.Y. Acad. Sci. 999: 302–303 (2003). © 2003 New York Academy of Sciences.
doi: 10.1196/annals.1284.040

presentation by Horatiu Radulescu demonstrates, composition theory may be only indirectly related to the perceived sonic result. Composition theory is by its nature subjective and fragmentary and may not attempt to account for every detail in a finished work. Maestro Radulescu describes how he approximated the effect of ring modulation spectra by imposing scordatura tunings on conventional instruments. Yet this discourse yields few clues as to how his music is finally constructed. This is not a fault. Very few composers have ever explained their working methods or narrative designs in any detail. Indeed, this is why we create music, which is by definition beyond words!

Dr. Minciacchi, who is both a composer and a neuroscientist, brings a unique perspective to this topic. In his compositions, the musical structure is a sonification of brain data derived from his own neurological research. While such a dual use of data by one person is unique, the translation of scientific data into music is not in itself new. A classic example is Charles Dodge's "Earth's Magnetic Field" (1970), which applied geophysical data to sound. The most important challenge in translation from scientific data to musical structure is finding an effective mapping between them. That is, if the scientific data have intrinsically interesting structural properties, how can these properties be made audible? As Dr. Minciacchi points out, the process of translating these data into sound objects required a profound reassessment in order to find appropriate compositional parameters.

The intervention of Dr. Molinari and his colleagues on "Neurobiology of rhythmic motor entrainment: a neurorehabilitation perspective" exposes a fascinating line of research in the service of rehabilitation. Through a careful argument, the authors prove that auditory information can guide human movement and can therefore be used in therapy. This research is an interesting foil to that of Dr. Bigand, in that it explains in specific terms how trained musicians listen differently.

Dr. Thaut's "Neural basis of rhythmic timing networks in the human brain" follows a similar vein of research, with a focus on localizing, for the first time, the neural circuitry underlying rhythmic synchronization. An interesting connection between the research of both Molinari and Thaut is an emphasis on the role of subliminal coupling mechanisms, or perception without consciousness.

If nothing else, we are learning more precisely what we do not know. As the composer Karlheinz Stockhausen once observed:

> Sonic vibrations do not only penetrate ears and skin. They penetrate the entire body, reaching the soul, the psychic center of perception. The esoteric only involves what cannot be described by means of existing scientific laws and rules. So the next step, time and again, is reinterpretation of the human body as a complicated instrument for perception. That is why every genuine composition makes conscious something of this esoteric realm. This process is endless, and there will be more and more esotericism as knowledge and science become increasingly capable of revealing human beings as perceivers. Neurologists have been seeking for years for the pilot that must obviously exist in the human brain but are unable to find it. They can turn human beings upside down but are unable to discover how the human system is coordinated and centered. That will be discovered step by step, and the profundity of what remains unexplained will also gradually become apparent. I believe that the ratio between the amount concealed, the mysterious, and how much is known always remains largely the same. The further your knowledge extends, the more you discover that you cannot explain.

More About the Musical Expertise of Musically Untrained Listeners

EMMANUEL BIGAND

Laboratoire d'Étude de l'Apprentissage et du Développement, UMR 5022, Université de Bourgogne, Dijon, France

ABSTRACT: Several behavioral experiments that were designed to compare the abilities of musicians and nonmusicians to process subtle changes in musical structures are surveyed. These experiments deal with different aspects of music perception including the processing of melodic and harmonic structures, the processing of large-scale structures, and implicit learning. In all these experiments, the so-called nonmusician listeners behaved in a very similar way as did highly trained students from music conservatories and music departments. This outcome suggests that when the experimental setting requires participants to process *musical structures* (in contrast to musical tones), the large audience of untrained listeners exhibits sophisticated musical abilities that are similar to those of musical experts. It has been suggested that musically untrained listeners are "experienced listeners" who use the same principles as musical experts in organizing their hearing of music.

KEYWORDS: musical expertise; music cognition; musicians; nonmusicians; harmonic processing; large-scale structure; musical brain

INTRODUCTION

The effects of expertise have been studied in many different domains of activity, sometimes leading to surprising results.[1–3] For example, it has been shown that professional cricketers do not differ greatly from novice players in predicting the point at which a ball will drop[4] and that bar waiters have more difficulty in considering the surface of liquid in a tilted container to be horizontal than does a control population.[5] In this paper we consider the effects of musical expertise. Musicians learn to play an instrument and to describe the musical structures they perceive in explicit terms. To what extent can this dual competence be said to modify the process of listening to music and the type of organizations perceived?

The answer to this question is less straightforward than has been generally supposed. First, methodological caution is required concerning the control of variables confounding expertise and the type of sound stimuli used. When experimental tasks require the judgment of an aspect of musical structure in which musicians have been

Address for correspondence: Emmanuel Bigand, Laboratoire d'Étude de l'Apprentissage et du Développement, UMR 5022, Université de Bourgogne, Dijon, France. Voice: +33 3 8039 5782; fax: +33 3 8039 5767.

bigand@u-bourgogne.fr

Ann. N.Y. Acad. Sci. 999: 304–312 (2003). © 2003 New York Academy of Sciences.
doi: 10.1196/annals.1284.041

explicitly trained (detecting a change in pitch or timbre, singing back melodies, identifying meter, tapping in time with the music, etc.), important differences between the two groups are frequently observed. Effects of expertise are thus confused with familiarity with a particular experimental task. The case is similar when experimental instructions employ technical musical terms, not explicitly understood by nonmusicians, or use notions so ill defined and so ambiguous that only musicians can grasp the purpose of the study.[6–8] In addition, the music used in experiments (classical music) is frequently more familiar to musicians than to nonmusicians,[9] thereby confounding expertise and familiarity with the stimulus.

What then are the preferred methods for comparing the two groups of listeners? The first consists of exploring the elementary perceptual intuitions experienced by all in their daily experience with music, which are not subject to overtraining in musicians. Examples of this include judgment of similarity between musical materials and the degree of completion of a piece of music, or the identification of the musical emotion expressed by musical extracts (see below). The second method consists of measuring the sensitivity of listeners to musical structures to which the investigator does not explicitly draw the participants' attention (an implicit measure). The use of implicit methods, such as the priming technique outlined below, can determine the structures "naturally" treated by the musical ear, that is, without a conscious effort underpinned by explicit response strategies. The use of implicit tasks is further justified given that musical structures are not devised for explicit perception.

Equally, the scientific study of musical expertise requires an appropriate definition of the term "musical perception." The perception of music is an infinitely rich experience that is not reducible to a sequence of simple situations consisting in the perception of rudimentary qualities of musical sound (pitch, timbre, duration). Elementary experimental tasks of this nature tell us more about the auditory abilities of listeners than their strictly *musical* abilities. That there may be differences between the two groups of listeners at early stages of comparison is not surprising, but it has few implications. Musical perception implies far more complex cognitive operations (categorization, memorization, integration, etc.) that are not necessarily more developed for music students than for students specialized in another field. In this paper we summarize empirical work focusing on four aspects of music listening (processing melody, harmony, large-scale musical structures, and musical learning). This work shows that differences between musicians and nonmusicians diminish, and sometimes disappear, when the requisite experimental tasks require higher cognitive processes. In these studies the term "musician" refers to students at national music conservatories who have studied musical and instrumental techniques for many years (10 years on average) and whose abilities have been confirmed by regular formal examination. Nonmusicians are students of the same age who have not had any specific musical training. As we will see, the differences observed between these two groups, when they exist, are negligible given on the scale of differences in their musical knowledge.

WHAT ARE THE EFFECTS OF EXPERTISE ON THE STRUCTURING OF MELODY?

A melody is a dynamic structure whose perceptual identity is defined by the integration of all its constitutive parameters. This integration leads to the perception of

(a)

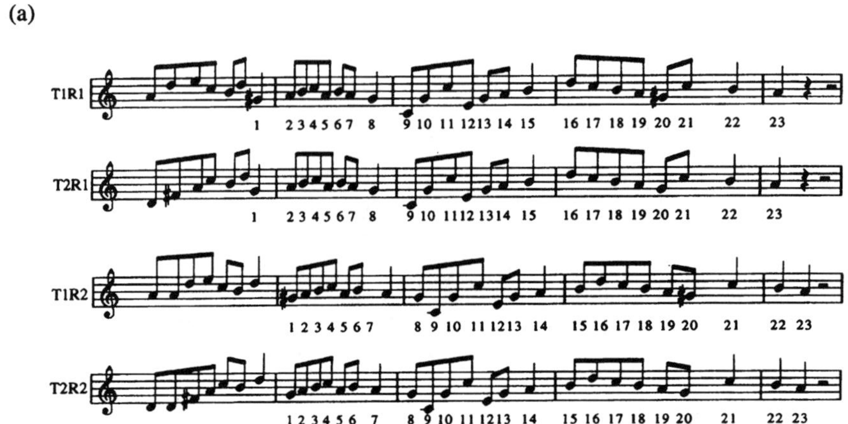

(b)

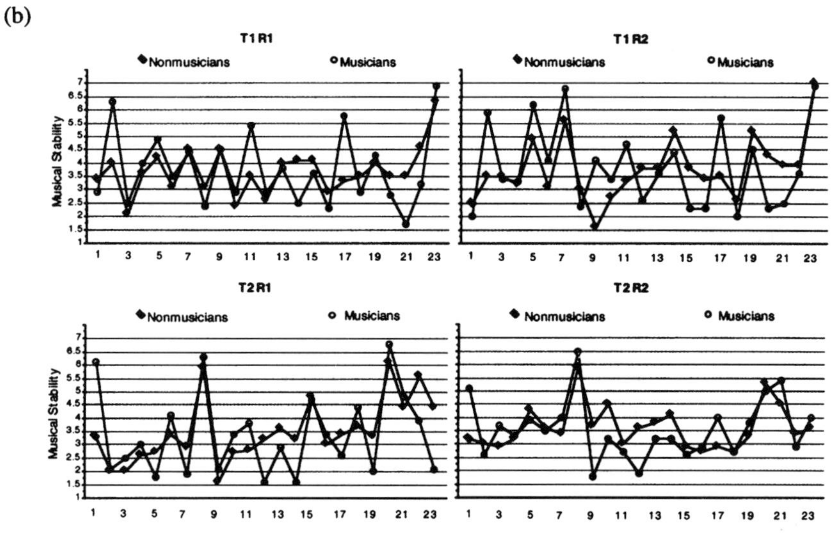

FIGURE 1.

melody as a dynamic form comprising musical tensions and resolutions that span several levels of time.[1,10] Different studies have been undertaken to specify how the different parameters of sound contribute to the definition of these dynamic forms. Consider the melodies in FIGURE 1a. The aim of this study was to determine the extent to which tonality and rhythm contribute to a definition of the dynamic form of the melody for musicians and nonmusicians.[11] The T1 melodies contain almost the same notes as the T2 melodies, the same rhythm, and the same contour. The tonal functions of these melodies are nevertheless greatly different because the T1 melodies are heard in A minor and the T2 melodies in G major. Therefore, the notes numbered 2, 5, and 7 (tonic notes) function as perceptual anchor points in T1, but as subtonic notes that are perceptually less salient in T2. The situation is reversed for notes 1 and 8, which are tonic in T2 but not in T1 (subtonic). Careful examination of the score reveals that the tonal function of the T1 and T2 melodies is inversely correlated. The experimental task used to investigate the sensitivity of listeners to changes in tonal function consisted of judging, on a 7-point scale, the degree of "musical tension" perceived for each melody note. The resulting profile would therefore be considered as an approximation of the perceived dynamic form.

The profiles obtained from the experiment (FIG. 1b) indicate that listeners were indeed sensitive to the contextual changes of tonal function between T1 and T2 and to changes in rhythm between R1 and R2. As can be seen, results from musicians and nonmusicians are strongly correlated in each of these situations, suggesting that common principles of melodic structuring are at work for these groups of listeners. We have also observed that these dynamic profiles are essential in the memorization of melodies.[12] Accordingly, musicians and nonmusicians estimate that more than 50% of the notes changed between T1 and T2 melodies, as well as between R1 and R2 (no note was changed in this case), and up to 64% of notes changed when the two types of change were combined (T1R1 versus T2R2).

Overall, the results we have obtained for the perception of melody support the conclusion that musically expert and novice listeners represent Western tonal melodies mentally as an abstract structure, which typifies the principal attentional trajectories developed during listening. For the two groups of listeners, these trajectories seem to be punctuated by the relationshops of musical tension and resolution that span the different levels of musical time.[1]

WHAT ARE THE EFFECTS OF EXPERTISE ON THE STRUCTURING OF HARMONY?

Western tonal harmony achieves a skillful balance between psychoacoustic constraints and the cultural conventions formed across several centuries of scientific, esthetic, and spiritual reflection. How does the contemporary Western musical ear apprehend these structures? What is the influence of musical education? We have undertaken two types of study in order to address these questions. In the first, we asked listeners to evaluate the musical tensions created by certain harmonic relations. We showed that the perception of musical tension in short musical sequences[13] and long musical sequences[14] is strongly correlated for musicians and nonmusicians. Thus, it seems that the perception of harmonic tension does not differ as a function of musical expertise.

Many other studies have been performed using the paradigm of harmonic priming. Using this technique, attention is drawn to an elementary perceptual task which the subject performs on a *target* chord, while the musical context in which the target chord is presented is manipulated without the subject's knowledge. The task may consist of deciding as quickly as possible whether a target chord is in tune or out of tune, whether or not it contains a clearly dissonant note, whether its constituent notes are played exactly together (15 for a review), or of identifying the phoneme on which the chord is sung.[16] The critical point is to determine how the performance of this perceptual task is influenced by the harmonic context in which the target chord appears.

The initial studies showed that judgment of a target chord's consonance is faster and more accurate when the chord is preceded by a harmonically related chord (C–G, for example) than a nonrelated chord (C–F#). The harmonic context leads the listener to implicitly anticipate compatible chords in this context according to the rules of Western harmony. The anticipated chords are thus cognitively present *before* being heard, which facilitates their perceptual processing when they actually occur. In the last five years, we have used this paradigm to show that Western adult listeners have an implicit sensitivity to very fine differences in harmonic function. Consider, for example, the musical sequences in FIGURE 2. The final two chords in these sequences are identical in all three contexts. However, their harmonic function changes from one context to the other. In the "highly expected" context, the final chord is a tonic chord (I). In the "unexpected" context, it acts as a subdominant chord (IV) following a perfect cadence in the key of D. In the "moderately expected" context, the harmonic function of this chord is more ambiguous. Perceived in relation to the second part of the sequence (which is identical to that of the unexpected condition), the target chord is a subdominant (IV) one. On the other hand, if it is heard in relation to the first part of the sequence (which is identical to the highly expected condition), this target chord may function as a tonic chord, marking a return (albeit rapid) to the principal key primed by the opening chords. In other words, the target chord is "primed" in the (moderately expected) condition by the first part of the sequence, which is not the case in the unexpected condition. In spite of the small scale of the perceptual difference between these three conditions, response for the target chord

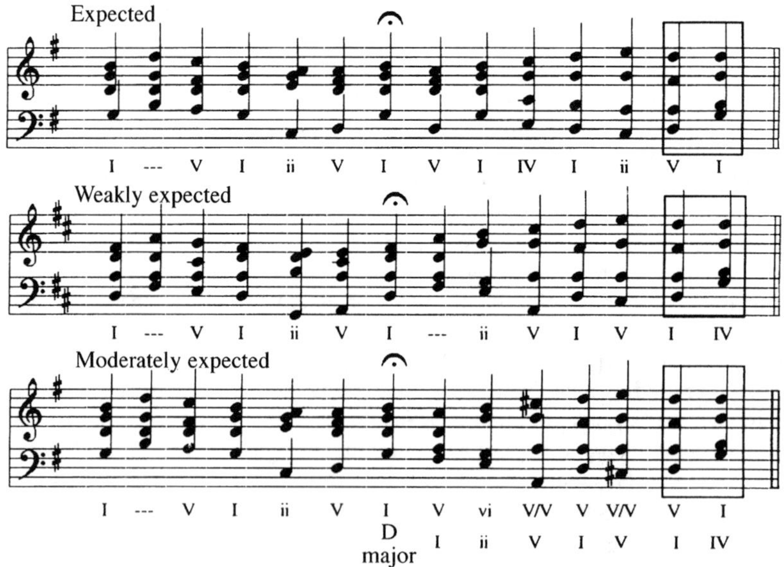

FIGURE 2.

is faster and more accurate in the highly expected and moderately expected conditions than in the unexpected condition. Given the subtlety of the harmonic manipulation used, it is remarkable that the nonmusicians accomplish this task similarly to musicians with a differentiated response to the unexpected and moderately expected conditions.[17]

Harmonic priming is at the root of perceptual expectations that develop automatically during listening and that play an important role in musical expression and emotion. For the present subject, the existence of these perceptual expectations for nonmusicians is important in view of the knowledge that an essential characteristic of expertise (in any domain) is the ability to anticipate events. The fact that the perceptual expectations of listeners—be they musicians or nonmusicians—do not differ or barely differ, either on a behavioral level[15] or on a neurophysiological level,[18–20] suggests that listeners without musical training are musical experts in spite of their inability to describe explicitly what they hear.

THE EFFECTS OF EXPERTISE IN MORE COMPLEX MUSICAL SITUATIONS

The preceding studies use musical sequences that illustrate in miniature the musical structures that can be observed in Western music. To what extent can the weak differences observed between the two groups of listeners be explained by the reductional nature of the musical stimuli used? Several experimental studies have been conducted with the help of pieces stemming from the existing musical repertoire in order to address this question. They led to results that reflect the weak effects of expertise. In this way, we have shown by using short minuets by Bach, Haydn and Mozart that musicians demonstrate the same sorts of difficulty as nonmusicians in resolving musical puzzles, with both groups experiencing great difficulty in integrating local harmonic structures with global structure.[21,22]

In a recent study, we aimed to better understand this result. It is known that the coherent presentation of the material improves considerably the process of memorization for experts, but not for novices.[23] By the same reasoning, we asked musicians and nonmusicians to memorize 20 extracts taken from the exposition sections of four sonatas by Haydn. These extracts were presented in either a coherent or an incoherent fashion. In the first instance, the 5 extracts corresponding to the exposition of a sonata were presented in sequential order, before turning to 5 extracts from another sonata, and so forth. In the incoherent condition, the 20 extracts from these sonatas were presented in a totally random way. Following this learning phase, subjects heard 44 musical extracts (20 from the learning phase, plus 24 new extracts taken from other sonatas by the same composer) and indicated on a 6-point scale their belief that the extract they heard was new or old. Contrary to the classical expertise effects reported in cognitive psychology, musicians do not perform better in the coherent condition. In other words, they do not benefit more than nonmusicians from the coherent presentation of the material. Based on these results, it can be argued that musicians do not have a better "comprehension" of large musical structures than do nonmusicians. This result, which agrees with that of many others, will come as no surprise to teachers of music analysis who observe the extent to which expert musi-

FIGURE 3.

cians, though good instrumentalists, experience great difficulties in hearing musical structures and forms.[24]

Finally, I present an example noteworthy for its lack of difference between musicians and nonmusicians. The study can be classed as research into the implicit learning of new musical systems.[25] We asked a composer to write 40 canons in the style of Webern, all based on a dodecaphonic series (FIG. 3a). In a learning phase, we presented 20 canons two times. In a test phase, we presented 20 pairs of canons. Each pair contains a canon composed from the same series but not yet heard by the subjects, and its foil (FIG. 3b). The subjects' task was to indicate which canon of the pair had been composed in the same fashion as that from the initial phase of the study. Though perceived as being extremely difficult, this task was accomplished at a rate above chance by musicians (63%) and nonmusicians (60%), with no significant difference between the two groups. This result is in accordance with work on implicit learning that shows that the listener is able to internalize complex statistical regularities (in this case, the same series of 12 sounds) through simple, passive exposure to environmental stimuli.

CONCLUSION

The study of expertise in the domain of music presents a certain number of scientific, pedagogical, and sociological interests. For scientists, the concern is an understanding of how expertise leads to changes in the processing of a specific type of information, by what learning process (implicit or explicit) does this expertise develop, and what mental changes might this learning entail as much from an anatomical as a functional perspective. At the present time, conclusions differ from one study to another. The choice of an experimental method plays a role in explaining this divergence. Equally, it seems that the theoretical orientation of researchers bears on the results that are reported in the literature. In this way, the work on brain plasticity tends to place an exaggerated importance on anatomical and/or functional differences associated with intensive musical training.[26–28] Of course, this work is essential on a neurophysiological level because it demonstrates that intensive perceptual learning can afford a functional reorganization in the brain of musician subjects. However, it would be wrong to conclude that these differences have any repercussions for the general cognitive and neurophysiological structure that allows musical processing in all its complexity.

The results reported above show strong similarities between the two groups of listeners that suggest that this structure is not strongly affected by explicit musical training. On the contrary, the mere repeated exposure to music seems to be sufficient for the development of a sophisticated auditory expertise in the absence of any form of explicit learning. This result is fairly commonplace when viewed alongside the many studies in implicit learning that have been conducted over the last few years, which attest to the extraordinary ability of the human mind to internalize highly complex structures. Mental reorganization resulting from these learning processes is undoubtedly more complex to observe than that resulting from elementary perceptual learning. No doubt these learning processes confer a real plasticity on the human mind. In the domain of music, we believe that this plasticity is translated more by the weak differences observed between musically trained and untrained listeners, than by neuroanatomical and/or functional differences associated with very elementary perceptual learning.

ACKNOWLEDGMENTS

I would like to thank all those who collaborated in this research—in particular, Marion Pineau, Barbara Tillmann, Bénédicte Poulin, François Madurell, Daniel A. D'Adamo, and Philippe Lallitte.

REFERENCES

1. LERDAHL, F. & R. JACKENDOFF. 1983. A Generative Theory of Tonal Music. MIT Press. Cambridge, MA.
2. MCKLOSKEY, M., A. WHASBURN & L. FLECH. 1983. Intuitive physics: the straight-down belief and its origin. J. Exp. Psychol. Learn. Mem. Cognit. **9:** 635–649.
3. CHOLLET, S. & D. VALENTIN. 2000. Expertise level and odour perception: What can we learn from red burgundy wines? Année Psychol. **100:** 11–36.

4. HOULSTON, D.R. & R. LOWES. 1993. Anticipatory cue-utilization processes amongst expert and nonexpert wicketkeeper in cricket. Int. J. Sport Psychol. **24:** 59–73.

5. HETCH, H. & D.R. PROFFITT. 1995. The price of expertise: effects of experience on the water-level task. Psychol. Sci. **6:** 90–95.

6. RADVANSKY, G.A., K.J. FLEMING & J.A. SIMMONS. 1995. Timbre reliance in nonmusicians' and musicians' memory for melodies. Music Percept. **13:** 127–140.

7. SMITH, J.D., D.G.K. NELSON, L.A. GROHSHKOPF & T. APPLETON. 1994. What child is this? What interval was that? Familiar tunes and music perception in novice listeners. Cognition **52:** 23–54.

8. HEBERT, S., I. PERETZ & L. GAGNON. 1995. Perceiving the tonal ending of tune excerpts: the role of preexisting representation and musical expertise. Can. J. Exp. Psychol. **49:** 193–210.

9. OHNISHI, T., H. MATSUDA, T. ASADA, *et al.* 2001. Functional anatomy of musical perception in musicians. Cerebr. Cortex **11:** 754–760.

10. JONES, M.R. & M. BOLTZ. 1989. Dynamic attending and responses to time. Psychol. Rev. **96:** 459–491.

11. BIGAND, E. 1997. Perceiving musical stability: the effect of tonal structure, rhythm and musical expertise. J. Exp. Psychol. Hum. Percept. Perform. **23:** 808–812.

12. BIGAND, E. & M. PINEAU. 1996. Context effects on melody recognition: a dynamic interpretation. Curr. Psychol. Cognit. **15:** 121–134.

13. BIGAND, E., R. PARNCUTT & F. LERDAHL. 1996. Perception of musical tension in short chord sequences: the influence of harmonic function, sensory dissonance, horizontal motion, and musical training. Percept. Psychophys. **58:** 125–141.

14. BIGAND, E. & R. PARNCUTT. 1999. Perception of musical tension in long chord sequences. Psychol. Res. **62:** 237–254.

15. TILLMANN, B., J. BHARUCHA & E. BIGAND. 2000. Implicit learning of tonality: a self-organizing approach. Psychol. Rev. **107:** 885–913.

16. BIGAND. E., B. TILLMANN, B. POULIN, *et al.* 2001. The effect of harmonic context on phoneme monitoring in vocal music. Cognition **81:** 11–20.

17. BIGAND, E., F. MADURELL, B. TILLMANN & M. PINEAU. 1999. Effect of global structure and temporal organization on chord processing. J. Exp. Psychol. Hum. Percept. Perform. **25:** 184–197.

18. REGNAULT, P., E. BIGAND & M. BESSON. 2001. Different brain mechanisms mediate sensitivity to sensory consonance and harmonic context: evidence from auditory event related brain potentials. J. Cognit. Neurosci. **13:** 241–255.

19. KOELSCH, S., T. GUNTER & A.D. FRIEDERICI. 2000. Brain indices of musical processing: "nonmusicians" are musical. J. Cognit. Neurosci. **12:** 3, 520–541.

20. MAESS, B., S. KOELSCH, T.C. GUNTER & A.D. FRIEDERICI. 2001. Musical syntax is processed in Broca's area: an MEG study. Nat. Neurosci. **4:** 540–545.

21. TILLMANN, B., E. BIGAND & F. MADURELL. 1998. Influence of global and local structures on solution of musical puzzles. Psychol. Res. **61:** 157–174.

22. TILLMANN, B. & E. BIGAND. 1998. Influence of global structure on musical target detection and recognition. Int. J. Psychol. **33:**107–122.

23. POULIN, B., E. BIGAND, W.J. DOWLING, *et al.* 2001. Do musical experts take advantage of global musical coherence in a recognition test? Society for Music Perception and Cognition. August 9–11, Queen's University, Kingston, Ontario, Canada.

24. LEVINSON, J. 1997. Music in the Moment. Cornell University Press. Ithaca, NY.

25. REBER, 1992.

26. SCHLAUG, G., L. JÄNCKE, Y. HUANG & H. STEINMETZ. 1995. In vivo evidence of structural brain asymmetry in musicians. Science **267:** 699–701.

27. ELBERT, T., C. PANTEV, C. WIENBRUCH, *et al.* 1996. Increased cortical representation of the fingers of the left hands in string players. Science **270:** 305–307.

28. PANTEV, C., R. OOSTENVELD, A. ENGELIEN, *et al.* 1998. Increased auditory cortical representation in musicians. Nature **392:** 811–814.

Neurobiology of Rhythmic Motor Entrainment

MARCO MOLINARI,[a,b] MARIA G. LEGGIO,[a,c] MARTINA DE MARTIN,[d]
ANTONIO CERASA,[a] AND MICHAEL THAUT[a,e]

[a]I.R.C.C.S. Santa Lucia Foundation, Rome, Italy

[b]Institute of Neurology, Catholic University, Rome, Italy

[c]Department of Psychology, University of Rome "La Sapienza," Rome, Italy

[d]National Institute of Geophysics and Vulcanology, Rome, Italy

[e]Molecular, Cellular, and Integrative Neuroscience Program,
Colorado State University, Ft. Collins, Colorado 80532, USA

ABSTRACT: Timing is extremely important for movement, and understanding the neurobiological basis of rhythm perception and reproduction can be helpful in addressing motor recovery after brain lesions. In this quest, the science of music might provide interesting hints for better understanding the brain timing mechanism. The report focuses on the neurobiological substrate of sensorimotor transformation of time data, highlighting the power of auditory rhythmic stimuli in guiding motor acts. The cerebellar role of timing is addressed in subjects with cerebellar damage; subsequently, cerebellar timing processing is highlighted through an fMRI study of professional musicians. The two approaches converge to demonstrate that different levels of time processing exist, one conscious and one not, and to support the idea that timing is a distributed function. The hypothesis that unconscious motor responses to auditory rhythmic stimuli can be relevant in guiding motor recovery and modulating music perception is advanced and discussed.

KEYWORDS: timing; rhythm perception; time discrimination; cerebellum; motor rehabilitation; music perception

INTRODUCTION

Rhythmicity is an essential temporal component of movement strictly embedded in many motor control functions. In different disorders of the central nervous system (CNS), motor symptoms have been linked to alteration in central timing mechanisms. Rehabilitation potentials of rhythmicity have been explored in different contexts. Rhythmic training has been investigated as potential treatment in Parkinson's disease,[1–4] Huntington's disease,[5] or stroke[6] and in patients with traumatic brain injury and other movement disorders.[7] Nevertheless, because it is currently unclear

Address for correspondence: Dr. Marco Molinari, Experimental Neurorehabilitation Lab, I.R.C.C.S. Santa Lucia Foundation, Via Ardeatina 306, 00179 Rome, Italy. Voice: +39-6-515012600; fax: +39-6-51501679.
m.molinari@hsantalucia.it

Ann. N.Y. Acad. Sci. 999: 313–321 (2003). © 2003 New York Academy of Sciences.
doi: 10.1196/annals.1284.042

which circuits and central mechanisms mediate the positive effects of rhythmic training, study of the neurobiological basis of rhythm may provide interesting data.

In the last decade, the neuroscience approach to music perception has greatly improved our understanding of how brain processes temporal information.[8] Thus, motor and musical sciences are discovering a common field in which they can positively interact. A multidisciplinary approach that links old methods, such as lesion studies, with more modern perspectives, such as the study of interrelations between music and brain functioning, may represent an interesting and fruitful tool.[9] The definition of the timing mechanisms in perception and motor control, and the knowledge of the circuits that allow auditory rhythmic stimuli to entrain motor acts will provide interesting cues to implement recovery of motor functions.

Both motor control and music perception highly depend on the capacity to detect time intervals and to process time information. In our laboratory, timing abilities were recently analyzed psychophysically and with functional neuroimaging in patients with motor disturbances due to cerebellar damage and in healthy subjects with different musical expertise. The choice of studying subjects with lesions of the cerebellar circuits was derived from the convergence of different lines of data in demonstrating a cerebellar role in timing. Patients with cerebellar pathologies show timing errors in muscle contraction[10,11] and increased variability during repetitive finger tapping.[12] Difficulties are also reported in perceptual tasks that require precise timing.[13,14] However, little is known about how the cerebellum contributes to timing, and different theories have been proposed. Ancillary non-timing functions required by timing tasks such as sensorimotor processing,[15] attention, or memory[16] have been considered to explain the timing deficits in patients with cerebellar disorders and to negate a central position for cerebellar processing in timing. Conversely, the cerebellum has been suggested as the site of primary temporal representation[17,18] or the site where temporal information from sensory inputs is extracted and funneled to the motor system.[15] However, the debate is still open.

The vast majority of human studies on timing use tasks requiring conscious responses, thus strongly depending on attention and memory abilities. Evidence that it is possible to vary synchronous tapping in response to consciously undetectable or not reliably detectable variations in rhythm frequency[19] provides the opportunity to test central time mechanisms for translating sensory time information into motor coding without depending on conscious detection of rhythmic tempo modulations.

TAPPING ABILITIES IN PATIENTS WITH CEREBELLAR DISORDERS

To test the role of the cerebellum in timing and to analyze the possible influences of consciousness, a consciously and unconsciously modulated synchronized tapping paradigm was recently used in subjects with cerebellar damage.[20] It was postulated that if the decoding of time information is an ability that requires an internal clock localized within the cerebellum, patients with cerebellar damage would be impaired in their ability to adjust their tapping rhythm to changes in the leading rhythm. Conversely, good timing adjustment would favor a different localization of the central clock for sensorimotor transduction of time information or a more distributed localization of this function.

Detection of rhythm changes

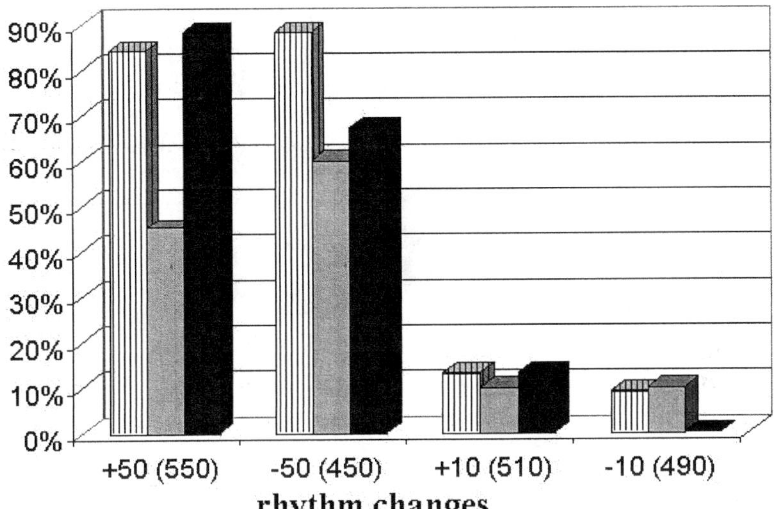

FIGURE 1. Capacity to detect rhythm changes in subjects with cerebellar damage. Detection rates of changes at 2% of IRI (+10,510 and −10,490) are very low (below chance) in subject with cerebellar damage as well as in healthy controls. Conversely, most large-step changes are detected by healthy controls and by patients with focal cerebellar lesions, except at −50 where the latter group performed less accurately. The detection rate of the atrophic patients is considerably lower, indicating degraded perception of frequency modulation in this group.

When tested in the capacity to detect frequency changes in an auditory rhythmic stimulus, subjects with cerebellar atrophy but not those with focal damage displayed significantly less accurate responses than did controls. The capacity to detect small changes, 2% (± 10 ms) variations of the base interval, was very low (below chance) in all subjects, indicating no conscious detection of the perturbation. Large-step changes, 10% of base interval (± 50 ms), were similarly detected by healthy subjects and by patients with focal cerebellar lesions. Degraded perception of frequency modulation, with detection rate below chance, was conversely observed in atrophic patients (Fig. 1). Therefore, cerebellar functioning appears to be linked with the capacity to consciously perceive changes in the rhythm.

The same subjects were also tested in their capacity to tap in synchrony with an auditory rhythmic stimulus. At a baseline frequency of 500 ms/2 Hz despite the differences in the motor coordination, healthy subjects and subjects with cerebellar atrophic or focal damage all tapped in synchrony. The interresponse interval (IRI) as well as synchronization errors (SE) was similar among the three groups. As expected, differences were present in motor implementation and were reflected by the high variability in the cerebellar group responses.

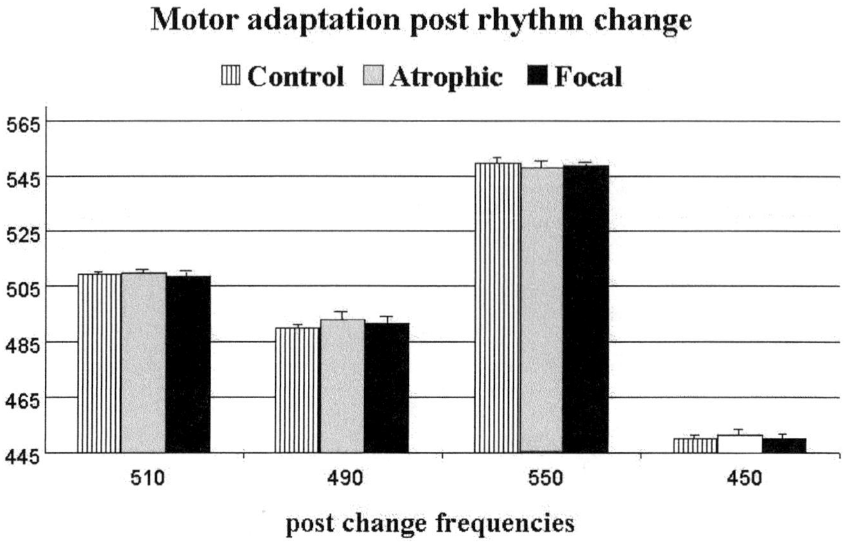

FIGURE 2. Averaged results of synchronization of the motor response to changes of the stimulus frequencies from the baseline (ISI = 500 ms/2 Hz) to 510, 490, 550, and 450 ms. Motor responses are synchronized to the stimulus frequencies in normal subjects as well as in the cerebellar, atrophic, and focal patient groups.

Interestingly, for all subjects, coupling between stimulus and motor response also continued without significant differences after both consciously perceived and not perceived changes in the leading rhythm (FIG. 2). Averaged synchronization errors, indicating positional timing between stimulus and motor response, were similar for size and positional errors independently of cerebellar damage. The good capacity of subjects with cerebellar disorders to modify their tapping to variations in the leading stimulus was also confirmed by analyzing the timing behavior of rhythmic motor traces before, during, and after the step change of the cue frequency, using a sigmoid function for modeling. Comparison across groups showed no difference in stability of frequency entrainment between the healthy group and the two cerebellar groups. Also, no differences in tracking stability were noted between pre-change and post-change cue frequencies. Conversely, as for the baseline rhythm, cerebellar damage appeared critical in influencing the variability of the motor response. In fact, standard deviations (SD) of SE values were significantly higher in persons with cerebellar disorders than in healthy subjects.

It is well established that synchronization strategies are different for small and large interstimulus interval (ISI) step changes.[19] Small step adaptations are made by rapid IRI rescaling to the new ISI value. Large-step changes are accomplished by a temporary (1–2 intervals) over- or under-correction, depending on the ISI increase or decrease before matching the new ISI interval. These strategies were similarly used by all tested subjects, indicating that the cerebellar damage does not affect post-step IRI rescaling. Furthermore, in general, rhythmic synchronization behavior was not modified. Stability of rhythmic frequency tracking, synchronization error, and speed of adaptation patterns to frequency modulations above and below conscious perception were similar in cerebellar and healthy subjects.

CEREBELLUM AND MOTOR ENTRAINMENT

The foregoing data indicate that cerebellar pathologies may impair the capacity to consciously detect rhythmic variation and the stability of the motor response but not the motor entrainment effect of rhythmic stimuli. It derives that it is possible to unconsciously detect frequency modulations and to extract time information from the sensory stimuli, making it available for the motor output without cerebellar processing. In other words, motor entrainment to rhythmic auditory stimuli can be induced by directly influencing motor effectors either in the cortex or at spinal levels. This is relevant for general theories on central timing and motor control and may provide interesting approaches to the rehabilitation of cerebellar motor defects.

The idea that sensorimotor timing can be processed in the periphery is in keeping with theoretical models.[21–23] It has been proposed that direct computing of timing information can be achieved in the auditory nerve by neural excitation patterns that provide precise physiological coding. This time coding can be transferred directly into adjacent motor structures, entraining neural motor codes and allowing synchronization between auditory stimulus and motor response. Further support of the peripheral processing theory is derived from the speed of step-change adaptations observed in subjects with cerebellar disorders. In fact, early adaptation suggests that frequency/interval-tuned modulation detectors are contained in early auditory processing and that coded temporal discharge patterns extend rapidly into neighboring motor regions. The fast processing and the optimal performances after cerebellar damage indicate that specially dedicated "ring circuitries" such as the cerebellum and/or the basal ganglia do not intervene in the processing of motor entrainment.

The peripheral processing theory, however, does not imply contradiction with the central mechanisms hypothesis, maintaining that core timing functions are located in the basal ganglia[24] or the cerebellum.[25] It is highly conceivable that when conscious processing of temporal information is required, as in temporal discrimination tasks, basal ganglia or cerebellar processing is specifically active, possibly at different stages of processing.[16] Also, studies in patients with cerebellar disorders support this interpretation. Deficits in time processing in cerebellar subjects are reported in conscious time discrimination tasks[26,27] but not in subconscious timing.[20]

MUSIC AND THE BRAIN

Skill acquisition is associated with changes in brain structure and function. Knowledge of the neurobiological mechanisms that allow such plastic changes is of extreme importance not only for understanding brain functioning in general but also for favoring recovery of function after CNS damage. Music professionals, who train with extraordinary intensity, may represent apt subjects for investigating the potential of training for shaping brain circuits.[9] In the last decade, contributions from different fields, spanning cognitive sciences to experimental psychology, behavioral neurology to neuroimaging, have significantly enlarged the scientific understanding of music.[28] Peculiarities in the brains of musicians have been reported on the basis of morphometric analysis of postmortem brains since the beginning of the last century.[29,30] Modern brain-imaging techniques have provided further demonstrations of anatomic differences between musicians and nonmusicians in different brain areas,[31–34] and the relevance of these findings in understanding musical abilities has recently been discussed.[28]

The simplest rhythmic act that we each often perform when listening to music is to tap in synchrony with the beat of the melody. To make the task simpler, think of following not a tune but a metronome. During such a task the motor system can adjust to a predictable pace by programming a movement sequence in advance according to an internal rhythm closely matching the external pacing. It may be that the generation of this isochronous tapping sequence is additionally controlled in a closed-loop manner, relating the visual or acoustic external stimuli with the tactile feedback of the fingers to evaluate how far the precise timing of the tapping sequence has been achieved. Tapping abilities have often been used to investigate neural circuits associated with time and rhythmic processing,[35–37] and a complex circuit involving different cortical areas, cerebellum, and basal ganglia has been identified as specifically active.

The capacity of musical and thus rhythmic training to affect isochronous tapping performances and the associated brain activation pattern was recently investigated in our laboratory.[38] Professional musicians and nonmusician subjects with no specific musical training were required to perform an acoustically triggered rhythmic finger-tapping task. The task was extremely simple and all subject were tapping with a high degree of accuracy. The average IRI was similar in the two groups with a larger dispersion of the data in the nonmusician group. This good level of performance was achieved in the two groups with different strategies. Analyses of the shift between the beat and the response as indicated by synchronization errors showed that lay subjects tended to anticipate the beat, whereas musicians' responses were quite delayed. Therefore, an "anticipatory" predictive technique was adopted by nonmusicians and a "posticipatory" reactive technique by musicians. The frequency distribution of the SE values allowed further highlighting of the differences. A peculiar bimodal distribution of SE frequencies was present in the musicians, with two peaks falling into two narrow ranges significantly after the beat (20–30, 90–110 ms). Nonmusician subjects showed a trend to a gaussian distribution, always centered before zero. A few subjects presented a mixed distribution characterized by a gaussian part and one or two peaks. When present, the peaks fell roughly in the same ranges as those of the musicians.

Functional neuroimaging data showed task-related activations in similar structures in both groups. Significant foci of activation were seen in the right cerebellum (lobule HV anterior lobe) and vermis (lobule VI posterior lobe), right putamen, left thalamus, and left primary motor (MI) and somatosensory (SI) cortices. Differences between groups were observed in the degree of activation among these structures. In particular, musicians presented greater cerebellar activation, whereas nonmusicians tended towards greater activation of cortical areas. The greater activation of the corticobasal ganglia loop in nonmusicians favors a highly automated sensorimotor process sustaining rhythmic tapping. Conversely the principal activation of a cerebello-cortical loop, as observed in musicians, can be interpreted as the result of continuous feed-forward control of the cerebellar processing over the S1-M1 cross-talking required for controlled rhythmic tapping.

CONCLUSION

The two studies reported are somehow in conflict. The first indicates that cerebellar damaged subjects are impaired in the capacity to consciously discriminate rhythms, but not in the sensorimotor transduction required in synchrony tapping. The second focuses on the cerebellocortical loop as a key circuit for rhythmic tapping, particularly in rhythmically highly trained subjects such as professional musicians. The apparent contradiction can be solved by taking into account the suggested differences in the cerebellar role in tapping during conscious versus subconscious sensorimotor time transduction. Along these lines, the tapping performed in a motor entrainment framework does not require cerebellar processing and fits well with the low activation observed in nonmusician subjects. Conversely, musicians, usually highly conscious of the nature of rhythmic patterns and structures, present high cerebellar activation that coupled with a specific delayed strategy supports the idea of continuous control over incoming sensory information for checking the synchrony between motor performance and auditory rhythm.

Different levels of rhythm perception may also be important for the neurobiology of music and the comprehension of how music is perceived. The demonstration that our body can react to rhythmic variations without their overt perception heralds the idea that components of a piece of music that are not overtly detectable might well affect listener reaction. In this context, musical studies like those carried out by Curtis Roads (this volume) on microsound might provide interesting cues to be tested in neuroscience settings.

In conclusion, the present data support the idea that timing is a distributed function in which different neural circuits can process time information, and their activation and role can vary according to the specific timing requirements of a given task. In particular, the present data demonstrate that cerebellar impairment does not abolish timing processing in the motor domain, suggesting a direct unconscious drive between auditory and motor structures that may be relevant for the development of rehabilitation strategies and for the understanding of music perception.

ACKNOWLEDGMENTS

The continuous encouragement and support of Prof. Carlo Caltagirone is gratefully acknowledged. The present work was supported by MURST, CNR, and Italian Ministry of Health grants (to M.M.).

REFERENCES

1. MCINTOSH, G.C., S.H. BROWN, R.R. RICE & M.H. THAUT. 1997. Rhythmic auditory-motor facilitation of gait patterns in patients with Parkinson's disease J. Neurol. Neurosurg. Psychiatry **62:** 22–26.

2. THAUT, M.H., G.C. MCINTOSH, R.R. RICE, *et al.* 1996. Rhythmic auditory stimulation in gait training for Parkinson's disease patients. Mov. Disord. **11:** 193–200.

3. MILLER, R.A., M.H. THAUT, G.C. MCINTOSH & R.R. RICE. 1996. Components of EMG symmetry and variability in parkinsonian and healthy elderly gait. Electroencephalogr. Clin. Neurophysiol. **101:** 1–7.

4. RICHARDS, L.J., T.J. KILPATRICK & P.F. BARTLETT. 1992. De novo generation of neuronal cells from the adult mouse brain. Proc. Natl. Acad. Sci USA **89:** 8591–8595.

5. THAUT, M.H., R. MILTNER, H.W. LANGE, *et al.* 1999. Velocity modulation and rhythmic synchronization of gait in Huntington's disease Mov. Disord. **14:** 808–819.

6. THAUT, M.H., G.C. MCINTOSH & R.R. RICE. 1997. Rhythmic facilitation of gait training in hemiparetic stroke rehabilitation. J. Neurol. Sci. **151:** 207–212.

7. THAUT, M.H. & G.C. MCINTOSH. Music therapy in mobility training with the elderly: a review of current research. Care Manag. J. **1:** 71–74.

8. OHNISHI, T., H. MATSUDA, T. ASADA, *et al.* 2001. Functional anatomy of musical perception in musicians. Cereb. Cortex **11:** 754–760.

9. MÜNTE, T.F., E. ALTENMÜLLER & L. JANCKE. 2002. The musician's brain as a model of neuroplasticity. Nat. Rev. Neurosci. **3:** 473–478.

10. DIENER, H.C., J. HORE, R. IVRY & J. DICHGANS. 1993. Cerebellar dysfunction of movement and perception. Can. J. Neurol. Sci. **20 (Suppl 3):** S62–S69.

11. TIMMANN, D., S. WATTS & J. HORE. 1999. Failure of cerebellar patients to time finger opening precisely causes ball high-low inaccuracy in overarm throws. J. Neurophysiol. **82:** 103–114.

12. VRY, R.B., S.W. KEELE & H.C. DIENER. 1988. Dissociation of the lateral and medial cerebellum in movement timing and movement execution. Exp. Brain Res. **73:** 167–180.

13. IVRY, R.B. 1996. The representation of temporal information in perception and motor control. Curr. Opin. Neurobiol. **6:** 851–857.

14. NAWROT, M. & M. RIZZO. 1995. Motion perception deficits from midline cerebellar lesions in humans. Vis. Res. **35:** 723–731.

15. PENHUNE, V.B., R.J. ZATORRE & A.C. EVANS. 1998. Cerebellar contributions to motor timing: a PET study of auditory and visual rhythm reproduction. J. Cognit. Neurosci. **10:** 752–765.

16. RAO, S.M., A.R. MAYER & D.L. HARRINGTON. 2001. The evolution of brain activation during temporal processing. Nat. Neurosci. **4:** 317–323.

17. IVRY, R.B. 1996. The representation of temporal information in perception and motor control. Curr. Opin. Neurobiol. **6:** 851–857.

18. MANGELS, J.A., R.B. IVRY & N. SHIMIZU. 1998. Dissociable contributions of the prefrontal and neocerebellar cortex to time perception. Cognit. Brain Res. **7:** 15–39.

19. THAUT, M.H., R.A. MILLER & L.M. SCHAUER. 1998. Multiple synchronization strategies in rhythmic sensorimotor tasks: phase vs period correction. Biol. Cybern. **79:** 241–250.

20. MOLINARI, M., M. THAUT, C. GIOIA, *et al.* 2001. Motor entrainment to auditory rhythms is not affected by cerebellar pathology. Soc. Neurosci. Abstr. **27:** Program No. 950.2.

21. BUONOMANO, D.V. & M.M. MERZENICH. 1995. Temporal information transformed into a spatial code by a neural network with realistic properties. Science **267:** 1028–1030.

22. BUONOMANO, D.V. 2000. Decoding temporal information: a model based on short-term synaptic plasticity. J. Neurosci. **20:** 1129–1141.

23. TECCHIO, F., C. SALUSTRI, M.H. THAUT, *et al.* 2000. Conscious and preconscious adaptation to rhythmic auditory stimuli: a magnetoencephalographic study of human brain responses. Exp. Brain Res. **135:** 222–230.

24. MALAPANI, C., B. RAKITIN, R. LEVY, *et al.* 1998. Coupled temporal memories in Parkinson's disease: a dopamine-related dysfunction. J. Cognit. Neurosci. **10:** 316–331.

25. IVRY, R. & S. KEELE. 1989. Timing functions of the cerebellum. J. Cognit. Neurosci. **1:** 136–152.

26. NICHELLI, P., D. ALWAY & J. GRAFMAN. 1996. Perceptual timing in cerebellar degeneration. Neuropsychologia **34:** 863–871.

27. CASINI, L. & R.B. IVRY. 1999. Effects of divided attention on temporal processing in patients with lesions of the cerebellum or frontal lobe. Neuropsychology **13:** 10–21.
28. BAECK, E. 2002. The neural networks of music. Eur. J. Neurol. **9:** 449–456.
29. AUERBACH, S. 1911. Beitrag zur Lokalisation des musikalischen Talents im Gehirn und am Schädel. Arch. Anat. Physiol. (Anat. Abt.) 1–10.
30. KLOSE, R. 1920. Das Gehirn eines Wunderkindes (des Pianisten Goswin Sökeland). Mschr. Psychiat. Neurol. **48:** 36–102.
31. MAZZIOTTA, J.C., M.E. PHELPS, R.E. CARSON & D.E. KUHL. 1982. Tomographic mapping of human cerebral metabolism: auditory stimulation. Neurology **32:** 921–937.
32. KEENAN, J.P., V. THANGARAJ, A.R. HALPERN & G. SCHLAUG. 2001. Absolute pitch and planum temporale. Neuroimage **14:** 1402–1408.
33. ZATORRE, R.J., P. BELIN & V.B. PENHUNE. 2002. Structure and function of auditory cortex: music and speech. Trends Cogn. Sci. **6:** 37–46.
34. PARSONS, L.M. 2001. Exploring the functional neruoanatomy of music performance, perception and comprehension Ann. N.Y. Acad. Sci. **930:** 211–229.
35. GERLOFF, C., J. RICHARD, J. HADLEY, *et al.* 1998. Functional coupling and regional activation of human cortical motor areas during simple, internally paced and externally paced finger movements. Brain **121 (Pt 8):** 1513–1531.
36. HOLDROYD, T., M. NIELSEN, S. MIYAUCHI, *et al.* 2001. An MEG investigation of rhythmic tapping comparing self-paced and externally paced modes. Neuroimage **13:** Abstr. 11099.
37. MAYVILLE, J.M., K.J. JANTZEN, A. UCHS, *et al.* 2002. Cortical and subcortical networks underlying syncopated and synchronized coordination revealed using fMRI. Functional magnetic resonance imaging. Hum. Brain Mapp. **17:** 214–229.
38. CERASA, A., M. DEMARTIN, U. SABATINI & M. MOLINARI. 2002. Music expertise influences rhythmic behaviour brain activation. Neuroimage **16:** Abstr. 20390.

Brain and Sound Resonance

The World of Self-Generative Functions as a Basis of the Spectral Language of Music

HORATIU RADULESCU

International Lucero Academy in Paris and Blonay, CH-1815 Clarens/Montreux, Switzerland

ABSTRACT: The author discusses the "preferential phenomenology" of sound spectra. Most interesting have been the sound relations that result from special filtering according to "rings" of resonance. Mathematical operations are required to describe this filtering of frequency multiples—spectral components—producing sum and difference tones. With new harmonic formats, a new phenomenological vocabulary of music is achieved that evolves far beyond its historical language.

KEYWORDS: brain; sound resonance; spectral language; music

INTRODUCTION

Since 1969, I have worked on what I term the "preferential phenomenology" of sound spectra. Most interesting have been sound relations that result from special filtering according to "rings" of resonance. Mathematical operations are required to describe this filtering of frequency multiples—spectral components—producing sum and difference tones (which are also multiples of the same fundamental, that is, they belong to the same spectrum).

Given that a ring modulation of frequency is perceived by our brain in a condition of high intensity (e.g., we hear four pitches while playing only two), we can emancipate these virtual functions, even at low dynamics, and make them, for musical reasons, into "real" pitches. Thus, we arrive at completely unfamiliar simultaneous pitch functions (formants), resulting from specific spectral self-generative processes. For example, primal functions 4 and 7 produce in sum 11 and in difference 3; primal functions 16 and 21 produce in sum 37 and in difference 5.

If this type of self-generation, using both sums and differences of frequencies, is proved to exist in both the acoustic and the perceptual domain, a more specific self-generative behavior of sounds based on the multiplication of the multiples is acous-

Address for correspondence: Dr. Horatiu Radulescu, International Lucero Academy in Paris and Blonay, 24, Rue du Lac, CH-1815 Clarens/Montreux, Switzerland. Voice: +41 21 964 6877; fax: +41 21 964 6884.

lucero@bluewin.ch

Ann. N.Y. Acad. Sci. 999: 322–363 (2003). © 2003 New York Academy of Sciences.
doi: 10.1196/annals.1284.043

tically possible, but has not yet been demonstrated as a common situation in the field of brain resonance. For example, bispectrality, that is, when harmonics 7 and 11 produce 77, can both be virtual fundamentals. With these harmonic formats, issuing from both sound resonance and brain resonance, we achieve a new phenomenological vocabulary of music, which evolves far beyond its historical language.

EXPLICIT SPECTRA OF A COMPACT ROW OF FUNCTIONS

One of the oldest scores to use the spectral technique is *Credo,* op. 10 (Paris, 1969) for nine celli, in which a compact spectrum of 45 theoretical harmonics of the cello's low C become the new fundamentals of 4170 micromusic events evolving over a duration of 55 minutes. The logarithmically unequal intervals in between these 45 are used over a range from low C_0 to a high F#.[4] (Throughout this paper, I use the German system of octave numbering.) These new spectrally organized pitches are stable plateaus of frequency, fundamentals of a new mode of 45 microtonal pitches; in other words, the 45 harmonics of low C_0 (64 Hz) have become new fundamentals for a random "spectrum pulse" orchestrated in 4170 "micro-volcanoes" of timbral processes. This process of treating what were formerly considered harmonics as "real" pitches I call the "emanation of the immanence."

The nine celli "read" a "fresco" score of nine macro-music large structures: α, β, γ, δ, ε, λ, τ, φ, and ω. Each of these large nine macro-music structures is built on an ever-increasing number of new fundamentals (corresponding to "ex"-harmonics 1 to 45):

> α on 1, β on 2 and 3, γ on 4, 5 and 6, δ on 7, 8, 9 and 10, and so on, until ω on 37, 38, 39, 40, 41, 42, 43, 44 and 45.

The first cello plays nine macro-music structures in 55 minutes :

> τ β ε δ λ α γ ω φ

Simultaneously, the second cello plays only eight in retrograde order; they are dilated in time:

> β τ φ ω γ α λ δ

The third cello plays seven macro-music structures again in direct order:

> φ τ β ε δ λ α

and so on, until the ninth cello, which plays only one macro-music during the whole 55 minutes:

> ω

Thus, the odd-numbered celli perform in direct movement and the even-numbered in retrograde movement. These multiple layers of music result from the gradual loss of specific macro-musics, and the dilation in time of the others that are performed simultaneously. The whole resembles the "reading" of a fresco from nine distances at the same time by nine celli; for the odd-numbered celli time runs from left to right, and for the even-numbered celli from right to left (i.e., simultaneously "direct" and "reversed" time).

The "chorale" of macro-music structures containing the 4170 micromusic events reaches a climax of only eight, as the maximum of the simultaneous different macro-music structures, the vertical density of nine being felt only as a premonition or a remembrance, since the missing ninth macro-music exists in the previous macro-form column or will appear in the next column (FIG. 1, macroform of *Credo*; FIG. 2, Processing Matrix). Intuitively, this is perceived as the "Divine Ninth Heaven," never to be reached by Man. The subconscious, conscious, and hyperconscious "floating"' within the Primal Chord of 45 different pitches (from which 23 are totally unique–the 23 odd harmonics) gives us the impression of a cosmic chord of Beginning and Ending, reminding us of the words of Guillaume de Machaut, "ma Fin est mon Commencement" (c.1364).

An Explicit Compact Spectrum as the Modus of "New Fundamentals" against Implicit Spectra Projected by False Fundamentals (with a Simultaneous Upper Cluster of Harmonics 28-65)

A Doini, op. 24α (1974) and *Alt A Doini,* op. 24β (1980) for 17 players with *sound icons* (vertical grand pianos bowed with rosined threads). The piano strings are retuned according to a "spectral scordatura." The principal compact zone covers a range of a twelfth, from harmonic 8 to harmonic 24 for version α, and of two and a half octaves for version β, from harmonic 6 to harmonic 33 (17 unique harmonics, with no repetition at octaves). The score symbols are micromusic processes, combining Jung's compass of the psyche (thought-feeling-intuition-sensation) with my 'sound plasma' compass: noise-sound-element-width (FIG. 3, matrix score of *Alt A Doini*; FIGS. 4, 5, and 6, performance symbols). If the principal zone has a fundamental of B natural (*A Doini*), the false virtual fundamentals are A, B flat, and C. The principal zone covers the same range as the interval between harmonics 1 and 3, that is, the interval between a fundamental and its first "new" harmonic, which is now filled by the scale of theoretical harmonics 8 to 24. The three "crude" implicit spectra of the false fundamentals fight against the explicit spectrum of B. The four spectra "shadow" one another, being slightly apart from each other in pitch but all "acting" in the same register.

Similar explicit compact spectra, harmonics 7 to 26 and 27 to 113, are set in opposition against an "inverse" spectrum, from harmonic 1 to –333, in *Thirteen Dreams Ago,* op. 26 (1978) for 33 strings (3 × 11), the higher group and the very low group of strings using spectral scordatura.

A special "galactic" uphill of spectral functions (7 to 57) is deployed in the evolution of the 139 micromusic events in *Ecou Atins* [e′kou a′tins], *Touched Echo* (1979). Bass flute (and grand flute), horn, soprano, cello, sound icon, and 29 bowed and spectrally tuned monochords figurate a "chorale" on the resonance of an E, where the "emanation of the immanence" renders the sound sources quite indistinguishable by increasing the spectral energy of the stable frequency-plateaus. We confuse them and mistake one for the other (FIGS. 7 and 8, score pages on colored slides). Related is the slow "chorale," of 24 minutes, also on an E spectrum, in *Frenetico il longing di amare,* op. 56 (1985), where bass voice, contrabass flute, and sound icon exchange elements of that spectrum to form special chords, mostly compact formants evolving as "expanding constellations."

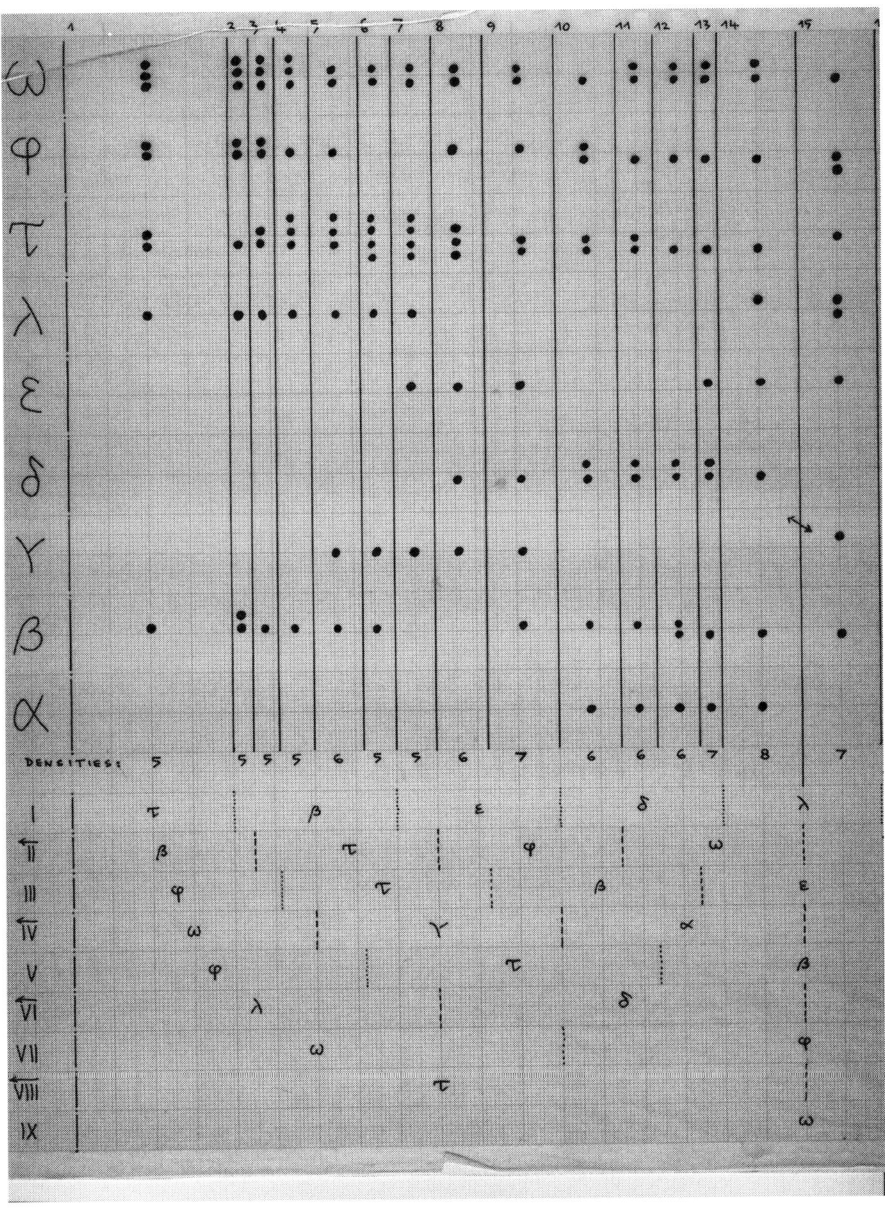

FIGURE 1.

14 15 16 17 18 19 20 21 22 23 24 25 26 27 28
ω 5 28 (50)
φ 4 26 (36)
τ 6 28 (47)
λ 2 19 (24)
ε 0 17
δ 1 19 (27)
γ 0 12
β 1 22 (24)
α 1 14 (15)
8 7 8 8 8 7 7 7 8 8 7 6 7 8 7 7
λ α γ ω φ
γ δ α λ δ
ε δ λ α
β λ δ ε
ε β δ
τ β
φ τ
ω φ
HORATIU RADULESCU
PARIS 1969
CREDO's
MACRO FORM

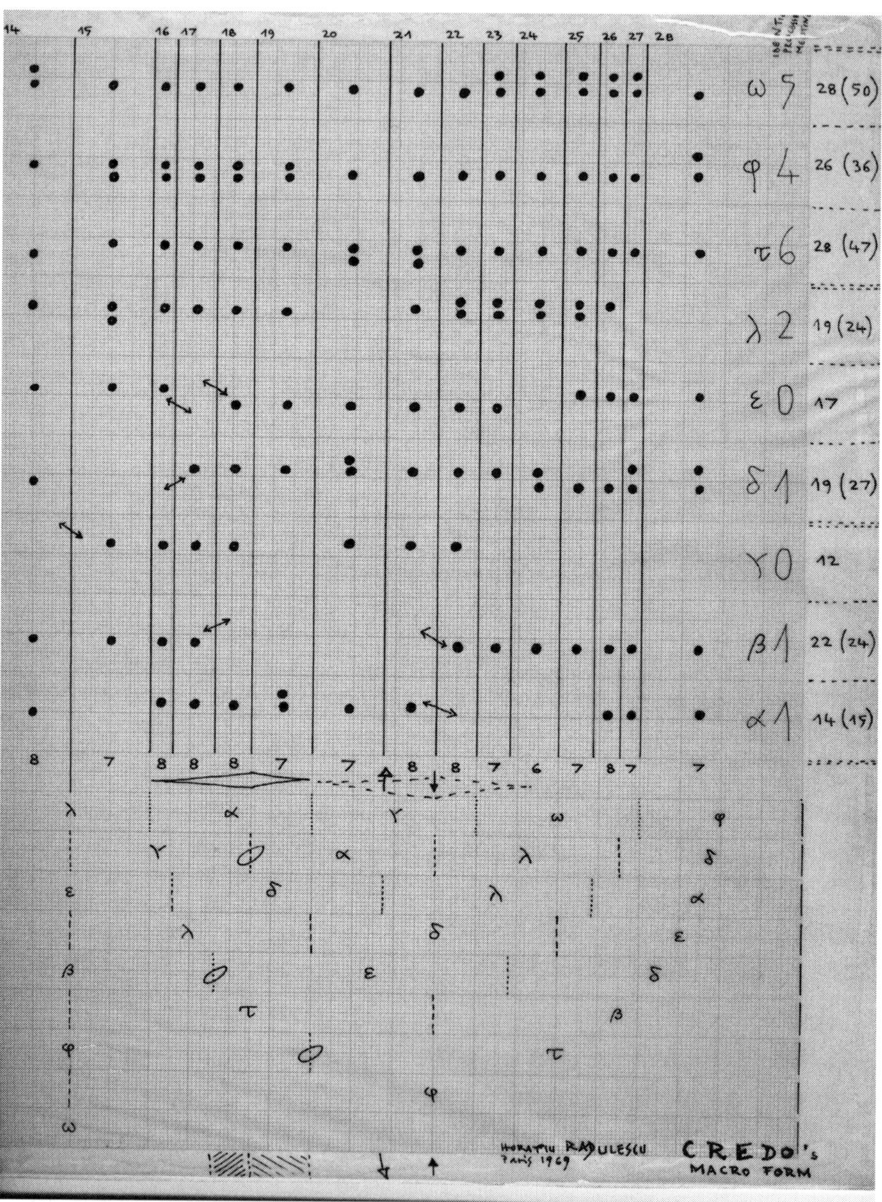

FIGURE 2.

Horatiu Radulescu: « ALT A DOÏNI » opus 24 nr. 2   for 17 sound icons   - 1980 NÜRNBERG
1 2 3 4 5 6 7 8 9 10 11 12 13 14 15 16 17 18 19
FREQUENCIES of the principal ZONE
h#~ 33° XVII   33° f#~
a## 31° XVI   31° e#.
a↑ 29° XV   29° ↓d## "(↑dis" tes)
~gis" 27° XIV   27° d"
↓g" 25° XIII   25° ~cis" or ~des
f↑" 23° XII   23° ↓h# (↑h')
e↓" 21° XI   21° ~a# (↓b')
~d" 19° X   19° ~as' or gis'
°c" 17° IX   17° fis' or ges'
_ais' 15° VIII   15° _e'
g#'_ 13° VII   14° tes' or dis'
e#' 11° VI   13° c#' or ↓d'
_dis' 10° V   12° c'····
~cis' 9° IV   11° a## or b♭°
h° 8° III   10° a°_
a↑° 7° II   9° ~g°
fis° 6° I   8° f°
VERSIONS: β   α
Timing of the "DRUNK" MATRIX
0 00 1 17 2 28 3 49 4 56 6 21 7 24 8 53 9 52 11 25 12 20 13 57 14 48 16 29 17 16 19 01 19 44 21 33 22 12 24 05
↓24° 12"
FUNDAMENTALS:
H   F
(B natural)   (FA)
(SI)
micro-tonal alterations:
3/4 of TONE
1/4 of TONE
↑ or ↓ 1/8
+ or − 1/16
• or ° 1/32
~ or ~ 1/64
···· or···· 1/128
© 1980 by Horatiu Radulescu - Versailles, France   LUCERO PRINT / SACEM PARIS

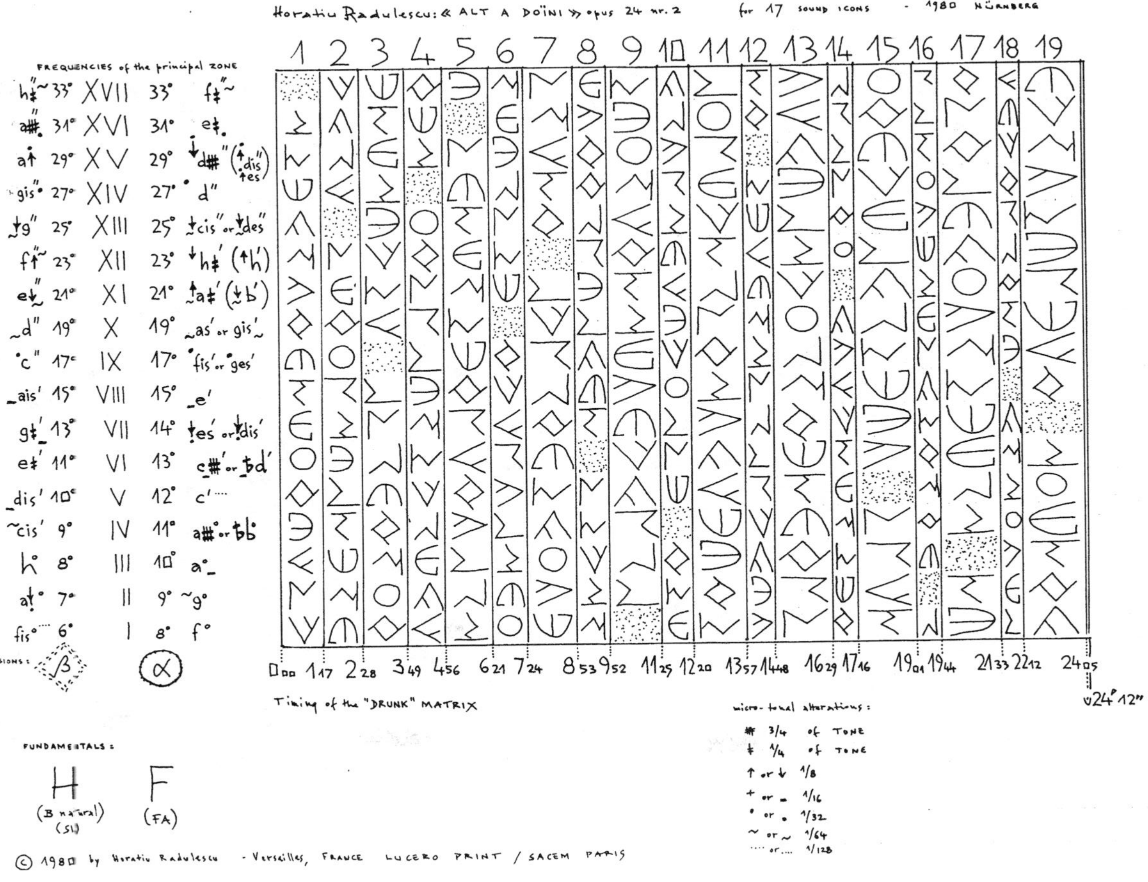

FIGURE 3.

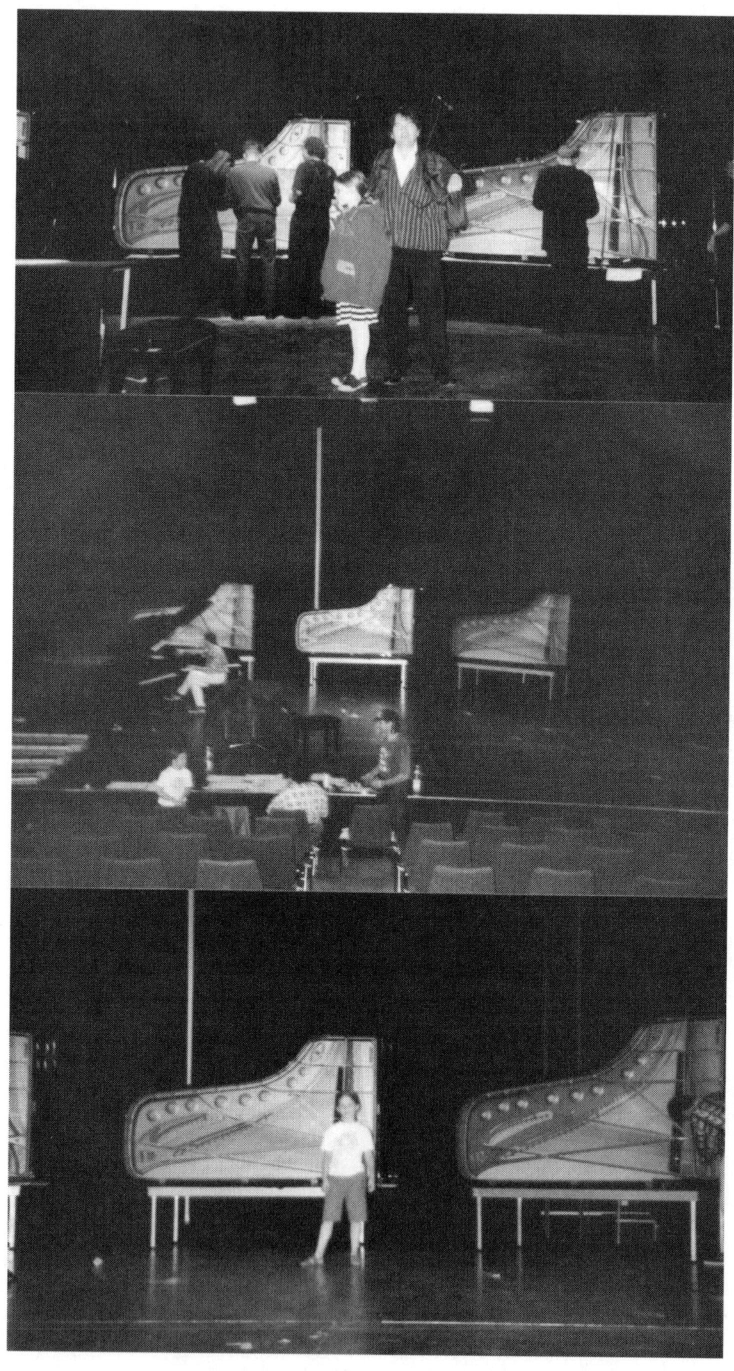

FIGURES 4 (top), 5 (middle), and 6 (bottom).

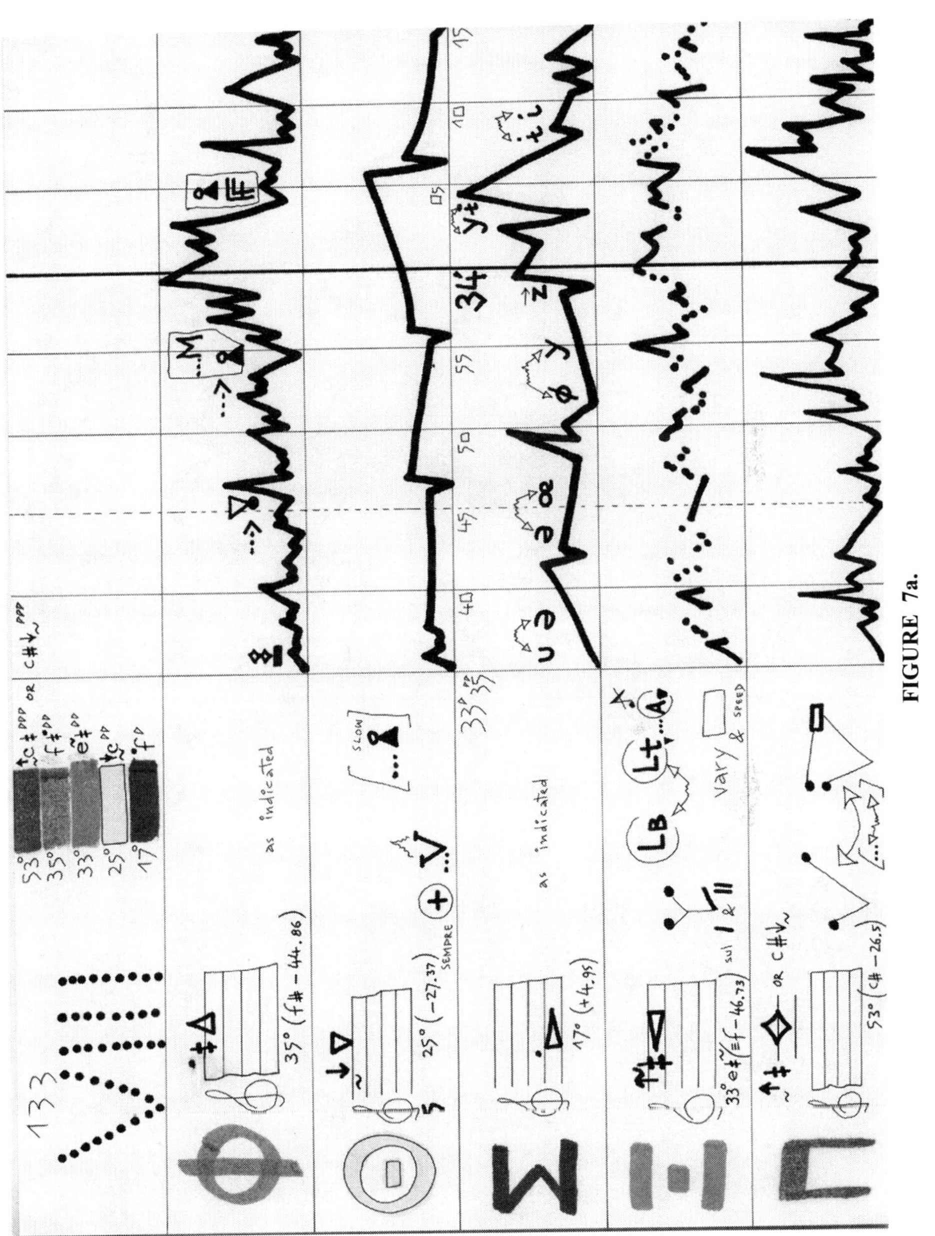

FIGURE 7a.

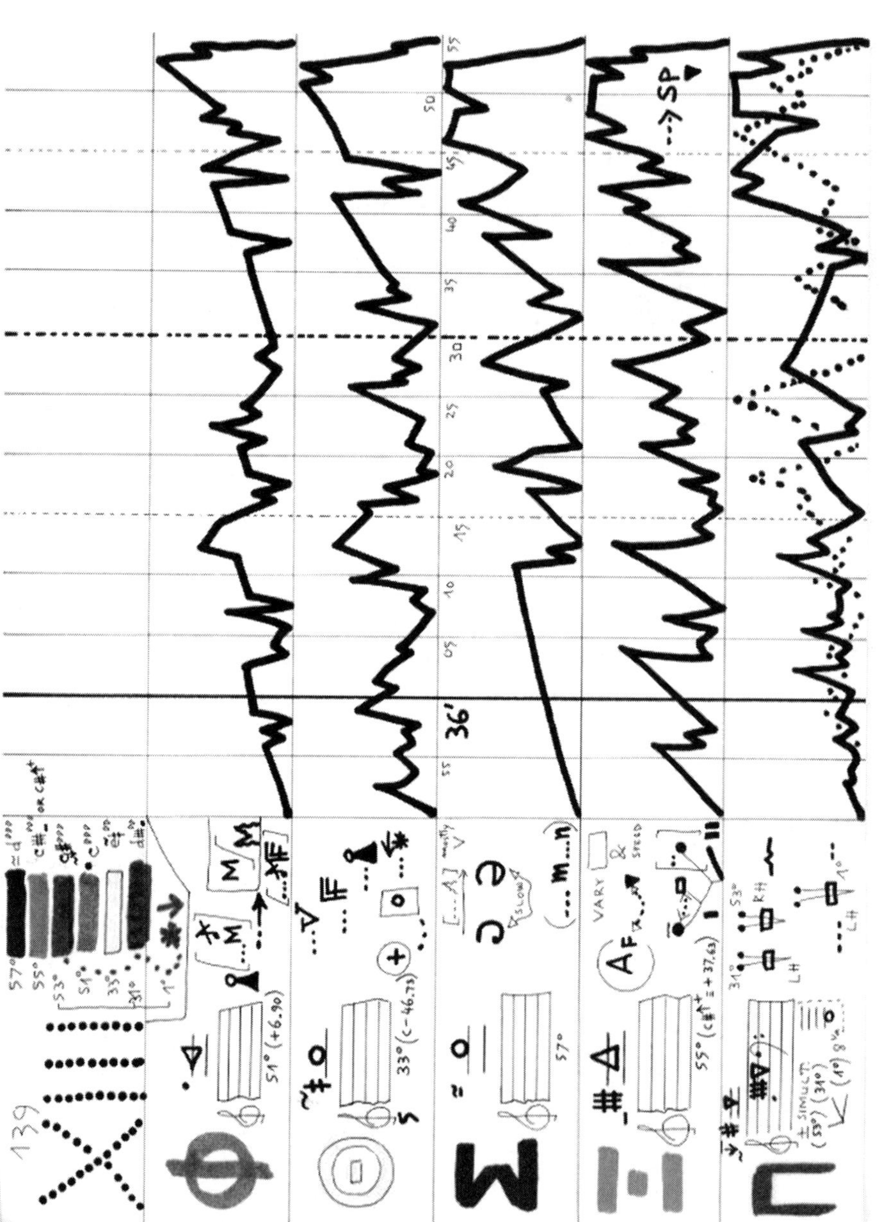

FIGURE 7b.

① DIRECTIONAL PROCESSES
TIMBRE - PSYCHE PROCESSES
N_t
W_s E_i
S_f
• aleatory performance as an AMINTIRE (remembrance) on all the three zones: high, middle (PRINCIPAL) & low REGISTERS
towards Noisy thought
" Elementary intuition
" Soundy feeling
" Width - sensation

② DENSE SILENCE with "drops" of SOUND - PHENOMENA
only on → high & low REGISTERS (NO PRINCIPAL frequencies)

③ GLIDING PROCESSES
N_f
W_i E_s
S_t
ON THE MIDDLE (PRINCIPAL) ZONE ONLY [*]
from Noisy feeling through Elementary sensation towards Soundy thought & VICE-VERSA
from Width-intuition through N_f towards E_s & VICE VERSA
from Soundy thought through W_i towards N_f & VICE VERSA
from E_s through S_t towards W_i & VICE VERSA

④ MULTI-DIRECTIONAL PROCESS
[*]
EVO - INVOLUTION sum of all processes // at least IRREGULAR PULSE of evolving neutral SOUND/NOISE

⑤ LOCATED PROCESSES
N_i
W_t E_f [*]
S_s
in the Noisy intuition
" the Elementary feeling
" the Soundy sensation
" the Width - thought

⑥ OPPOSING PROCESSES
N_s
W_f E_t [*]
S_i
(fast & irregularly)
all processes against W_f N_s E_t S_i
" " " W_f N_s E_t S_i
" " " W_f S_i N_s E_t
" " " W_f N_s S_i E_t

⑦ NEUTRAL PROCESSES
the same space as for [*]
any Noisy intuition or Soundy sensation
" Width-thought or Elementary feeling

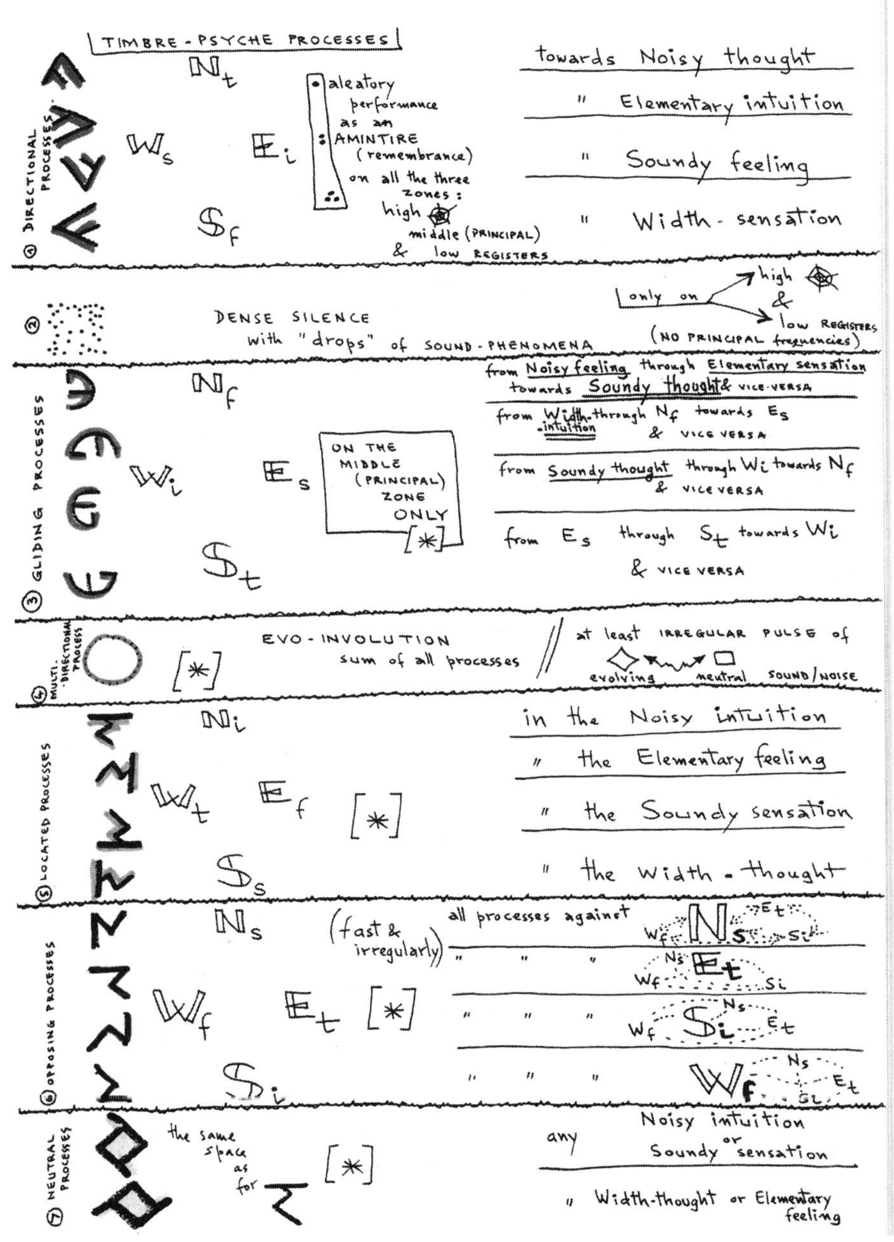

FIGURE 8.

Compact Spectrum of Totally Unique Functions

"<u>infinite to be cannot be infinite, infinite anti-be could be infinite</u>": 4th String Quartet, op. 33 (1976–1987):

The spectral scordatura of the 128 open strings of the imaginary "viola da gamba" built around the audience by eight quartets, the ninth being placed in the center of the space:

One hundred twenty-eight unique functions of pitch chosen within the harmonics 36 to 641 of a C (1 Hz), cover a range of over four octaves, from D_{-1} to e^2. The "geology" of these functions consists of various densities of pitch, the steps in Hz using different powers of the number 2: 4, 2, 8, and 16 (from low to high register) in order to respect the historic tuning, the tradition of those strings.

By contrast, the central ninth quartet modulates within 27 spectra. It uses ρ, μ, and δ micromusic, that is, ring modulation, multiplication of multiples, and dilation of formants, respectively, the music of "preferential" and "self-generative" functions. Each micromusic is tuned rigorously according to its duration; for example, a music of 13″ uses a spectrum of D 3/4 tone$^+$ sharp (or E 40 cents lower, approximately): this is historically the first scientifically strict interdependence between *time* and *pitch*. Again, in the correlation between time and pitch, brain and sound are two aspects of the same unique reality (cf. part 2, ring spectra) (FIG. 9, spectral scordatura of the 128 open strings in op. 33).

Many other scores use this type of scordatura based on compact spectra of totally unique functions, some as specific situations:

In *Forefeeling Remembrances,* op. 64 (1985), 14 identical voices pulsate the following 14 functions proportionally to their spectral role: 6, 7, 8, 9, 10, 11, 13, 15, 17, 19, 21, 23, 25, and 27.

In *Sensual Sky,* op. 62 (1985), an explicit spectrum is figurated by nine instruments—alto flute, clarinet, alto saxophone, trombone, sound icon, violin, viola, cello, and double bass—while, as background, two contrabass flutes project various implicit spectra on more or less unstable fundamentals via overblowing sound production.

Subconscious Sound Trajectory Inside a Compact Spectrum

Do Emerge Ultimate Silence, op. 30 (1974–1984): A 34-pitch spectral chain describes a 28-minute unique "cantus firmus" of a deeply Buddhist time: a *melopée* (chant) of uphill-downhill profile where each of the 34 spectral sounds appears only once in one of the two "sides" of this melodic mountain. This very slow, mountain-shaped "cantus firmus" is often accompanied by emanations of parts or the whole of the 34 pitch functions. The whole evolves on six concentric groups of children's voices supported by bowed monochords in just tuning: harmonics 11 to 44 of a low D of 18 Hz (g monesis $_0$ - g monesis^2). The children sing 34 different pitches within an ambit of two octaves, with steps becoming progressively smaller towards the high register, from the "neutral second" g monesis $_0$ – a $_0$ to an interval smaller than a quartertone, g^{+2} – g monesis ^{2} (monesis = half-sharp).

Do Emerge Ultimate Silence can be realized with 34 soloists or even with 340 voices, that is, 34 groups of 10 voices each, for example, in the Pantheon in Rome.

FIGURE 9a.

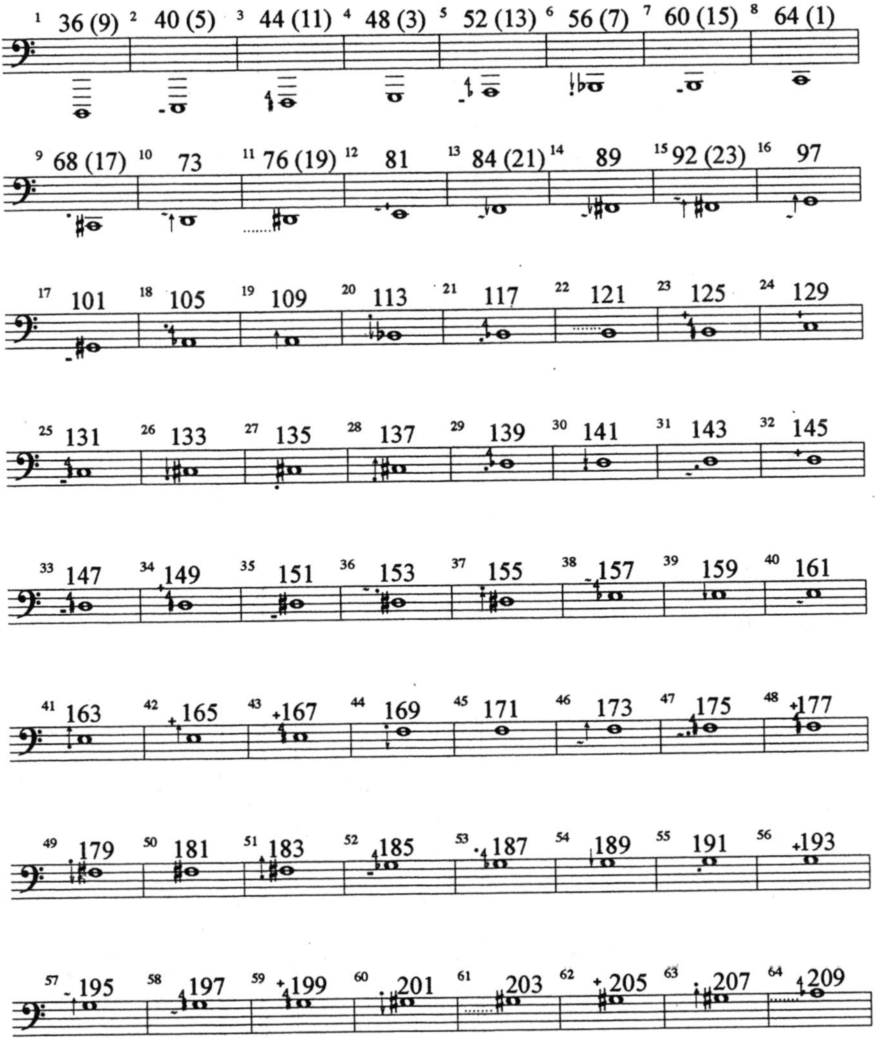

infinite to be cannot be infinite, infinite anti-be could be infinite
HORATIU RADULESCU

FIGURE 9b.

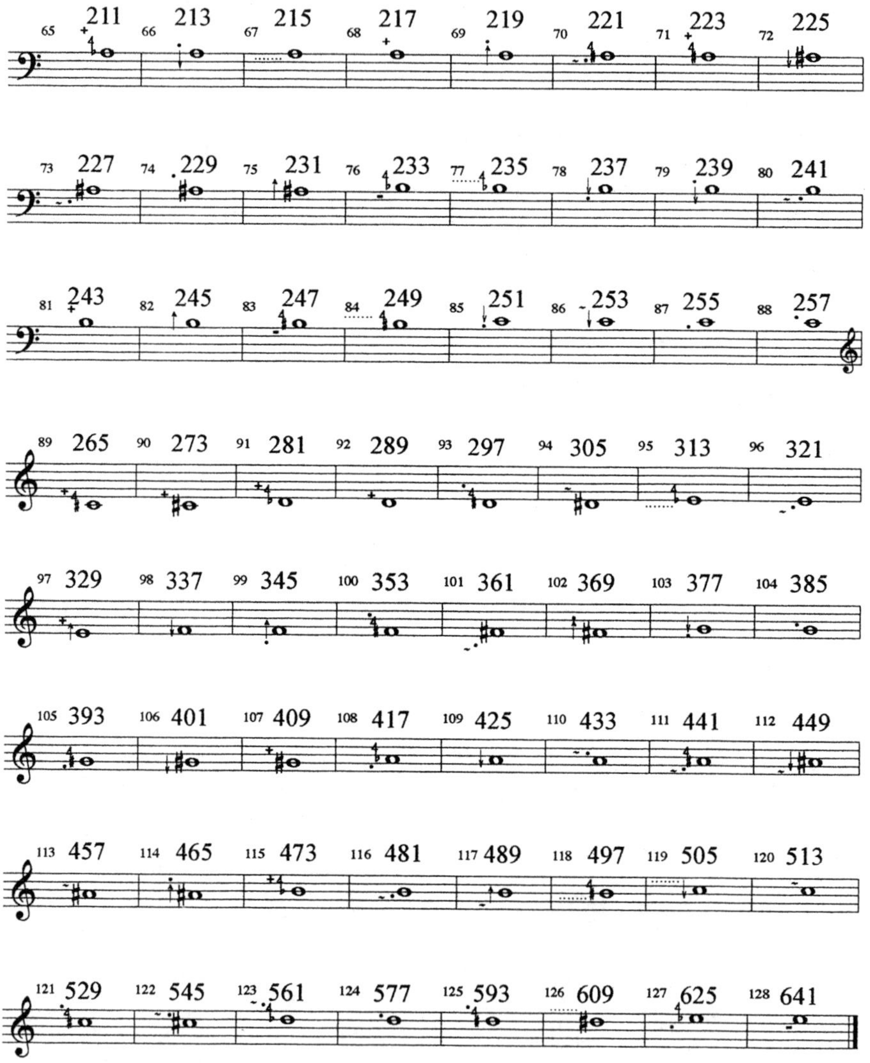

The psychoacoustic tension is intense throughout the concert hall or church, the 34 points being symmetrically distributed thoughout the audience: the result is a circular movement of sound or the eruption of chords, with different pitches or in unison, all voices contributing the private sound of their monochord or imitating other voices in unison, and each voice singing, at least once, each of the 34 harmonics. Sometimes the movements of sound on different concentric groups can revolve contrariwise and at a very different speed and with varying densities of the voices. Each group of voices uses a unique selection of functions without any repetition inside the

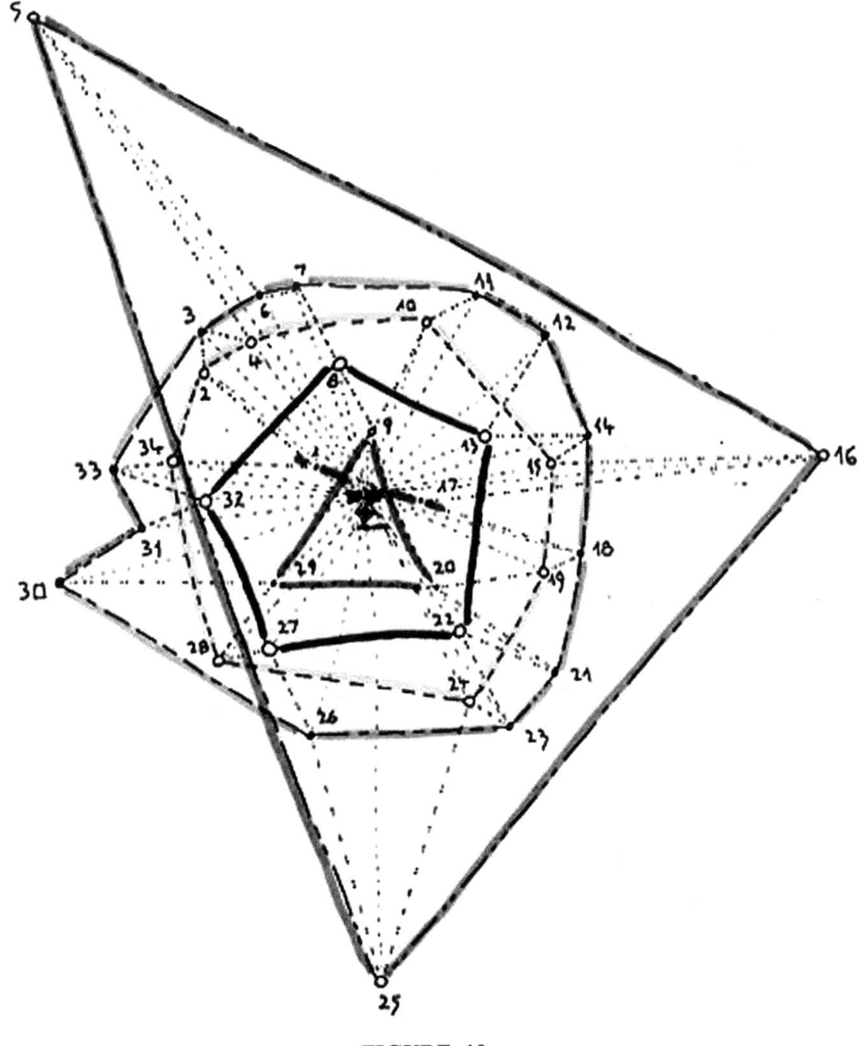

FIGURE 10.

1984
Horatiu Radulescu
(PLAYERS')
NR.
SINGERS
TUTTI
GROUPS OF CHILDREN
(13 years old ≈)
in Do EMERGE ULTIMATE SILENCE

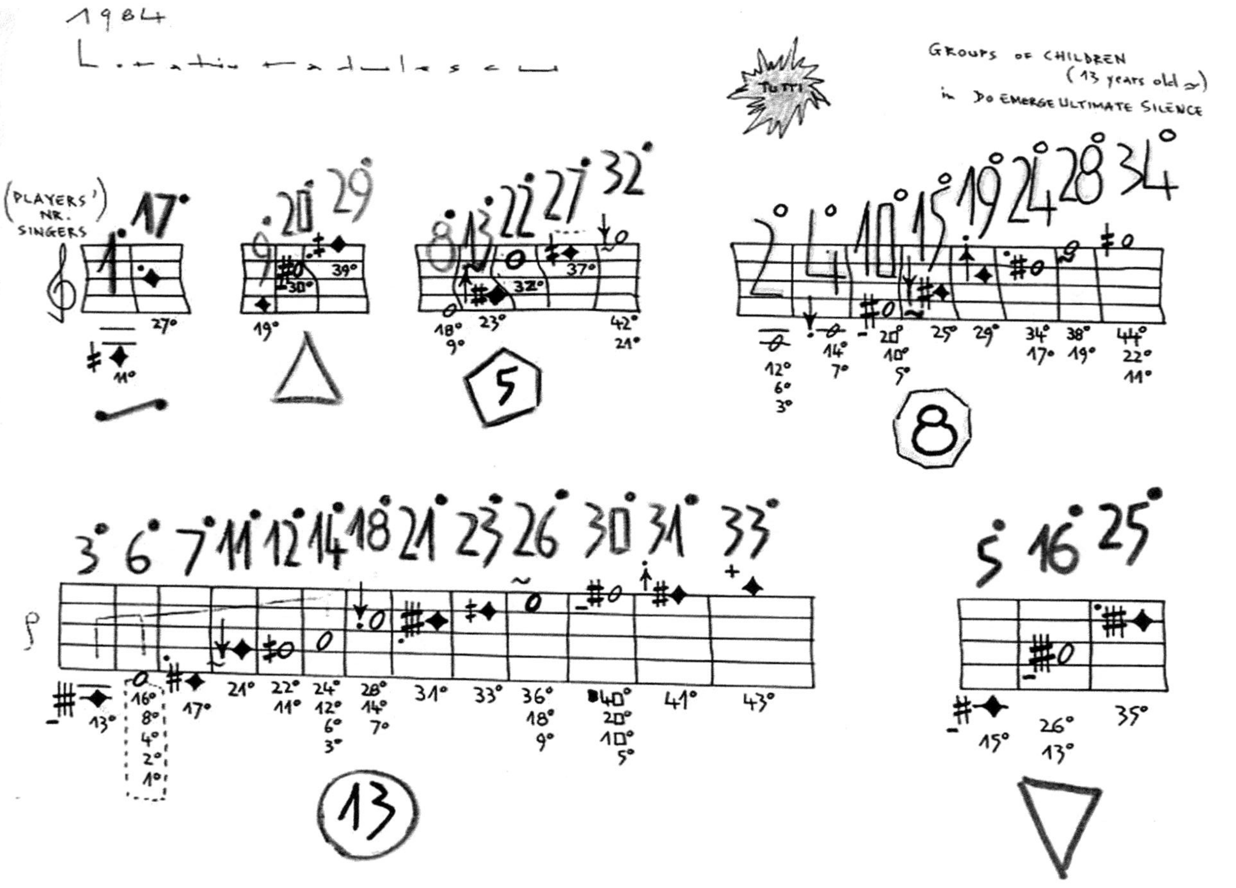

FIGURE 11.

same group. There are 2, 3, 5, 8, 13, and 3 voices per group, respectively. The seventh group is the "zigzag star" of tutti (all 34 voices) (FIGS. 10 and 11).

Double and Complementary "Bridges" of Compact Spectra

Incandescent Serene, op. 35 (1982): Twelve quartets, each consisting of one contrabass flute, one viola, one horn, and one double bass, figurate filters of aural information ("plages de l'audible") of a horizontal "sand clock" and of rhombus shapes that integrate a chain of 137 micromusic "mobiles." The upper "bridge" is built on a virtual scordatura from harmonic 12 (G_0) to 83 (e monesis[1]) and is played by flutes, violas (specially retuned according to elements of a C spectrum), and horns. The contrary lower "bridge," built on a spectral scordatura from harmonic 16 to 95 (same order numbers being also the actual Hz), is played by the double basses. The unique frequency plateaus of the open strings of the double basses correspond per string to harmonics 16 to 27 (string IV); 28, 29, 30, 31, 33... 45, 47 (string III); 49, 51... 69, 71 (string II); 73, 75...93, 95 (string I).

Compact Spectra of Unique Functions (No Repetition at Just Octaves), Each Revolving at a Different Speed

The family of compositions op. 42α, *Outer Time,* and op. 42β, *Inner Time* (1980–2003): In *Outer Time,* for 42 Thai gongs spectrally tuned (harmonic 6 to 83 or 11 to 83), the virtual spectral scordatura uses harmonics 6 to 11 or 11 to 21 as a compact zone and then only odd harmonics up to the 83rd in order to obtain a row of 42 unique spectral functions, as unique as the 12 pitches of a Webern serial row.

Four layers of time/speed are simultaneously perceived:

MMT – Macro Macro Time, e.g., 62′ 13″ (whole macroform)

mMT – micro Macro Time, e.g., 5″ up to 55″ (the 137 mobiles)

MmT – Macro micro Time, e.g., one attack per ± 1″ (the 81 "rhythm characters")

mmT – micro micro Time, e.g., aural information of about 1/333 of a second (2–3 ms).

On the 42 spectral "orbits," the pitch functions revolve at a speed ranging from 11 circles for the lowest function to 83 circles for the highest one, over a total duration of 62′13″ (FIG. 12, the 81 rhythm characters [microagogic rhythm] used in various scores such as "practicing eternity," String Quartet no. 6, op. 91 (1992); FIG. 13, one of the 56 pages of *Outer Time*).

Reverse and Compact Spectra

[ɩ] inverse spectrum and [ε] explicit/compact spectrum (e.g., the scordatura of the 11 strings in *Unde Incotro* ['un de în ko 'tro] (and *Where Beyond*), op. 55 (a and b) (1984, 1992): From left to right: double bass, two celli, two violas, and six violins - a direct display of sources in a semicircle on stage (plus alto flute in the b version, *Where Beyond*): left/low – right/high, according to the specialization of our stereo hearing. The lowest functions are the secondary and tertiary ones, unique theoretical harmonics 6 to 47 as the scordatura of strings, while the highest are the primary ones, harmonics 1, 2, and 3, each on a great number of strings (FIG. 14, scordatura of unique strings in *Unde Incotro*).

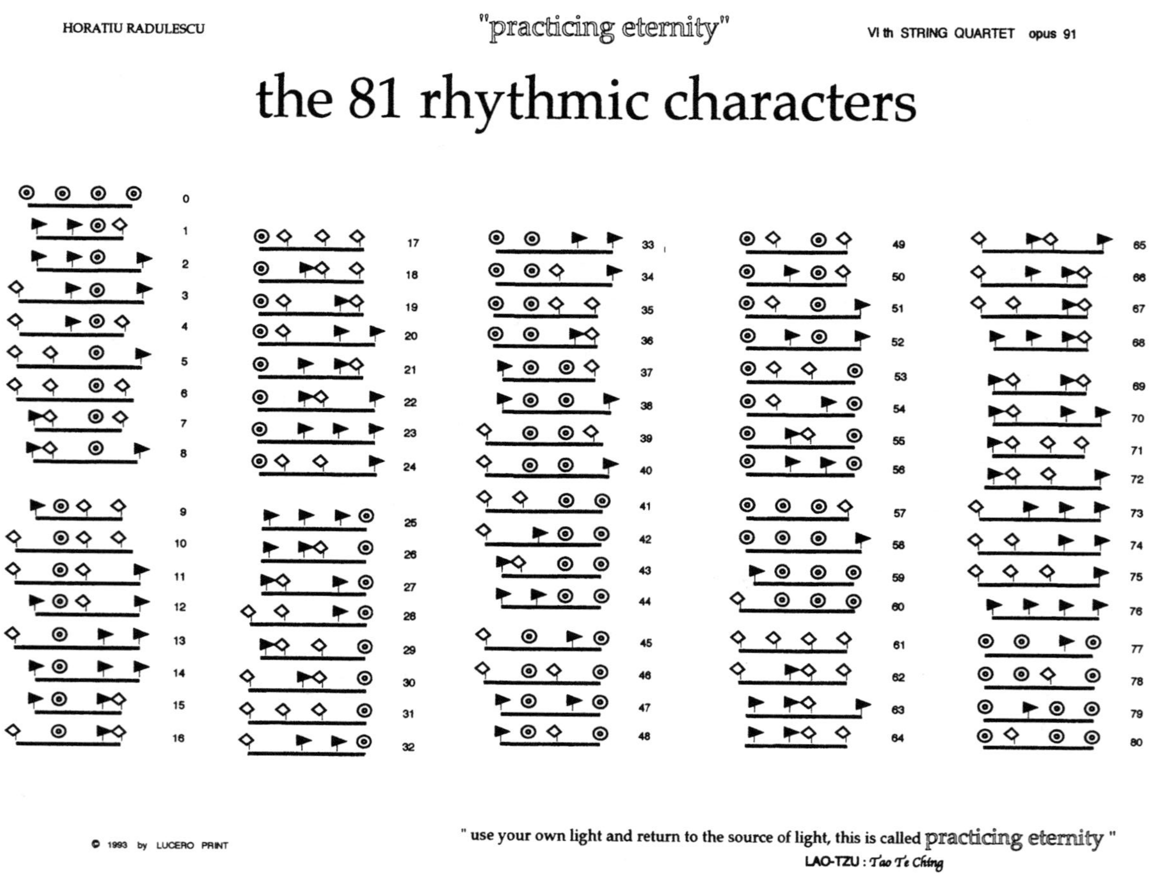
HORATIU RADULESCU
"practicing eternity"
VI th STRING QUARTET opus 91
the 81 rhythmic characters
© 1993 by LUCERO PRINT
" use your own light and return to the source of light, this is called practicing eternity "
LAO-TZU : Tao Te Ching

FIGURE 12.

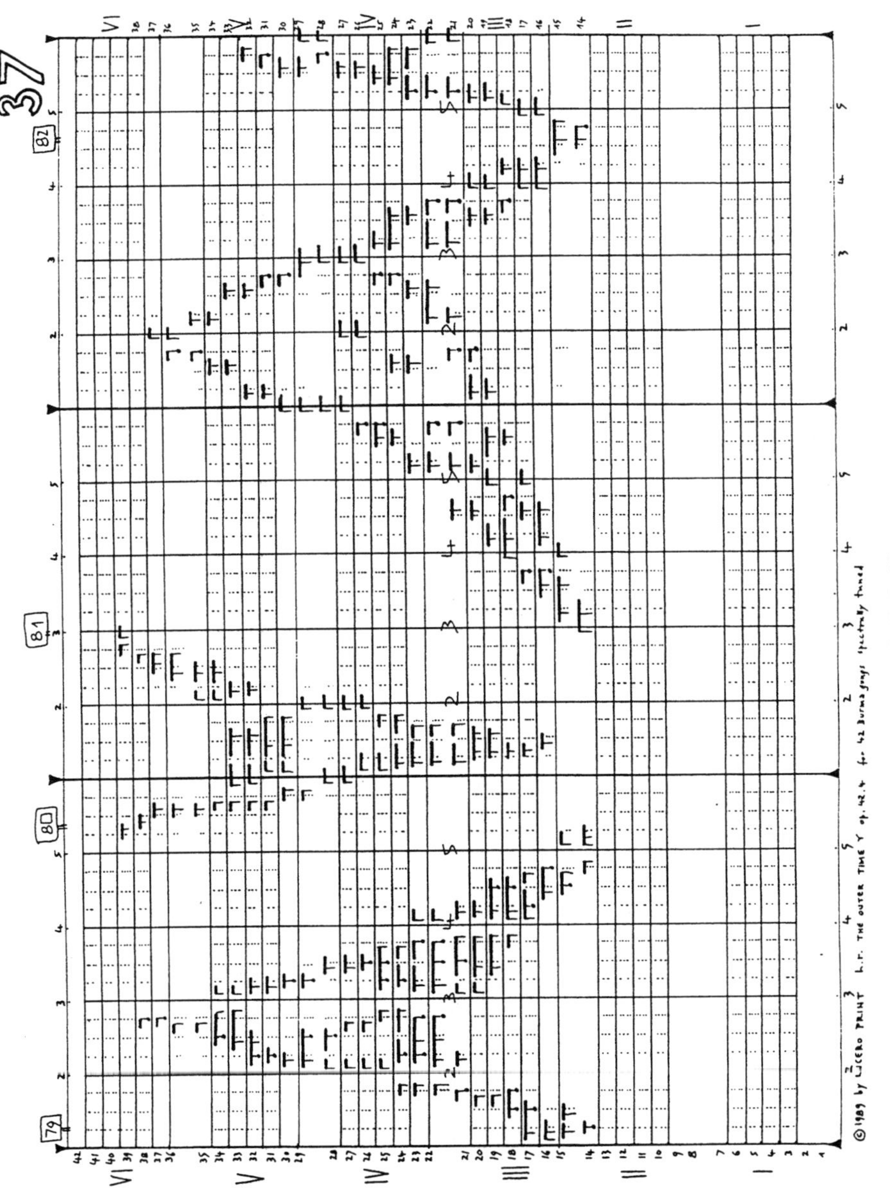

FIGURE 13.

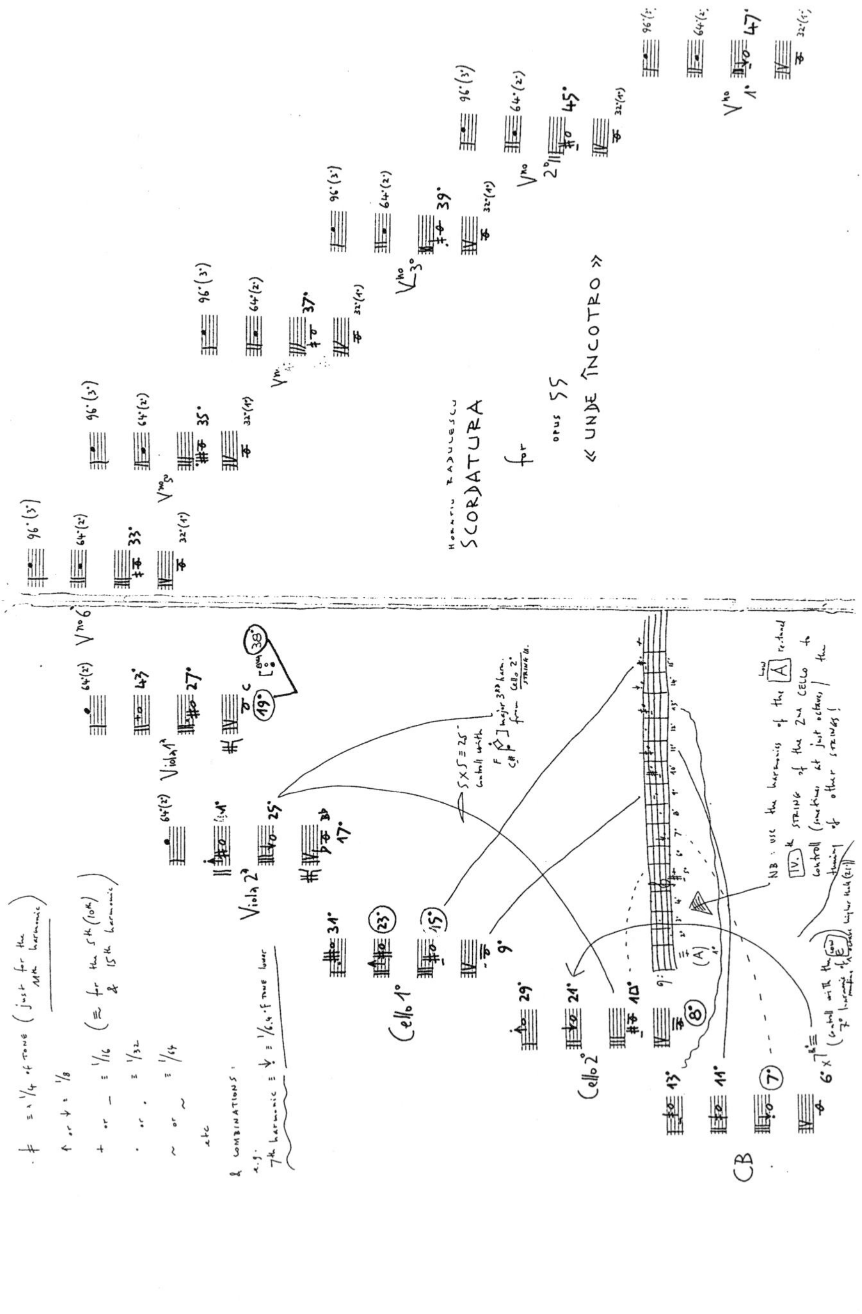

Horatiu Radulescu
SCORDATURA
for
opus 55
« UNDE ÎNCOTRO »
Cello 1°
Cello 2°
Viola 1ª
Viola 2ª
CB

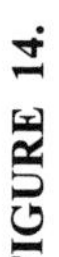

FIGURE 14.

RING SPECTRA: PREFERENTIAL PHENOMENOLOGY USING SPECTRAL FILTERING ACCORDING TO RINGS OF RESONANCE

Proto Ring Spectra

The interference of frequency multiples (spectral components) produces additional and differential sounds (sum and difference tones), the latter also being multiples of the same fundamental, that is, belonging to the same spectrum. Such a ring modulation of functions is perceived by our brain in a condition of high intensity (e.g., we hear four pitches while playing only two). We can emancipate these virtual functions, even at low dynamics, and make them, for musical reasons, into "real" pitches. Thus, we arrive at completely unfamiliar simultaneous pitch functions (formants), resulting from specific spectral self-generative processes.

For example, primal functions 16 and 21 produce functions 5 and 37. This type of ring function is encountered in very early works such as *Omaggio a Domenico Scarlatti,* op. 2 (1967), for piano or harpsichord, approximated by tempered elements: e_0 c^2 f^2 e flat^3. In particular, the lower part of the formant contradicts the rules of classic harmony and counterpoint: its simultaneity of major third (harmonic 5), tonic (harmonic 16), and perfect fourth (harmonic 21) is not allowed and is considered dissonant in the history of music, but now becomes an element of a self-generative chain, a genealogic tree of frequencies, for both our brain and our musical language, most healthy and "consonant." This psychoacoustic triad is now one of my musical signatures.

Another proto ring spectrum approximated by tempered elements is that of the end of *Taaroa,* op. 7 (1968–1969), for orchestra, humanized by the singing of all of the players, where the theoretical harmonic 15 (leading tone or "sensible") and 17 (Phrygian second or Neapolitan minor second) produce in sum 32 (tonic), which easily is allowed to have its perfect fifth (dominant), this resulting also from the further ring modulation as additional 47 (32+15) and 49 (32+175), giving in the upper register the 96 (3) [47+49], a just octave of the dominant. This formant 15-17 – 32-48 is a chord that Sergiu Celibidache appreciated and loved: acoustically and spiritually "our imperfect step," a slightly wider major second as in some Hindu and Byzantine music, towards—in the high—"the Eternal," the perfect fifth of Pythagoras, Guillaume de Machaut, and Josquin Des Prez.

Proto ring modulation of spectral functions produces a "Brancusi infinite column," a modal chain based on only two intervals (harmonics 6-7 and 13-16, in a condensed register [m3rd – p5th, - m3rd – p5th, and so on]). This can be further transposed via pivot functions until the whole chromatic set of 12 different pitches is reached. I used this approach in *Madrigal,* op. 3 for children's voices and orchestra (1967), *Being and non-being create each other,* Piano Sonata no. 2, op. 82 (1991), the Piano Concerto *The Quest,* op. 90 (1996), and other works.

Another proto ring spectral function produces a similar but different type of Brancusi infinite column, a modal chain based also on only two intervals (harmonics 7-8 and 12-13, in a condensed register [M2nd – p5th, - M2nd – p5th, and so on]). Further transposition via pivot functions gives the defective chromatic set of eight different pitches. *Agnus Dei,* op. 84 (1991), for two violas is entirely built on those two columns. Related also is the fourth movement of the 3rd Piano Sonata, op. 86, where the "Dance of the Eternal" is based on the first column (of 12 different pitches) and only the conclusion, surprisingly, brings the acoustic freshness of the second column (of eight different pitches).

Advanced Ring (ρ), Dilation (δ), and Multiplication (μ) Spectra

As we saw in the section on the 128 open strings of the imaginary viola da gamba built by the eight surrounding string quartets of op 33, the 128 unique functions of pitch permit these three types of spectral life to materialize. The ninth quartet in the center is based mostly on ρ and only exceptionally on δ and μ as well as very intense and "multiple-stringed" unisons. As an example of how music ρ can create a historically new language, we look at the very first micromusic where *three unique and different types of f are presented against* C: harmonic 21 (f^3 1/7 tone lower), 22 (f^3 monesis) as generators, harmonic 43 (f^4 1/17 tone higher) as additional, and 1 (the fundamental C_0) as differential.

We know that in the recent history of serial, modal, and polytonal music only two different types of a same step of the mode were used more or less simultaneously (Webern, Bartók, Stravinsky, Enescu, Hindemith, Stockhausen, Berio, Boulez, Nono, *et al.*), for example, b natural $_0$ (c flat ^{1}) – f ^{1} – b flat ^{1} (Webern), e $_0$ – c ^{1} – e flat ^{1} (Enescu).

Music δ operates a special filtering on a spectrum. A chord of several harmonics, for example, 3, 4, 7, and 11 (vicinity +1, +3, +4), is imagined in other formants of the same spectrum, keeping constant the vicinity proportions around an axis [α] or starting from the bottom [β] or from the top [γ]. If we consider the inner axis 4 – 8 – 16 – 32, then the different formants, ranged vertically, describe the following chain of chords. To feel the dilation/contraction of the same formant/chord keeping the vicinity proportions stable, we are obliged to make all of these chords dance around the axis with their different modal content as represented by colors: red (the most spread chord) green, blue, and finally (the most condensed one) violet. To actually feel the process, the axis was kept on place, stable, as if the fundamental were jumping on different octaves to obtain all of those harmonics with different order numbers. This means that we perform just the violet functions, that is, we are on very high formant strata operating various filtering on the same spectrum region. In any case, the actual spectral functions are the left colored numbers, whereas the division by two is no longer possible at the given rank of functions. We notice that in this example the ambit of nearly two octaves is progressively contracted to an augmented octave, then further to a slightly augmented perfect fifth, and finally to a major third.

The rich content in various microtonal pitches of the spectrally tuned 128 open strings enables music δ to be produced. The players of the central quartet realize a chorale of dancing formants of δ music too, but they are obliged to use special microtonal functions produced by unusual fingerings, because their instruments are tuned normally, using an a ^{1} of 432 Hz.

The very complex δ music events of this imaginary 128-string "viola da gamba," sometimes combined simultaneously with ρ or being crossed by τ music events, were computed in Paris by Vincent Bourgue using 16 dimension-calculus and probabilistic approach with chance variables.

Example of music δ with inner axis 4 – 8 – 16 – 32; the resultant formants/chords are ranked among the differently colored columns.

NB! We keep constant the vicinity proportions (vicinity +1, +3, +4):

[a] around an inner axis (see previous page) or

11	14	20	32
7	10	16	28
4	7	13	26

[b] starting from a stable bottom 3 - 6 - 12 - 24
or [g] from a stable top 11 - 22 - 44 - 88

7	18	40	84
4	15	37	81
3	14	36	80

AdvancedExplicit/Compact [ɛ], Ring [ρ], and Inverse Spectra [ʊ]

Seven spectra in *Iubiri* (Amours), op. 43 (1981), consist of 343 unique micro-music events that integrate a registral shape of an enormous horizontal "sand clock" of 46'–from a seven-octave range to a focus on a central d' and then a faster widening back to seven octaves, but giving a constant feeling of ascension because the fundamentals of the seven modulating spectra-regions are the first seven different theoretical harmonics of the primal spectrum: 1, 3, 5, 7, 9, 11, and 13. These functions are reinforced exactly where their original ascending places are located within the primal spectrum (FIG. 15, registral evolution of the *Iubiri* macroform; FIG. 16: (a) fragment of score; (b) fragment of a flute part).

A related concept is that on which *Awakening Infinity,* op. 53 (1983), is based, where 25 players are placed on stage in a circle, reversed and augmented in the auditorium via slight amplification. A simultaneous double complementary stereophony is perceived, for example, a diagonal on and from the stage heard as SW-NE is heard at the same time from and in the concert hall spatially augmented and reflected as NW-SE. These two circles of stereophony build a figure of 8: two superimposed circles constituting the subconscious upright symbol of an "awoken infinity."

Implicit [ʊ–ρ] Ring Modulations Through Sums (Σ) of High Harmonics' Biphonies (Harmonics 7 to 20) Producing at High Intensity Various Differential Sounds; Explicit [ɛ–ρ] Ring Modulations as Arpeggios (A) of Chords Built on Spectral Components of Virtually Very Low Fundamentals

Das Andere, op. 49, for viola or cello (1984) (NB: performed by the cellist Catherine Marie Tunnell in Venice at the Neuroscience and Music Congress 2002) and *You - tree kalotrope,* op. 57 (1985), for tuba and double bass with background of eight shakuhachi (those of *Starriness,* op. 52, based on the random spectra of "whistle sounds" of the eight flutes) (FIG. 17 (a and b): macroform with S and A patterns in *Das Andere;* FIG. 17(c), fragment of page 13 of the score).

1188 f monesis 3
1560 b_(h_) 2
756 b flat 2 1/6.4 tone lower
2346 f sharp 2 1/7 tone higher
1144 f monesis 2
39 e half flat 2 (& 1/27 tone lower)
1938 e flat 2 (& 1/80 tone lower)
35 c 3/4 tone sharp 2 (+ 1/39 tone higher)
axis 432 - 832 - 1632 - 32 c 2
31 b_(h_) monesis 1 (& 1/40 tone lower)
1530 b_(h_) 1
728 b flat 1 1/6.4 tone lower
324 g 1 slightly higher (+ 1/102 tone higher, i.e. 1,95396 cents)

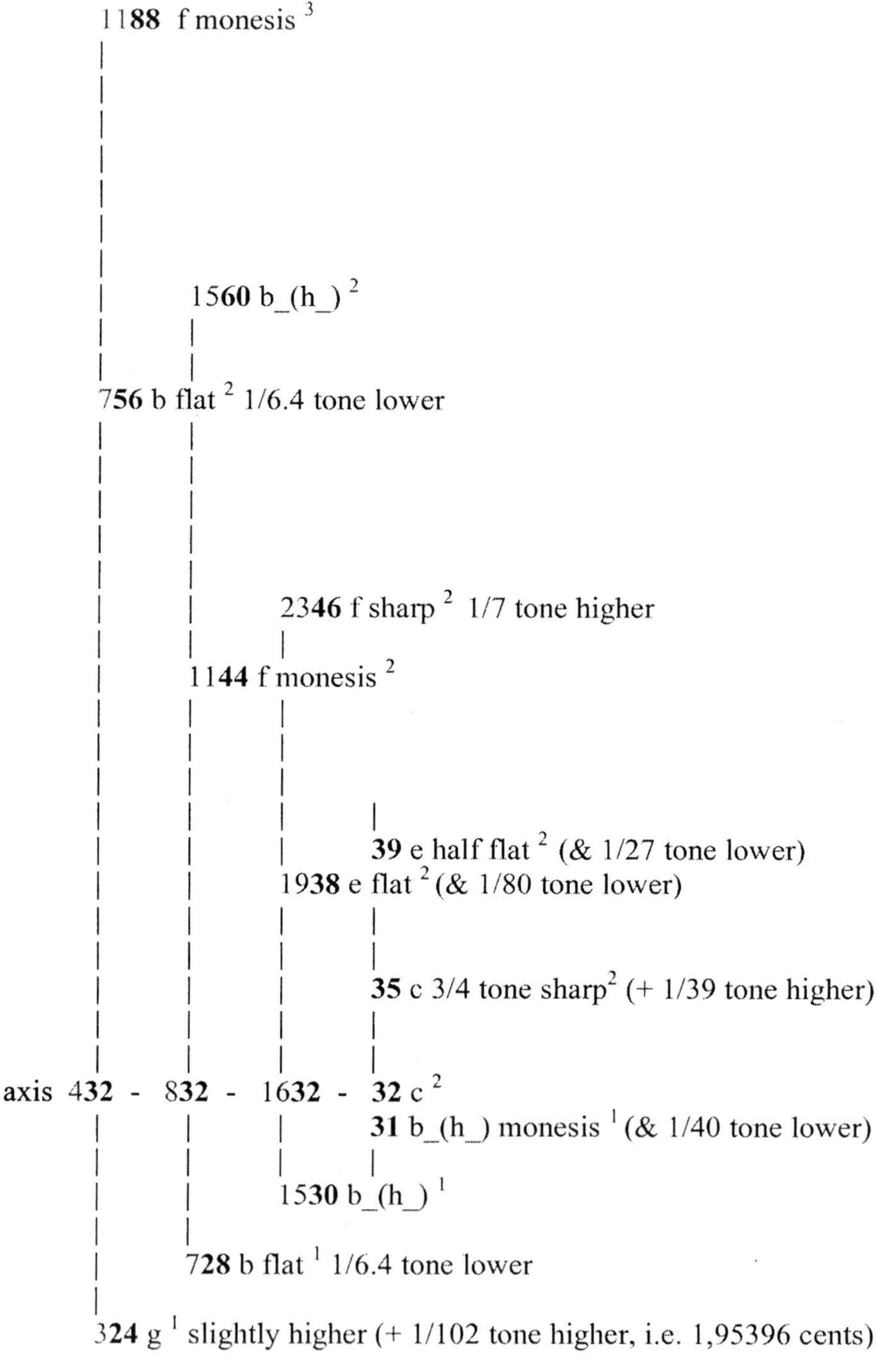

FIGURE 15a.

I. There are 343 orchestration-impulses fighting against and within
   a periodicity which itself has the time proportion : 6 5 4 2
   7 1 3 . In other words the 7 regions of the work each contain
   7 bridges (each of which itself contains 7 orchestration-impulses
   The first 49 orchestration-impulses happen every 6 beats (where
   1cm=2sec.). In the second region, the orchestration-impulses
   occur every 5 beats, etc...
   This macro-pulse of the tempi could be shown :

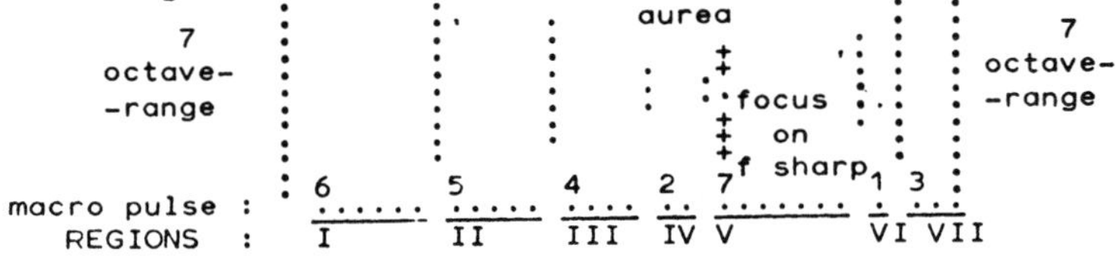

(accel.) (rall.) (accel.) (rall.)

   The strict proportional writing of these tempi uses the unit of
   1cm for a light signal pulsating at 2 second-intervals. The device built for this purpose gives colour-light-signals for a mesure of 5 centimeters (lasting 10 seconds) :
       RED (left wall) & VIOLET,VIOLET,VIOLET,VIOLET (right wall)
   as a mesure of $\frac{5}{2}$ where each half note $\flat$ lasts 1 centimeter and
   equals $\quarternote$ =30 MM.
   Therefore the minutes are indicated in the score: $1'$, $2'$, etc
   and for each 10 seconds:

       O.oo .   1___.___2___.___3___.___4___.___5____.____I starting
        1 10"   10"   idem   I minute 2

II. The frequencies (pitches) used describe the shape of a horizon-
    tal hour-glass   sectio
                              aurea
         7   7
       octave-   focus   octave-
       -range   on   -range
                                   f sharp
                  6   5   4   2 7   1 3
    macro pulse :
    REGIONS   :   I   II   III IV V   VI VII

Each region is built on "fundamentals" taken from SPECTRA of dif
ferent pitches:
Istregion of C ; IInd of G ; IIIrd of _E ; IVth of B$\natural$flat ; Vth
of D ; VIth of F$\sharp$; VIIth of G$\sharp\sharp$.
The fundamental of each region is itself the NEXT (ODD) NEW HAR-
MONIC of the ORIGIN-SPECTRUM of C.
It should be noted that sometimes the pitches were deduced from
very high layers of those SPECTRA, which contain more and more
compressed (minute) micro-tonal divisions of the octave, the
logarithmical enrichment per octave increasing according to the
row O, 1, 2, 4, 8, 16, 32, 64,etc..., which is equivalent to the
number of new frequency-functions found per octave.

FIGURE 15b.

HORATIU RADULESCU   IUBIRI   OPUS 43
SILENCE
[45' 44'']
(≈ 16'')
46'
28·IV·81 Versailles

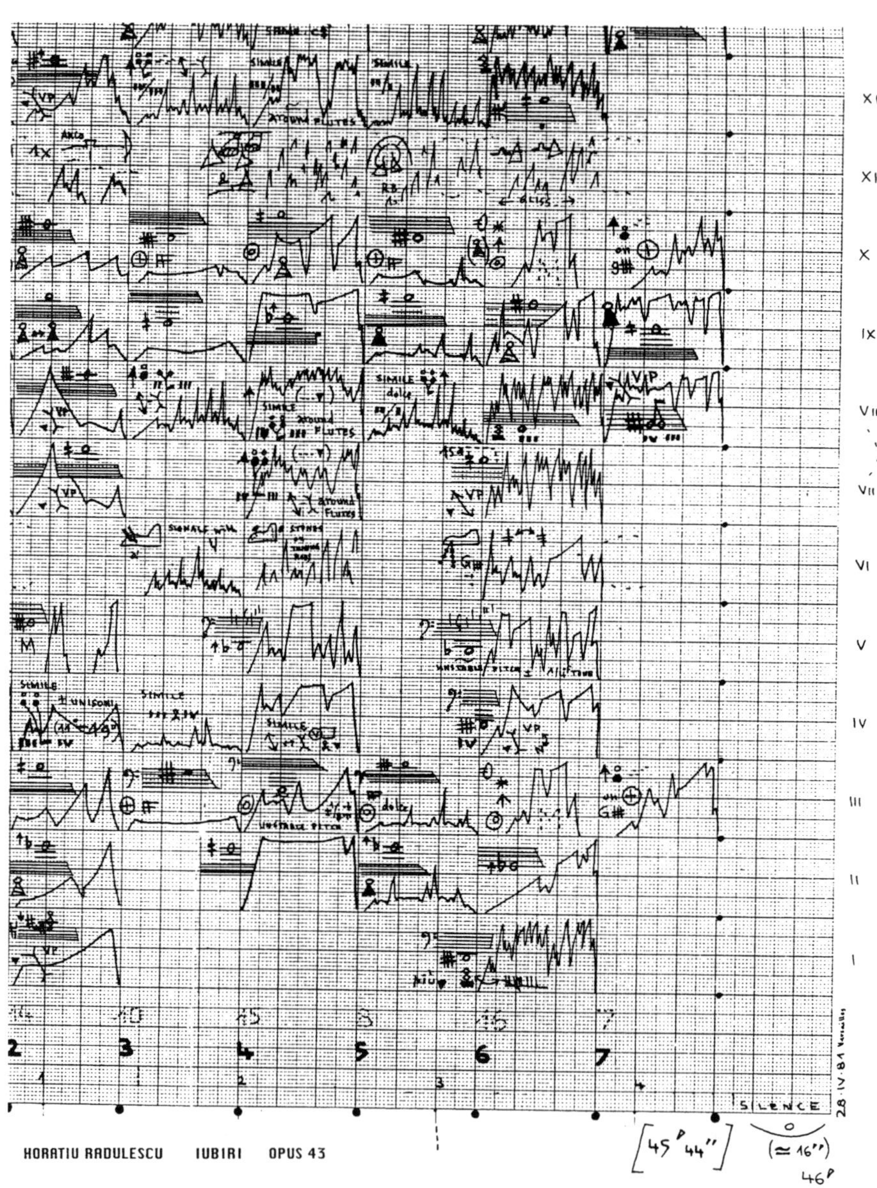

FIGURE 16a.

1ST REGION on C
FLAUTO 1° (PICCOLO & CONTRABBASSO)
HORATIU RADULESCU
IUBIRI
op.43
[ju'birj]
"AMOURS"
Versailles
1980/81
PICCOLO
FAST, tra.
Fa# T.G.
1/3
FLAUTO 1°
PAGE 1
SLOW ALTERNATING
END
of
PAGE
1
1 measure from
page 2
©1981 by LUCERO PRINT

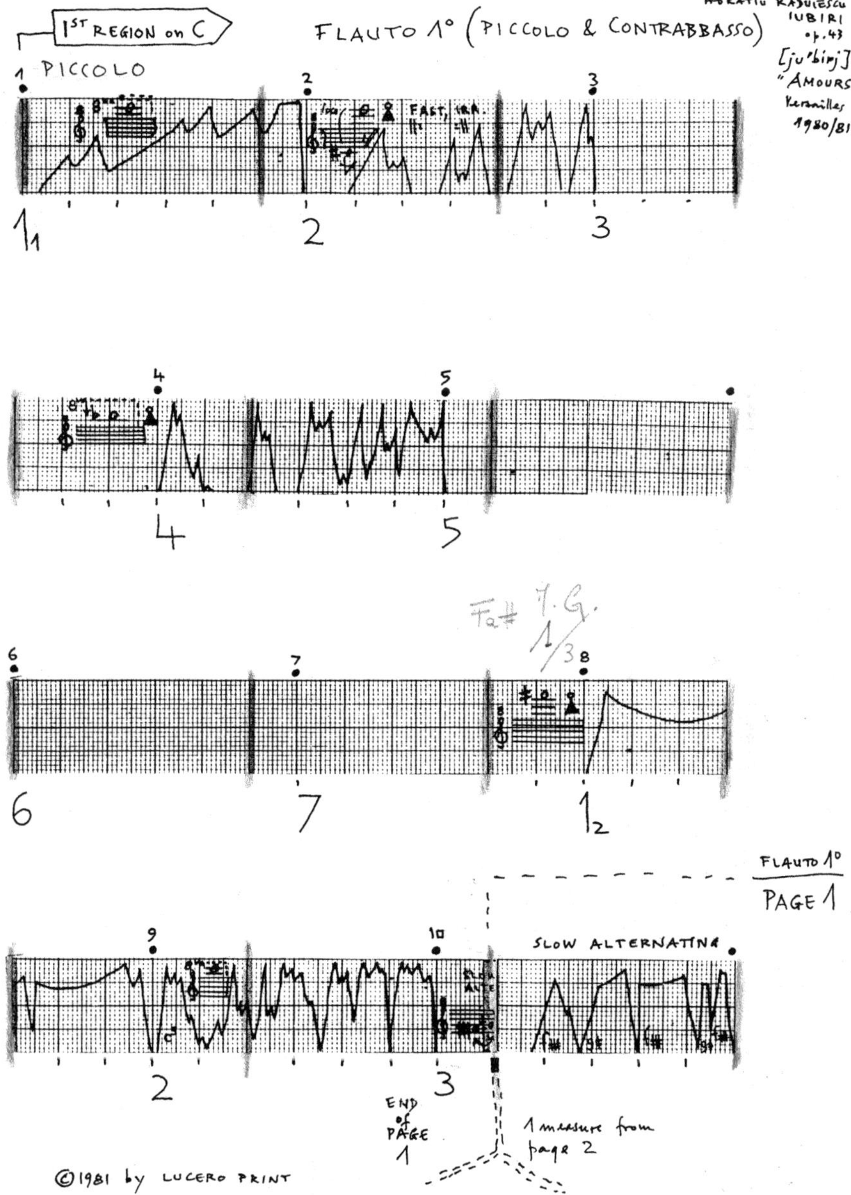

FIGURE 16b.

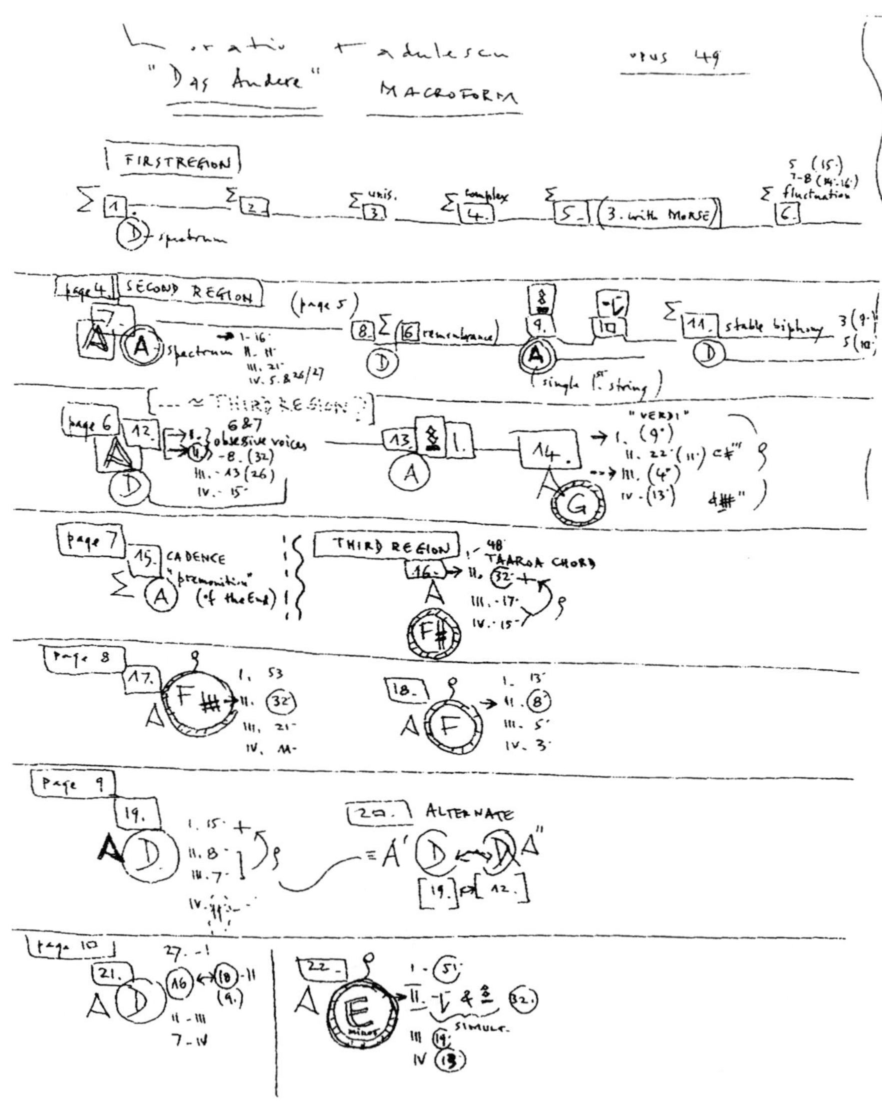

FIGURE 17a.

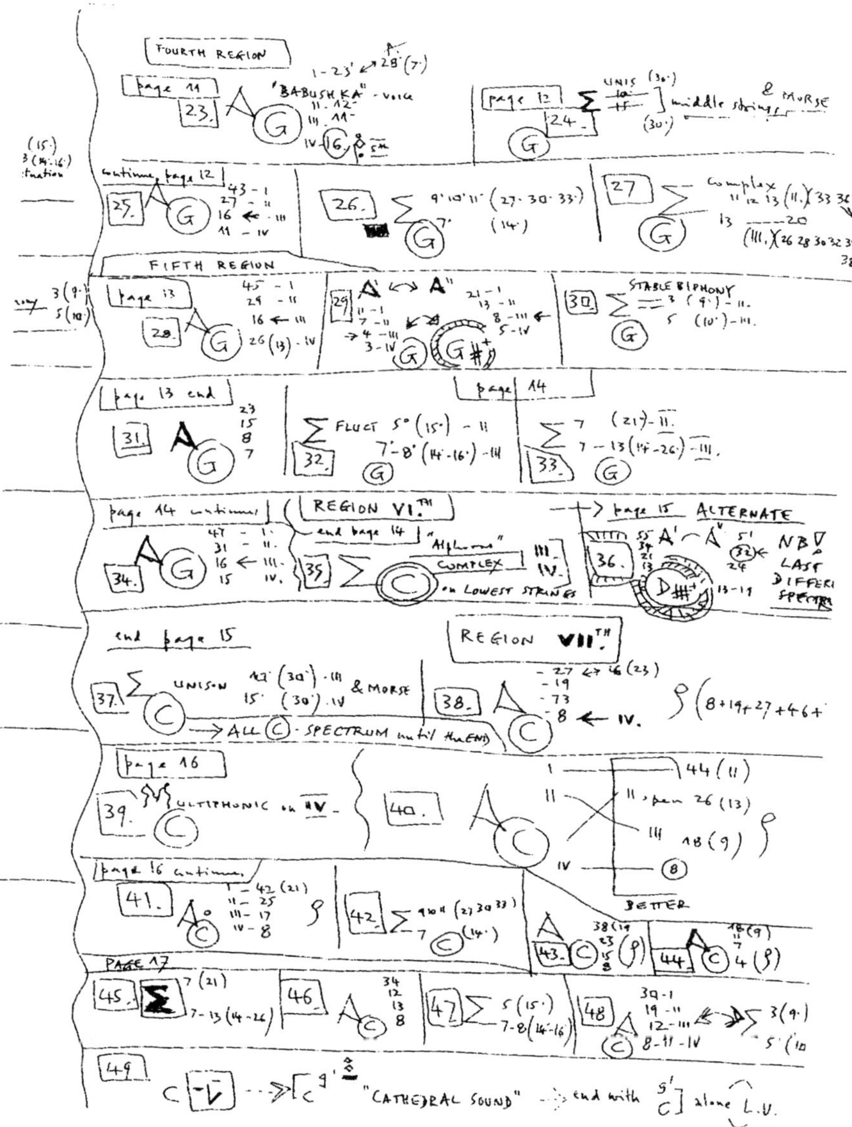

FIGURE 17b

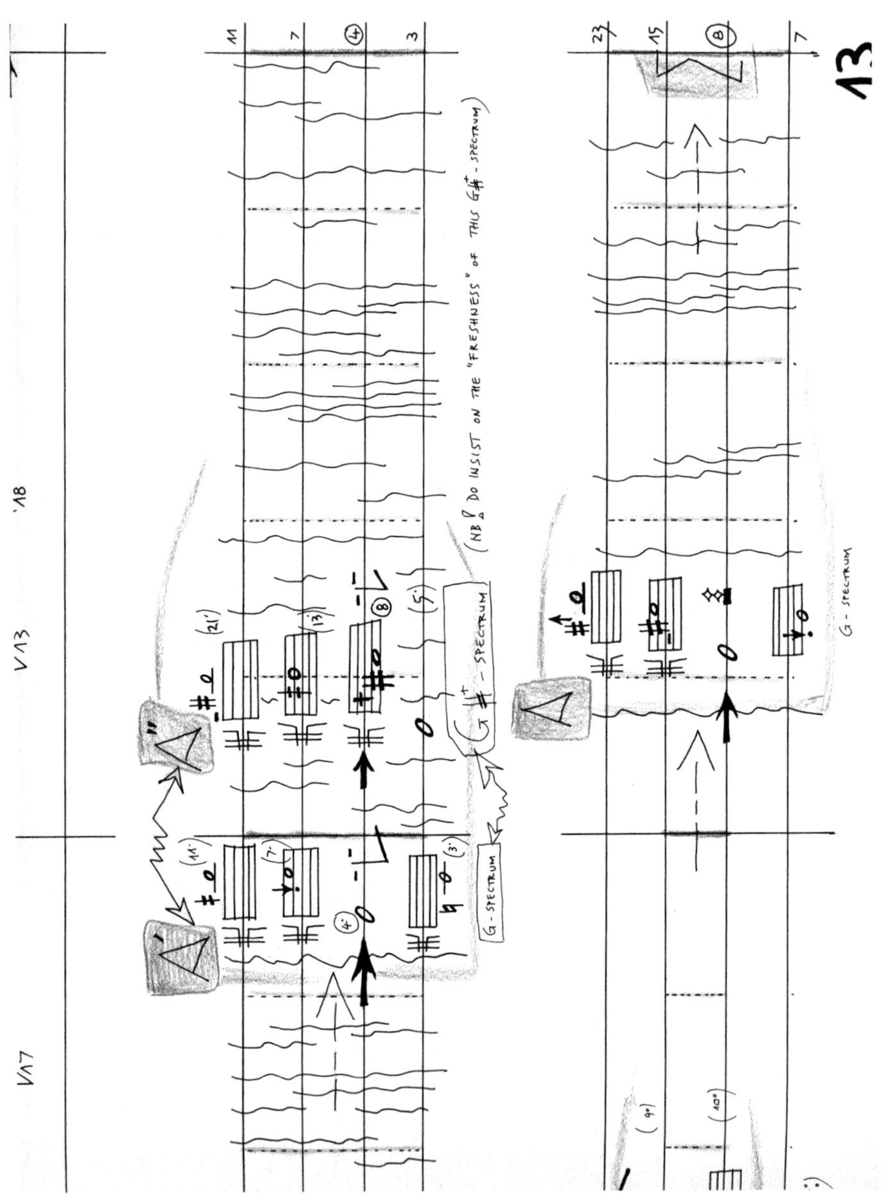

(NB: I DO INSIST ON THE "FRESHNESS" OF THIS G# - SPECTRUM)
G# - SPECTRUM
G - SPECTRUM
G - SPECTRUM
G - SPECTRUM

FIGURE 17c.

Explicit [ρ] Ring Formants as Slowest Arpeggio Arches

Sums and differences of harmonics up to the 23rd built on all fundamentals are implemented by the solo trombone in *Trombe d'oro della solarità,* op. 65 (1986). The whole should be realized in a very resonant medium (a Romanesque church, e.g., Speyer Funeral Basilica of the German Emperors, about 1030–1061) with a background music of sound icons performing low random spectra as "thunder"-sound, with rosined thick bow-threads on the lowest strings of concert grand pianos (preferably Imperial Bösendorfers) (FIG. 18, a page of the score, with logarithmic spectral natural intervals).

Erupting [ρ] Ring and [Π] Random Spectra with Increased Spatial Movement

Mirabilia Mundi, op. 70, for 7 large ensembles (1986, ca. 88 players): 12 strings, 17 flutes, organ, 8 trombones, 12 players with gold and silver crotales, 13 very large tam tams, 25 bass voices of monks (moving sound source), is a ritual for the same resonant Speyer Basilica. The resonance of 14 seconds in the Speyer Basilica inspired us to use a special display of the sound sources in concentric circles, as close as possible to the audience, and at "sacred" points, resembling imaginary knots of the total space vibration.

Similarly, the 40 flautists with 72 flutes in *Byzantine Prayer–Requiem for Giacinto Scelsi,* op. 74 (1988), read a score on two different but simultaneous pages: the whole 40-point "star" of flutes and/or the dialogue in between the 8 concentric groups of flutes materialize "Venetian" cori spezzati or various circular sound movement on elements of erupting ring [ρ] and of random spectra [Π].

For the psychoacoustic "volcano" created by the great number of flutes in a high-ly resonant room, again authentic old sacred spaces are preferred, as in the case of *Do emerge ultimate silence* with the children's choirs; the Rome Pantheon, for example, might be an ideal location for this ritual.

Genuine [ρ] Ring Scordatura and [π] Microrandom Spectra

Solo instruments such as the violin (with the singing voice of the performer) in *Dr. Kai Hong's Diamond Mountain* VI and VII, op. 77 (2001–2002), the double bass in *Azzurro profondo dello sguardo,* op. 78 (2002), the cello or viola in *Lux Animae,* op. 97 (1996), or even the percussion with seven wooden monks' plates in *Eterno,* op. 103 (2002), all with ring spectral scordatura, are all reinforced by the energy of the self-generative resonance.

A unique case with [ρ] ring scordatura, [π] microrandom spectra, and [μ] multi-plication of multiples (bispectral language) is the duo for oboe d'amore and spectral piano (12 pitches retuned according to harmonics of some spectra) *Animae morte carent* (after Ovid), op. 85 (1995).

Some of the most remote "shadows" of pure harmony and intrinsic spectrality issue from the authentic logic of sound and brain phenomenology. The self-generativity of functions seems to make music itself (in German, the sound phenom-ena "musizieren").

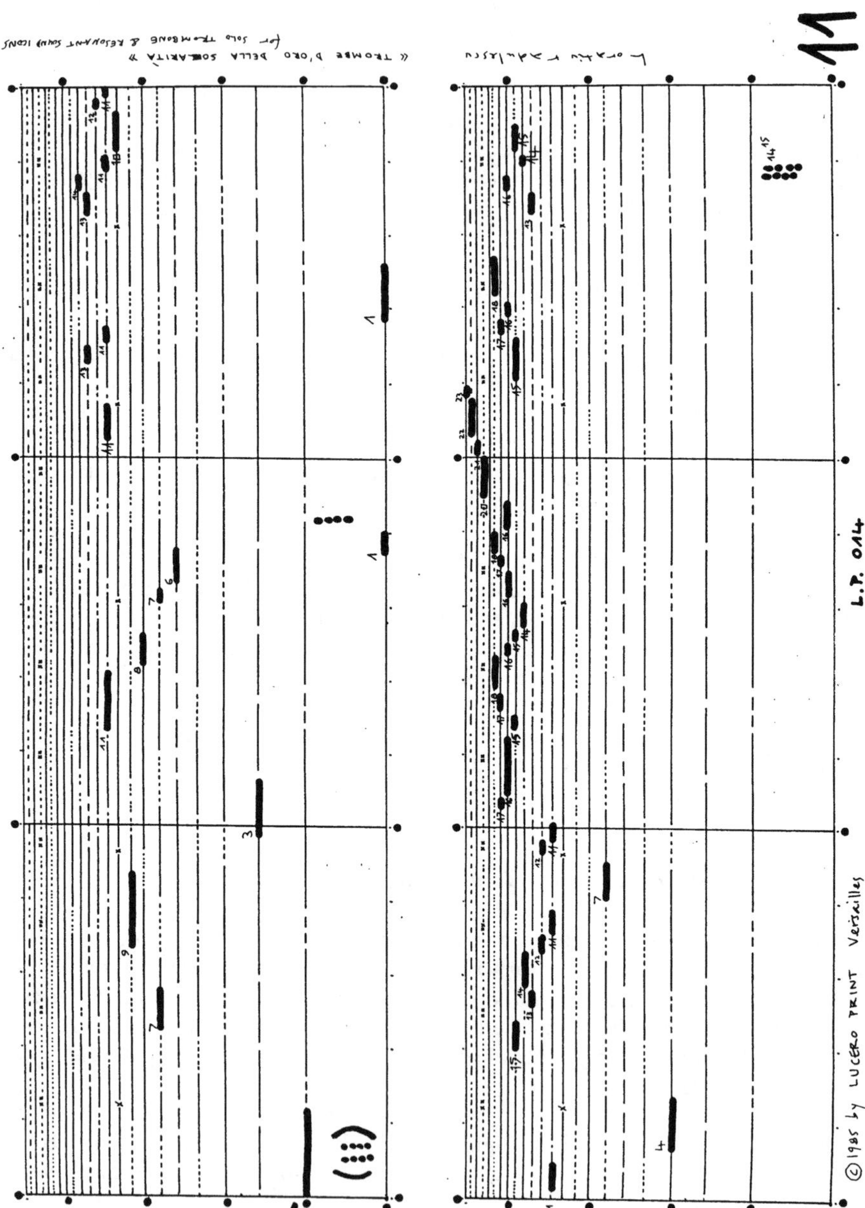

FIGURE 18.

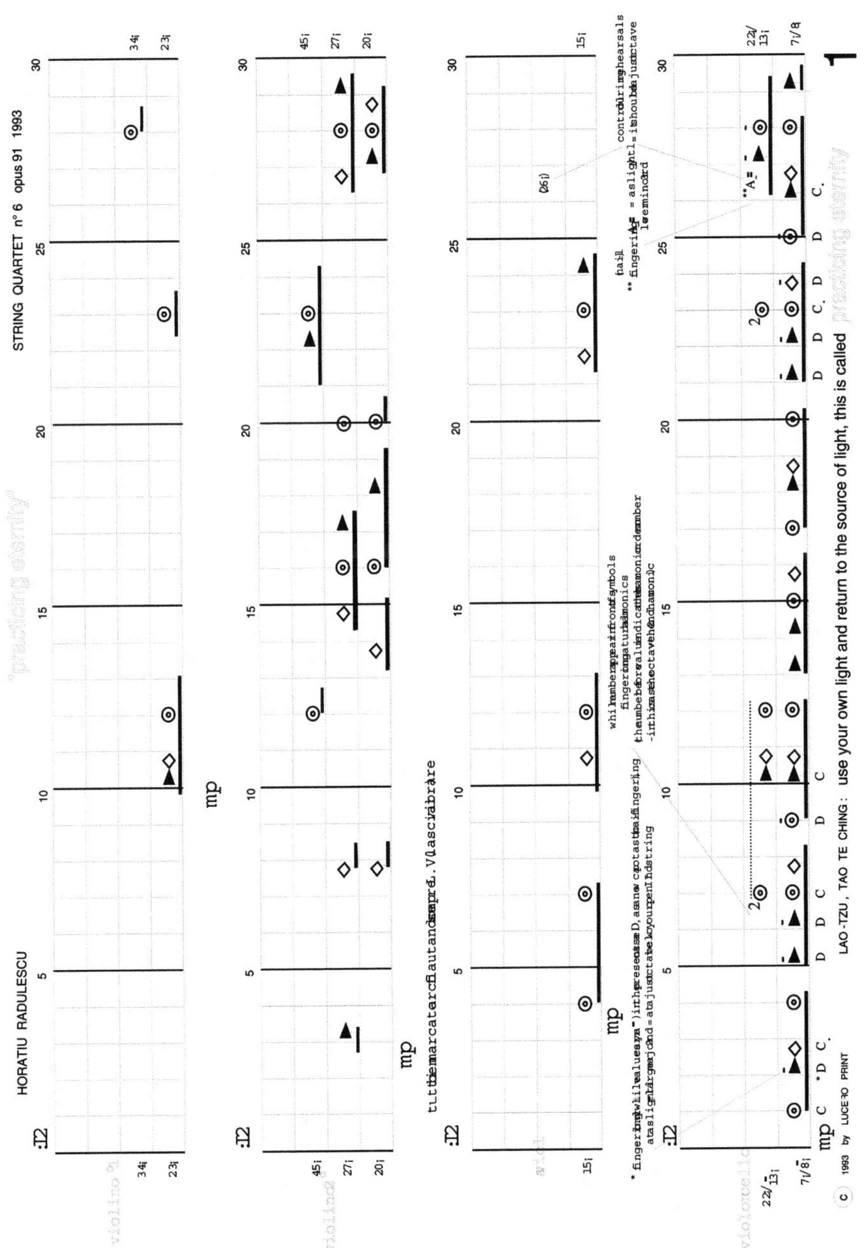
HORATIU RADULESCU
STRING QUARTET n° 6 opus 91 1993
"practicing eternity"
LAO-TZU, TAO TE CHING : use your own light and return to the source of light, this is called practicing eternity
© 1993 by LUCERO PRINT

FIGURE 19.

Complex [ρ] Ring Spectra and Scordatura with Semantic Magic Text Implication

'*Before the universe was born,*' String Quartet no. 5, op 89 (1990–1995), uses scordaturae covering seven spectra; in addition, the phonetic, notional, and magical significance of the text fragments of Lao-tsu's *Tao te Ching* determine some sound production and performance techniques.

Complex [ρ] Ring Spectra and Scordatura, Plus Explicit/Compact Spectrum [ε], with Microagogic Rhythm, Macrorandom Spectra [Π], and Multiplication of Multiples [μ]

'*practicing eternity*' (after Lao-tsu), String Quartet no. 6, op. 91 (1992), uses a single scordatura on a D spectrum, with 16 different functions as open strings, producing over 333 pitches: harmonics of those 16 ex-theoretical harmonics, representing 207 different steps of pitch, with a content of 89 totally unique basic spectral functions. The harmonic inquiry reaches utmost phenomenological "alchemy" in between heterohomophony of quasi-birds' chirping and plurimelodic neobyzantine spectral modes of fractal polyphony, the latter being difficult to analyze or reducible to monody, polyphony, homophony, or heterophony (FIG. 19, a page from '*practicing eternity*').

A ritual of psychoacoustic trance is *Angolo Divino,* op. 87 (1994), for large orchestra where the same 81 microagogic characters of rhythm vivify the sober deployment of macroregister "mobiles" coming from the highest "heaven" and progressively opening on an "Eternal Sight" embracing both highest and lowest "skies" through most secretly murmured and most luminous powerful waves.

TEMPERED ("NONHARMONIC") FORMANTS AS THEORETICAL BASIS OF FUNDAMENTALS WITH INCREASED SPECTRAL CONTENT ("SPECTRUM PULSE")

Already in '*Vies pour les cieux interrompus,*' String Quartet no. 2, op. 6 (1966), *Taaroa* for orchestra, op. 7 (Bucharest 1968–1969), especially in its second movement, "Rivelazione" based on various cluster geology, and *Everlasting longings* for string orchestra: 8, 6, 4, 4, and 2, op. 13 (Paris, 1971), I developed an intense treatment of timbre, highly detailed and producing rich aural information in the second formant: "spectrum pulse" (cf. the theory of composition booklet *Sound plasma—music of the future sign,* written in Paris in 1973, published by Edition Modern Munich in 1975).

Several compositions, op. 8, 9, 11, 12, 14, 15, and 18 (Paris 1969–1973), use "*global sound sources*": I, instrument/object sound; H, human abstract source; N, natural phenomena; E, electronic sound; L, articulated language, concrete human source, and prepare conceptually the venue of *Capricorn's nostalgic crickets* for seven identical woodwind op. 16h (Grasse 1972–1973) and of *Lamento di Gesù* for large orchestra, op. 23 (Paris, 1973; Munich, 1975). *Capricorn's* pseudo-random reading of a quartertone-based diffracted canon builds 96 rich columns with a very active second formant. This unique sound source reaches the characteristics of all

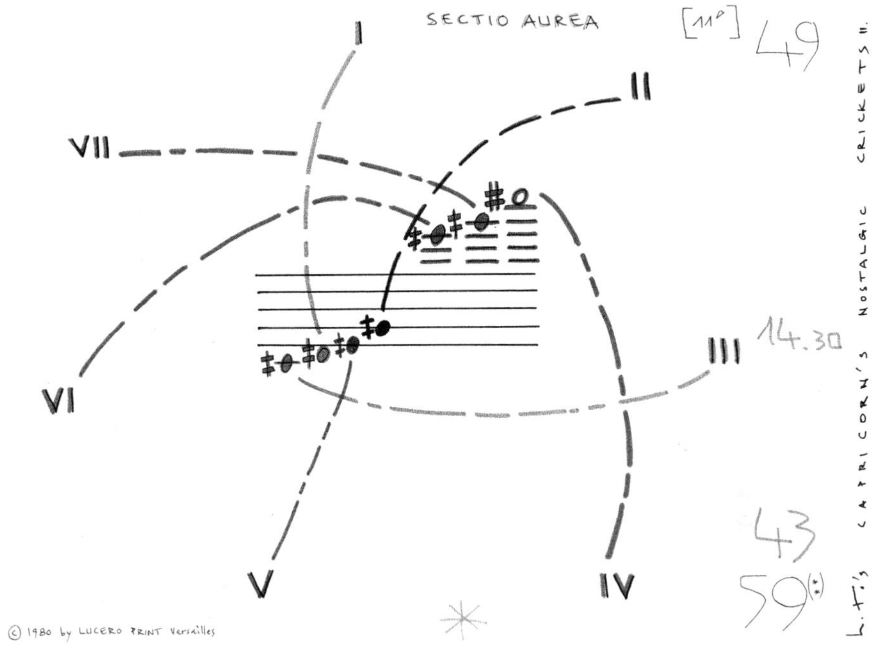

FIGURE 20.

other global sources not present (FIG. 20, 1 of the 96 "eruptions" in *Capricorn's Nostalgic Crickets*).

Lamento di Gesù, op. 23, is a "music on music on silence" where the large aggregates, each of 24 unique functions in quartertones, integrate a horizontal, lying "trinity" from a 7-octave range to a progressively narrowing ambit until the final cluster of the same 24 different pitch plateaus (FIG. 21, a page of *Lamento di Gesù).* Three simultaneous roles of actual, hiding/concealing, and sustaining/echoing source on each of the 24 function orbits make the 91 instruments of the orchestra quite "undetectable." Cause and effect are concealed, as is the Divine, the Eternal.

Other scores use these types of "nonharmonic," tempered, or microtempered fundamentals.

Their "spectrum pulse" is such an intense harmonic sonic life that the "nonharmonic" character (in French, "inharmonique") of these fundamentals' temperament seems to be mostly a very peculiar filtering operating on a harmonic content. One can imagine logarithmic paper being visually filtered in order to become millimetric paper.

In this timbral process of the "emanation of the immanence" we can assume to act conceptually on very high formants of the primal spectrum in order to dispose of enough functions to be able to operate that filtering. The rich "spectrum pulse" of these new fundamentals contributes to unifying all the aural information into a complex multilayered "spectrality."

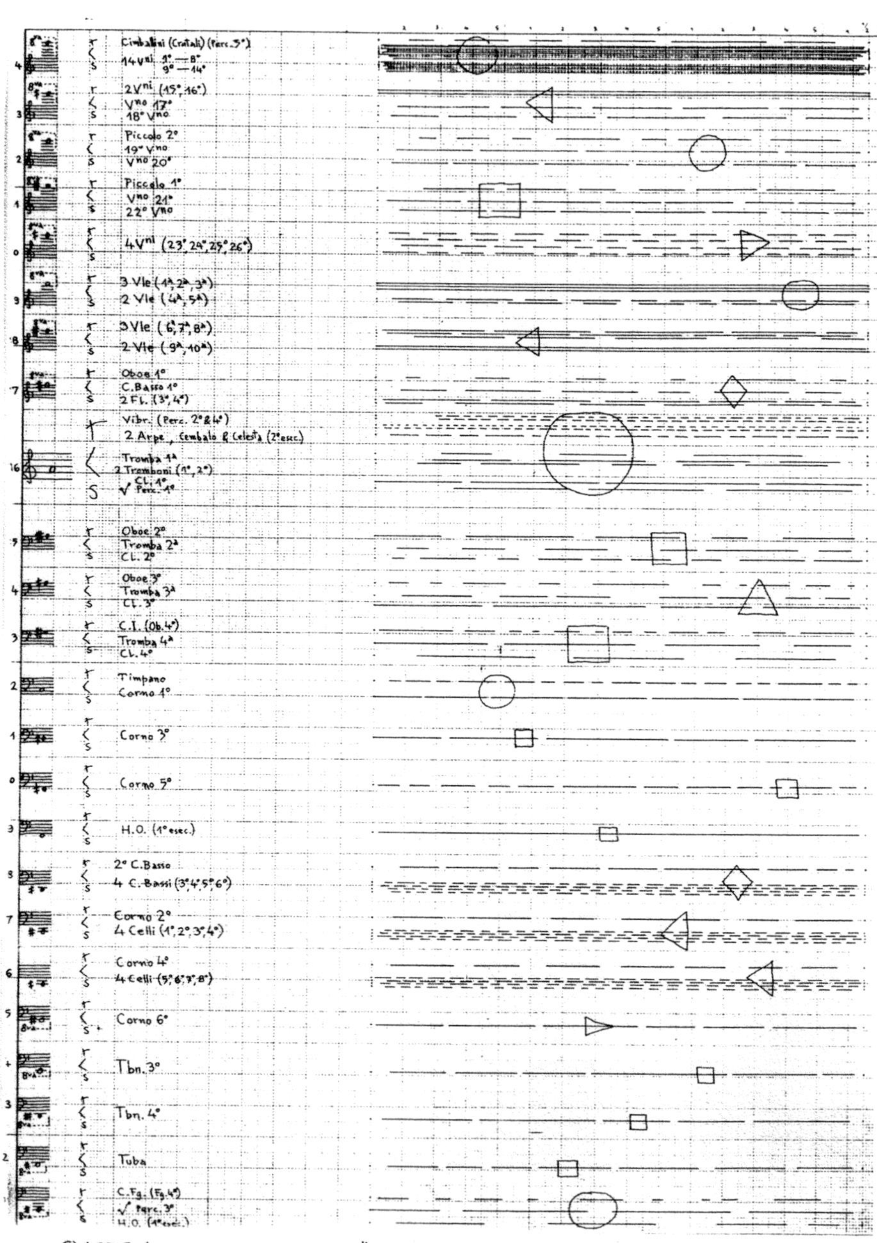

FIGURE 21a.

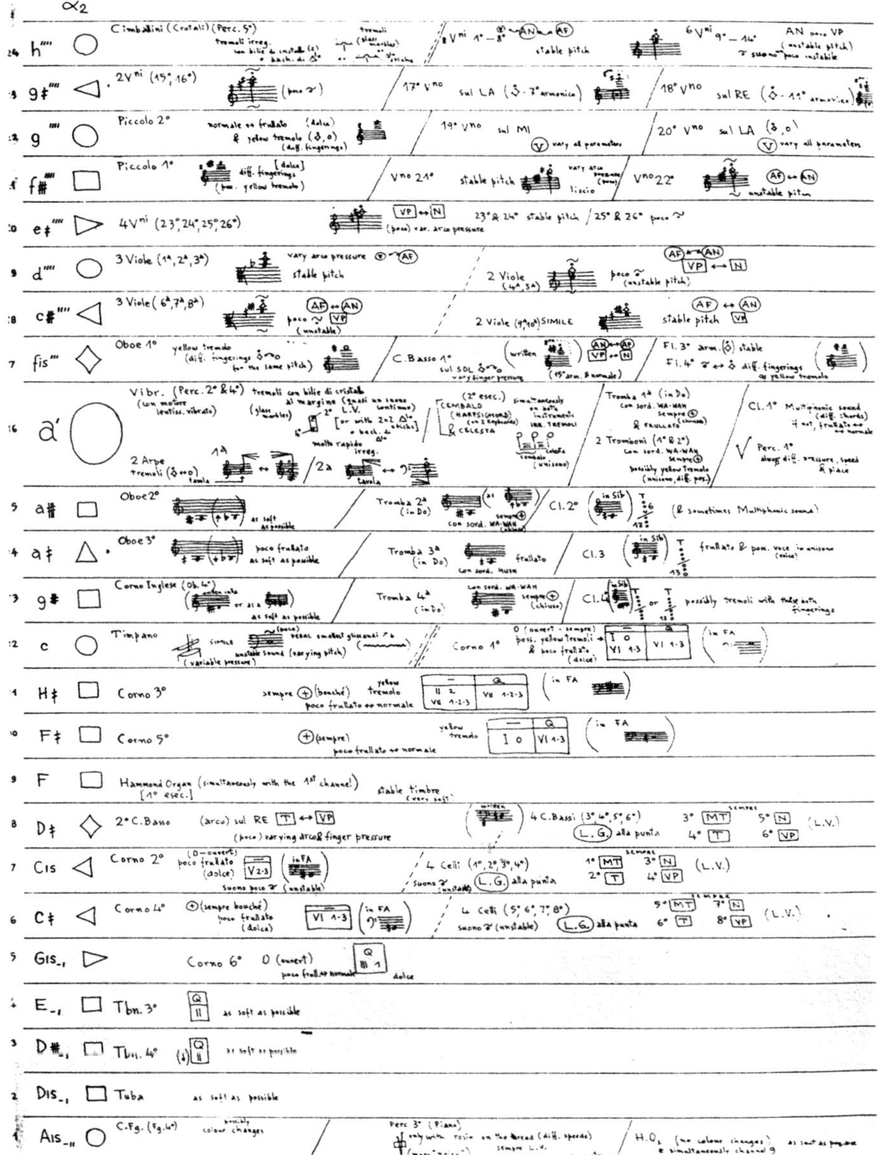

FIGURE 21b.

© 1976 by EDITIONS JOBERT PARIS
[page 7 of 44]
HORATIU RADULESCU DORUIND opus 27
(48 actual parts)
for 48 voices a cappella
(14 S, 10 A, 10 T, 14 B)

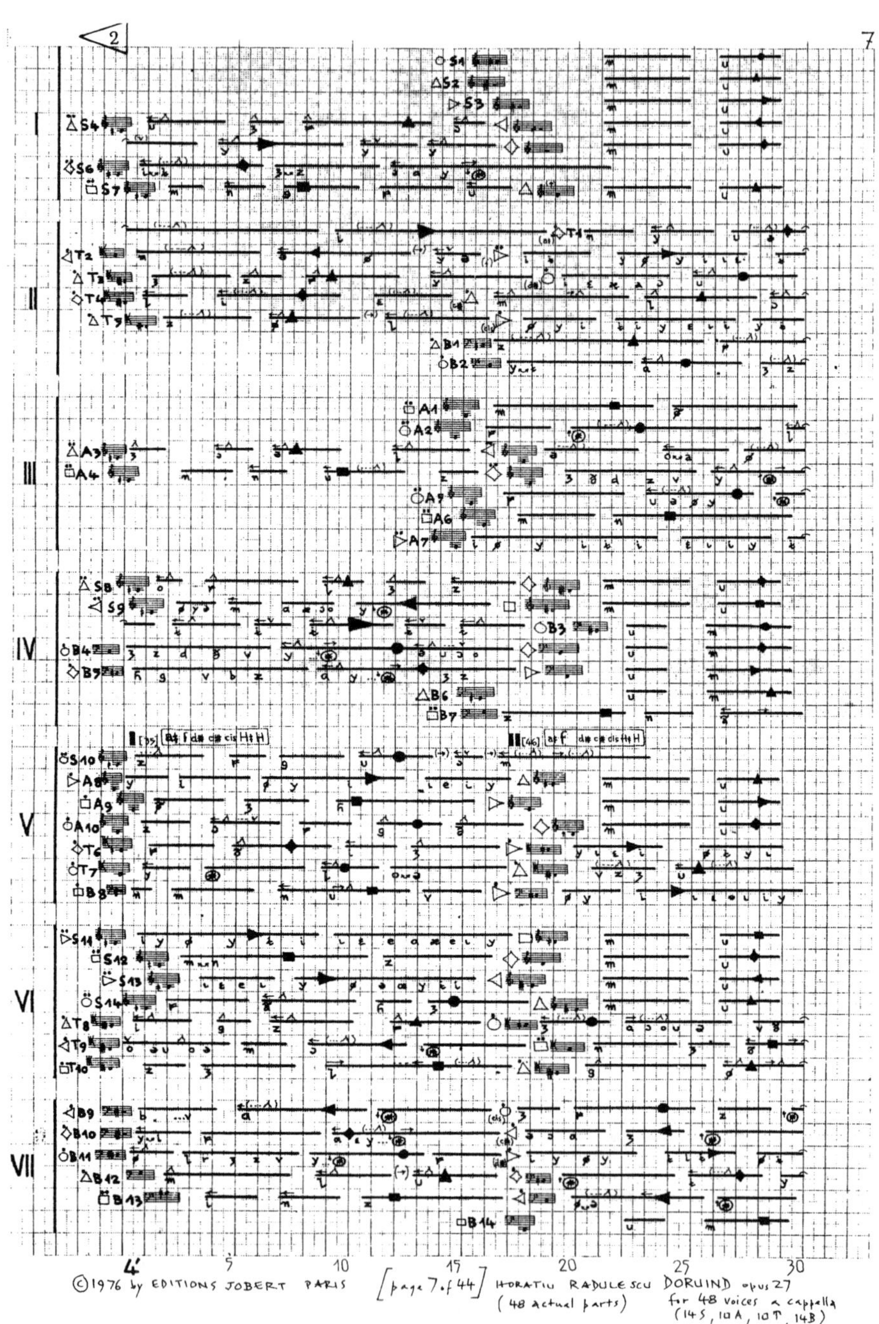

FIGURE 22.

A special case is *Doruind* for 48 voices—14, 10, 10, and 14, op. 27 (1976), where the slowest cantus firmus, a row of 24 quartertones deployed only once in 22′, has a unique downhill-uphill registral trajectory and constitutes, with its seven groups of generative sounds, the basis of an advanced ring process (r). For example, one of the middle macroform regions, being based on only two generators, will reach, on the seventh layer of ring modulation. about 250 pitches, simulated through simultaneous singing and whistling of slightly unstable "pitch plateaus." All sounds are live, all being performed *a cappella*—the ring modulation (ρ) of elements of a microtonal tempered scale forces the same problem as in *Lamento di Gesù* to arise in another way: nonharmonic fundamentals (extremely high functions) as the basis of harmonic spectra. However, the preferential self-generative ρ functions attain progressively through successive "auras" of ρ, the perfect, ultimate, and primal state of ε–compact explicit spectrum (FIG. 22, a page of *Doruind*).

ELEMENTS OF TEMPERED ("NONHARMONIC") SCALES IN "PLANETARY" DISPLAY SIMULATE OR APPROXIMATE ACTUAL NONTEMPERED FORMANTS

Self-generative spectral functions (ρ) and explicit compact spectra (ε) are non-tempered. Their natural, infinitely varied, microtonal deviation from the tempered scale can be expressed by the tempered pitches of a piano (for example) only through a very strict distribution of the approximate spectral functions at well-respected registral distances.

See the r starting the 2nd Piano Sonata op. 82 "being and non being create each other" (Lao-tzu):

Bb spectrum: 3(6) 4(8) ...[7] is not used because too out of tune! ...10, 11, ... 21 (FIG. 23a.)

Simulation of a compact spectrum (ε) of E (8 functions) and bispectrality through the sudden intervention of some important elements of an F spectrum (4 functions) constitue the opening of the 4th Piano Sonata op. 92 "like a well...older than God" (Lao-tzu):

E spectrum 1, 9, 5, 15, 11, 27, 13, 19

F spectrum 1, 5, 11, 13 (or 27)

... on the late resonance, the E-elements change into F-elements...

(E spectrum 1, 9, 5, 15, 11, 27, 13, 19)

(F spectrum 15, 17, 19, 7, 21, 5, 3, 9)

(See FIG. 23b, first page of "like a well... older than God.")

One of the most advanced approximations of spectral r and e can be found in the second movement, "Sacred Sound, the Second," of *The Quest,* Piano Concerto op. 90 (Freiburg 1996) and in the other versions of "Sacred Sound" in the 4th Piano Sonata and in the Cello Sonata *Exil Intérieur* op. 98 (1997) (FIG. 24, fragment of the Cello Sonata).

I. Immanence

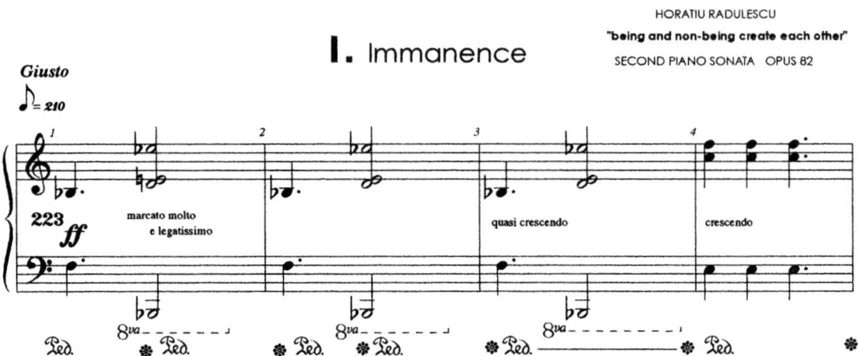

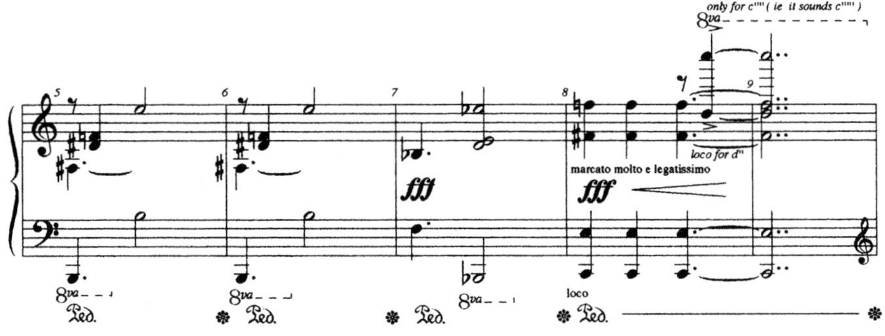

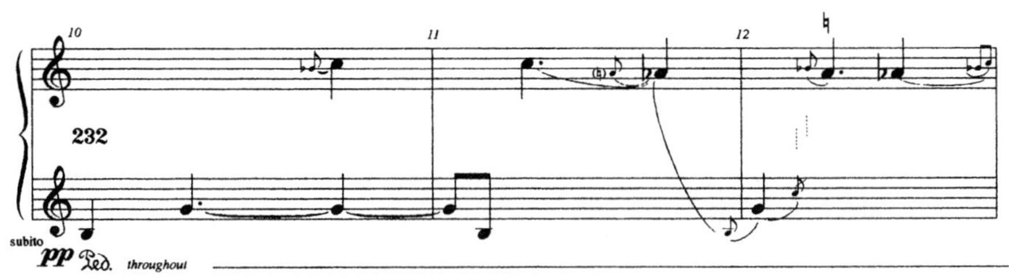

FIGURE 23a.

I.

trumpets of the eternal

FIGURE 23b.

HORATIU RADULESCU   CELLO SONATA OPUS 98
II. the sacred sound β
Grave
♩ = 90
8va
I.
II.
Cello
III.
IV.
ff
pp
fff
fff
fff
fff
loco
loco
8va
8va
8va
Piano
fff
fff
fff
fff
fff
ff
subito
loco
pp
ff
sempre Ped.
© 1997 by LUCERO PRINT
19

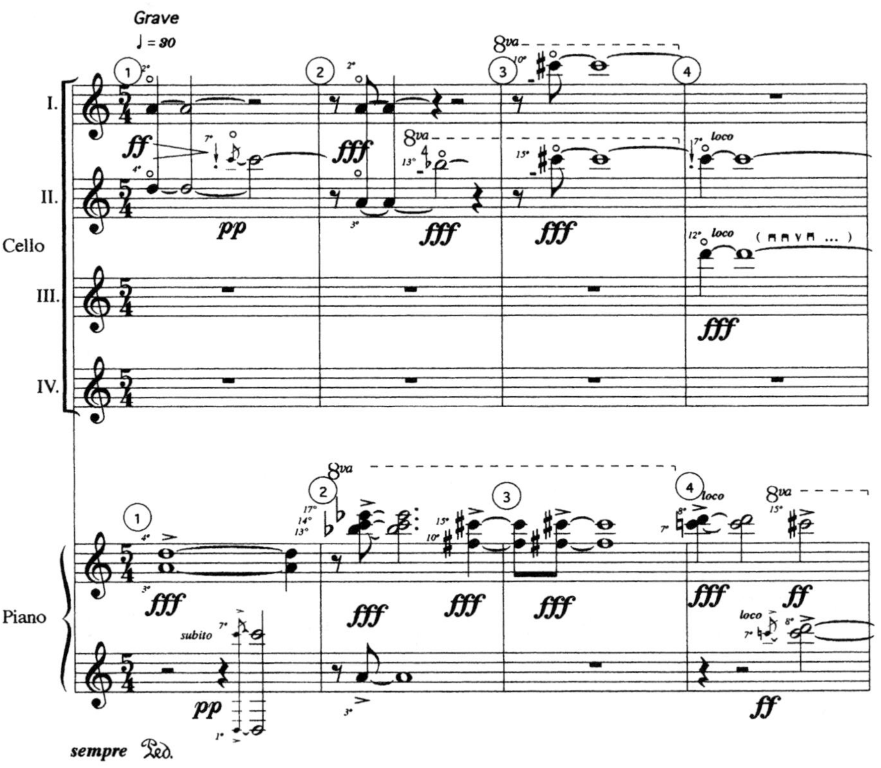

FIGURE 24.

RANDOM SPECTRALITY

Invented instruments and performances with three types of consciousness—composer, player, amateur (as in *Wild Incantesimo* for nine orchestras, op. 17b)—produce also savage and rich spectra, sometimes containing elements that are difficult to analyze, for example, the "invisible double bass" created by the resonance of sound icons. Forty-four laboratories of sound are encoded by the 44 signs taken from very old civilizations. The performance uses a score of over 4,000 colored slides projected onto 19 cinema screens encircling the audience. Over 4,000 micromusic events are realized by the 162 players (18 per orchestra) in 112 minutes.

Other cases of random spectrality are found in *Astray* for two duos of sound icons and saxophones, where the macroform is read at three speeds on two simultaneous layers, approaching psychoacoustically the problem of "presentiment of remembrances;" also, the recitation in 42 languages in *Hierophany,* op. 16r, with 42 children arrives at pitched abstract colored noise from the multilingual magic, notional, phonetic "alchemy" of a unique poem.

SPACE AS TIME OF SPECTRAL RITUALS

To combine the compass of sound (element, width, sound, and noise) and that of the psyche (thought, feeling, intuition, and sensation), the sound sources such as the 9 celli in *Credo* and *Ultimo Credo*, the 34 (or 340) children's voices in *Do Emerge Ultimate Silence,* the 9 orchestras in *Wild Incantesimo,* the 40 flautists with 72 flutes in *Byzantine Prayer*, the 9 string quartets in *'Infinite to be cannot be infinite, infinite anti-be could be infinite',* the 7 groups in *Mirabilia Mundi,* the revolving gongs in *Outer Time,* and the "ethereal harp" of the 7 clarinets in *Inner Time,* the sound icons of *A Doini* or of *Clepsydra,* are living in the temple of time, and bring us the movement and the vibration of light.

Neural Basis of Rhythmic Timing Networks in the Human Brain

MICHAEL H. THAUT

Center for Biomedical Research in Music,
Colorado State University,
Fort Collins, Colorado 80523, USA

ABSTRACT: The study of rhythmicity provides insights into the understanding of temporal coding of music and temporal information processing in the human brain. Auditory rhythms rapidly entrain motor responses into stable steady synchronization states below and above conscious perception thresholds. Studying the neural dynamics of entrainment by measuring brain wave responses (MEG) we found nonlinear scaling of M100 amplitudes generated in primary auditory cortex relative to changes in the period of the rhythmic interval during subliminal and supraliminal tempo modulations. In recent brain imaging studies we have described the neural networks involved in motor synchronization to auditory rhythm. Activated regions include primary sensorimotor and cingulate areas, bilateral opercular premotor areas, bilateral SII, ventral prefrontal cortex, and, subcortically, anterior insula, putamen, and thalamus. Within the cerebellum, vermal regions and anterior hemispheres ipsilateral to the movement became significantly activated. Tracking temporal modulations additionally activated predominantly right prefrontal, anterior cingulate, and intraparietal regions as well as posterior cerebellar hemispheres. Furthermore, strong evidence exists for the substantial benefits of rhythmic stimuli in rehabilitation training with motor disorders.

KEYWORDS: auditory rhythm; music; synchronization; entrainment; neural networks; brain imaging

INTRODUCTION

The capability for the perception and volitional production of rhythms is unique to the human brain and is dependent on the capacity for stable, precise, rapid, and complex time organization in the brain. Converging evidence[1] shows pulse salient models as underlying rhythm processing. Conceptually, these are models that require synchronization of acoustical events at various hierarchical levels of rhythmic organization into felt pulse patterns that function as isochronous temporal templates with a given period. The study of the neural substrates of rhythm involving synchronization tasks closely reflects the process of intrinsic temporal pattern

Address for correspondence: Dr. Michael H. Thaut, Center for Biomedical Research in Music, Colorado State University, Fort Collins, Colorado 80523, USA. Voice: 970-491-7384; fax: 970-491-7541.
mthaut@lamar.colostate.edu

Ann. N.Y. Acad. Sci. 999: 364–373 (2003). © 2003 New York Academy of Sciences.
doi: 10.1196/annals.1284.044

formation within synchronized pulse structures as the core effort in the perception and production of rhythm. Traditionally, preference in experimental designs has been given to simple tasks such as finger tapping in synchrony to metronome-like pulse beat sequences. Time-span reduction theory supports the validity of these designs for the study of rhythm because all rhythms can eventually be reduced to isochronic prototypes.[2,3]

RHYTHMIC SYNCHRONIZATION: EVIDENCE FROM PSYCHOPHYSICS AND BEHAVIORAL DATA

Several findings in our recent research have contributed to a systems understanding of control mechanisms of rhythm in the brain. First, as was shown previously,[4,5] steady-state coupling of rhythm and motor response is achieved quickly within 1 to 2 repetitions of the rhythmic stimulus interval. In a study employing random step changes in the tempo/frequency of metronome sequences we showed evidence for direct frequency entrainment. In small step changes, those below the level of conscious perception, period errors are adjusted first, and then synchronization errors (time difference between tap and onset of sound) are recovered gradually. During larger step changes, which are consciously perceived around step changes of 4–5% or larger of the period duration, synchronization error and period error are adjusted conjointly by temporary overcorrection of the response period.[6] Thus, a nonlinear controller can be proposed in the human brain that employs multiple synchronization strategies depending on the dynamic state of the synchronization system.[7–9]

These results were replicated in a recent study by using a syncopated synchronization design in which the beat and the motor response were 180 degrees out of phase.[10] A recursive mathematical model showed a predominance of 5:1 in weighting factor for adjustment of period versus synchronization error to resynchronize the motor response to the new tempo, providing strong support for oscillator coupling underlying entrainment mechanisms. Oscillators can only couple via frequency modulations.

Further evidence of subliminal coupling mechanisms was demonstrated in a study in which we employed a metronome pattern that was continuously modulated in tempo following a cosine wave function at levels above and below conscious perception.[11] Response curves followed precisely the temporal dynamics of the stimulus modulation with continuous period correction at lag 1.

In rhythmic synchronization the motor response exhibits considerable temporal variability in relation to the beat interval. In a recent study we investigated if the fluctuations were deterministically chaotic or truly random.[12,13] A direct test for low-dimensional determinism failed to demonstrate its presence. Fractal analysis showed the absence of long-range correlations for the periods of the motor response but the presence of long-range correlations for synchronization errors. Together, the synchronization mechanisms appear therefore to be characteristic of a self-correcting system with stochastic noise attributes,[14–16] however, with bounded phase errors that signal set-point limits of perceived simultaneity between motor rhythm and auditory rhythm.[17] Such system resembles directed brownian motion, which is random rather than chaotic but which also contains correlations.[13,18,19]

Rythmic Tapping

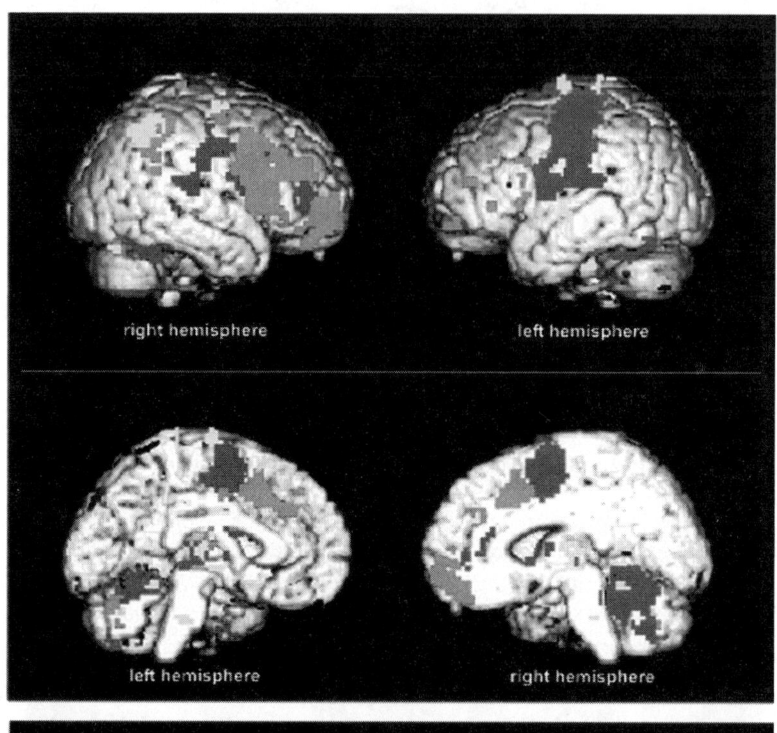

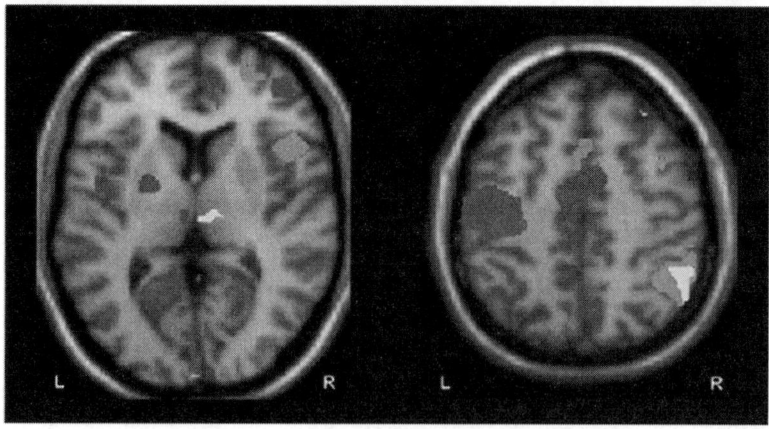

FIGURE 1. *See following page for legend.*

NEUROPHYSIOLOGICAL AND NEUROANATOMICAL EVIDENCE

We investigated the neural dynamics of rhythmic synchronization during tempo modulations by applying the step change experiment just described in conjunction with brain wave measures using MEG.[20] Random step change rates were 2% and 20% of a base interval of 500 ms. Results revealed that the amplitude of the M100 component of the brain magnetic field scaled proportionally to changes in rhythmic interval duration regardless if the tempo modulation was perceived consciously or not. Previous dependency of the M100 intensity on temporal stimulus characteristics has been shown,[21,22] but not in a time frame as brief as in our experiment. Dipole analysis revealed an M100 generator stably in localization and latency within the supratemporal Heschl's gyrus, indicating that the change in cerebral time coding response was driven by the synchronized firing of more activated neurons with a fixed neural pool. It is a natural assumption that subcortical relays (not measured in this study) may also contribute to or generate the cortical activity patterns, such as those via reticulospinal pathways.[23,24]

In a series of brain imaging experiments we used a variety of experimental tasks of rhythmic motor synchronization consisting of isochronous finger tapping, tapping to random rhythms, and tapping to rhythms whose tempo fluctuations were cosine wave modulated similar to experiments just described.[25] Using positron emission tomography (PET) we were able to identify for the first time the basic neural network underlying rhythmic synchronization (FIG. 1, TABLE 1). During isochronous right-handed finger tapping, the basic network consists of primary sensorimotor areas contralateral to the moving hand, bilateral SII, bilateral opercular premotor areas, and, subcortically, contralateral insula, putamen, and thalamus. Additional significant increases in regional cerebral blood flow were seen in right rostral ventrolateral prefrontal cortex. Within the cerebellum, right cerebellar vermis and predominantly right anterior hemispheres were activated. Implementation of temporal tracking, both to cosine-modulated or randomized rhythms, was associated with additional parietothalamic and premotor activity, predominantly ipsilateral to the movement. Finally, primarily right prefrontal, anterior cingulate, and intraparietal areas as a well as posterior cerebellar areas became active specifically during cosine-modulated tapping. Prefrontal activation showed an expanding regional network depending on the magnitude of the modulation. At 3% of base interval, medial frontal areas were activated. Ventrolateral areas were additionally activated during the 7% condition, and dorsolateral areas prominently during the 20% condition.

Further analysis of cerebellar patterns[26] of rCBF change revealed distinct corticocerebellar circuits underlying rhythmic synchronization (FIGS. 2 AND 3). Anterior

---

FIGURE 1. Neural maps for three conditions of rhythmic tracking: isochronous synchronization vs listening to rhythmic stimulus (*dark*); finger tapping to a rhythmic stimulus whose periods are cosine wave modulated at 20% of the base interval (*grey*) vs isochronous tapping; tracking a rhythm stimulus with randomized period durations (**light*) vs isochronous tapping. Significant group differences are given at $P < 0.005$, N = 9. Activations (in MNI space) up to 10 mm below the surface of the brain are projected onto medial and lateral surfaces. Same activations are shown horizontally at the thalamus and basal ganglia level ($z = 6$) and at the level of the anterior cingulate, primary sensorimotor area, and inferior parietal lobe close to the intraparietal sulcus ($z = 48$).

TABLE 1. Isochronous tapping versus listening—Summary table of activation areas during isochronous rhythmic auditory-motor synchronization, after subtracting areas activated during the condition of listening to the rhythmic stimulus only. The areas listed constitute the basic neural network mediating isochronous rhythmic auditory-motor synchronization, using finger tapping as the motor response.

	Left Hemisphere	Right Hemisphere
Right inferior frontal gyrus (ventrolateral prefrontal)		45 / 48 / 6 (3.92)*
Precentral gyrus (dorsal premotor areas)	−39 / −15 / 60 (6.41)	
Precentral gyrus and operculum (ventral premotor)	−48 / −3 / 15 (5.83)	63 / 6 / 18 (4.30)*
Anterior insula	−33 / 15 / 15 (3.77)*	
Cingulate motor areas	0 / −3 / 48 (5.66)	
Superior frontal gyrus (SMA)		6 / −3 / 69 (3.75)*
Primary sensorimotor	−36 / −21 / 48 (6.89)	
Opercular parietal (SII)	−48 / −24 / 18 (4.57)*	63 / −24 / 12 (4.07)*
Thalamus	−3 / −21 / 6 (3.64)*	
Putamen	−21 / 0 / 6 (3.95)*	
Cerebellar vermis		9 / −48 / −12 (5.66)
Anterior cerebellar hemisphere	−24 / −54 / −24 (3.96)*	24 / −42 / −21 (6.63)
Posterior cerebellar hemisphere	−21 / −63 / −24 (3.98)*	

Statistically significant differences using a significance level of $p<0.001$. Results were corrected for spatial extent ($p <0.05$). Several subcortical and even some cortical areas with very focal activation peaks were too small to 'survive' such a spatial correction, and are marked with an asterisk (*).

cerebellar lobe activation in vermal and hemispheric regions ipsilateral to the movement was present with similar strength during isochronous, randomized, and cosine-modulated tracking (FIG. 2a and b) and covaried with activations mostly in primary sensorimotor and thalamic areas. Within bilateral hemispheric parts of the posterior cerebellar lobe magnitude of rCBF, change covaried with the magnitude of change of the rhythmic stimulus interval: it was lowest for the isochronous rhythm, stepwise increases were relative to the 3% and 7% tempo change, and highest activations occurred at similar levels during the random and the 20% change condition (FIG. 2c and d). Changes in activation thus only correlated with mean degree of temporal change but not with tracking strategy (random vs regularly modulated). Secondary sensorimotor activations, especially in the right inferior parietal cortex close to the intraparietal sulcus mirrored the pattern of rCBF change in the cerebellum. Finally, during the 20% condition the only higher activation than in the other conditions was seen in more medial parts of the left posterior hemisphere, close to crus I/lobule VI and more caudally in crus VIIa (FIG. 2e), corresponding to regional activations in the right inferior parietal cortex in the depth of the intraparietal sulcus, lateral right prefrontal cortex (right hemispheric homologue to Broca's area), and dorsolateral prefrontal cortex. In conclusion, the observations suggest that the described cerebellar

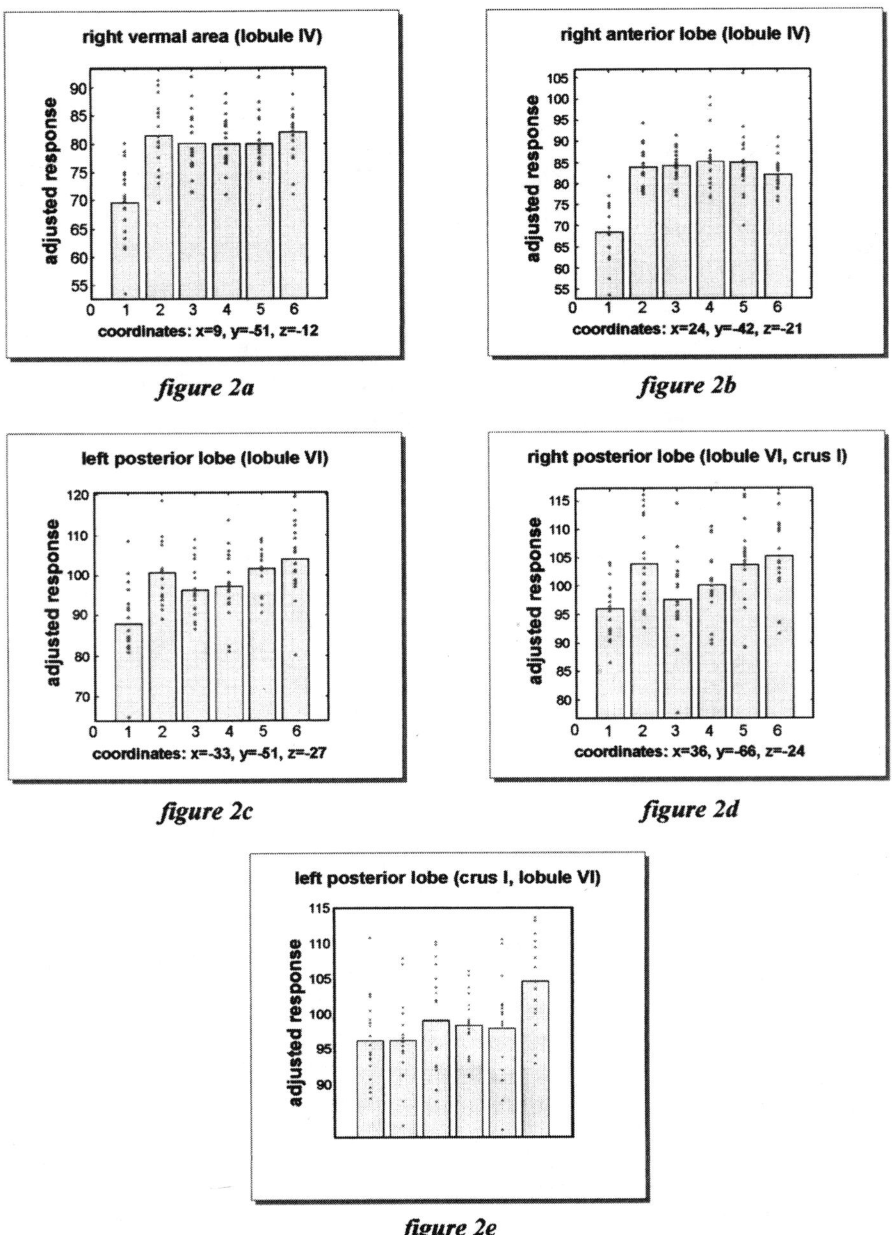

FIGURE 2. Adjusted rCBF mean and individual values for six experimental conditions of rhythmic synchronization: listening to rhythmic stimulus (1); finger tapping to random rhythm (2); isochronous finger tapping (3); finger tapping to rhythmic stimulus modulated at 3% of base interval (4), at 7% of base interval (5), and at 20% of base interval (6).

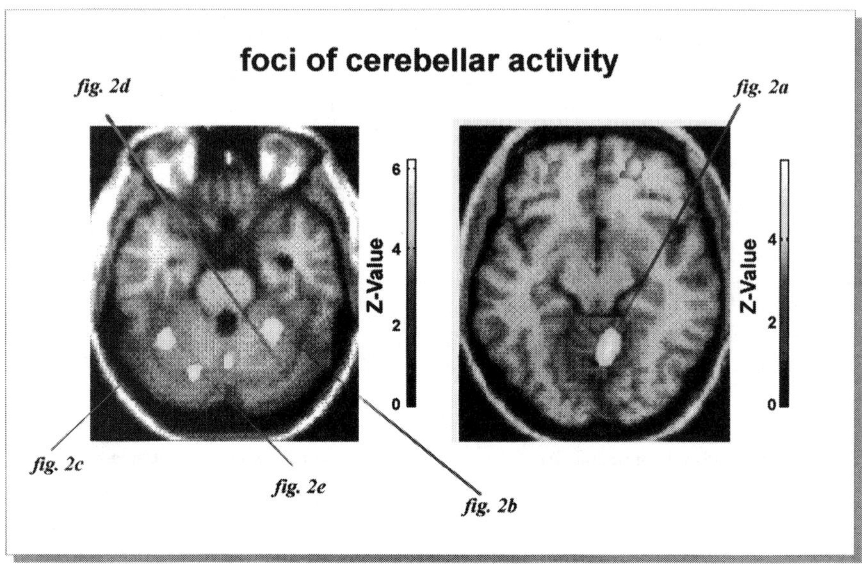

FIGURE 3. Foci of cerebellar activity during rhythmic motor synchronization corresponding to experimental conditions and activation patterns displayed in FIGURE 2.

regions form parts of distinct neural networks related to different components of temporal control during rhythmic synchronization.

Similar cortical topologies for isochronous rhythmic synchronization were also found in a study by colleagues Jerome Sanes and Martina DeMartin and I conducted recently[27] to compare isochronous with polyrhythmic (hemiola) synchronization at two different movement frequencies. The hemiola synchronization showed more activation strength than the isochronous condition. Faster auditory rhythm rates showed higher activation rates bilaterally in anterior portions of the superior temporal gyrus bordering on the opercular regions in the right superior parietal lobe. Faster movement did not show distinct brain activation patterns. The hemiola condition yielded more activation in the bilateral supplementary motor area (SMA) as well as two small clusters in the right supramarginal gyrus and left cerebellar hemisphere. Activation in basal ganglia structures of the right anterior caudate and putamen was reduced during hemiola synchronization versus isochronous tracking.

Finally, my colleague, Larry Parsons, and I studied[28] whether distinct neural activation patterns underly different components of rhythm. We separately studied the perception of tempo, duration, pattern, and meter in rhythmic discrimination tasks. We also tested nonmusicians and musicians. Results showed impressively that distinct and partially overlapping neural networks subserve each component of rhythm. A detailed description and interpretation of the study is given elsewhere.[29] However, the data of the study clearly underline the importance for neurobiological investigations into musical rhythm to be consistent with theoretical concepts of rhythm in musicology.

CONCLUSION

Research into the neurobiology of music suggests that music and rhythm can serve as a model of temporality of the human brain.[30,31] We would therefore concur with proposals that music is related to adaptive core functions of the biology of the human nervous system.[32]

A widely distributed cortical and subcortical network subserves the motor, sensory, and cognitive aspects of rhythm processing.[33–35] Consistent engagement of distinct neural circuits in the cerebellum across musical-rhythmic tasks suggests a central role of the cerebellum in the temporal organization of cognitive and perceptual processes in music.[36] However, cerebellar pathology does not affect the capacity of auditory rhythms to entrain rhythmic motor responses.[37] Furthermore, temporal information processing in rhythm appears to follow multiple parallel neural computation processes. Such processes may be coded on a cellular level in the oscillatory timing patterns of synaptic network coupling in the auditory system and subsequently entraining other brain areas via resonant physiological network functions.[38,39]

Lastly, clinical studies have shown striking evidence that auditory rhythm and music can be effectively harnessed for specific therapeutic purposes in the rehabilitation of movement disorders with different neuropathologies.[40] Such findings underscore further the complex ways in which music engages the human brain and in which the brain that engages in music, even in states of brain injury, can be changed by engaging in music.[41,42]

REFERENCES

1. PARNCUTT, R. 1994. A perceptual model of pulse salience and metrical accent in musical rhythm. Music Percept. **11:** 409–464.
2. VOS, P.G. & E.L. HELSPER. 1992. Tracking simple rhythms: on-beat versus off beat performance. *In* Proceedings of the NATO Advanced Research Worshop on Time, Action, and Cognition. F. Macar & V. Pouthas, Eds. :287–299. Kluwer Academic Publishers. Amsterdam.
3. SCHENKER, H. 1935. Der freie Satz. Universal Edition. Vienna.
4. MICHON, J.A. 1967. Timing in Temporal Tracking. Van Gorcum. Assen NL.
5. HARY, D. & P. MOORE. 1987. Synchronizing human movement with an external clock source. Biol. Cybern. **56:** 305–311.
6. THAUT, M.H. & R.A. MILLER. 1994. Multiple synchronization strategies in tracking of rhythmic auditory stimulation. Proc. Soc. Neurosci. 146.11.
7. THAUT, M.H., R.A. MILLER & L.M. SCHAUER. 1998. Multiple synchronization strategies in rhythmic sensorimotor tasks: phase vs period correction. Biol. Cybern. **79:** 241–250.
8. THAUT, M.H. & L.M. SCHAUER. 1997. Weakly coupled oscillators in rhythmic motor synchronization. Proc. Soc. Neurosci. 298.20.
9. RAND, R.H., A.H. COHEN & P.J. HOLMES. 1988. Systems of coupled oscillators as models of central pattern generators. *In* Neural Control of Rhythmic Movements in Vertebrates. A.H. Cohen, S. Rossignol & S. Grillner, Eds. :333–367. Wiley. New York.
10. KENYON G.P., A.E. IRWIN, G.C. MCINTOSH & M.H. THAUT. 2000. Fast motor adaptations to subliminal frequency shifts in auditory rhythm during syncopated sensorimotor synchronization. Proc. Soc. Neurosci. 63.9.
11. THAUT, M.H., B. TIAN & M. AZIMI. 1998. Rhythmic finger tapping to cosine wave modulated metronome sequences: evidence of subliminal entrainment. Hum. Mov. Sci. **17:** 839–863.

12. SALVINO, L.W. & R. CAWLEY. 1994. Smoothness implies determinism: a method to detect it in time series. Phys. Rev. Lett. **73:** 1091–1094.

13. ROBERTS, S., R. EYKHOLT & M.H. THAUT. 2000. Analysis of correlations and search for evidence of deterministic chaos in rhythmic motor control by the human brain. Phys. Rev. E **62:** 2597–2607.

14. MITRA, S., M.A. RILEY & M.T. TURVEY. 1997. Chaos in human rhythmic movement. J. Motor Behav. **29:** 195-198.

15. KELSO, J.A.S. 1995. Dynamic Patterns: The Self-Organization of Human Brain and Behavior. MIT Press. Cambridge, MA.

16. KUGLER, P.N. & M.T. TURVEY. 1989. Information, Natural Law, and the Self-Assembly of Rhythmic Movement: Theoretical and Experimental Investigations. Lawrence Erlbaum. New Jersey.

17. HASAN, M.A. & M.H. THAUT. 1999. Autoregressive moving average modeling for finger tapping with an external stimulus. Percept. Motor Skills **88:** 1331–1346.

18. HAUSDORF, J.M., C.K. PENG, Z. LADIN, *et al.* 1995. Is walking a random walk? Evidence for long-range correlations in stride interval of human gait. J. Appl. Physiol. **78:** 349-358.

19. CHEN, Y., M. DING & J.A.S. KELSO. 2001. Origins of timing errors in human sensorimotor coordination. J. Motor Behav. **33:** 3-8.

20. TECCHIO, F., C. SALUSTRI, M.H. THAUT, *et al.* 2000. Conscious and preconscious adaptation to rhythmic auditory stimuli: a magnetoencephalographic study of human brain responses. Exp. Brain Res. **135:** 222–230.

21. IMADA, T., M. WATANABE, T. MASHIKO, *et al.* 1997. The silent period between sounds has a stronger effect than the interstimulus interval on auditory evoked magnetic fields. Electroencephal. Clin. Neurophysiol. **102:** 37–45.

22. LU, Z.L., S.J. WILLIAMSON & L. KAUFMAN. 1992. Human auditory primary and association cortex have different lifetimes for activation traces. Brain Res. **572:** 236–241.

23. ROSSIGNOL, S. & G. MELVILL JONES. 1976. Audiospinal influences in man studied by the H-reflex and its possible role in rhythmic movement synchronized to sound. Electroencephal. Clin. Neurophysiol. **41:** 83–92.

24. PALTSEV, Y.I. & A.M. ELNER. 1967. Change in the functional state of the segmental apparatus of the spinal cord under the influence of sound stimuli and its role in voluntary movement. Biophysiology **12:** 1219–1226.

25. STEPHAN, K.M., M.H. THAUT, L.G. WUNDERLICH, *et al.* 2002. Conscious and subconscious sensorimotor synchronization: prefrontal cortex and the influence of awareness. Neuroimage **15:** 345–352.

26. STEPHAN, K.M, M.H. THAUT, G. WUNDERLICH, *et al.* 2002. Cortico-cerebellar circuits and temporal adjustments of motor behavior. Proc. Soc. Neurosci. 462.8

27. SANES, J.N., M. DEMARTIN, J. WECKEL & M.H. THAUT. 2001. Brain activation patterns for producing symmetrically and asymmetrically synchronized movement rhythms. Neuroimage **13:** 1249 (abstr.).

28. PARSONS, L.M. & M.H. THAUT. 2001. Functional neuroanatomy of the perception of musical rhythm in musicians and nonmusicians. Neuroimage **13:** 925 (abstr.).

29. PARSONS, L.M. 2001. Exploring the functional neuroanatomy of music performance, perception, and comprehension. Ann. N.Y. Acad. Sci. **930:** 211–231.

30. RAO, S.M., A.R. MAYER & D.L. HARRINGTON. 2001. The evolution of brain activation during temporal processing. Nat. Neurosci. **4:** 317–323.

31. HARRINGTON, D.L. & K.Y. HAALAND. 1999. Neural underpinnings of temporal processing: a review of focal lesion, pharmacological, and functional imaging research. Rev. Neurosci. **10:** 91–116.

32. BERLYNE, D E. 1971. Aesthetics and Psychobiology. Appleton, Century & Croft. New York.

33. PENHUNE, V.B., R.J. ZATORRE & A. EVANS. 1998. Cerebellar contributions to motor timing: a PET study of auditory and visual rhythm reproduction. J. Cognit. Neurosci. **10:** 752–765.

34. PLATEL, H., K. PRICE, J.C. BARON, *et al.* 1997. The structural components of music perception. Brain **120:** 299–243.

35. SCHLAUG, G. 2001. The brain of musicians: a model for functional and structural adaptation. Ann. N.Y. Acad. Sci. **930:** 281–299.
36. SCHMAHMANN, J.D, Ed. 1997. The Cerebellum and Cognition. Academic Press. New York.
37. MOLINARI, M., M.H. THAUT, C. GIOIA, *et al.* 2001. Motor entrainment to auditory rhythm is not affected by cerebellar pathology. Proc. Soc. Neurosci. 950.2.
38. BUONOMANO, D.V. 2000. Decoding temporal information: a model based on short-term synaptic plasticity. J. Neurosci. **20:** 1129–1141.
39. BUONOMANO, D.V. & M.M. MERZENICH. 1995. Temporal information transformed into a spatial code by a neural network with realistic properties. Science **267:** 1028–1030.
40. THAUT, M.H., G.P. KENYON, M.L. SCHAUER & G.C. MCINTOSH. 1999. The connection between rhythmicity and brain function: implications for therapy of movement disorders. IEEE Engin. Med. Biol. **18:** 101–108.
41. PASCUAL-LEONE, A., D. NGUYET, L.G. COHEN, *et al.* 1995. Modulation of muscle responses evoked by transcranial magnetic stimulation during the acquisition of new fine motor skills. J. Neurophysiol. **74:** 1037–1045.
42. THAUT, M.H. & D.A. PETERSEN. 2002. Plasticity of neural representations in auditory memory for rhythmic tempo: trial dependent EEG spectra. Proc. Soc. Neurosci. 373.8.

Effects of Relaxing Music on Salivary Cortisol Level after Psychological Stress

STÉPHANIE KHALFA,[a] SIMONE DALLA BELLA,[b] MATHIEU ROY,[c]
ISABELLE PERETZ,[c] AND SONIA J. LUPIEN[d]

[a]*Laboratory of Neurophysiology and Neuropsychology, INSERM EMI-U 9926,
Université de la Méditerranée, Faculté de Médecine Timone,
13385 Marseille cedex 05, France*

[b]*Department of Psychology, Ohio State University, Columbus, Ohio 43210, USA*

[c]*Laboratory of Music Neuropsychology and Auditory Cognition, University of Montreal,
and Department of Psychology CP 6128 Succursale Centre-Ville Montréal H3C 3J7
Quebec, Canada*

[d]*Laboratory of Human Psychoneuroendocrine Research, Douglas Hospital
Research Center, and McGill University, Department of Psychiatry,
Verdun H4H 1R3 Quebec, Canada*

ABSTRACT: The goal of the present study was to determine whether relaxing music (as compared to silence) might facilitate recovery from a psychologically stressful task. To this aim, changes in salivary cortisol levels were regularly monitored in 24 students before and after the Trier Social Stress Test. The data show that in the presence of music, the salivary cortisol level ceased to increase after the stressor, whereas in silence it continued to increase for 30 minutes.

KEYWORDS: cortisol; stress; music; emotion; relaxation; anxiety

INTRODUCTION

Music is a powerful tool in evoking emotions.[1] Musical experience in everyday life results in a more positive and happier disposition in many individuals.[2] By extension, listening to music can be effective in reducing the negative effects of stress. Supportive evidence can be found in the observation that listening to music resulted in a marked reduction in salivary cortisol levels in patients exposed to presurgical stress.[3] The purpose of this study was to replicate this finding in a nonmedical setting.

Address for correspondence: Dr. Stéphanie Khalfa, Laboratory of Neurophysiology and Neuropsychology, INSERM EMI-U 9926, Université de la Méditerranée, Faculté de Médecine Timone, 27, Bd Jean Moulin, 13385 Marseille cedex 05, France. Voice: (33)-04-91-29-98-15; fax: (33)-04-91-78-99-14.
skhalfa@skhalfa.com

Ann. N.Y. Acad. Sci. 999: 374–376 (2003). © 2003 New York Academy of Sciences.
doi: 10.1196/annals.1284.045

METHODS

Twenty-four francophone male university students were evaluated before and after the Trier Social Stress Test (TSST).[3] The test is a validated stress task known to induce significant increases in cortisol levels.[4] It consists of a speaking task and mental calculations performed in front of an audience. The sampling of salivary cortisol was undertaken 20 and 30 minutes after the subject's arrival (in the afternoon) and served as a baseline. Next, subjects prepared their speech for 10 minutes (anticipation phase). At the end of this phase and of the TSST, two other samples were collected. Afterwards, a sample was taken every 15 minutes until the end of the 45-minute stress recovery period (FIG. 1). During this time, the students were comfortably seated and were asked to relax in silence or with a musical tape being played. The tape was a concatenation of 10 relaxing excerpts from Enya, Vangelis, and Yanni and was delivered via loudspeakers. The salivary cortisol response to the stress was absent in seven subjects; they were eliminated from the analyses. The silence group was comprised of eight subjects (mean age 23.8 years ± 3.3), and the music group was comprised of nine subjects (mean age 24 years ± 4.2). The experiment was conducted with the full understanding and consent of each participant.

RESULTS

The variations in mean salivary cortisol levels for each group are shown in FIG-URE 1. There were no significant differences in cortisol levels between the silence

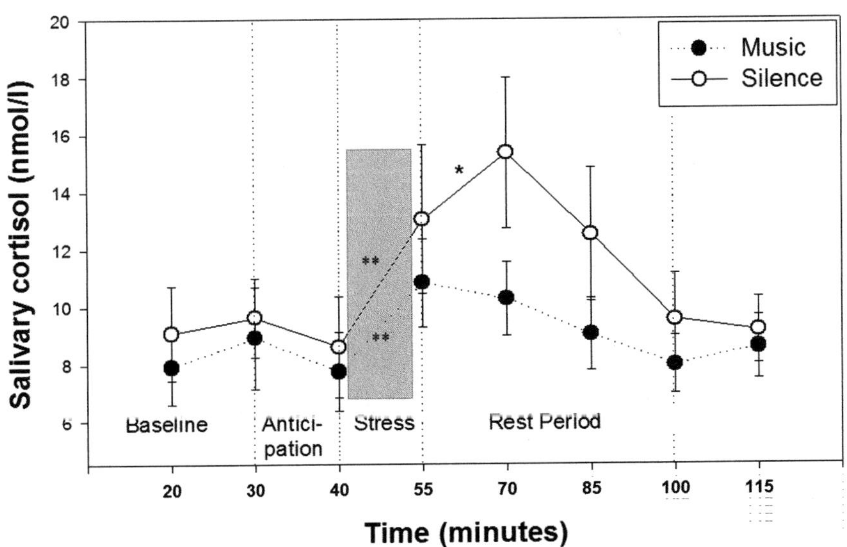

FIGURE 1. Means and standard errors of salivary cortisol concentration from 20 to 115 minutes after subjects' arrival; **P <0.01; *P <0.05.

and music groups during the baseline period. Also, both groups displayed a significant increase in salivary cortisol levels in response to the TSST (silence: t(8) = 3.3; P <0.01; music: t(7) = 8.2; P <0.001), and there was no group effect with regard to the magnitude of the cortisol response to the stressor. However, after the stressor and in the first part of the stress recovery period (time 70 of the protocol), salivary cortisol levels continued to increase in the silence group (t(8) = 2.3; P <05), whereas they did not in the music group (P >0.1). Following the rest period, salivary cortisol levels returned to baseline levels in both groups.

DISCUSSION

The psychological stressor provoked a strong emotion that was revealed by a sharp increase in cortisol levels within 15 minutes. The concentration of cortisol decreased more rapidly in the saliva of the subjects exposed to music than in the group recovering from stress in silence, suggesting that relaxing music after a stressor can act by decreasing the poststress response of the hypothalamic-pituitary-adrenal axis. This result replicates previous findings,[4,5] indicating that relaxing music is more effective than silence in decreasing cortisol levels after stress induction. Further studies are necessary to determine if the effect of relaxing music on the stress recovery period is specific to music or can be obtained with different stimuli.

ACKNOWLEDGMENTS

This research was supported by fellowships from the Canadian Institutes of Health Research and from the education government of Quebec (to S.K.) and the REPAR section of the Fonds de la Recherche en Santé du Québec (to M.R.) and by a grant from the Natural Science and Engineering Research Council (to I.P.).

REFERENCES

1. GOLDSTEIN, A. 1980. Thrills in response to music and other stimuli. Physiol. Psychol. **8:** 126–129.
2. SLOBODA, J.A. & S.A. O'NEILL. 2001. Emotions in everyday listening to music. *In* Music and Emotions. Theory and Research. P.N. Juslin & J.A. Sloboda, Eds. :415–429. Oxford University Press. Cambridge.
3. MILUK-KOLASA, B., Z. OBMINSKI, R. STUPNICKI & L. GOLEC. 1994. Effects of music treatment on salivary cortisol in patients exposed to pre-surgical stress. Exp. Clin. Endocrinol. **102:** 118–120.
4. KIRSCHBAUM, C. & D.H. HELLHAMMER. 1993. The "Trier Social Stress Test": a tool for investigating psychobiology stress responses in a laboratory setting. Neuropsychobiology **28:** 76–81.
5. KIRSCHBAUM, C. & D.H. HELLHAMMER. 1994. Salivary cortisol in psychoneuroendocrine research: recent developments and applications. Psychoneuroendocrinology **19:** 313–333.

Effect of Unilateral Temporal Lobe Resection on Short-Term Memory for Auditory Object and Sound Location

CÉLINE LANCELOT,[a,b] SÉVERINE SAMSON,[a,b] PIERRE AHAD,[c] AND MICHEL BAULAC[b]

[a]University of Lille 3, URECA, Villeneuve d'Ascq Cedex, France

[b]Epilepsy Unit, Salpêtrière Hospital, Paris, France

[c]McGill University, Department of Psychology, Montreal, Canada

ABSTRACT: To investigate auditory spatial and nonspatial short-term memory, a sound location discrimination task and an auditory object discrimination task were used in patients with medial temporal lobe resection. The results showed a double dissociation between the side of the medial temporal lobe lesion and the nature of the auditory discrimination deficits, suggesting that right and left temporal lobe structures are differently involved in auditory spatial and nonspatial short-term memory.

KEYWORDS: short-term memory; auditory object; sound location; temporal lobe

INTRODUCTION

Auditory short-term memory (STM) is necessary to process sound stimulus in order to regulate behavior accordingly. As already demonstrated in vision, differentiation between object quality ("what" pathway) and object location ("where" pathway) seems to exist in audition as well, even though relatively few studies have examined this issue. These two types of processing seem to depend on distinct memory mechanisms, as suggested in experimental[1,2] and brain-imaging studies.[3,4] In looking for neural substrate involved in the maintenance of nonverbal information in auditory STM, several investigators clearly outlined the importance of the right anterior temporal lobe structures when melodic information was used.[5,6] By contrast, a widely distributed network including temporal, parietal, and frontal lobe structures bilaterally seems to be involved in STM for auditory spatial information.[3,4,7] The goal of the present study was to assess the role of temporal-lobe struc-

Address for correspondence: Céline Lancelot or Séverine Samson, URECA, Université de Lille 3 Domaine universitaire du Pont de Bois, BP 149, 59653 Villeneuve d'Ascq Cedex, France. Voice: +33 3 20 41 64 43; fax: +33 3 20 41 63 24.
celinelancelot2@aol.com

Ann. N.Y. Acad. Sci. 999: 377–380 (2003). © 2003 New York Academy of Sciences.
doi: 10.1196/annals.1284.046

tures in auditory STM by directly contrasting auditory STM for spatial and non-spatial information within the same group of patients who had undergone a unilateral medial temporal lobe excision. For this purpose, we used bird songs as stimuli and hypothesized that patients with a unilateral right temporal lobe lesion will be impaired in nonspatial STM involving bird songs, whereas patients with either right or left temporal lobe excision will be disturbed STM for sound location.

METHODS

Participants. Patients who had undergone right (RT, $n = 9$) or left (LT, $n = 10$) temporal lobe resection for the relief of intractable seizures were tested. Resection consisted of unilateral medial temporal lobe removal including the amygdala, hippocampus, and various amounts of surrounding cortices (entorhinal, perirhinal, and parahippocampal). Twelve normal control subjects (NCs) who were matched to the patient groups with respect to age, sex, and education were also tested.

Procedure. The participants, who were blindfolded, sat in a chair positioned in the center of a horizontal semicircular array, with four loudspeakers on the left side ($-15°$, $-30°$, $-45°$, $-60°$) and four on the right ($15°$, $30°$, $45°$, $60°$). Each participant was tested using two discrimination tasks. In the *auditory object discrimination task,* the participant had to decide if the two bird songs were identical or different. Within a single trial, the stimuli came from the same loudspeaker (randomly chosen among the eight loudspeakers). In the *sound location discrimination task,* the participant had to determine whether the two locations were identical or different. Within a single trial, the exact same bird song was presented twice from either same or different loudspeakers. In different trials, the two loudspeakers were separated by an angle of 30 degrees within the same hemispace. For each discrimination task, a condition with a 4-second interstimulus interval (ISI) was used. The patients were also tested in a similar task, using a 1-second ISI.

RESULTS

An analysis of variance was carried out on the percentage of correct detection in the 4-second ISI condition with Task as the within factor and Group (NC, RT, LT) as the between factor. An effect of Task [$F(1, 28) = 4,46$; $P < 0.05$] as well as a Task by Group interaction [$F(2, 28) = 4,09$; $P < 0.05$] were found (FIG. 1). Posthoc statistical comparisons (Newman-Keuls) demonstrated that auditory object discrimination was disrupted in the RT group relative to the NC ($P < 0.05$) and LT ($P < 0.05$) groups. The LT group did not differ significantly from the NC group ($P > 0.05$). Conversely, sound location discrimination was disrupted in the LT group relative to the NC group ($P < 0.05$), whereas the RT group did not differ from the NC ($P > 0.05$) and LT groups ($P > 0.05$). To verify that all participants could discriminate auditory object and sound location with only 1-second delay, another ANOVA was carried out. The results showed a significant effect of Task [$F(1, 28) = 17,8$; $P < 0.05$], indicating that performance was lower in the sound location than in the auditory object discrimination task, but there was no effect of Group ($P > 0.05$) or Task by Group interaction ($P > 0.05$).

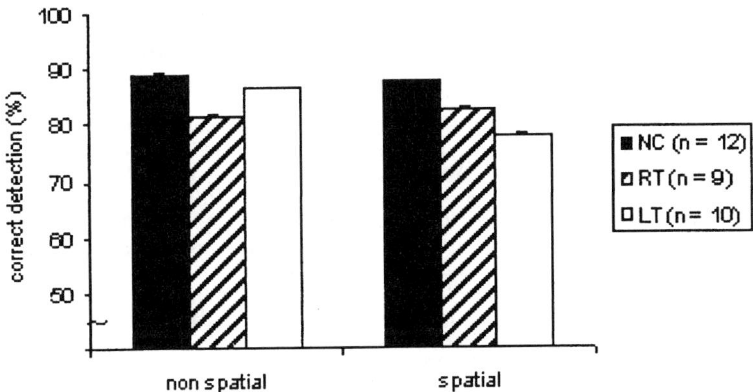

FIGURE 1. Mean percent of correct detection in spatial and nonspatial short-term memory for each group of subjects (LT, left temporal lobe lesion; RT, right temporal lobe lesion; NC, normal control). *Bars* indicate 1 standard error of the mean.

DISCUSSION

The main finding of this study suggests that auditory STM for spatial and nonspatial information can be differently affected by right or left temporal lobe lesions, revealing a double dissociation between the side of the temporal lobe lesion and the nature of the auditory STM deficits. This result confirms the functional dissociation between STM for auditory object and sound location reported in studies with normal subjects.[1,2] According to our predictions, patients with right but not left medial temporal lobe lesions were impaired in auditory object discrimination, suggesting the predominant role of the right temporal lobe in STM of bird songs. This result is consistent with previous works which suggest the contribution of right temporal structures in STM of tonal information.[5,6] Conversely, sound location discrimination was only affected by left but not right anterior temporal lobe excision, indicating that left temporal lobe structures are particularly important in auditory STM for spatial information. Such a cerebral asymmetry might be attributed to the role of left hemisphere function in categorical judgment[8] that might have been reinforced by the use of a limited number of loudspeakers.

CONCLUSION

This study demonstrates a double dissociation between the side of the lesion and the auditory discrimination deficit, suggesting the role of the left and the right medial temporal lobe structures in auditory spatial and nonspatial short-term memory, respectively.

REFERENCE

1. CLARKE, S., M. ADRIANI & A. BELLMANN. 1998. Distinct short-term memory systems for sound content and sound localization. NeuroReport **9:** 3433–3437.
2. ANOUROVA, I., P. RAMA, K. ALHO, *et al.* 1999. Selective interference reveals dissociation between auditory memory for location and pitch. NeuroReport **10:** 3543–3547.
3. ALAIN, C., S.R. ARNOTT, S. HEVENOR, *et al.* 2001. "What" and "where" in the human auditory system. Proc. Natl. Acad. Sci. USA **98:** 12301–12306.
4. MAEDER, P.P., R.A. MEULI, M. ADRIANI, *et al.* 2001. Distinct pathways involved in sound recognition and localization: a human fMRI study. NeuroImage **14:** 802–816.
5. SAMSON, S. & R.J. ZATORRE. 1988. Melodic and harmonic discrimination following unilateral cerebral excision. Brain Cognit. **7:** 348–360.
6. ZATORRE, R.J. & S. SAMSON. 1991. Role of the right temporal neocortex in retention of pitch in auditory short-term memory. Brain **114:** 2403–2417.
7. MARTINKAUPPI, S., P. RAMA, H.J. ARONEN, *et al.* 2000. Working memory of auditory localization. Cereb. Cortex **10:** 889–898.
8. KOSSLYN, S.M., O. KOENIG, A. BARRETT, *et al.* 1989. Evidence for two types of spatial representations: hemispheric specialization for categorical and coordinate relations. J. Exp. Psychol. Hum. Percept. Perform. **15:** 723–735.

Implicit and Explicit Emotional Memory for Melodies in Alzheimer's Disease and Depression

NOLWENN QUONIAM,[a] ANNE-MARIE ERGIS,[a,b] PHILIPPE FOSSATI,[b,f]
ISABELLE PERETZ,[c] SÉVERINE SAMSON,[d] MARIE SARAZIN,[e] AND
JEAN-FRANÇOIS ALLILAIRE[b,f]

[a]*Laboratoire de Psychologie Clinique et de Psychopathologie, Institut de Psychologie, Université Paris 5, Paris, France*

[b]*Service de Psychiatrie adulte, Groupe hospitalier Pitié-Salpêtrière, Paris, France*

[c]*Département de Psychologie, Université de Montréal, Montréal, Canada*

[d]*Laboratoire URECA, UFR de Psychologie, Université de Lille 3, Villeneuve d'Ascq Cedex, France*

[e]*Service de Gériatrie, Hôpital Bretonneau, Paris, France*

[f]*CNRS, UMR 7593, Paris, France*

ABSTRACT: The present study investigates the impact of emotional deficits on implicit and explicit memory for musical stimuli in patients with Alzheimer's disease and elderly depressed patients. Results showed that unlike Alzheimer's patients, depressed patients were unable to develop a positive affective bias of judgment for previously heard melodies.

KEYWORDS: implicit memory; explicit memory; emotions; exposure effect; Alzheimer's disease; depression.

INTRODUCTION

Patients with Alzheimer's disease (AD) and patients with depression present memory and emotional disorders that are different in nature. All aspects of explicit memory are severely impaired in Alzheimer's disease, whereas implicit memory (which does not require conscious recollection of previously presented information) is spared in many situations (for a review, see Ref. 1). Depressed patients show impaired performance in explicit memory tasks that involve strategic retrieval processes, whereas implicit memory is usually preserved.[2,3] Regarding emotional disorders, AD patients present less depressive symptoms than depressed patients, but they show more apathy.[4] Few studies have examined the interactions between emo-

Address for correspondence: Nolwenn Quoniam, Laboratoire de Psychologie Clinique et de Psychopathologie, Institut de Psychologie, Université René Descartes Paris 5, 71 avenue Edouard Vaillant, 92100 Boulogne-Billancourt, France. Voice: 33 1 55 20 59 53.
nolwenn.quoniam@free.fr

Ann. N.Y. Acad. Sci. 999: 381–384 (2003). © 2003 New York Academy of Sciences.
doi: 10.1196/annals.1284.047

TABLE 1. Demographic and clinical characteristics for the three groups

Groups	Controls ($n = 16$)	Alzheimer patients ($n = 10$)	Depressed patients ($n = 9$)
Age	75.75 (4.04)	79.20 (1.82)	75.20 (4.35)
Education	9.87 (2.36)	11.10 (3.47)	11.11 (3.51)

TABLE 2. Scores on emotional scales in the three subjects groups

	Controls	Depressed patients	Alzheimer patients	Group comparisons
AES[a]	72.50 (9.86)	44.11 (7.67)	32.30 (7.73)	MA > D > T
GDS[b]	3.18 (2.19)	20.88 (3.40)	5.80 (2.48)	D > MA > T
EHD[c]	2.25 (3.33)	46.55 (14.22)	14.10 (9.70)	D > MA > T
EBRS[d]	0.06 (0.25)	19.44 (5.93)	3.50 (3.02)	D > MA > T

[a] Apathy Evaluation Scale.
[b] Geriatric Depressive Scale.
[c] Echelle d'humeur dépressive.
[d] Emotional Blunting Rating Scale.

tional disorders and memory in AD and depression. In general, implicit and explicit memory for both positive and negative stimuli (words and pictures) are enhanced in AD.[5,6] On the other hand, most studies conducted with depressed patients revealed enhanced performance in explicit memory for words with negative valence, and no effects of emotional valence on implicit tasks. These data suggest impaired explicit processing of positive emotional stimuli in depression.[7]

To assess memory as it relates to emotions, we explored the mere exposure effect. According to Zajonc,[8] the "mere exposure effect" is the increase in positive affect that results from the repeated presentation of previously unfamiliar stimuli. Few studies have examined exposure effects in AD and none in depression. The present study is the first to investigate exposure effects of music on a preference task (implicit memory) and on recognition task (explicit memory) in both AD and depression.

MATERIAL AND METHODS

Ten AD patients, 9 depressed patients, and 16 control subjects, all matched for age and education, were included in the present study (TABLE 1). Exposure effects for unfamiliar melodies were explored with a task elaborated by Peretz and Gaudreau.[9] During the study phase, subjects listened to novel but conventional melodies, which were randomly presented 1, 5, or 10 times. During the preference task, subjects were asked to rate how much they liked each melody on a 10-point scale. A priming effect was evidenced when subjects preferred old melodies to new ones. During the recognition task, subjects were asked to recognize previously studied melodies among "new" ones.

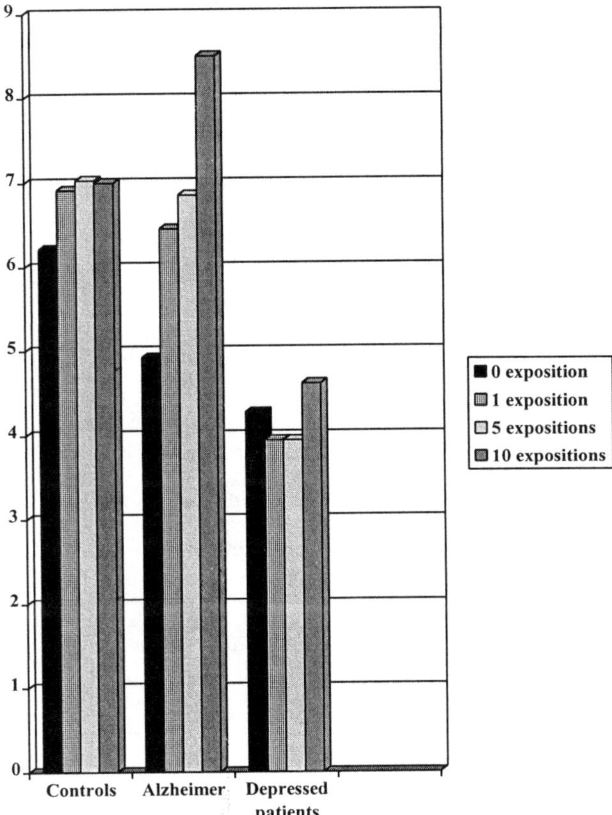

FIGURE 1. Effects of exposure in the three groups.

RESULTS AND DISCUSSION

As predicted, AD patients presented significantly less negative symptoms but more apathy than did depressed patients (TABLE 2). More importantly, a double dissociation emerged between the two groups: AD patients showed normal priming effects on preference judgments, whereas depressed patients showed a drastic impairment (the ANOVA revealed a group x condition interaction, with $F(6, 96) = 5,78$; $P < 0.0001$) (FIG. 1); recognition was severely impaired in AD, but was unaffected in depressed patients [$F(2, 32) = 24,64$; $P < 0.0001$]. Unlike Alzheimer patients, depressed patients were unable to develop a positive affective bias of judgment for previously heard melodies, suggesting that impaired emotional processing of positive stimuli in depression may affect processes functioning at an automatic level.

REFERENCES

1. CARLESIMO, G.A. & M. OSCAR-BERMAN. 1992. Memory deficits in Alzheimer's patients: a comprehensive review. Neuropsychol. Rev. **3:** 119–169.
2. NIEDEREHE, G. 1986. Depression and memory impairment in the aged. *In* Clinical Memory Assessment. L.W.L Poon, Ed. :226–237. A.P.A. Washington, DC.
3. NIEDEREHE, G. & C. YODER. 1989. Metamemory perceptions in depressions of young and older adults. J. Nerv. Ment. Dis. **177:** 4–14.
4. BUNGENER, C. *et al.* 1996. Affective disturbances in Alzheimer's disease. J. Am. Geriatr. Soc. **44:** 1066–1071.
5. HAMANN, S.B. *et al.* 2000. Memory enhancement for emotional stimuli is impaired in early Alzheimer's disease. Neuropsychology **14:** 82–92.
6. PADOVAN, C. *et al.* 2002. Evidence for a selective deficit in automatic activation of positive information in patients with Alzheimer's disease in an affective priming paradigm. Neuropsychologia **40:** 335–339.
7. WATKINS, P.C. *et al.* 1996. Unconscious mood-congruent memory bias in depression. J. Abnorm. Psychol. **105:** 34–41.
8. ZAJONC, R.B. 1968. Attitudinal effects of mere exposure. J. Pers. Soc. Psychol. **9:** 1–28.
9. PERETZ, I. *et al.* 1998. Exposure effects on music preference and recognition. Mem. Cognit. **26:** 884–902.
10. BRINK, T. L. *et al.* 1983. Development and validation of a geriatric depression screening scale. J. Psychiatr. Res. **17:** 37–49.
11. JOUVENT, R. *et al.* 1998. La clinique polydimensionnelle de l'humeur dépressive. Psychiatr. Psychobiol. **3:** 245–253.
12. ABRAMS, R. & M.A. TAYLOR. 1978. A rating scale for emotional blunting. A.P.A. **135:** 226–229.
13. MARIN, R.S. *et al.* 1991. Reliability and validity of the Apathy Evaluation Scale. Psychiatr. Res. **38:** 143–162.

Musicians Differ from Nonmusicians in Brain Activation despite Performance Matching

NADINE GAAB AND GOTTFRIED SCHLAUG

Department of Neurology, Music and Neuroimaging Laboratory,
Beth Israel Deaconess Medical Center and
Harvard Medical School,
Boston, Massachusetts 02215, USA

ABSTRACT: Brain activation patterns in a group of musicians and a group of nonmusicians (matched in performance score to the musician group) were compared during a pitch memory task using a sparse-temporal sampling functional magnetic resonance imaging experiment. Both groups showed bilateral activaton (left more than right) of the superior temporal gyrus, supramarginal gyrus, posterior middle and inferior frontal gyrus, and superior parietal lobe. Musicians showed greater right posterior temporal and supramarginal activation, whereas nonmusicians had greater activation of the left secondary auditory cortex.

KEYWORDS: musicians; pitch memory; fMRI; planum temporale; performance

INTRODUCTION

Functional brain differences between musicians and nonmusicians have been found mainly in perisylvian brain regions using various perceptual tasks including just listening to music or performing pitch, harmony, melody, or rhythm tasks.[1–5] These studies speculated that musicians and nonmusicians might process music in a different way: increased musical sophistication would be associated with more lateralized (mostly left) activation. However, it is unclear if group differences in the actual performance of these tasks (e.g., percentage of correct answers), in cognitive strategies, in specialized musical abilities (e.g., absolute pitch), or even in anatomic differences between musicians and nonmusicians account for these between-group functional differences.

Address for correspondence: Gottfried Schlaug, M.D., Ph.D., Department of Neurology, Music and Neuroimaging Laboratory, Beth Israel Deaconess Medical Center and Harvard Medical School, 330 Brookline Avenue, Boston, MA 02215. Voice: 617-632-8912; fax: 617-632-8912.
gschlaug@bidmc.harvard.edu

Ann. N.Y. Acad. Sci. 999: 385–388 (2003). © 2003 New York Academy of Sciences.
doi: 10.1196/annals.1284.048

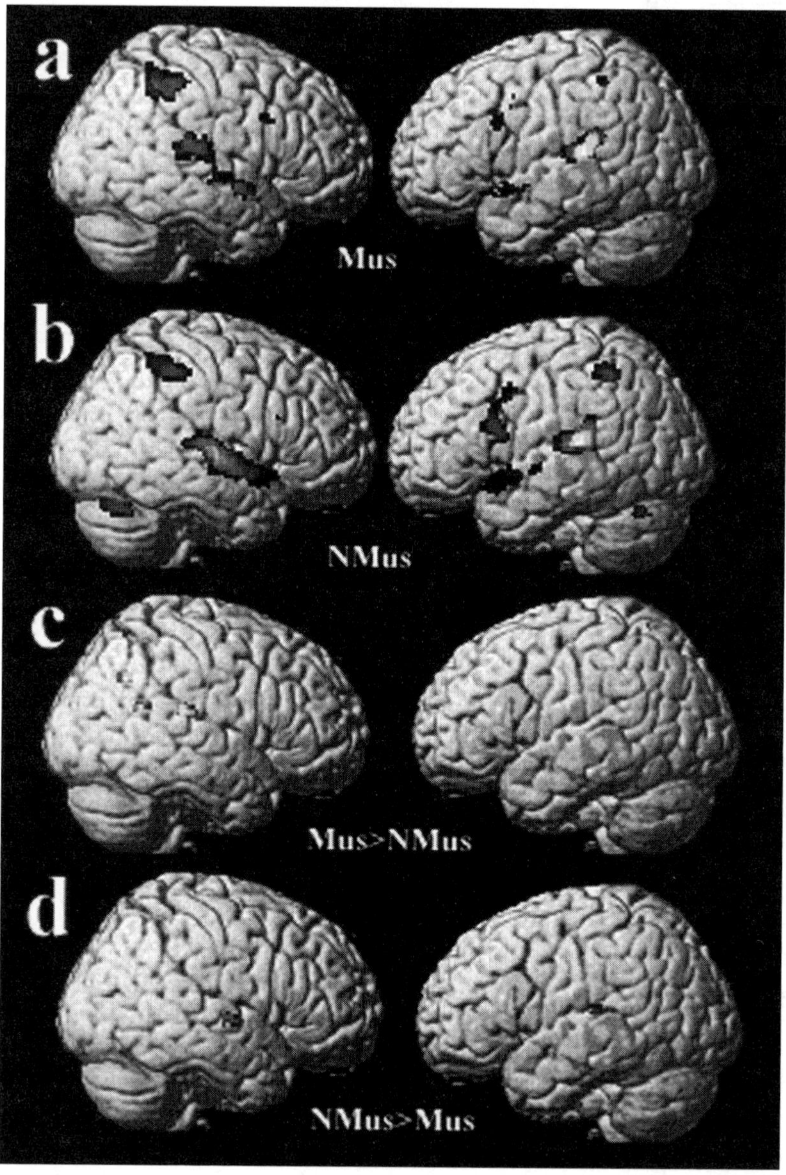

FIGURE 1. Group mean activation maps for musician (**a**) and nonmusician (**b**) groups (P <0.05, corrected for multiple comparisons); (**c**) the contrast "Musicians>Nonmusicians;" (**d**) the contrast Nonmusicians>Musicians. All activation maps are only for MR acquisitions obtained at 0–3 seconds after the end of auditory stimulation. Later time points (MR acquisitions obtained at 4–6 seconds after the end of auditory stimulation) are not shown.

METHODS

Ten right-handed musicians and 10 right-handed nonmusicians (age range 18–40 years; matched for age and gender) participated in this study. Subjects had to listen to a sequence of tones (6–7 tones with a total duration of 4.6 seconds) and were asked to decide if the last or second to the last tone (depending on a visual prompt) was the "same" as or "different" from the first tone, indicating the answer by a button press response. This task was contrasted with a motor control condition in which subjects pressed a right or left button, depending on a visual prompt. The non-musicians were matched with the musician group in their performance score (% correct responses) in the pitch memory task. For this experiment, we defined musicians as those who had a formal music education and played a musical instrument regularly. A nonmusician was defined as someone who had never played a musical instrument and did not have a formal musical education.

MR Data Acquisition and Image Analysis. Functional magnetic resonance imaging (fMRI) was performed on a Siemens Vision 1.5 Tesla whole-body MR scanner. To avoid interference with the MR scanner noise as well as auditory masking effects, a sparse temporal sampling fMRI method with a long repetition time (TR) of 17 seconds was used. This ensured that the clustered volume MR acquisition (over 2.75 seconds) was always separated from the actual auditory task. In addition, the stimulus-to-imaging delay time was varied between 0 and 6 seconds in a jitter-like fashion to explore the time course of brain activation in response to the perceptual and cognitive demands of this pitch memory task. FMRI data were analyzed using the SPM99 software package (Institute of Neurology, London, UK). Planum temporale measurements were done as previously described.[6]

RESULTS

The musician group had a mean correct response rate of 78% (SD = 6), whereas the mean of the nonmusician group was 76% (SD = 6) (P >0.05) after individually matching nonmusicians with musicians depending on task performance. Group mean activation images for musicians and nonmusicians showed involvement of the superior temporal gyrus, supramarginal gyrus, posterior middle and inferior frontal gyrus, and superior parietal lobe bilaterally in the pitch memory task (FIG. 1a and b). Contrasting both groups directly with each other for time points 0–3 (MR scan acquired 0–3 seconds after the end of auditory stimulation) revealed that musicians had more activation of the posterior planum temporale as well as the supramarginal gyrus on the right and the superior parietal lobe bilaterally compared to the non-musicians (FIG. 1c). Nonmusicians differed from musicians by greater activation of a region in the anterior part of the planum temporale (immediately posterior to the HG) on the left (FIG. 1d). Analysis for later imaging time points (MR scans acquired 4–6 seconds after the end of auditory stimulation) revealed greater activation in the right inferior parietal lobe in musicians than in nonmusicians. Planum temporale measurements did not indicate significant differences in hemispheric asymmetry between both groups.

CONCLUSIONS

We interpret our results as indicating perceptual and cognitive processing differences between musicians and nonmusicians, because performance scores and measures of anatomic hemispheric asymmetry were similar between both groups. Although musicians seemed to use more short-term auditory storage centers (e.g., supramarginal gyrus), nonmusicians relied more on early perceptual brain regions (e.g., primary and early secondary auditory areas) within the superior temporal lobe to solve the pitch memory task with similar performance.

REFERENCES

1. MAZZIOTTA, J.C., M.E. PHELPS, R.E. CARSON & D.E. KUHL. 1982. Tomographic mapping of human cerebral metabolism: auditory stimulation. Neurology **32:** 921–937.
2. ALTENMUELLER, E. 1986. Hirnelektrische Korrelate der zerebralen Musikverarbeitung beim Menschen. Eur. Arch. Psychiatr. Neurol. Sci. **235:** 342–354.
3. BESSON, M., F. FAITA & J. REQUIN. 1994. Brain waves associated with musical incongruities differ for musicians and non-musicians. Neurosci. Lett. **168:** 101–105.
4. PANTEV, C., R. OOSTENVELD, A. ENGELIEN, *et al.* 1998. Increased auditory cortical representation in musicians. Nature **392:** 811–814.
5. OHNISHI, T., H. MATSUDA, T. ASADA, *et al.* 2001. Functional anatomy of musical perception in musicians. Cereb. Cortex **11:** 754–760.
6. KEENAN, J., V.T HANGARAJ, A. HALPERN & G. SCHLAUG. 2001. Absolute pitch and planum temporale. NeuroImage **14:**1402–1408.

Part III: Music and Development

Introduction

JOHN SLOBODA

Department of Psychology, Keele University, Newcastle, Staffs ST5 5BG, UK

Several "big questions" are addressed by the contributions to this section. These include: What mechanisms for processing and understanding music exist in very young humans? How does musical training and experience change musical behavior and its underlying neural substrate? Can we distinguish which age-related changes in musical capacity are a result of developmental maturation, passively experienced acculturation, and/or active engagement in structured musical training regimens? Does musical training or engagement have effects on other nonmusical aspects of skill and development? Are there significant individual differences in musical capacity between individuals and if so, what causes them? And finally, what reliable and effective research methods are available for systematically investigating these questions?

The contributions to this section demonstrate that many of these questions are now at the center of productive research agendas. Less than 25 years ago, publications with titles such as "Music and the Brain" were filled with largely anecdotal, speculative, and exploratory studies in which developmental issues hardly figured. Developmental Music Science, of which this section contains some of the best exemplars, now consists of detailed, mutually reinforcing, and self-correcting investigatory programs in major science laboratories around the world.

I can identify several reasons for the recent consolidation of this area of research. First, music scientists have adapted and developed a range of effective behavioral methods for investigating early cognitive function. These are largely "borrowed" from broader developmental psychology and include "head turning" methods for probing infant cognition. Second, music scientists have been proactive at forging collaborations with medical researchers who have access to fMRI and other advanced technologies for investigating brain function. Third, educators and funders of education have increasingly been asking for scientific evidence to justify the place of music in the school curriculum for both itself and its potential benefit to general academic and personal development. And fourth, increasing concern with the failure of many young people to acquire significant music performance skills has prompted fundamental questions about the precursors of high musical achievement.

Several of this section's contributions focus on the devlopment of pitch processing and especially the relation between absolute and relative processing modes. Contrary to "received" opinion, the evidence marshalled here (particularly in the contributions of Saffran and Trehub) suggests that humans of all ages have the

Address for correspondence: Professor John Sloboda, Department of Psychology, Keele University, Newcastle, Staffs ST5 5BG, UK.

Ann. N.Y. Acad. Sci. 999: 389–391 (2003). © 2003 New York Academy of Sciences.
doi: 10.1196/annals.1284.057

capacity to process pitch materials in both absolute and relative ways and that both skills can be developed by appropriate training at any age.

These and several other contributions to this section stress the similarities between infant and adult cognition and the contributions of specific musical experience (rather than general developmental factors) to differences in performance between groups or individuals. The contributions of Krumhansl, Drake, Pantev, Besson, and Trainor all highlight the role of musical experience in shaping behavior and brain characteristics. What is particularly exciting about these studies is the integration of behavioral and neuropsychological data in the hands of scientists who are becoming increasingly expert in handling both types of data.

On the other hand, some authors (e.g., Costa-Giomi, Oerter, and Gruhn) review evidence which suggests that experience is not the only causal factor determining musical outcomes. Costa-Giomi provides evidence that a training program in harmonic awareness was not effective in children under the age of 8 years. She suggests that children below this age do not have the cognitive capacity to integrate attention to harmony with melodic and rhythmic information, because these more salient aspects of music capture their attention and divert their processing capacity.

However, such developmental blocks cannot be absolute, because there are numerous reliable reports of individual children handling harmonic information at a significantly younger age than 8. Some renowned composers were handling harmony in simple compositions considerably before that age. The "nature-nurture" debate still rages with respect to individual differences. However, it appears hard for science to provide decisive tests of the relative contribution of "innate musical talent," "innate general cognitive capacity," motivation, and learning to a final musical outcome.

Educators may well find the detail of some of these debates too far removed from their concerns, because the task in front of them is to find ways of developing whatever capacity exists in the students they work with. The emphasis on the cognitive aspects of musical development within this volume may also underrepresent the crucial importance of social, motivational, and emotional factors in underpinning and sustaining musical activity. For instance, Gruhn surmises that "drop out" from music programs may arise from frustration caused by reaching capacity limits. Social development research is beginning to suggest that mismatches between the cultural and social values embodied in music education programs and the cultural and social priorities of the young person may account for far more dramatic levels of frustration and disengagement than arrival at individual capacity limits.

The final major issue elucidated in this section is the effect of music engagement on other cognitive skills. Overy's paper is a clear contribution to the growing body of knowledge that demonstrates significant "transfer effects" between musical activity and nonmusical cognition. This study is particularly interesting in that it shows specific effects on some, but not all, aspects of language performance in dyslexic children. This is a clear demonstration that music's effects are not the result of some general overall "motivation" or "priming" effect, but are a consequence of specific overlaps between cognitive components of music processing and related components of nonmusical tasks.

Where should developmental music sciences be heading next? My hope is that the *Annals* of the proceedings of a future conference in this series would be able to report significantly more research into developmental aspects of emotional, esthetic, and creative aspects of musical development. Allied with this is a hope that researchers will increasing find ways of addressing these issues through meaningful and complete musical objects or acts. Most of us engage music because of its inherent and inescapable beauty and value of its best products. The "exemplar" stimulus of most music cognition experiments is still the disembodied brief tone sequence to which respondents make a discrete time-limited response (e.g., yes/no). This may be methodologically convenient, but it is experientially and thus ultimately scientifically impoverished. Finally, I would hope for more strategic alliances to be built between music educators (particularly general educators in the school system) and researchers. There is still too much of a sense that research is driven by idiosyncratic and largely uncoordinated impulses of individual researchers, largely independent of, and tangential to, the concerns of the music education community and those who fund it.

Introductory Remarks on Musical Beginnings: Ten Years Later

CARMINE FAIENZA AND GIUSEPPE COSSU

University of Parma, Parma, Italy

ABSTRACT: Over the last 10 years, the neurosciences have witnessed an exponential increase of research in the neuropsychology of music. Consequently, old and new questions on the ontogeny of musical skills can be refined and reformulated with greater accuracy. Three interrelated issues are discussed.

KEYWORDS: music; language; phylogenesis; ontogeny; infants

PREMISE

Ten years ago, in October 1992, we organized a meeting in Parma on "Music, Speech and the Developing Brain." It is worth a reminder that the subtitle of the meeting dubbed the emergence of musical beginnings as a case for "the modularity of mind." Indeed, a shared concept among most of the invited speakers was that musical skills are deeply embedded in the biology of our species and are governed by a highly specific neural architecture.

Ten years later, the Fondazione Mariani offered us a privileged opportunity to introduce, in this international meeting, a session devoted to the neurobiology of musical beginnings. By viewing the field in retrospect, we realize that over the past 10 years, the neurosciences have witnessed an exponential increase of research in the neuropsychology of music. As a consequence, old and new questions on the ontogeny of musical skills can be refined and formulated with greater accuracy.

In our introductory remarks, we would like to address three interrelated issues. The first issue deals with the links between music and biology from a phylogenetic perspective. Although this session is devoted to the ontogeny of music, the theoretical assumption of a biologically specific system devoted to the emergence of musical skills requires that some words be devoted to clarifying in which sense we can reasonably assume a biological predisposition for music in infancy. The second issue is concerned with the links between the emergence of musical and linguistic competencies. The third (and final) issue is concerned with the neurophysiological evidence for a dedicated neural system underlying the acquisition of musical skills.

Address for correspondence: Dr. Carmine Faienza, University of Parma, Parma, Italy.

Ann. N.Y. Acad. Sci. 999: 392–396 (2003). © 2003 New York Academy of Sciences.
doi: 10.1196/annals.1284.049

A PHYLOGENETIC PERSPECTIVE

It is now well documented that infants' responsiveness to maternal music is stable and protracted in time.[1,2] Indeed, infants seem to be mesmerized by their mother's singing.[3] But where does infants' fascination for music come from? Ultimately, enjoying a lullaby does require some cognitive skills, but infants are unlikely to generate their musical competence overnight. Hence, the assumption of musical predispositions in infants implies a prenatal biological basis for music.[3] In this volume, Sandra Trehub addresses such an issue. She outlines those features of the auditory system that are shared by humans with many other species, ranging from monkeys to fish (as documented by Large and Crawford[4]). Nonetheless, the claim of a dedicated neural system for musical cognition, which implies a phylogeny of music, is likely to be received with skepticism within many scientific circles. Pinker,[5] for instance, maintains the neuropsychological irrelevance of music and claims that "as far as biological cause and effects are concerned, music is useless."

Yet, it was Darwin[6] who, in *The Descent of Man,* speculated that the emergence of language *might have been fostered* by preexisting musical abilities in the human species. Darwin reasoned that "we have every reason to suppose that articulate speech is one of the latest of the arts acquired by man and as the instinctive power of producing musical notes and rhythms is developed low down in the animal series, it would be altogether opposed to the principle of evolution, if we were to admit that man's musical capacity has been developed from the tones used in impassioned speech. …We may go even further than this and believe that musical sounds afforded one of the bases for the development of language."

Indeed, as Marler has repeatedly indicated,[7] learning of bird songs reveals fascinating similarities to language learning and to the underlying neural architecture thereof.

To some extent, these results are corroborated by experimental evidence on the primordial mechanisms of language processing in infants. In 1994, Mehler and his coworkers[8] discovered that suprasegmental features provide infants with a key for telling the difference between their future mother tongue and a foreign idiom. On the fourth day after birth, French infants responded preferentially to sentences produced in the French language compared to Russian sentences. Clearly, it is the infant's ability to detect the prosodic contours that allows them to tell the two languages apart. Indeed, such a basic skills turns out to be a key connection between language and music. As Patel and Daniele[9] have documented, there is an empirical basis for the claim that "spoken prosody leaves an imprint on the music of culture."

Another approach to the biological roots of music points to cross-cultural evidence for universals. Temporal processing, for instance, appears as a likely candidate for the status of temporal universals in music.[10]

MUSIC AND LANGUAGE IN ONTOGENY

Infants are compellingly attracted by music no less than by language. When 6-month-old infants viewed their own mothers (videotaped while singing to their infants), they showed more sustained attention to mother's singing episodes than to their speaking episodes.[11] Hence, when the lullaby melts the two ingredients (words and melody), newborn babies rest as if they were mesmerized. Saffran deals in detail with these intriguing connections between music and language in ontogeny. Music unfolds a very special power in regulating infant moods.

These recent findings suggest that infants come to the task of decoding language and music already equipped with highly sophisticated tools. It has been shown, for instance, that infants can remember words from stories over 2 weeks later.[12] Because music, like language, requires outstandingly complex computation, it is possible that infant memory for musical stimuli is similarly powerful. Indeed, experimental evidence shows that infants can retain familiar music for 2 weeks in long-term memory.[13] As Trehub[11] states in her forthcoming chapter: "Infants exhibit a propensity for relational processing of pitch and timing patterns and for enhanced processing of consonant intervals, which are at the heart of music processing across cultures. It is tempting to accept infants as musical beings on the basis of relatively *mature* pattern processing skills." Infants are sophisticated musical listeners indeed. They can link musical passages together, thus recovering a coherent musical event.[14]

In a similar vein, although from a slightly different perspective, the infant's musical sophistication is conceived of as the outcome of general principles in perception, namely, the Principles of Sameness and Difference.[15] It is relevant to note that such principles apply to both children and adult's performance in music discrimination. There is, in fact, an instinctive tendency of our species to cluster into perceptual units those events that have similar physical characteristics or that occur close in time. Another candidate is our predisposition towards better processing of regular versus irregular sequences. As a matter of fact, we do tend to hear as regular sequences ones that are not really regular.

NEUROPHYSIOLOGY OF EARLY BEGINNINGS IN MUSIC

These findings deepen our astonishment at the infants' prodigious capacity for processing musical passages. Furthermore, infants' musical skills raise a number of questions concerning the architecture of the underlying neurofunctional systems. A fundamental issue in this regard concerns the functional architecture of the auditory cortex. Recent neurophysiological studies demonstrate independent processing streams for sound localization and identification, analogous to the "what" and "where" stream in the visual cortex, although the modular arrangements are specific.[16] These findings suggest that the concomitant tasks of localizing auditory "where" and analyzing auditory "what" are carried out by two functionally independent neural systems. Infants are likely to be equipped with such refined machinery from birth.

Besides these intrinsic architectural aspects, a more dynamic dimension must be taken into account. It concerns the role of activity-dependent mechanisms in controlling the development of synaptic connectivity. Indeed, a recent critical review on such an issue by Cohen-Cory[17] documents a role "for both activity-independent and activity-dependent processes in regulating early synaptogenic events in the developing brain, from the level of individual synapses to the level of topographically organized neuronal maps." Further support for activity-dependent synaptogenesis comes from research on the effect of prenatal auditory enrichment on the developmental expression of synaptic proteins (synaptophysin and syntaxin1).[18]

Furthermore, as suggested by Pantev's stimulating contribution, the dynamic architecture of cortical tonotopic maps provides an optimal adaptation to a changing acoustic environment.

A second crucial issue concerns the relation between music and speech. It is widely recognized that speech and musical sounds exploit different acoustic cues and that these differences are reflected in the underlying neural architecture.[19] In particular, the temporal resolution is processed more efficiently in the left auditory cortical areas, whereas spectral resolution is better processed in the right auditory cortical areas. Such functional specializations optimize the efficiency in processing the acoustic environment. As suggested by Zatorre *et al.*,[19] such cortical asymmetries are likely to have emerged as a general solution to optimize processing of the acoustic environment in both the temporal and the frequency domains.

As we can foresee from this brief review, there is plenty of grist for the mill of scientific inquiry for the decade to come as well as plenty of reasons to expect exciting new discoveries on the astonishing performances of musical infants.

REFERENCES

1. MATASAKA, N. 1999. Preference for infant-directed singing in 2-day-old hearing infants of deaf parents. Dev. Psychol. **35:** 1001–1005.
2. TREHUB, S.E. & BERGSON. 1999. Mother's speech and song for infants. *In* R.J. Zatorre & I. Peretz, Eds. The Biological Foundations of Music. Ann. N.Y. Acad. Sci. **930:** 19.
3. TREHUB, S.E. 2001. Musical predisposition in infancy. *In* R.J. Zatorre & I. Peretz, Eds. The Biological Foundations of Music. Ann. N.Y. Acad. Sci. **930:** 1–16.
4. LARGE AND CRAWFORD.
5. PINKER, S. 1994. The Language Instinct. Allen Lane. The Penguin Press. London.
6. DARWIN, C. 1896/1964. The Descent of Man and Selection in Relation to Sex. New York, The Modern Library.
7. MARLER, P. 1975. On the origin of speech from animal sounds. *In* The Role of Speech in Language. J.F. Kavanagh & J.E. Cutting, Eds. :11–37. MIT Press. Cambridge, MA.
8. MEHLER, J. & E. DUPOUX. 1994. What Infants Know. Blackwell. Oxford.
9. PATEL, A.D. & J.R. DANIELE. 2003. An empirical comparison of rhythm in language and music. Cognition **87:** B35–B45.
10. DRAKE, C. & D. BERTRAND. 2001. The quest for universals in temporal processing in music. *In* R.J. Zatorre & I. Peretz, Eds. The Biological Foundations of Music. Ann. N.Y. Acad. Sci. **930:** 17–27.
11. TREHUB, S. 2003. Infant musical abilities and their implications (abstr.). *In* The Neurosciences and Music. International Conference. Venice. Ann. N.Y. Acad. Sci. **999:** this volume.
12. JUSCZYK, P.W. & E.A. HOHNE. 1997. Infant's memory for spoken words. Science **277:** 1984(abstr.)1986.
13. SAFFRAN, J.R., M.M. LOMAN & R.R.W. ROBERTSON. 2000. Infant memory for musical experiences. Cognition **77:** B15–B23.
14. TREHUB, S., G. SCHELLENBERG & D. HILL. 1997. The origins of music perception and cognition: a developmental perspective. *In* Perception and Cognition of Music. I. Deliège & J. Sloboda, Eds. :103-128. Psychology Press. East Sussex.
15. MÉLEN, M. 1999 Les principes du même et du différent comme organisateurs du groupement rhythmique chez le nourisson: une étude empirique. Musicae Scientiae **3:** 41–66.
16. READ, H.L., J.A. WINER & C.E. SCHREINER. 2002. Functional architecture of auditory cortex. Curr. Opin. Neurobiol. **12:** 433–440.
17. COHEN-CORY, S. 2002. The developing synapse: construction and modulation of synaptic structures and circuits. Science **298:** 770–776.

18. ALLADI, WADHWA & SINGH. 2002.
19. ZATORRE, R.J., P. BELIN & V.B. PENHUNE. 2002. Structure and function of auditory cortex: music and speech. Trends Cognit. Sci. **6:** 37–46.
20. FAIENZA, C., Ed. 1994. Music, Speech and the Developing Brain. The Case of the Modularity of Mind. Guerini e Associati. Milan.
21. COSSU, G., C. FAIENZA & C. CAPONE. 1994. Infant's hemispheric computation of music and speech. *In* Music, Speech and the Developing Brain. The Case of the Modularity of Mind. C. Faienza, Ed. :181-202. Guerini e Associati. Milan.

Musical Learning and Language Development

JENNY R. SAFFRAN

Waisman Center and Department of Psychology, University of Wisconsin – Madison, Madison, Wisconsin 53706, USA

ABSTRACT: How do infant learners acquire structure, given complex environments? In this chapter, we consider the role played by statistical learning—tracking patterns in the environment—in the acquisition of language and music. The results from a series of experiments suggest that similar learning mechanisms may operate in both domains, but that these mechanisms are also influenced by domain-specific perceptual biases and input structure.

KEYWORDS: infant learning; statistical learning; music; language

How do infants begin to make sense of their world? Filled with patterned stimuli and replete with sensory information of all kinds, infants must somehow begin to derive structure amidst all of the noise—both figuratively and literally—that characterizes their environments. Innate predispositions are one way to sort through the noise to find the signal. Learning is another. In this chapter we consider one type of learning—statistical learning, or the detection of patterns in the environment—and the possible role played by statistical learning mechanisms in the acquisition of complex structure.

In particular, we focus on the acquisition of language and music. These two domains share much in common structurally: both are auditory (with the exception of signed languages), highly patterned, and internally consistent. Music and language also share the distinction of being two of the stimuli that are most interesting to developing humans. Along with faces, young infants are most consistently engaged by speech and by music (making singing a particularly welcome combination of face, speech, and music). Linguistic and musical knowledge are arguably the most complex systems universally acquired by humans early in life. Finally, both language and music consist of some structures that are present cross-culturally, along with other structures that vary across cultures. This means that young learners must be capable of learning in these domains, or else they could not acquire the features of their native language that are not universal (e.g., the particular sounds, words, and grammatical devices of English versus Thai) or the aspects of their native musical system that are not universal (e.g., the scale tones of Western tonal music versus Javanese music).

Address for correspondence: Jenny R. Saffran, Department of Psychology, University of Wisconsin – Madison, Madison, WI 53706. Voice: 608-262-9942; fax: 608-262-4029.
jsaffran@wisc.edu

Ann. N.Y. Acad. Sci. 999: 397–401 (2003). © 2003 New York Academy of Sciences.
doi: 10.1196/annals.1284.050

How might this learning occur? To answer this question, we have begun to explore the potential contributions of statistical learning: the ability to track consistent patterns in the input to discover units and structures. In particular, we have explored the role of statistical learning in the discovery of consistent patterns that lie within continuous streams of sound. For example, consider the problems faced by learners exposed to a novel language. In order to acquire the language, learners must find the sound sequences corresponding to words in speech; learners can then map those sound sequences to meanings and discover their syntactic relationships. However, the initial task of finding words is nontrivial, because fluent speech does not contain pauses or other consistent acoustic cues marking word boundaries.[1] Nevertheless, infants can rapidly extract words from continuous speech by 7.5 months of age.[2]

There are multiple cues that infants might exploit to discover word boundaries in fluent speech.[3] The statistical properties of language might be particularly useful to infants engaged in segmenting words.[4,5] For example, consider the two-word sequence *pretty baby*. The syllable *pre* precedes a small set of other syllables, including *ty*, *tend*, and *cede*; the probability that *pre* is followed by *ty* is thus quite high (roughly 80% in speech to young infants). However, because the syllable *ty* occurs word-finally, it can be followed by any syllable that can begin a word in the English language. Thus, the probability that *ty* is followed by *ba*, as in *pretty baby*, is extremely low (roughly 0.03% in speech to young infants). Given the statistical properties of the input language, the ability to track sequential probabilities could be an extremely useful tool for infant language learners.

To ask whether infants possess statistical learning mechanisms usable for word segmentation, we exposed 8-month-olds to a "nonsense" language in which the only cues to word boundaries were the statistical properties of the syllable sequences.[5,6] The infants first listened to a 2-minute continuous sequence of syllables, in which "words" were present but word boundaries were not marked, for example, *golabup-abikututibubabupugolabu* etc. We then tested the infants to determine if they could discriminate the words in the language from sequences spanning word boundaries (part-words); these were sequences that the infants had heard, but that did not contain the statistical properties of words. To succeed in this task, the infants had to track the statistical properties of the input. We tested the infants using the Headturn Preference Procedure,[7] in which discrimination of the two types of test items (words versus part-words) was assessed by calculating how long infants listened to each test item. Our results confirmed that infants can indeed use sequential statistics to find word boundaries, despite the brevity of their exposure to the novel language and the lack of any other cues to word boundaries.

Our language experiments as well as those from other research groups demonstrate that infants have access to powerful statistical learning mechanisms that are deployed in the absence of any training or explicit reinforcement and that may be useful for many aspects of language acquisition.[8–10] One immediate question raised by these results is whether the learning mechanism in question is dedicated solely to the acquisition of language or whether it might be used by infants for learning in other domains. Although the issue of domain-specificity versus domain-generality (modularity) plays a prominent role in discussions of adult psychological and neural structure, studies of learning have typically not directly addressed this issue directly (for discussion, see Elman *et al.*, Ref. 11).

The domain of music provided a natural arena in which to extend our work on statistical language learning. We thus developed a pseudo-musical analog of our language task by translating each syllable of our nonsense language into a sine-wave tone; for example, *golabu* became CFE. We then exposed both infants and adults to a statistical learning task in which "tone-words" were discoverable solely by virtue of their statistical properties.[12] As in our linguistic tasks, we discovered that learners were adept at tracking the probabilities with which particular tones co-occurred to locate the boundaries between tone-words. It is thus likely that learners can use the same statistical learning mechanisms in both linguistic and musical tasks; subsequent studies have since demonstrated that the same mechanism can also be used for visuospatial and visuomotor learning.[13–15]

The musical nature of the tone sequence task we used raised an interesting issue concerning the nature of musical learning. Consider the tone sequences used in our experiment, for example, CFEGAD#F#A#B etc. What are the units over which learners computed statistics? There are multiple perceptual primitives that might have entered into learners' computations. Given a sequence like CFE, learners might have tracked the probabilities with which the absolute pitches C, F, and E co-occurred, an analogy with tracking the probabilities with which the syllables *go*, *la*, and *bu* co-occurred. Alternatively, infants might have tracked the probabilities with which the relative pitches, or the intervals between those pitches, co-occurred: what was the likelihood that an ascending perfect fourth (the interval between C and F) was followed by a descending minor second (the interval between F and E)?

Given the design of our original tone-learning experiment, it was impossible to determine which of these perceptual primitives was tracked during learning; did learners detect the statistical properties of absolute pitch sequences or relative pitch sequences? Learners at various ages appear to have access to both types of cues, but they may use one type of pitch information preferentially in some tasks relative to others.[16,17] To address this question, we manipulated our tone sequence statistical learning task so that in some conditions, successful test discrimination required the use of absolute pitch cues during learning, whereas in other conditions, successful test discrimination required the use of relative pitch cues during learning.[18] The results suggested that learners of different ages capitalized on different perceptual primitives to perform our task: 8-month-old infants tracked absolute pitch cues, while adults tracked relative pitch cues, given exactly the same stimulus materials.

Because infants in other types of tasks are skilled at detecting relative pitch information,[17] we hypothesized that the lack of musicality in our tone sequence stimuli might have influence infants' performance. The materials used in the Saffran and Griepentrog[18] experiment were atonal and did not conform to any of the conventions of Western tonal music. It is therefore possible that infants' reliance on absolute pitch cues in these experiments was a function of how the infants processed our stimuli; the lack of melodic structure may have led infants to focus on the nongeneralizable absolute pitch information rather than on the melodic information carried by relative pitch. To test this hypothesis, we exposed infants and adults to a tone sequence segmentation task in which the materials conformed to the key of C major.[19] We reasoned that the increment in musicality afforded by the inclusion of some tonal structure might lead infants to begin to track relative pitch in our materials. Interestingly, however, the infants continued to depend on absolute pitch cues for the test discrimination rather than relative pitch cues.

Why might some tasks lead infants to show one type of perceptual processing whereas other tasks suggest the use of a different set of perceptual cues? One possibility concerns an interaction between learners' existing perceptual sensitivities and the structure of the tasks themselves. This hypothesis was previously explored in the domain of pitch processing by birds.[20] When European starlings were presented with a pitch discrimination task that could be performed using either absolute or relative pitch cues, the birds initially solved the task using absolute pitches. However, when the task was changed to require transfer of the pitch sequences, the birds began to use relative pitch cues. These results suggest that the birds had access to both types of pitch cues, but that the demands of the task determined which pitch dimensions they used to perform the test discriminations.

It is possible that a similar process occurs during pitch learning in human infants. That is, infants may have access to both types of pitch cues, and the structure of the learning task and/or test discrimination affects which aspects of pitch are tracked. If this hypothesis is correct, then when absolute pitch cues are rendered unreliable, infants should begin to track relative pitch cues instead. We thus created tone sequences that no longer contained consistent absolute pitch sequences; this was accomplished by continually transposing the "tone words."[21] Relative pitch sequences, however, remained highly predictable. Given these materials, infants were able to capitalize on relative pitch cues, which they failed to do when absolute pitch information was also available. By essentially removing absolute pitch cues as a signal of structure in the input, infants began to reliably discriminate the test materials using relative pitch cues.

The results of this line of research, taken together, support a nuanced view of infant learning capabilities. On the one hand, infant learners are powerful: they can detect structure using statistical cues, this learning mechanism operates rapidly and in the absence of reinforcement, and similar learning occurs across domains. On the other hand, learning is constrained, both by the infants' perceptual capabilities and by the structure of the material to be learned. Infants may avoid William James' "blooming buzzing confusion," at least in part, by virtue of a perceptual system that weights some cues more highly than others and by a learning system that can flexibly adapt to the task at hand. Little is yet known about the role played by such systems in acquiring the native musical system and how the infants' inherent preferences and biases interact with information to be learned. We are hopeful, however, that comparisons of learning across domains, and careful consideration of the tasks facing learners within a given domain, will lead us to understand how the developing brain so masterfully acquires the intricacies of the native environment.

ACKNOWLEDGMENTS

The preparation of this chapter was funded by a grants to J.R.S. from the International Foundation for Music Research, National Institute of Child Health and Human Development (RO1HD37466), and National Science Foundation (BCS-9983630). Thanks to Erik Thiessen for helpful comments on the manuscript.

REFERENCES

1. COLE, R. & J. JAKIMIK. 1980. A Model of Speech Perception. Lawrence Erlbaum. Hillsdale, NJ.
2. JUSCZYK, P.W. & R.N. ASLIN. 1995. Infants' detection of the sound patterns of words in fluent speech. Cognit. Psychol. **29**: 1–23.
3. JUSCZYK, P.W. 1999. How infants begin to extract words from fluent speech. Trends Cognit. Sci. **3**: 323–327.
4. SAFFRAN, J.R., E.L. NEWPORT & R.N. ASLIN. 1996. Word segmentation: the role of distributional cues. J. Mem. & Lang. **35**: 606–621.
5. SAFFRAN, J.R., R.N. ASLIN & E.L. NEWPORT. 1996. Statistical learning by 8-month-old infants. Science **274**: 1926–1928.
6. ASLIN, R.N., J.R. SAFFRAN & E.L. NEWPORT. 1998. Computation of conditional probability statistics by 8-month-old infants. Psychol. Sci. **9**: 321–324.
7. KEMLER NELSON, D.G., P.W. JUSCZYK, D.R. MANDEL, *et al.* 1995. The Headturn Preference Procedure for testing auditory perception. Infant Behav. Dev. **18**: 111–116.
8. SAFFRAN, J.R. & D.P. WILSON. 2003. From syllables to syntax: multi-level statistical learning by 12-month-old infants. Infancy **4**: 273–284.
9. THIESSEN, E.D. & J.R. SAFFRAN. 2003. When cues collide: statistical and stress cues in infant word segmentation. Dev. Psychol. **39**: 706–716.
10. MAYE, J., J.F. WERKER, & L. GERKEN. 2002. Infant sensitivity to distributional information can affect phonetic discrimination. Cognition **82**: B101–111.
11. ELMAN, J.L., E.A. BATES, *et al.* 1996. Rethinking Innateness: A Connectionist Perspective on Development. MIT Press. Cambridge, MA.
12. SAFFRAN, J.R., E.K. JOHNSON, R.N. ASLIN & E.L. NEWPORT. 1999. Statistical learning of tone sequences by human infants and adults. Cognition **70**: 27–52.
13. HUNT, R.H. & R.N. ASLIN. 2001. Statistical learning in a serial reaction time task: simultaneous extraction of multiple statistics. J. Exp. Psychol. Gen. **130**: 658–680.
14. FISER, J. & R.N. ASLIN. 2002. Statistical learning of new visual features by infants. Proc. Natl. Acad. Sci. USA **99**: 15822–15826.
15. KIRKHAM, N.Z., J.A. SLEMMER & S.P. JOHNSON. 2002. Visual statistical learning in infancy: evidence for a domain general learning mechanism. Cognition **83**: B35–B42.
16. TAKEUCHI, A.H. & S.H. HULSE. 1993. Absolute pitch. Psychol. Bull. **113**: 345–361.
17. TREHUB, S.E., G. SCHELLENBERG & D. HILL. 1997. The origins of music perception and cognition: a developmental perspective. *In* Perception and Cognition of Music. I. Deliège & J. Sloboda, Eds. :103–128. Psychology Press. East Sussex, UK.
18. SAFFRAN, J.R. & G.J. GRIEPENTROG. 2001. Absolute pitch in infant auditory learning: evidence for developmental reorganization. Dev. Psychol. **37**: 74–85.
19. SAFFRAN, J.R. 2003. Absolute pitch in infancy and adulthood: the role of tonal structure. Dev. Sci. **6**: 37–45.
20. MACDOUGALL-SHACKLETON, S.A. & S.H. HULSE. 1996. Concurrent absolute and relative pitch processing by European starlings. J. Comp. Psychol. **110**: 139–146.
21. SAFFRAN, J.R., K. REECK, A. NIEBUHR & D. WILSON. 2003. Changing the tune: the structure of the input affects infants' use of absolute and relative pitch. Manuscript submitted.

Toward a Developmental Psychology of Music

SANDRA E. TREHUB

*Department of Psychology, University of Toronto at Mississauga,
Mississauga, Ontario, Canada*

ABSTRACT: Research on music perception has revealed numerous parallels
between infants and adults, but these findings have had little influence on adult
research. Studies of pitch memory in infants, children, and adults are presented to illustrate potential gains from a developmental approach. Although the
prevailing wisdom is that absolute pitch processing dominates in early life until
it is supplanted by relative pitch processing, recent research offers no support
for that view. After a week of exposure to English folk melodies, infants remember the melodies, but they do not distinguish the original versions from transposed versions. Relative pitch processing dominates later on, but it does not
occur at the expense of absolute pitch processing. For example, adults can identify the pitch level of familiar musical recordings in the context of foils that are
pitch shifted by one or two semitones. Children 5–9 years of age can identify
the pitch level of familiar recordings when the foils are pitch shifted by two
semitones but not by one semitone. By contrast, Japanese children are successful in the context of one-semitone shifts. In short, a developmental
approach can provide insights of comparable importance on many issues in
music cognition.

KEYWORDS: pitch memory; absolute pitch; relative pitch; infants; children;
adults

INTRODUCTION

In recent years, it has become clear that we begin life with predispositions for musical engagement.[1–4] The concern here is with receptive rather than productive abilities, the discrepancy between the two being considerably greater in music than in
other domains such as language. For example, toddlers understand words well before
they produce them, the lag being several months at the earliest stages of language
acquisition.[5] The corresponding gap between receptive and productive abilities is
considerably greater in music, often on the order of years rather than months.

INFANTS ARE INHERENTLY MUSICAL

The claim that infants are inherently musical[4] is based on their sensitivity to critical features of music. For example, they engage in relational processing of pitch[1,2]

Address for correspondence: Sandra E. Trehub, Department of Psychology, University of
Toronto at Mississauga, Mississauga, Ontario, Canada L5L 1C6. Voice: 905-828-5415; fax: 905-
569-4326.

Sandra.Trehub@utoronto.ca

Ann. N.Y. Acad. Sci. 999: 402–413 (2003). © 2003 New York Academy of Sciences.
doi: 10.1196/annals.1284.051

and temporal patterns,[6] which is essential for the appreciation of music. Specifically, they recognize the invariance of melodic patterns across changes in pitch level[7,8] and tempo.[9] Infants' detection of invariant pitch or rhythmic patterning does not reflect poor discrimination. On the contrary, they are sensitive to changes of a semitone or less in the context of multitone sequences.[10–12] Similarly, infants are sensitive to changes in temporal grouping,[9,13] meter,[14,15] tempo,[16] duration,[17] and timbre.[18]

ORIGINS OF CONSONANCE AND DISSONANCE

Infants are also sensitive to the consonance and dissonance of musical patterns.[19–21] They categorize intervals on the basis of their consonance and dissonance,[22] and they retain more fine-grained information from patterns with consonant intervals—melodic or harmonic—than from those with dissonant intervals.[20] Infants also exhibit rudimentary esthetic preferences for consonant over dissonant music. For example, 6-month-olds are more attentive while listening to Mozart minuets than to altered versions in which dissonant intervals replace many of the consonant intervals.[19] From as early as 2 months of age (earlier, no doubt, if we could test them), they prefer sequences of consonant, harmonic intervals to those with dissonant intervals.[23] By 4 months of age, they listen contentedly to European folk melodies, but they squirm, fuss, and turn away when presented with dissonant versions of those melodies.[21]

DOMAIN-GENERAL OR DOMAIN-SPECIFIC SKILLS?

These are but a few of the ways in which infants' processing of music or music-like patterns parallels that of adults. Can we assume that this preparation for music listening is uniquely human? Traditionally, relative pitch processing has been regarded as an exclusively human disposition, with absolute pitch processing being the *modus operandi* of nonhuman primates[24] and songbirds.[25] However, recent work by Wright and his associates[26] casts doubt on this widely held view. They trained rhesus monkeys to make nonverbal judgments of "same" or "different" on a series of trials involving novel melody pairs (i.e., new melodies presented on each trial). Subsequently, they tested these monkeys on other pairs in which the second melody was identical to the first (including its pitch level), transposed by one or two octaves (i.e., same melody at a different pitch level), or entirely different. Monkeys displayed a pattern of responses that one might expect from human listeners. They judged octave-transposed melodies as "same" when the melodies were tonal but as "different" when they were atonal. For example, "Happy Birthday" was recognizable in transposition (i.e., response of "same"), but randomly generated melodies were not (i.e., response of "different" for octave transpositions). On the one hand, these findings confirm the biological basis of relational processing in music. On the other hand, they indicate that relational pitch processing is not unique to human listeners but may be part of a heritage that is shared with many other species. Indeed, there is considerable evidence to support the contention that relational processing or perceptual invariance operates across age and species for pitch structure, spectral structure, and temporal structure.[27]

MOTIVATIONAL FACTORS

If species-specific predispositions do not account for the pattern processing that characterizes human music listening, they must account for the attraction to music from the earliest days of life,[28,29] for caregivers' universal disposition to sing to infants,[30,31] for young children's spontaneous music-making,[32,33] and for the worldwide prominence of music in cultural rituals.[34,35] Common threads across these diverse realms of activity include the emotional impact of music on listeners and practitioners[36–38] and the facilitation of social bonding.[31,39,40]

Music is highly effective in regulating or optimizing mood in listeners as diverse as infants,[29,41] adolescents,[42,43] and adults,[44] including those with psychotic tendencies[45] or late-stage dementia.[46] Aside from pharmacologic agents, music may be the most effective regulator of mood. As a result, it has been used extensively for lofty goals such as promoting physical or emotional well-being[47] and for less lofty goals such as influencing consumer behavior.[48] No doubt, the emotional impact of music and its ability to enhance social bonds have fueled the development of uniquely human musical behaviors.

THE GREAT DIVIDE: INFANTS AND ADULTS

Despite the fact that research on music perception has revealed intriguing parallels between infants and adults,[1,2,19] the infant findings have had little influence on adult research. The questions investigated with infants may seem elementary or obvious for adults. We know, for example, that adults encode consonant intervals more readily than dissonant intervals and that they prefer consonant to dissonant music, but we do not know *why* that is the case. The presumption is that music processing biases arise largely from extended exposure to culture-specific music,[49] but this presumption is incorrect in at least some respects. A related belief is that listening skills become increasingly refined or differentiated with increasing age and experience. At times, however, naïveté about cultural conventions leads infants to outperform adults on specific speech and music tasks. For example, infants exhibit more differentiated perception of some non-native speech contrasts,[50] melodic changes,[10] and atypical meters.[15] In short, developmental research can inform adult research by revealing the initial state of the organism, age-related changes in domain-general processing, and age- and experience-dependent changes in domain-specific processing.

A SAMPLE DEVELOPMENTAL AGENDA: ABSOLUTE AND RELATIVE PITCH PROCESSING

To illustrate potential gains from a truly developmental approach to music perception, consider the case of absolute pitch processing. The tiny minority of individuals (1 in 10,000) who possess absolute pitch (AP) can identify or produce isolated pitches in the absence of a reference pitch. Everyone else is thought to have very poor memory for absolute aspects of pitch.[51–53] Unique structural asymmetries or patterns of cortical activation have been identified in AP possessors,[54–56] but these differences may be a consequence of AP rather than its cause.

AP is generally attributed to early and prolonged musical training[57,58] acting in concert with genetic predispositions.[59,60] In one large sample of musicians, 40% who began lessons before age 4 had AP, in contrast to 26% who began at 4-6 years and 8% who began at 6–9 years.[57] For early lessons to result in AP, it may be necessary for training in note identification to precede the acquisition of relative pitch skills.[53] The prevalence of early training among AP possessors is consistent with a critical period for AP,[53,61] but the retrospective status of such information raises questions about its reliability. It is possible, for example, that parents of future AP possessors initiate musical training in response to their children's precocious musical interests or talents.[62] Beyond the uncertain early history of AP possessors is the typical incidence of earlier and more extensive training in musicians with AP compared to those without AP.[63] In principle, cumulative training along with enhanced quality of training could be as important as, or more important than, age of first lessons.

Views of AP and its acquisition have implications for the developmental course of pitch processing more generally. Takeuchi and Hulse,[61] among others,[64] consider absolute pitch processing as the dominant strategy in the preschool period, which could account for the relative ease of acquiring arbitrary pitch labels at that time. Saffran and her associates[65,66] go further, arguing that absolute pitch processing dominates from early infancy. By the end of the preschool period, however, children are thought to shift from absolute pitch processing to relative pitch processing.[61,66] One account of this shift implicates *unlearning* resulting from exposure to specific tunes in different keys.[53] Only AP possessors are thought to maintain absolute as well as relative modes of pitch processing. Just as relative pitch processing is thought to occur at the expense of absolute pitch processing,[53,61] AP possession is thought to have negative consequences for relative pitch processing.[63,67–69]

There are numerous problems with the proposed developmental timetable for absolute and relative pitch processing. For example, relative pitch processing is well documented in infancy,[1,2] but reliable evidence of absolute pitch processing in this period is lacking.[70] Claims of absolute pitch processing in infancy[65,66] are based on short-term memory for stimuli whose pitch relations are difficult for adults to remember. Ideally, questions regarding the priority of absolute or relative pitch processing should be settled with reference to long-term representations rather than short-term performance on novel materials.[70]

PITCH MEMORY IN INFANTS

Platinga and Trainor[71] examined infants' long-term retention of absolute and relative aspects of familiar melodies. They exposed 6-month-olds to one of two English folk tunes for 6 minutes daily over the course of 1 week. After the end of the familiarization phase (i.e., 1 day later), infants were able to distinguish the familiar tune from the unfamiliar tune, as reflected in greater attention to the novel tune. There was no indication, however, that infants remembered the original pitch level, because they accorded comparable attention to renditions in the original key and to those in a novel key (i.e., transposition of a perfect fifth). These findings are consistent with the priority of relative pitch processing in infancy. It is possible, however, that relational processing has priority for pitch but not for other dimensions of musical patterns. After comparable familiarization with one of two folk melodies, infants exhibited greater attention to renditions with novel tempo or timbre than to the original versions, which indicates that their memory for these melodies was tempo or timbre specific.[72] Studies such as these may underestimate infants' retention of pitch level because of relatively limited exposure to the "familiar" melodies. Perhaps infants would more readily remember the pitch level of their mothers' stereotyped performances of nursery songs.[29,73]

PITCH AND TEMPO MEMORY IN ADULTS

In general, adults with little or no musical training have reasonably good relative pitch, which enables them to recognize familiar tunes played at a novel pitch level and to detect pitch errors in performance.[74] Nevertheless, they perform at chance levels on conventional AP tests, which require identification or production of isolated pitches. Musicians without AP have comparable difficulties with isolated tones except when a reference tone is available, in which case they can capitalize on their knowledge of pitch intervals.[53]

In contrast to their poor memory for isolated pitches, adults exhibit surprisingly good memory for the pitch level of familiar musical materials. For example, mothers' repeated performances for infants (i.e., same songs on different occasions) are nearly identical in pitch level.[73] Consistency in pitch production is also evident in college students' repeated renditions of folk songs[75] and their attempts to sing along with imagined versions of favorite hit songs.[76] Their productions in the latter instance are within two semitones of the canonical recording. In these cases of song production, motor memory may play a substantial role, obscuring the independent contribution of pitch memory.

Schellenberg and Trehub[77] evaluated college students' recognition of the pitch level of instrumental excerpts from popular television programs: *E.R., Friends, Jeopardy, Law & Order, The Simpsons*, and *X-Files*. On each trial, participants heard two excerpts, one of which was shifted one or two semitones upward or downward by means of professional editing software that preserve the tempo and timbre. Students were required to identify the "real" version (first or second presented), that is, the one that matched the version heard on the television program in question. Performance significantly exceeded chance levels on one- and two-semitone shifts, with better performance on the latter (FIG. 1, Experiment 1). However, increasing exposure to pitch-shifted excerpts over the course of the test session reduced performance accuracy relative to that shown on the initial comparisons. Although the altered versions generated progressive interference with memory for the original details, adults still performed above chance levels. Other adults who were tested on the same task but with unfamiliar recordings (Experiment 2) performed at chance levels, ruling out the possibility of extraneous cues from the pitch-shifting manipulation.

These findings indicate that adults with minimal musical training remember the pitch level of music heard incidentally. Their ability to detect one-semitone shifts is noteworthy in view of the common occurrence of semitone errors in individuals with AP.[78] Musically untrained adults do not have deficient pitch memory, as alleged.[51,53] Instead, their memory for pitch rivals that of musicians as long as the test context features familiar, ecologically valid materials. It is clear, then, that listeners

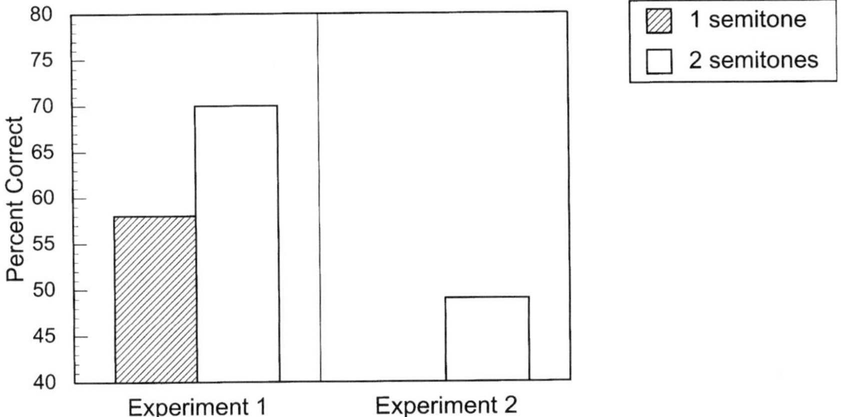

FIGURE 1. Adults' accuracy in identifying the pitch level of familiar (Experiment 1) and unfamiliar (Experiment 2) musical excerpts in the context of foils altered by one or two semitones. Chance level is 50%. Data from Schellenberg and Trehub.[77]

encode surface details of music, such as its pitch level, along with relational features such as melodic and temporal structure. Adults' identification of popular recordings from excerpts as brief as 100 ms[79] indicates that they encode timbral or spectral information as well. They encode comparable surface details from speech,[80] which enables them to recognize familiar voices.[81,82]

What about tempo? In ongoing research,[83] adults are being tested on their memory for the tempo of familiar musical excerpts. Test trials consist of the original excerpt paired with excerpts whose tempo is altered by 10% (faster or slower). On a separate occasion, participants are being tested on two-semitone pitch shifts involving the same excerpts. Performance to date reveals that adults can identify the original tempo, their accuracy being at least as good as that on two-semitone pitch shifts. Does recognition of these features depend on the complexity or engaging quality of the auditory patterns or merely on their familiarity? To provide answers to this question, participants in the tempo/pitch study are attempting to identify a conventional telephone dialtone—an unengaging but highly familiar stimulus—in the context of a pitch-shifted foil. Preliminary results reveal more accurate performance on the dialtone than on television theme music. This finding is consistent with exposure as the principal contributor to pitch memory, as is AP possessors' superior memory for pitches corresponding to white rather than black piano keys.[78,84]

PITCH MEMORY IN CHILDREN

The findings on pitch memory in adults are inconsistent with the view that relative pitch processing supplants absolute pitch processing in early childhood. Relative pitch processing may be the dominant or most relevant mode of musical processing in adulthood but, as noted, adults are clearly capable of encoding and retaining information about absolute pitch. It is likely that they can direct their attention to absolute or relative aspects of pitch, as necessary for the task at hand, just as they can focus on the content, voice quality, or prosody of a spoken message. There is no means of quantifying the relative efficiency of processing these various aspects of speech or music. We can ask, however, whether adults perform better or worse than children whose musical experience is much more limited. If increasing competence in relative pitch processing occurs at the expense of absolute pitch processing, as suggested,[53,61] then children, who are less efficient than adults in relative pitch processing, may outperform adults on memory tasks that depend on absolute aspects of pitch.

Such comparisons are currently in progress.[85] Children between 5 and 9 years of age are being tested on excerpts from the soundtracks of four television programs or movie videos that they watch regularly. The test procedure is the same as that used with adults except that the musical materials include vocal portions, as do most programs on the television diet of local children. Although the pitch-shifting manipulation does not have perceptible consequences with instrumental materials, it may generate subtle voice quality distortions, providing potential cues to the pitch change. Testing to date has revealed above-chance performance on two-semitone shifts but chance-level performance on one-semitone shifts and no age-related changes in performance (FIG. 2). Thus, there is no support for the view that children are better than adults at remembering the pitch level of familiar musical materials. On the contrary, they perform more poorly than adults in this respect. No doubt, these children have had less exposure than adults to music in general and, perhaps, to the musical excerpts in particular. Although older children have had the benefit of greater musical exposure than younger children, their experience in this regard does not seem to enhance their memory for the pitch level of familiar music. Tests of adults who are unfamiliar with this music will indicate whether vocal cues are responsible for children's success on the two-semitone shifts.

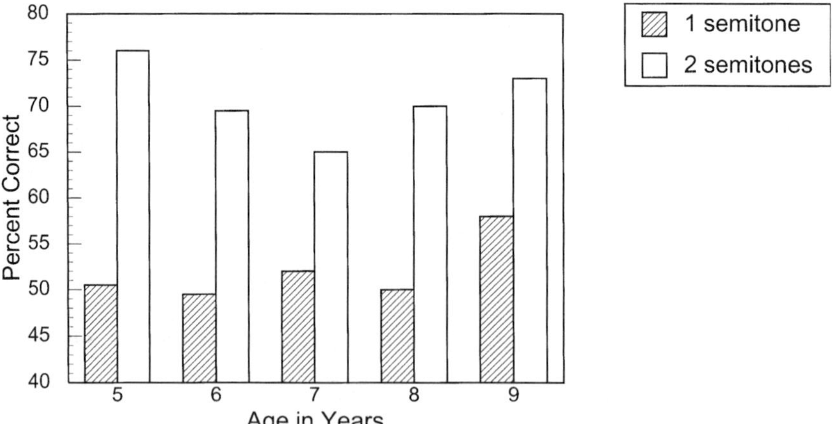

FIGURE 2. Children's accuracy in identifying the pitch level of familiar musical excerpts in the context of foils altered by one or two semitones. Chance level is 50%.

A parallel study is currently underway in Japan.[86] Japanese children provide an interesting comparison group because, on average, they have earlier and more extensive musical training than do North American children. Moreover, popular Japanese programs for children feature instrumental theme music in some cases and accompanied vocal music in others. The set of test materials includes four excerpts from programs that are familiar to each participant, two featuring instrumental themes and two featuring vocal-plus-instrumental themes. Results to date indicate that 5- and 6-year-old children perform above chance levels on one-semitone comparisons, and they perform no differently on instrumental than on instrumental-plus-vocal materials. In short, Japanese children are clearly superior to their North American peers in retaining the pitch level of familiar music. Information about their musical background is being examined with a view to identifying potential causal factors.

IMPLICATIONS OF DEVELOPMENTAL RESEARCH ON PITCH MEMORY

Findings from the research on pitch memory with infants, children, and adults offer little support for the conventional view that absolute pitch processing prevails in early life, being replaced by relative pitch processing sometime thereafter.[61,65,66] Instead, there is every reason to believe that absolute and relative pitch processing are operative from the beginning of life and that they remain so throughout life. What may differ as a function of age and musical experience are the specific circumstances that elicit one mode of processing rather than the other, creating the illusion that one mode functions in a dominant or exclusive manner.

Only musicians with AP are credited with dual processing. Obviously, they would be unable to function as musicians unless they had considerable expertise in relative pitch processing. Their identification of intervals[69] and keys[63] indicates that AP reduces processing efficiency on such tasks, but it does not impair accuracy. By the same token, non-AP musicians use absolute cues when making key judgments, and many of them have a single internalized reference tone such as A_4.[53] Even the untrained children and adults who identified the pitch level of television theme music[77,85,86] were making intuitive key judgments despite their ignorance of note names and the concept of key.

In the context of relatively unfamiliar materials, novice listeners remember a great deal more than global features or the "gist" of music. There are indications that they automatically encode surface features while listening to music[87] or speech.[80] Tests of explicit memory often imply that surface details fade with the passage of time, but their persistence is evident in implicit memory tasks.[80,87] The automatic processing of music-specific information (e.g., intervals) is also reflected in electrical activity in the brain.[88] In arguing for automatic processing of surface features in music, Dowling et al.[87] make the case with respect to interval processing. On the basis of the research presented here, that argument can be extended to absolute pitch processing in the context of music or speech.

If some form of absolute pitch processing is universal, then why is AP so rare? Perhaps it is because the defining criteria necessitate specialized knowledge of note names, which must be applied to decontextualized, essentially meaningless, tones. When AP occurs in the absence of early musical training, it is often associated with disability. For example, many blind school-age children develop AP after limited musical training.[89] Obviously, auditory cues play an important role in the lives of blind children, which may account for their disposition to attend carefully to absolute as well as relative aspects of pitch. AP also occurs with greater than expected frequency in individuals who are autistic[90,91] or developmentally delayed.[92,93] In cases of developmental delay or autism, deficits in cognitive flexibility may be relevant.[93] Very young children may have comparable generalization deficits, which may increase their "susceptibility" to AP.

Absolute pitch processing is but one of many skills for which a developmental perspective can yield important insights. Whether the questions of interest concern structural asymmetries in the brain, patterns of cortical activation, emotion, esthetic preferences, therapeutic interventions, cognitive consequences of musical training, or pedagogy, a developmental perspective has much to offer. Researchers engaged in these endeavors are invited to join the developmental enterprise.

ACKNOWLEDGMENT

The preparation of this paper was assisted by a grant from the Natural Sciences and Engineering Research Council of Canada.

REFERENCES

1. TREHUB, S.E. 2000. Human processing predispositions and musical universals. *In* The Origins of Music. N.L. Wallin, B. Merker & S. Brown, Eds. :427–448. MIT Press. Cambridge, MA.
2. TREHUB, S.E. 2001. Musical predispositions in infancy. Ann. N.Y. Acad. Sci. **930:** 1–16.
3. TREHUB, S.E. 2003. The developmental origins of musicality. Nature Neurosci. **6:** 669–673.
4. TREHUB, S.E. 2002. The musical infant. *In* Conceptions of Development: Lessons from the Laboratory. D.J. Lewkowicz & R. Lickliter, Eds. :231–257. Psychology Press. New York.
5. BLOOM, L. 1993. The Transition from Infancy to Language: Acquiring the Power of Expression. Cambridge University Press. New York.
6. DRAKE, C. & D. BERTRAND. 2001. The quest for universals in temporal processing in music. Ann. N.Y. Acad. Sci. **930:** 17–27.
7. CHANG, H.W. & S.E. TREHUB. 1977. Auditory processing of relational information by young infants. J. Exp. Child Psychol. **24:** 324-331.
8. TREHUB, S.E., L.A. THORPE & B.A. MORRONGIELLO. 1987. Organizational processes in infants' perception of auditory patterns. Child Dev. **58:** 741–749.
9. TREHUB, S.E. & L.A. THORPE. 1989. Infants' perception of rhythm. Categorization of auditory sequences by temporal structure. Can. J. Psychol. **43:** 217–229.
10. TRAINOR, L.J. & S.E. TREHUB. 1992. A comparison of infants' and adults' sensitivity to Western musical structure. J. Exp. Psychol. Hum. Percept. Perform. **18:** 394–402.
11. TREHUB, S.E., E.G. SCHELLENBERG & S.B. KAMENETSKY. 1999. Infants' and adults' perception of scale structure. J. Exp. Psychol. Hum. Percept. Perform. **25:** 965–975.
12. TREHUB, S.E., A.J. COHEN, L.A. THORPE & B.A. MORRONGIELLO. 1986. Development of the perception of musical relations: semitone and diatonic structure. J. Exp. Psychol. Hum. Percept. Perform. **12:** 295-301.
13. CHANG, H.W. & S.E. TREHUB. 1977. Infants' perception of temporal grouping in auditory patterns. Child Dev. **48:** 1666-1670.

14. BERGESON, T.R. 2002. Perspectives on music and music listening in infancy. Ph.D. thesis, University of Toronto. Canada.

15. HANNON, E.E. & S.E. TREHUB. 2003. Metrical categories in infancy and adulthood. Presented at the 25th Annual Meeting of the Cognitive Science Society, Boston, MA, July/August.

16. BARUCH, C. & C. DRAKE. 1997. Tempo discrimination in infants. Infant Behav. Dev. **20:** 573–577.

17. THORPE, L.A. & S.E. TREHUB. 1989. Duration illusion and auditory grouping in infancy. Dev. Psychol. **25:** 122–127.

18. TREHUB, S.E., M. ENDMAN & L.A. THORPE. 1990. Infants' perception of timbre: classification of complex tones by spectral structure. J. Exp. Child Psychol. **49:** 300–313.

19. TRAINOR, L.J. & B.M. HEINMILLER. 1998. The development of evaluative responses to music: infants prefer to listen to consonance over dissonance. Infant Behav. Dev. **21:** 77–88.

20. SCHELLENBERG, E.G. & S.E. TREHUB. 1996. Natural musical intervals: evidence from infant listeners. Psychol. Sci. **7:** 272–277.

21. ZENTNER, M.R. & J. KAGAN. 1996. Perception of music by infants. Nature **383:** 29.

22. SCHELLENBERG, E.G. & L.J. TRAINOR. 1996. Sensory consonance and the perceptual similarity of complex-tone harmonic intervals: tests of adult and infant listeners. J. Acoust. Soc. Am. **100:** 3321–3328.

23. TRAINOR, L.J., C.D. TSANG & V.H.W. CHEUNG. 2002. Preference for sensory consonance in 2- and 4-month-old infants. Music Percept. **20:** 187–194.

24. D'AMATO, M.R. 1988. A search for tonal pattern perception in cebus monkeys: why monkeys can't hum a tune. Music Percept. **5:** 453-480.

25. HULSE, S.H., S.C. PAGE & R.F. BRAATEN. 1990. An integrative approach to auditory perception by songbirds. *In* Comparative Perception, Vol **2:** Complex Signals. W.C. Stebbins & M.A. Berkley, Eds. :3–34. Wiley. New York.

26. WRIGHT, A.A., J.J. RIVERA, S.H. HULSE, *et al.* 2000. Music perception and octave generalization in rhesus monkeys. J. Exp. Psychol. Gen. **129:** 291–307.

27. HULSE, S.H., A.H. TAKEUCHI & R.F. BRAATEN. 1992. Perceptual invariances in the comparative psychology of music. Music Percept. **10:** 151–184.

28. MASATAKA, N. 1999. Preference for infant-directed singing in 2-day-old hearing infants of deaf parents. Dev. Psychol. **35:** 1001–1005.

29. TREHUB, S.E. & T. NAKATA. 2001–2002. Emotion and music in infancy. Musicae Scientiae Special Issue: 37–61.

30. TREHUB, S.E. & E.G. SCHELLENBERG. 1995. Music: its relevance to infants. Ann. Child Dev. **11:** 1–24.

31. TREHUB, S.E. & L.J. TRAINOR. 1998. Singing to infants: lullabies and play songs. Adv. Infancy Res. **12:** 43–77.

32. CAMPBELL, P.S. 1988. Songs in their Heads: Music and its Meaning in Children's Lives. Oxford University Press. Oxford.

33. KARTOMI, M.J. 1991. Musical improvisations of children at play. The World of Music **33:** 53–65.

34. DISSANAYAKE, E. 1992. Homo Aestheticus: Where Art Comes From and Why. Free Press. New York.

35. MERRIAM, A.P. 1964. The Anthropology of Music. Northwestern University Press. Evanston, IL.

36. JUSLIN, P.N. 2001. Communicating emotion in musical performance. *In* Music and Emotion: Theory and Research. P.N. Juslin & J.A. Sloboda, Eds. :309–337. Cambridge University Press. Cambridge.

37. SCHERER, K.R. & M.R. ZENTNER. 2001. Emotional effects of music: production rules. *In* Music and Emotion: Theory and Research. P.N. Juslin & J.A. Sloboda, Eds. :361–392. Cambridge University Press. Cambridge.

38. STORR, A. 1992. Music and the Mind. The Free Press. New York.

39. BROWN, S. 2000. Evolutionary models of music: from sexual selection to group selection. *In* Perspectives in Ethology, Vol. **13:** Evolution, Culture, and Behavior. F. Tonneau & N.S. Thompson, Eds. :231–281. Kluwer Academic. New York.

40. KOGAN, N. 1997. Reflections on aesthetics and evolution. Crit. Rev. **11:** 193–210.

41. SHENFIELD, T., S. TREHUB & T. NAKATA. Maternal singing modulates infant arousal. Psychol. Music **31:** 365–375.
42. SLOBODA, J.A. & S.A. O'NEILL. 2001. Emotions in everyday listening to music. *In* Music and Emotion: Theory and Research. P.N. Juslin & J.A. Sloboda, Eds. :415–429. Oxford University Press. Oxford.
43. ZILLMANN, D. 1988. Mood management: using entertainment to full advantage. *In* Communication, Social Cognition, and Affect. L. Donohew *et al.*, Eds. :147–171. Erlbaum. Hillsdale, NJ.
44. KONECNI, V. 1982. Social interaction and music preference. *In* The Psychology of Music. D. Deutsch, Ed. :497–516. Academic Press. San Diego, CA.
45. RAWLINGS, D., F. TWOMEY, E. BURNS & S. MORRIS. 1998. Personality, creativity and aesthetic preference: comparing psychoticism, sensation seeking, schizotypy and openness to experience. Empiric. Stud. Arts **16:** 153–178.
46. GOETELL, E., S. BROWN & S.-L. EKMAN. 2002. Caregiver singing and background music in dementia care. West. J. Nurs. Res. **24:** 195–216.
47. BUNT, L. 1997. Clinical and therapeutic uses of music. *In* The Social Psychology of Music. D.J. Hargreaves & A.C. North, Eds. :249–267. Oxford University Press. Oxford.
48. NORTH, A.C. & D.J. HARGREAVES. 1997. Experimental aesthetics and everyday music. *In* The Social Psychology of Music. D.J. Hargreaves & A.C. North, Eds. :84–103. Oxford University Press. Oxford.
49. KRUMHANSL, C.L. 1990. Cognitive Foundations of Musical Pitch. Oxford University Press. New York.
50. WERKER, J.F. & R.C. TEES. 1999. Influences on infant speech processing: toward a new synthesis. Ann. Rev. Psychol. **50:** 509–535.
51. BURNS, E.M. 1999. Intervals, scales, and tuning. *In* The Psychology of Music, 2nd Ed. D. Deutsch, Ed. :215–264. Academic Press. San Diego.
52. KRUMHANSL, C.L. 2000. Rhythm and pitch in music cognition. Psychol. Bull. **126:** 159–179.
53. WARD, W.D. 1999. Absolute pitch. *In* The Psychology of Music, 2nd Ed. D. Deutsch, Ed. :265–298. Academic Press. San Diego.
54. HIRATA, Y., S. KURIKI & C. PANTEV. 1999. Musicians with absolute pitch show distinct neural activities in the auditory cortex. Neuroreport **200:** 999–1002.
55. OHNISHI, T., H. MATSUDA, T. ASADA, *et al.* 2001. Functional anatomy of musical perception in musicians. Cereb. Cortex **11:** 754–760.
56. SCHLAUG, G., L. JAENCKE, Y. HUANG & H. STEINMETZ. 1995. In vivo evidence of structural brain asymmetry in musicians. Science **267:** 699–701.
57. BAHARLOO, S., P.A. JOHNSTON, S.K. SERVICE, *et al.* 1998. Absolute pitch: an approach for identifying genetic and non-genetic components. Am. J. Hum. Genet. **62:** 224–231.
58. SERGEANT, D. 1969. Experimental investigation of absolute pitch. J. Res. Music Ed. **17:** 135–143.
59. BAHARLOO, S., S.K. SERVICE, N. RISCH, *et al.* 2000. Familial aggregation of absolute pitch. Am. J. Hum. Genet. **67:** 755–758.
60. GREGERSEN, P.K. 1998. The genetics of pitch perception. Am. J. Hum. Genet. **62:** 221–223.
61. TAKEUCHI, A.H. & S.H. HULSE. 1993. Absolute pitch. Psychol. Bull. **113:** 345–361.
62. BROWN, W.A., H. SACHS, K. CAMMUSO & S.E. FOLSTEIN. 2002. Early music training and absolute pitch. Music Percept. **19:** 595–597.
63. MARVIN, E.R. & A.M. BRINKMAN. 2000. The effect of key color and timbre on absolute pitch recognition memory. J. Exp. Psychol. Learn. Mem. Cognit. **22:** 1166–1183.
64. CROZIER, J.B. 1997. Absolute pitch: practice makes perfect, the earlier the better. Psychol. Music **25:** 110–119.
65. SAFFRAN, J.R. 2003. Absolute pitch in infancy and adulthood: the role of tonal structure. Dev. Sci. **6:** 35–43.
66. SAFFRAN, J.R. & G.J. GRIEPENTROG. 2001. Absolute pitch in infant auditory learning: evidence for developmental reorganization. Dev. Psychol. **37:** 74–85.

67. MIYAZAKI, K. 1992. Perception of musical intervals by absolute pitch possessors. Music Percept. **9:** 413–426.
68. MIYAZAKI, K. 1993. Absolute pitch as an inability: identification of musical intervals in a tonal context. Music Percept. **11:** 55–71.
69. MIYAZAKI, K. 1995. Perception of relative pitch with different references: some absolute–pitch listeners can't tell musical interval names. Percept. Psychophysics **57:** 962–970.
70. TREHUB, S.E. 2003. Absolute and relative pitch processing in tone learning tasks. Dev. Sci. **6:** 44–45.
71. PLANTINGA, J. & L.J. TRAINOR. 2002. Long–term memory for pitch in 6-month-old infants. Conference presentation, The Neurosciences and Music. Venice, Italy, October.
72. TRAINOR, L.J., L. WU & C.D. TSANG. 2003. Long-term memory for music: infants remember tempo and timbre. Dev. Sci. In press.
73. BERGESON, T.R. & S.E. TREHUB. 2002. Absolute pitch and tempo in mothers' songs to infants. Psychol. Sci. **13:** 71–74.
74. DRAYNA, D., A. MANICHAIKUL, M. DE LANGE, *et al.* 2001. Genetic correlates of musical pitch recognition in humans. Science **291:** 1969–1972.
75. HALPERN, A.R. 1989. Memory for the absolute pitch of familiar songs. Mem. & Cognit. **17:** 572–581.
76. LEVITIN, D.J. 1994. Absolute memory for musical pitch: evidence from the production of learned melodies. Percept. Psychophysics **56:** 414–423.
77. SCHELLENBERG, E.G. & S.E. TREHUB. 2003. Good pitch memory is widespread. Psychol. Sci. **14:** 262–266.
78. MIYAZAKI, K. 1988. Musical pitch identification by absolute pitch possessors. Percept. Psychophysics **44:** 501–512.
79. SCHELLENBERG, E.G., P. IVERSON & M.C. MCKINNON. 1999. Name that tune: identifying popular recordings from brief excerpts. Psychon. Bull. Rev. **6:** 641–646.
80. GOLDINGER, S.D. 1996. Words and voices: episodic traces in spoken identification and recognition memory. J. Exp. Psychol. Learn. Mem. Cognit. **22:** 1155–1183.
81. PAPCUN, G., J. KREIMAN & A. DAVIS. 1989. Long-term memory for unfamiliar voices. J. Acoust. Soc. Am. **85:** 913–925.
82. VAN LANCKER, D., J. KREIMAN & T. WICKENS. 1985. Familiar voice recognition: patterns and parameters. II. Recognition of rate-altered voices. J. Phonetics **13:** 39-52.
83. TREHUB, S.E. & E.G. SCHELLENBERG. 2003. Good memory for tempo is widespread. In preparation.
84. TAKEUCHI, A.H. & S.H. HULSE. 1991. Absolute-pitch judgments of black- and white-key pitches. Music Percept. **9:** 27–46.
85. SCHELLENBERG, E.G., S.E. TREHUB & T. NAKATA. 2003. Musical pitch memory: developmental and cross-cultural perspectives. Presented at the Auditory Perception, Cognition, and Action Meeting. Vancouver, Canada, November.
86. NAKATA, T., S.E. TREHUB & E.G. SCHELLENBERG. 2004. Cross-cultural perspectives on pitch memory. Presented at the International Congress on Acoustics. Kyoto, Japan, April.
87. DOWLING, W.J., B. TILLMAN & D.F. AYERS. 2002. Memory and the experience of hearing music. Music Percept. **19:** 249–276.
88. TRAINOR, L.J., K.L. MCDONALD & C. ALAIN. 2002. Automatic and controlled processing of melodic contour and interval information measured by electrical brain activity. J. Comp. Neurosci. **14:** 430–442.
89. WELCH, G.F. 1988. Observations on the incidence of absolute pitch (AP) ability in the early blind. Psychol. Music **16:** 77–80.
90. HEATON, P., B. HERMELIN & L. PRING. 1998. Autism and pitch processing: a precursor for savant musical ability. Music Percept. **15:** 291–305.
91. HEATON, P., L. PRING & B. HERMELIN. 1999. A pseudo-savant: a case of exceptional musical splinter skills. Neurocase **5:** 503–509.
92. LENHOFF, H.M., O. PERALES & G. HICKOK. 2001. Absolute pitch in Williams syndrome. Music Percept. **18:** 491–503.
93. MOTTRON, L., I. PERETZ, S. BELLEVILLE & N. ROULEAU. 1999. Absolute pitch in autism: a case study. Neurocase **5:** 485–501.

Experimental Strategies for Understanding the Role of Experience in Music Cognition

CAROL L. KRUMHANSL

Department of Psychology, Cornell University, Ithaca, New York 14853, USA

ABSTRACT: Research in music cognition has assessed the role of experience by investigating the effects of development, training, and cross-cultural differences. A fourth approach is to apply statistical techniques to identify patterns that are relatively frequent in the listeners' musical experience. As an example of this approach, a functional magnetic resonance imaging (fMRI) study was conducted on melodic expectancy. Two patterns that frequently began musical themes in classical and folk music were identified. A final continuation tone was added to these opening patterns. Some of the continuation tones were frequent in the musical styles; others were infrequent. Listeners judged how well the continuation tones fit with their expectations. A sparse-sampling (event-related) design was used in the fMRI study, with image acquisition at varying delays after the sequence. Early acquisitions showed activation in the auditory cortex, and late acquisitions showed effects of the motor response. These results suggest the timing of the image acquisitions was appropriate. Contrasting melodies and monotonic controls showed right inferior frontal activation, similar to that found in other studies. However, no differences were found as a function of whether or not the continuation tone was frequent or infrequent in the statistical style analysis. Methodological differences of this study from other recent fMRI studies on harmonic expectations are discussed.

KEYWORDS: music cognition; melodic expectation; fMRI; statistical learning

INTRODUCTION

Research in music cognition offers a rich set of possibilities for investigating the role of experience. A frequently taken approach is to look at the development of musical abilities from infancy, through early childhood, and continuing through the school years. Despite the extensive developmental literature now available, there is the disadvantage that maturational changes co-vary with musical experience. Another approach is to look at effects of musical training. Unlike other domains such as vision and language, musical experience includes years of explicit instruction for some individuals. The effects of training, however, cannot be distinguished unambiguously from preexisting differences that might have inclined the individual toward musical activities. A third approach is to look at music and listeners from dif-

Address for correspondence: Dr. Carol L. Krumhansl, Department of Psychology, Uris Hall, Cornell University, Ithaca, NY 14853, USA. Voice: 607-255-6351; fax: 607-255-8433.
clk4@cornell.edu

Ann. N.Y. Acad. Sci. 999: 414–428 (2003). © 2003 New York Academy of Sciences.
doi: 10.1196/annals.1284.052

ferent cultures. In this case, it is possible to equate musical expertise and look quite directly at the effects of prior experience with the specific musical style. This requires understanding the various musical styles and their cultural context sufficiently well to construct appropriate materials and experimental methods.

A fourth approach, which is taken in the present paper, is to conduct an ecological survey of regularities within a musical style using statistical techniques. This identifies patterns that are frequent in the listeners' experience, offering insights into patterns that may influence how they perceive, organize, and remember music. Psychological research on music, summarized briefly below, has suggested that a cognitive principle that is basic to acquiring musical knowledge is sensitivity to statistical regularities in music. That is, listeners internalize patterns that are frequent in their experience. Recent research has indicated that statistical learning may also play a role in language acquisition.[1–3] In this research, infants are presented with sequences of syllables, and later tests suggest that they have learned which syllables frequently co-occured. A learning process such as this may lead to the identification of combinations of syllables as words. Recently, the paradigm has been extended to tones.[4] Thus, early in development humans appear to be sensitive to frequent successions of sounds, and this sensitivity may encompass both language and music.

There is now extensive research showing convergence between statistical regularities in music and perceptual and cognitive responses. It has been proposed that, among other effects, these regularities help establish tone centrality; in Western music, this is called key or tonality. Regularities include repetition of tones and tone sequences, melodic and rhythmic emphasis, durational and metric stress, and positioning of central tones at or near beginnings and endings of phrases. Through repeated exposure to music, listeners implicitly develop a mental representation that captures these regularities. This representation can then be used to encode and remember musical patterns in the future and generate expectations while listening. Sensitivity to these regularities may also enable listeners to adapt relatively easily to novel musical styles. According to Meyer, "musical styles are complex systems of probability relationships, and the meaning of any given musical event depends on its relationship to other possible events within the style."[5]

Early empirical evidence pointing to these conclusions came from a cross-cultural study of tonal hierarchies. Castellano *et al.*[6] used contexts from 10 North Indian *ragas*. North Indian music has a greatly expanded set of scales (called *thats*, all with the same tonic) compared to the major/minor system of Western music. Theoretical treatments of Indian music describe a hierarchy of the importance of tones. Short excerpts from *ragas* with different underlying scales were followed by probe tones that listeners rated as to how well they fit with the excerpts. The ratings largely confirmed theoretical predictions. Surprisingly, this was true for both Western and Indian listeners. Both groups gave high ratings to the first (*Sa*) and fifth (*Pa*) scale tones, which are described as most structurally significant and which were sounded in the drone accompanying the melody. They also gave relatively high ratings to the *vadi* tone, which is designated for each *rag* and is given emphasis in the melody. The agreement between the two groups of listeners could be attributed to the finding that the ratings generally reflected the relative durations of tones in the musical contexts preceding the probe tones. That is, tones that were sounded for longer durations (and more frequently) were judged as fitting better with the contexts. This suggested that the distribution of tones conveys the tonal hierarchy to listeners unfamiliar with the style. Beyond this, only the Indian listeners were sensitive to the scales (*thats*) underlying the *ragas*.

The correspondence between musical statistics in Western tonal-harmonic music and perceptual judgments was then examined.[7,8] The tonal hierarchy measured in a probe tone task with Western key-defining contexts[9] was compared with various statistical analyses of musical pieces.[10–12] These analyses tabulated the frequency of occurrence or total duration of each tone of the chromatic scale in pieces by Schubert, Mozart, Hasse, Strauss, Mendelssohn, and Schumann. These correlated strongly with the probe-tone profile for the corresponding key, with correlations ranging from $r(10) = 0.86$ to 0.97. The extent to which the probe-tone ratings resembled the tone distributions supports Meyer's proposal that listeners have internalized the statistical properties of music. Tones high in the tonal hierarchy tend to be sounded more frequently and for longer duration. Sensitivity to these tones would serve the purpose of initially establishing and then maintaining the listeners' sense of the tonal reference points. Subsequent statistical analyses have considered a variety of styles and more specific issues in how tone distributions relate to tonality, harmony, and meter.[13–15]

The question of how rapidly and completely listeners adapt to novel distributions of pitch was investigated by Oram and Cuddy.[16] They constructed melodic sequences of 20 pure tones. Each sequence was generated from either a diatonic tone set, C, D, E, F, G, A, B, or a nondiatonic (novel) tone set that did not conform to any major or minor key. Within each sequence, the frequency of occurrence of each tone was determined according to the following rule: one tone of the tone set occurred eight times, two tones occurred four times each, and the remaining four tones occurred just once. The tones selected to occur most frequently never formed the simple pattern of a major triad in root position. Both musically trained and untrained listeners heard the sequences in a probe-tone paradigm. The results for both levels of musical training were straightforward. Probe tone ratings were systematically related to the frequency of occurrence of the tones in the sequence.

Musical knowledge, however, did play a role. First, as would be expected, the data for the musically trained listeners revealed some degree of assimilation to the Western tonal hierarchy for diatonic sequences. Second, the effect of frequency of occurrence was more pronounced for diatonic sequences than for nondiatonic sequences. Third, in a somewhat surprising result, the effect of frequency of occurrence was more pronounced for musically trained than untrained listeners. The important conclusion from this study is that listeners are flexible and adaptable to novel pitch distributions and that extensive music training does not appear to interfere with this sensitivity; if anything, it appears to enhance it.

A subsequent study used composed melodies.[17] As before, sequences contained either a seven-tone diatonic or non-diatonic tone set. Professional composers wrote 20-second flute melodies under several constraints. They were to allot 8 seconds of the melody to one tone of the tone set, 4 seconds to each of two different tones, and only 1 second to the remaining four tones of the set. They could vary the frequency of occurrence and tone duration of each tone so long as they kept strictly to the allotted time, and the tones could be distributed across octaves within the flute range. In a standard probe-tone paradigm, the results closely replicated those of the first study by Oram and Cuddy. Listeners gave highest ratings to the tone sounded for eight seconds. Ratings for other tones followed, in order, the total duration for which the probe tone had been sounded in the flute melody.

A pair of articles considered how tone distributions influenced melodic expectations.[18,19] These tabulated the frequency of occurrence of tones in two musical styles: Finnish spiritual folk hymns and yoiks (melodies sung by the Sami of Northern Scandinavia). In the first of these styles, the tone distribution corresponded closely with those of Western tonal-harmonic music. In the second of these, it showed a pentatonic, rather than a diatonic, pattern. In both cases, the tone distributions were evident in the judgments of what constituted an appropriate continuation of melodic fragments. That is, listeners judged those tones frequent in the musical style as good continuations. These studies also considered the frequency of two-tone and three-tone transitions. The two-tone statistics record the frequencies of pairs of tones, and the three-tone statistics record the frequencies of three-tone patterns. The results showed that the higher-order statistics (two- and three-tone transitions) influenced expectation judgments more for listeners experienced with the musical culture. These higher-order statistics were also shown to be important for classifying short excerpts from these two styles.[20] Thompson and Stainton have also compiled statistics on three-tone sequences in folk melodies to study regularities in melodic patterns.[21]

ECOLOGICAL METHODOLOGY FOR AN FMRI STUDY OF MELODIC EXPECTANCY

The fMRI study, done in collaboration with Robert Zatorre at the Montreal Neurological Institute, investigated the neural systems involved in processing melodic patterns. A sparse sampling procedure was chosen in which the acquisition of BOLD (blood-oxygen-level–dependent) responses occurred during a period of silence after the melodic pattern ended. This meant that the melody was heard without the simultaneous noise of the scanner. To maximize the number of observations during the experimental session, short but tonally unambiguous sequences were needed.

With these considerations in mind, the study began with an ecological survey of frequent initial patterns in a corpus of Western melodies in order to identify initial segments that clearly specify the tonality or key of the music. The search tools in Themefinder.com[22] were used for this purpose. Two patterns appeared relatively frequently: the progression with scale degrees 5 1 2 3 (with intervals +P4 +M2 +M2) and the progression with scale degrees 5 1 7 6 (with intervals +P4 −m2 −M2). In the themes included in the corpus, the first of these appeared 248 times in the classical music corpus and 396 times in the folk music corpus. The second of these appeared 294 times in the classical music corpus and 302 times in the folk music corpus. Vos[23] has also noted the prevalence of ascending fourth (and descending fifth) openings in tonal music.

The same corpus was used to determine what tones followed these patterns when they were the initial segments of the music. The results are shown in FIGURE 1. As can be seen, the pattern 5 1 2 3 was most often followed by the scale degree 4, sometimes the scale degree 5, and rarely the scale degree 6. Nondiatonic continuations essentially never occurred. The pattern 5 1 7 6 was almost always followed by scale degree 5, and rarely scale degree 3, and again nondiatonic continuations were

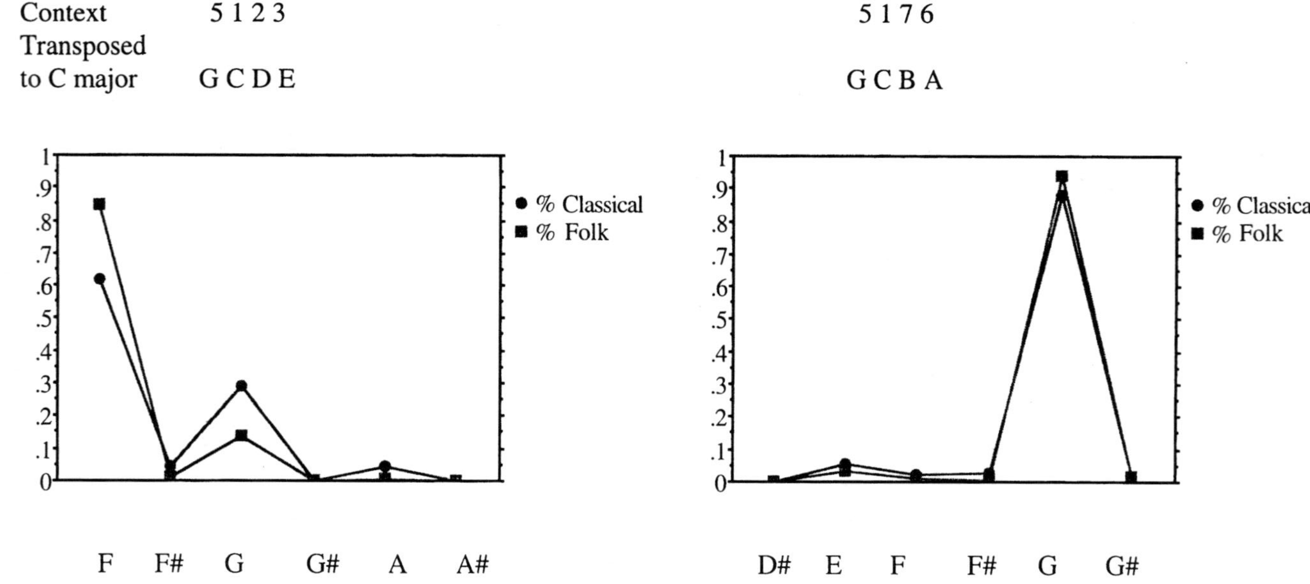

FIGURE 1. Shows the probability of different tones following the openings of 5 1 2 3 (transposed to C major: G C D E) and the 5 1 7 6 (G C B A) in the corpus of classical and folk music.

FIGURE 2. Two example sequences for the 5 1 2 3 context (*top*) and the 5 1 7 6 context (*bottom*) used in the experiment.

extremely rare. It is interesting to note the high degree of consistency across the two musical styles considered.

In the experiment, these initial patterns were followed by tones that either were or were not frequent continuations in the musical corpus. The continuation tones are shown in FIGURE 2. The continuation tones were always in the melodic direction of the end of the context sequence (ascending for the 5 1 2 3 context and descending for the 5 1 7 6 context). The rhythmic pattern was the same for all melodies. It emphasized the tonic as the downbeat of the first measure with the duration of a half-note. The critical (to-be-judged) continuation tone also appeared on the downbeat and with half-note duration. The sequence 5 1 2 3 was presented in the key of C major; the sequence 5 1 7 6 was presented in the key of B major, a key that is distantly related to C major. This was intended to prevent a sense of key developing across multiple trials, thus requiring listeners to respond to the continuation tone relative to the context on each trial. The melodic sequences were sounded in a piano timbre.

The temporal structure of each trial is shown in FIGURE 3. The context plus continuation tone lasted for a total duration of 2.0 seconds. This was followed at varying delays by the brain image acquisition. The choice of delays (the onset of scanning ranging from 3.33 to 6.0 seconds after the onset of the continuation tone) was based on the fMRI results of Belin *et al.*[24] They found that the blood-oxygen-level–dependent (BOLD) response to a single speech-like sound reached its peak in the auditory cortex in the range of 2.5–5 seconds after the onset of the sound. The slightly longer delays used here were chosen with two considerations in mind: to cover times when later, more cognitive processes might occur and to coincide with times when the pre-

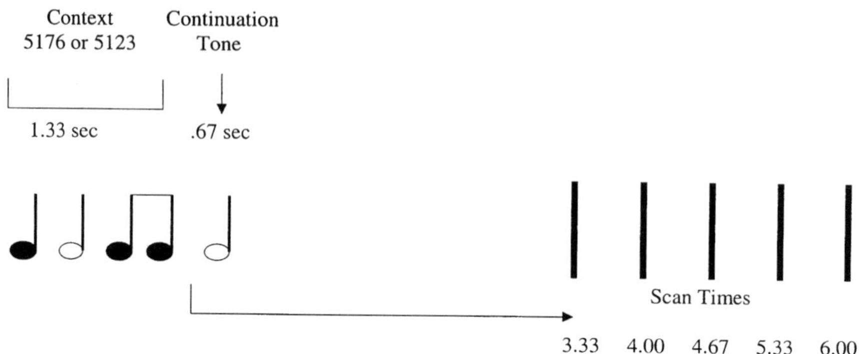

FIGURE 3. Timing of the brain acquisition scans relative to the sequences in the experiment. The times correspond to the onset of the image acquisition, which lasted approximately 2 seconds.

dicted response to the continuation tone was relatively strong compared to the response to the other tones in the sequence.

Twelve musically trained, right-handed adult subjects participated in the fMRI study. On average, the subjects had 10.3 years of musical instruction and 14.9 years of experience playing music. The parameters for the echo planar imaging (EPI) included 20 slices of 5-mm thickness with 100-ms acquisition time each. The slices were positioned to be parallel to the Sylvian fissure and did not include the cerebellum. In addition, high resolution T1-weighted scans were obtained for each participant. During the experiment, they judged how well the continuation tone fit with their expectations as to how they thought the melody would continue. The response was made on a scale of 1–4 (1 = not at all well, 4 = very well) by pressing one of four buttons on an MRI-compatible response pad with the fingers of their right hand. The sequences were presented over MRI-compatible headphones (SONY MDR-V900) at a level of approximately 82 dB; the headphones were held in place with an air-filled pillow, which also stabilized the position of the head.

In addition to the 120 melody trials, there were 60 monotone trials and 60 trials with silence. The monotone trials consisted of tones presented with the same piano timbre and rhythm as the melodic sequences. The tone chosen was a quarter tone between F4 and F#4, selected so that it would not reinforce either of the two keys of the sequences, C major and B major. On the monotone trials, subjects were required to press randomly one of the four response buttons. The silence trials were included as a general check on equipment and statistical analyses.

FIGURE 4. (*Left*) The brain areas more active for early scan times (*top*) and late scan times (*bottom*). (*Right*) The brain areas more active in the melody-monotone contrast (*top*) and the brain area less active in the melody-monotone contrast (*bottom*).

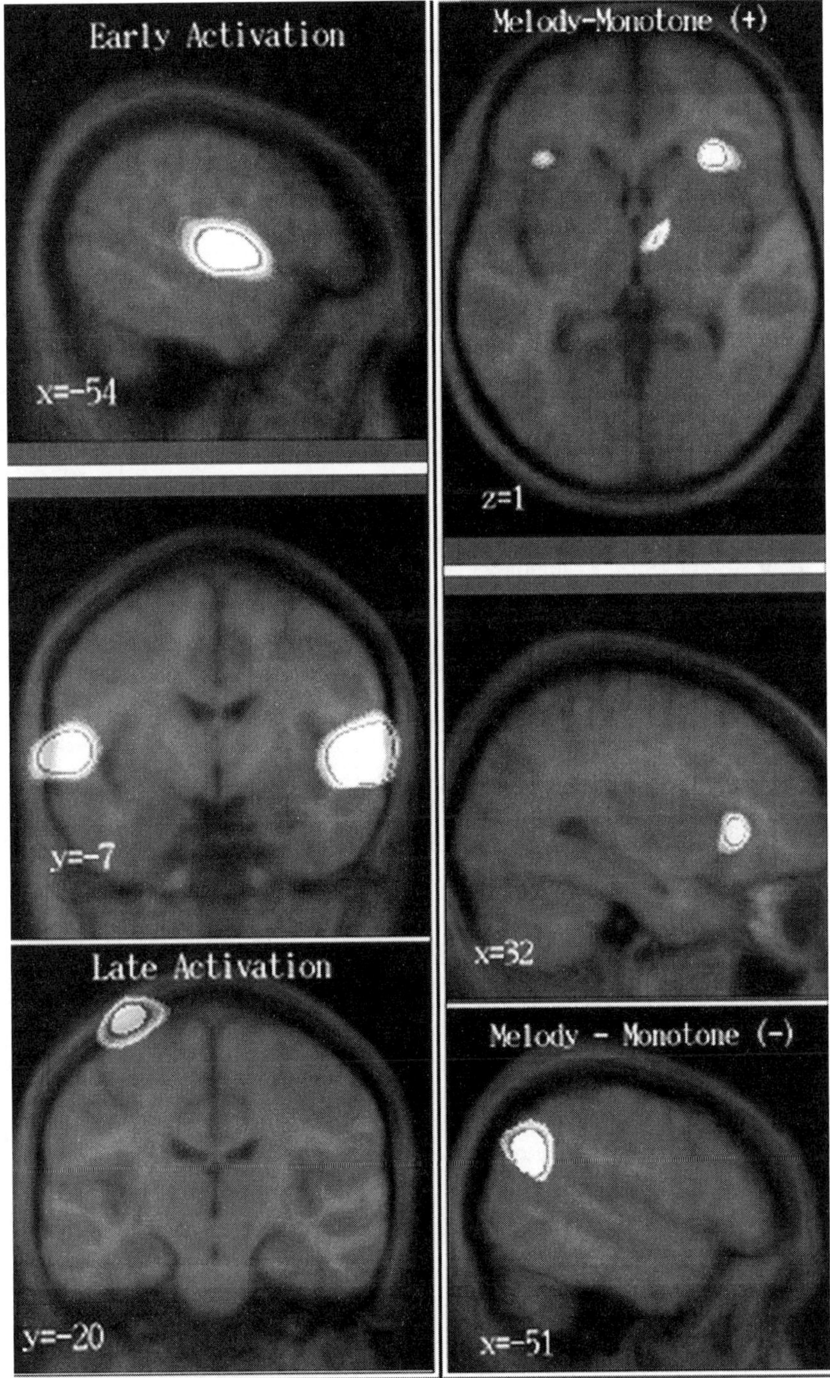

FIGURE 4. *See previous page for legend.*

TIMING AND TASK EFFECTS ON BRAIN ACTIVATIONS

The statistical analysis determined brain regions with *t*-statistics that were significant at $P < 0.05$. These are shown as white areas on the averaged normalized anatomical scans in FIGURE 4; the coordinates shown refer to Talairach coordinates. The first statistical analysis considered the effect of the time of image acquisition relative to the melodic sequences. As seen in the left column of FIGURE 4, regions in the superior temporal gyrus bilaterally showed much stronger activation at early scan times than at late scan times. These results are in accordance with the predictions based on Belin *et al.*[23] The area with activation that was stronger at late scan times is shown in the bottom of the left column of FIGURE 4. This locus in the left primary motor cortex suggests it may be related to the right-hand button response required by the task. These results indicate that the scan times fell during the interval between initial processing of the stimulus and the judgment of the quality of the melodic continuation. Similar patterns of activation were found when the same analysis was done on the monotonic sequences (which also required a motor response).

The next analysis of interest compared BOLD activity for the melody trials with that for the monotone trials. Activations in this contrast would relate to processing the tonal structure in the melody as contrasted with processing the monotone controls (which had the same rhythm and timbre as the melodies). The results are shown in the right column of FIGURE 4. As seen at the top, a positive activation was found in the inferior frontal gyrus and insula. It was significant only in the right hemisphere, but approached significance in the left hemisphere. This result is consistent with a variety of previous studies, as will be discussed. As shown at the bottom of the right column of FIGURE 4, a decrease in activation was found in the inferior parietal lobe, an area with connections to the frontal lobe and implicated in attentional processes. The area of significant decrease in activation was almost exclusively on the left.

Similar activation in the right inferior frontal cortex and insula has been found in a number of music studies using a variety of methods. It was found, for example, in the positon emission tomography (PET) study of musical imagery by Halpern and Zatorre[25] in their Cue/Imagery condition. In that condition, listeners heard the first few notes of well-known tunes and were asked to imagine the tunes. This area was also found in a PET study of musical memory by Zatorre *et al.*[26] In one condition, subjects had to compare the first note of a melody with the last note, and judge which was higher. This is called the High Load condition. In another condition, subjects had to compare the first and second tones. This is called the Low Load condition. Subtracting the Low Load condition from the High Load condition showed right inferior frontal activation. Similar activation appeared in two fMRI studies by Tillmann *et al.*[27] and Koelsch *et al.*[28] These studies are especially relevant to the results to be presented and will be described in detail. Finally, Steinke *et al.*[29] reported a case study of patient K.B. who has a widespread right frontal lesion. This condition is accompanied by severe impairments in the recognition of instrumental melodies and poor performance on a variety of tasks, including pitch discrimination, probe-tone judgments, and contour discriminations. These results all point to right frontal involvement in the processing of tonal relationships.

MELODIC EXPECTANCY EFFECTS ON BRAIN ACTIVATIONS

During the scanning sessions, the subjects also made behavioral judgments of the continuation tones. Intersubject agreement was strong, and the average results are shown as the circles in FIGURE 5. As can be seen, judgments for the 5 1 7 6 context sequences showed the highest rating for the continuation tone on scale degree 5, followed by scale degree 3, followed by scale degree 4. Nondiatonic tones received low ratings. Judgments to the 5 1 2 3 context sequences showed highest ratings for scale degrees 4 and 5, followed by scale degree 6. Again, nondiatonic tones received low ratings. The factors correlating with these behavioral responses were assessed using multiple and stepwise regressions. The best predictors of the behavioral responses were the tonal hierarchy from the probe-tone study of Krumhansl and Kessler[9] and the log frequency of occurrence in the classical and folk corpus in Themefinder.com. Together, these two factors produced a multiple regression of $R(2,9) = 0.99$, $P = 0.0001$, with each factor contributing significantly. The model fit is shown as the squares in FIGURE 5.

The next analysis asked whether brain activity correlated with behavioral judgments. Were activations in some brain areas greater for tones that were expected in the contexts? Alternatively, were greater activations found for tones that were not expected in the contexts? A statistical contrast was formed based on the behavioral judgments. It used the behavioral judgments (shifted to have a mean of 0) as weights in the contrast. In essence, this uses the different continuation tones as controls for the others. However, no brain area had activity that was significantly correlated (either positively or negatively) with the behavioral results. That is, no specific effect appeared in the fMRI data of how expected the continuation tone was in the context preceding it.

To pursue this question further, two additional statistical contrasts were formed based on the tonal hierarchy and log frequency of occurrence, separately. Recall that these two factors together predicted the behavioral data almost perfectly. Each of these factors separately correlated significantly with the behavioral judgments, $r(10) = 0.94$ and 95, $P = 0.0001$, respectively. It might be that one of them correlated with brain activity, but not the other. The tonal hierarchy (closely related to both scale and harmonic structure) may be more explicitly theorized by these musician subjects than frequency of occurrence, which is likely to be implicit knowledge. Thus, differences might appear between these two factors. Again, however, there were no brain areas whose activity correlated with either of these factors.

Four other factors were also tried. These had been found to influence judgments in previous studies of melodic expectancy mentioned earlier. The first was scale membership, which correlated with the behavioral judgments, $r(10) = 0.86$, $P = 0.0003$. The last three were more of an acoustical nature. The first of these was the consonance of the continuation tone with the tonic, which correlated significantly with the behavioral judgments, $r(10) = 0.90$, $P = 0.0001$. The next was the proximity of the continuation tone to the last tone of the context. The last was the pitch height of the continuation tone. Neither of the last two correlated with the behavioral judgments. And no brain areas were found whose BOLD response correlated significantly with any of these four factors.

In contrast to the present results, a number of studies have succeeded in finding brain responses to unexpected musical events. For example, Besson and Faïta[30]

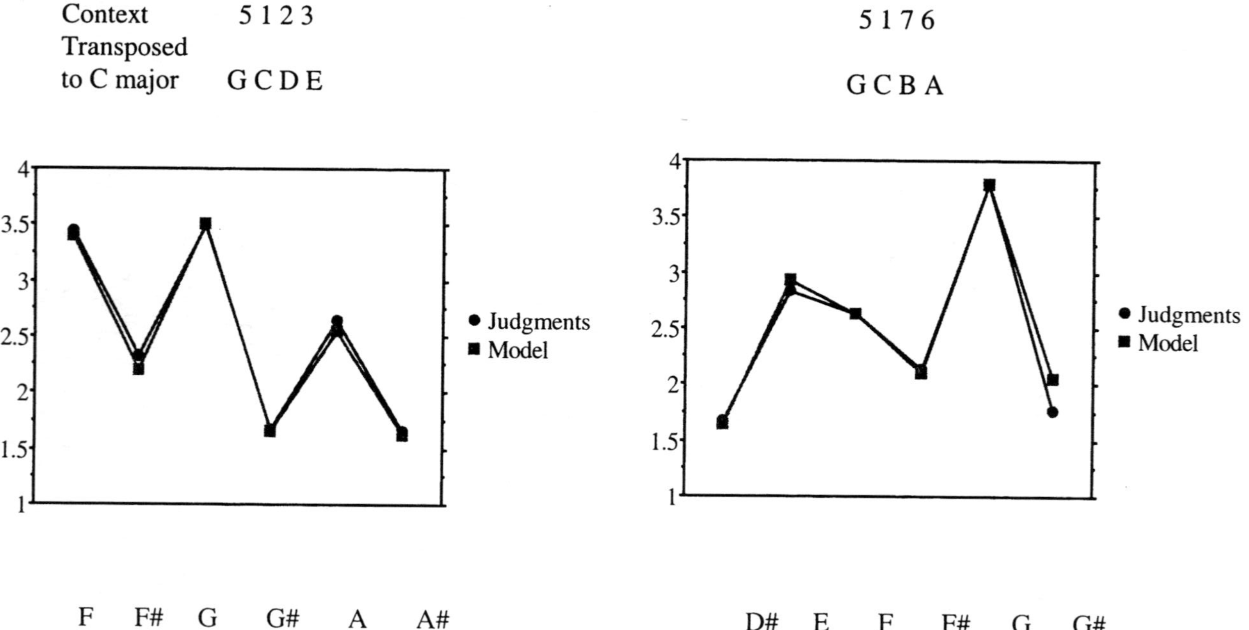

FIGURE 5. The melodic continuation judgments (shown as *circles*) for the 5 1 2 3 context (*left*) and the 5 1 7 6 context (*right*). These are compared with the fit of a multiple regression model (shown in *squares*) that had two factors: major key tone profiles[9] and the log frequency of occurrence of the continuation tones.

found a late positive component in event-related potentials (ERPs) when the final tone of a melody was changed to a tone out of key. This was found whether or not a semantically incongruous word was sung to that tone. Using both ERPs and EMFs (event-related magnetic fields), Koelsch and collaborators[31,32] found brain responses elicited by an unexpected chord in a chord sequence. This was manifested as an early right anterior negativity (ERAN). It may be that ERP and EMF methods are better suited to studying brain responses at the level of specific violations of melodic and harmonic expectancy, because of their greater temporal sensitivity than the BOLD signal. However, two fMRI studies recently showed brain areas with different responses to expected and unexpected musical events. Given their relevance, some methodological details will be described.

The first study by Tillmann *et al.*[27] looked at brain activations in a harmonic priming paradigm. In this paradigm, a harmonic sequence of seven chords served as the priming context. The eighth and final chord was the target, which was either expected or unexpected given the tonality of the priming sequence. The unexpected chord was a half-step away from the tonic of the context key (a B major chord if the context key was C major and vice versa). Half the target chords of each type were made dissonant by adding a raised fifth. The task was to judge whether the final chord was consonant (a major chord) or dissonant (a major chord with a raised fifth added). Thus, the task was independent of the tonality manipulation (expected versus unexpected target chord). Trials were grouped into blocks in which the key of the priming context was constant; chord sequences within blocks were separated by a 2500 ± 500-ms period of silence. Scanning occurred continuously during the runs, which also included periods of rest.

The basic result from the behavioral data (both in and out of the scanner) was that consonant chords were identified more quickly and more accurately when they were expected given the tonality of the context than when they were unexpected. Unlike most previous behavioral studies using this paradigm, the data acquired in the scanner found that related dissonant targets were also identified more quickly and accurately than were unrelated dissonant targets. Often, this paradigm has found that unrelated chords tend to be judged as more dissonant. This difference from experiments outside the scanner raises the possibility that the scanner noise may have affected the processing of the stimulus manipulation.

The main interest in this paradigm, however, is the effect of the related versus unrelated prime-target relationship on the processing of consonant targets. For these targets, the BOLD response was greater for unexpected targets than expected targets in a number of regions, notably in a region of interest around the frontal operculum (including inferior frontal gyrus, pars opercularis, and anterior insula). In addition, activations in this region were generally greater in the right hemisphere than the left hemisphere and for dissonant targets than consonant targets. This was apparent in both the number of voxels reaching significance and in the magnitude of the statistic (z-score).

The second highly relevant study is the fMRI experiment by Koelsch *et al.*[28] As in the other study, sequences of chords were presented in rapid succession, and scanning occurred continuously during stimulus presentation. The key of the sequences changed approximately every three or four sequences. In-key sequences contained five chords in a single key. The study included two different stimulus manipulations that might be described as related to tonality. In the first, called modulations, the last three chords were in a key different from the key of the in-key sequences (a shift of two semitones, from C to D major); this is most clearly a manipulation of tonality. In the second, called clusters, the final tonic chord of the five-chord sequence was replaced by a dissonant tone cluster that included out-of-scale tones. Although this manipulation introduced out-of-key tones, it might be considered to be more comparable to the dissonance manipulation of Tillmann *et al.*[27] A third manipulation changed the timbre of one or two in-key chords from piano to an instrument other than piano. The subjects' task was to respond whenever they heard the deviant instrument. However, they also had to detect the clusters, because the appearance of a cluster signaled that they should change the button used to indicate that a timbre deviant had occurred. Thus, as in the Tillmann *et al.*[27] study, the task-relevant variables were independent of the stimulus manipulation of tonality.

In the analysis of the fMRI data, the in-key sequences were used as controls for the other three conditions: modulations, clusters, and deviant instrument. Significant activations were found in the superior temporal gyrus and inferior frontal gyrus, including Brodmann's areas 44/6 and insula as well as thalamus and other areas. Activations in these conditions (compared to in-key sequences) were stronger for deviant instruments (directly response relevant) and clusters (indirectly response relevant) than modulations. For the modulations, the activation in inferior frontal gyrus was stronger in the right hemisphere than in the left hemisphere. At a general level, the pattern of activations was similar to that of Tillmann *et al.*[27]

The present study, like these, found right inferior frontal activations when melodies were compared to monotone controls. However, no differences were found as a function of whether the final element of the sequence was or was not expected in the tonal context. A number of methodological differences may account for these discrepancies. First, the present study used melodic sequences as opposed to harmonic (chord) sequences. Harmonic sequences may be stronger stimuli for establishing deviations from tonality. The tonal violations in the present experiment were obvious, as evidenced by the consistent behavioral responses, but they may be relatively weak compared to the harmonic violations of the other studies. Second, the previous studies presented sequences blocked by key, whereas the present study randomly intermixed sequences in two distantly related musical keys. The behavioral data, however, suggested listeners had no difficulty adjusting to the shifting tonal center, so this seems an unlikely possibility.

Another explanation for the differing results is that the present study used sparse sampling as opposed to continuously acquiring data, while sequences were presented without interruption as was done in the other studies. It may be that with the sparse sampling, the image acquisition was mistimed to pick up the hemodynamic response to the continuation tone. Arguing against this possibility is the analysis that showed activation in auditory areas for early acquisition times and activation attributable to responses at later scan times. This suggests that the range of delays was appropriate to the task. Alternately, it may be that the activation observed in the Tillmann *et al.*[27] and Koelsch *et al.*[28] studies is in part a by-product of the scanning noise produced by the continuous sampling. Given that the noise is also present in the control conditions, this would not be a problem unless the noise interacted with the stimulus manipulation. For example, the scanning noise might shift the listeners' strategy to a more top-down process where they relied more on harmonic knowledge, making the unexpected events more salient.

A final reason for the divergent results to consider is the relation between the task and the stimulus manipulation of tonality. In the present study, the task given listeners was directly relevant to the stimulus manipulation: they were to judge how well the final tone of the sequence fit with their expectations for how the melodic sequence would continue. This had the advantage that it provides a behavioral measurement that could be directly related to the hemodynamic responses. In the other two studies, the overt task was independent of the tonality manipulation in the experimental design. In the Tillmann et al.[27] experiment, it was to judge the consonance or dissonance of the last chord. The advantage for the related targets (the priming effect) shows, however, that this judgment is not independent of the tonality manipulation (whether the target chord is or is not related to the priming context). In the Koelsch et al.[28] experiment, the primary task was to judge whether notes in a deviant timbre had occurred, with a secondary task of detecting dissonant tone clusters. Again, it is possible that the tasks interacted with the tonality manipulation, which was whether or not a modulation had occurred. This raises the possibility that in both studies the findings reflect the need to filter out a tonality manipulation that is irrelevant to the task, rather than an effect of processing tonality itself.

As the literature on music cognition using brain imaging techniques expands, some of these methodological issues will become more clear. Studies are currently being conducted that should provide relevant comparative data. For the present, I conclude by reiterating the main advantages of an ecological approach to music experimentation. This approach can guide the selection of stimulus materials to be representative of listeners' musical experience. Patterns that are typical of a style are also likely to have been codified in music theory, thus bringing to bear predictions and interpretations from another disciplinary perspective. Finally, the approach makes contact with the research on statistical learning in language, which may lead to a greater understanding of basic psychological principles that are common to the domains of language and music.

ACKNOWLEDGMENTS

The research was supported by a grant (Award # 0121839) to Carol L. Krumhansl from the National Science Foundation and a Fellowship from the John Simon Guggenheim Memorial Foundation. The discussion of statistical learning and music was formulated in collaboration with Lola Cuddy. Comments on an earlier version of this manuscript by Stefan Koelsch and Barbara Tillmann are gratefully acknowledged.

REFERENCES

1. SAFFRAN, J.R. et al. 1996. Statistical learning by 8-month-old infants. Science **274:** 1926–1928.
2. SAFFRAN, J.R. et al. 1996. Word segmentation: the role of distributional cues. J. Memory & Lang. **35:** 606–621.
3. SAFFRAN, J.R. et al. 1997. Incidental language learning: listening (and learning) out of the corner of your ear. Psychol. Res. **8:** 101–105.
4. SAFFRAN, J.R. et al. 1999. Statistical learning of tone sequences by human infants and adults. Cognition **70:** 27–52.

5. MEYER, L.B. 1956. Emotion and Meaning in Music. University of Chicago Press. Chicago.
6. CASTELLANO, M.A. *et al.* 1984. Tonal hierarchies in the music of North India. J. Exp. Psychol.: General **113**: 394–412.
7. KRUMHANSL, C.L. 1985. Perceiving tonal structure in music. Am. Scientist **73**: 371–378.
8. KRUMHANSL, C.L. 1990. Cognitive Foundations of Musical Pitch. Oxford University Press. New York.
9. KRUMHANSL, C.L. & E.J. KESSLER. 1982. Tracing the dynamic changes in perceived tonal organization in a spatial representation of musical keys. Psychol. Rev. **89**: 334–368.
10. YOUNGBLOOD, J. 1958. Style as information. J. Music Theory **2**: 24–35.
11. HUGHES, M. 1977. A quantitative analysis. *In* Readings in Schenker analysis and other approaches. M. Yeston, Ed. :144–164. Yale University Press. New Haven.
12. KNOPOFF, L. & W. HUTCHINSON. 1983. Entropy as a measure of style: the influence of sample length. J. Music Theory **27**: 75–97.
13. VOS, P.G., F.M. GREMMEN & J. DAUTZENBERG. 2003. On the frequency distribution of the 12 pitch classes in music by J.S. Bach. In revision.
14. HURON, D. 1993. Chordal-tone doubling and the enhancement of key perception. Psychomusicology **12**: 154–171.
15. JÄRVINEN, T. 1995. Tonal hierarchies in jazz improvisation. Music Percept. **12**: 415–437.
16. ORAM, N. & L.L. CUDDY. 1995. Responsiveness of Western adults to pitch-distributional information in melodic sequences. Psychol. Res. **57**: 103–118.
17. CUDDY, L.L. 1997. Tonal relations. *In* Perception and Cognition of Music. I. Deliège & J. Sloboda, Eds. :329–352. Taylor & Francis. Hove, Sussex.
18. KRUMHANSL, C.L. *et al.* 1999. Melodic expectation in Finnish spiritual folk hymns: Convergence of statistical, behavioral, and computational approaches. Music Percept. **17**: 151–195.
19. KRUMHANSL, C.L. *et al.* 2000. Cross-cultural music cognition: cognitive methodology applied to North Sami yoiks. Cognition **76**: 13–58.
20. KRUMHANSL, C.L. 2000. Tonality induction: a statistical approach applied cross-culturally. Music Percept. **17**: 461–479.
21. THOMPSON, W.F. & M. STAINTON. 1998. Expectancy in Bohemian folk song melodies: evaluation of implicative principles for implicative and closural intervals. Music Percept. **15**: 232–252.
22. HURON, D. *et al.* Themefinder is a non-profit collaborative project of the Center for Computer Assisted Research in the Humanities (CCARH), Stanford University, and the Cognitive and Systematic Musicology Laboratory at the Ohio State University.
23. VOS, P.G. 1999. Key implications of ascending fourth and descending fifth openings. Psychol. Music **27**: 2–18.
24. BELIN, P. *et al.* 1999. Event-related fMRI of the auditory cortex. NeuroImage **10**: 417–429.
25. HALPERN, A.R. & R.J. ZATORRE. 1999. When that tune runs through your head: a PET investigation of auditory imagery for familiar melodies. Cereb. Cortex **9**: 697–704.
26. ZATORRE, R.J., A.C. EVANS & E. MEYER. 1994. Neural mechanisms underlying melodic perception and memory for pitch. J. Neurosci. **14**: 1908–1919.
27. TILLMANN, B. *et al.* 2003. Activation of the inferior frontal cortex in musical priming. Cognit. Brain Res. **16**: 145–161.
28. KOELSCH, S. *et al.* 2002. Bach speaks: a cortical "language-network" serves the processing of music. NeuroImage **17**: 956–966.
29. STEINKE, W.R. *et al.* 2001. Dissociations among functional subsystems governing melody recognition after right-hemisphere damage. Cognit. Neuropsychol. **18**: 411–437.
30. BESSON, M. & F. FAÏTA. 1995. An event-related potential (ERP) study of musical expectancy: comparison of musicians with nonmusicians. J. Exp. Psychol.: Human Percept. Perf. **21**: 1278–1296.
31. KOELSCH, S. *et al.* 2000. Brain indices of music processing: "nonmusicians" are musical. J. Cognit. Neurosci. **12**: 520–541.
32. MAESS, B. *et al.* 2001. Musical syntax is processed in Broca's area: an MEG study. Nature Neurosci. **4**: 540–545.

Synchronizing with Music:
Intercultural Differences

CAROLYN DRAKE AND JAMEL BEN EL HENI

*Laboratoire de Psychologie Expérimentale, CNRS, Université René Descartes,
92774 Boulogne-Billancourt, France*

ABSTRACT: The way in which listeners perceive music changes throughout
childhood, but little is known about the factors responsible for these changes.
One factor, explicit music training, has received considerable attention, with
studies indicating that musicians demonstrate a more complex hierarchical
mental representation for music and superior temporal organizational skills.
But does acculturation—the passive exposure to a particular type of music
since birth—also influence the acquisition of these skills? We compared the
music synchronization performance of Tunisian and French subjects with
music from these two contrasting musical cultures. Twelve musical excerpts
were selected from the two popular music cultures, matched for perceived
tempo, complexity, and familiarity, and subjects were asked to tap in time with
the music. Tapping mode (rate and hierarchical level) varied with subjects'
familiarity with the musical idiom, as evidenced by an interaction between
musical culture and type of music: participants synchronized at higher hierarchical levels (and over a wider range) with music from their own culture than
with an unfamiliar type of music. Thus, passive acculturation as well as explicit
music tuition influence our perception and cognition of music.

KEYWORDS: music synchronization; temporal organization; hierarchical structure; intercultural studies; musical training

INTRODUCTION

Turn on your radio to your local music station and tap along with the first piece
of music you hear. Now tune in to a station playing music from a culture that you are
unfamiliar with, and tap along again. You may notice a major difference in the way
with which you tap; you may be surprised to notice that you tap slower with the music from the familiar than with the unfamiliar culture. In this paper we demonstrate
this phenomenon experimentally and present a theoretical framework to account for
it. We therefore investigate whether we perceive music from our own culture in the
same way as we do music from a different, unfamiliar culture. We suggest that certain fundamental processes involved in music perception and production function

Address for correspondence: Carolyn Drake, Laboratoire de Psychologie Expérimentale,
CNRS UMR 8581, Université René Descartes, Institut de Psychologie, Centre Universitaire de
Boulogne, 71, avenue Edouard Vaillant, 92774 Boulogne-Billancourt, France. Voice: (33) 01 55
20 59 30.

drake@idf.ext.jussieu.fr

Ann. N.Y. Acad. Sci. 999: 429–437 (2003). © 2003 New York Academy of Sciences.
doi: 10.1196/annals.1284.053

differently from music from an idiom we know compared with music from an idiom we do not know. We first present the processes under investigation and then describe ways in which their application could vary.

PSYCHOLOGICAL PROCESSES INVOLVED IN SYNCHRONIZING WITH MUSIC

What psychological mechanisms may be involved in synchronization with musical excerpts? The Dynamic Attending Theory proposed by Jones[1–5] suggests one possible set of processes. We used this theoretical framework to evaluate how listeners are able to synchronize with our musical excerpts. Jones considers that each individual has an internal rhythm or rate at which event processing is optimal (*referent period*). When listening to a complex sequence such as music (composed of multiple hierarchical metric levels), listeners *attune* their internal rhythms to one of those in the sequence. Attunement usually occurs with events at the hierarchical level closest to a listener's referent period, called the *referent level* (FIG. 1). Once listeners have "latched onto" the music at this intermediate rate, they can direct their attention towards other hierarchical levels by a process of *focal attending*. They can direct their attention down the metric hierarchy towards events occurring at faster rates (*analytic attending*) or up the metric hierarchy towards events occurring at slower rates (*future-oriented attending*). These focal listening strategies do not exclude the possibility of adopting a global listening strategy whereby the multiple metrical levels are perceived simultaneously. Here, we examine the functioning of three processes: (1) The *referent period* is measured by the rate that participants spontaneously tap in the absence of all sound sequences (spontaneous motor tempo). (2) The *referent level* is measured by the hierarchical metric level corresponding to the rate at which participants spontaneously synchronize (spontaneous synchronization tempo). (3) *Focal attending* is measured by asking listeners to synchronize with music at rates faster (or slower) than their referent level and then faster (or slower) still, until they can no longer accomplish the task. This method indicates the ease with which individuals can move away from their referent level and the range of hierarchical

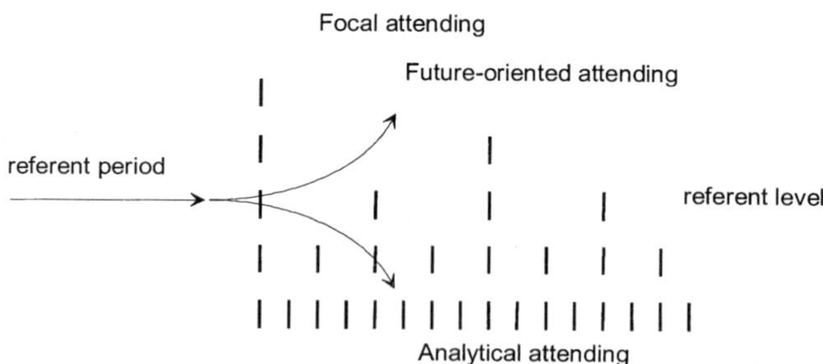

FIGURE 1. Schematic representation of dynamic attending to a hierarchical sequence: guided initially by their referent period, listeners attend spontaneously to one particular hierarchical level (referent level) and may then direct their attention to longer (future-oriented attending) or shorter (analytic attending) periodicities within the stimulus sequence by a process of focal attending.

levels accessible to them (number of hierarchical levels) (see Refs. 5 and 6 for full details of these dependent measures).

ORIGIN OF THESE PROCESSES

Previous studies have investigated how these processes vary with two factors.[7,8] First, the role of *explicit musical training* has been highlighted.[6] The performance of a group of adult musicians was compared with that of an equivalent group of adult nonmusicians. They found that musicians spontaneously synchronized with slower hierarchical levels (slower referent level) and had access to more hierarchical levels (greater focal attending). Second, the role of *age* has been investigated.[5] The performances of 4-, 6-, 8-, and 10-year-old children and adults were compared. It was found that: (1) referent period slowed with age, (2) referent level slowed with age, and (3) focal attending increased with age. The additional influence of musical training was observed even from an early age: 6-year-old musicians behaved like the 8-year-old nonmusicians for all three dependent measures (a shift of about 2 years was observed up to the age of 10 years). Thus, these processes appear to be influenced by both explicit musical training and age.

How can music training and age influence these three processes? In *explicit music training*, musical tuition usually focuses on the musical structure, familiarizing the musicians with the metric structure, thus enhancing the complexity of the mental representation and facilitating focal attending to particular levels. Also, musicians learn to organize events over longer time spans by focusing on higher hierarchical levels. Other processes must be investigated to understand how *age* may intervene. It could be purely a question of maturation, whereby the system only reaches maturity during adolescence. Appropriate data are not yet available to disqualify this hypothesis. However, we favor another hypothesis, that of passive acculturation by implicit learning. Even before birth, infants are very frequently immersed in music, and exposure to music remains high throughout life (nursery rhymes at primary school, "pop music" in teenage years, a variety of music later in life, etc.). The idea is that we pick up regularities in the music we hear and learn to extract the typical musical structure. We therefore create expectations about what usually happens in our music. The role of such implicit learning in music perception has now been demonstrated.[9] It should be noted that we usually grow up in a particular musical culture, hearing repeated exemplars of a particular musical idiom. It is therefore likely that this implicit learning is specific to the particular musical culture in which people grow up.

In accordance with the preceding logic, we should therefore be able to extract the typical musical structure of music from the culture in which we grew up, but we should not be able to do so (or at least less successfully) in music from a culture with which we are not familiar. We therefore adapt the previous tasks to compare synchronization performance of French and Tunisian subjects with French and Tunisian music. We predict that participants will better extract the musical structure of the music with which they are familiar and therefore will organize events over longer time spans at higher hierarchical levels and so will synchronize at a slower rate with music from their own culture than with music from a culture with which they are not familiar.

MATERIAL AND METHOD

Participants. Forty-eight adults who reported normal hearing took part in this experiment. Their mean age was 27 years (ranging from 20–40 years). Half the participants were Tunisian and the other half were French. In each nationality group, half were musicians (who had received formal music tuition in both an instrument and music theory for at least 5 years and played at least twice a week at the time of the experiment) and half were nonmusicians (who had never received any formal music tuition, did not play an instrument, and rarely listened to music). The Tunisian musicians (TM) did the experiment in Tunisia at the Music Conservatory in Sousse, and the Tunisian nonmusicians (TN) did the experiment in Tunisia at the Human Science University in Tunis. The French musicians (FM) and French nonmusicians (FN) did the experiment at the University of René Descartes in Paris, France.

Stimuli. Six French and six Tunisian songs were selected as stimuli from a large sample of typical popular music from each culture. Pretests were run in which listeners from each culture judged all the songs from their own culture in the entire sample for "familiarity" and "complexity" on independent scales of 1–7. Mean IOI (inter-onset interval; duration of interval between onsets of successive beats) was calculated in order to obtain an indication of tempo (medium = 500–700 ms IOI, fast <300 ms IOI, and slow >800 ms IOI). Six songs were then selected from each culture so as to equalize across these parameters. One familiar and one unfamiliar song was chosen for each singer to balance across styles. Each excerpt lasted 1 minute. TABLE 1 presents the characteristics of the selected songs.

Apparatus. Creation and presentation of the stimuli and recording of responses were controlled by a personal computer equipped with a Creative SoundBlaster 16 Value sound card. Stimuli were presented binaurally through headphones (Sennheiser HD 580). Finger taps were recorded on a specially designed touch-sensitive

TABLE 1. Details of the songs selected from the full sample

Culture	Song	Singer	Tempo	Familiarity	Time signature
Tunisian	Maftoun	Hédi Jouini	Fast	F - 7	2/4
	Ya Kassi	Hédi Jouini	Slow	U -3	4/4
	Ain	Lotfi Bouchnak	Fast	F - 7	4/4
	Ya meziana	Lotfi Bouchnak	Medium	U - 3	6/8
	Sultan	Saber Rebaï	Slow	F - 7	4/4
	Aghrab	Saber Rebaï	Medium	U - 2.5	4/4
French	Milord	Edith Piaf	Fast	F - 7	4/4
	On cherche un Auguste	Edith Piaf	Slow	U - 2	3/4
	Flamandes	Jacques Brel	Medium	F - 6	4/4
	Comme	Jacques Brel	Fast	U - 1	2/4
	Besoin de personne	Véronique Sanson	Slow	F - 6	4/4
	Comédie	Véronique Sanson	Medium	U - 1	4/4

pad that coded the time of onset of each tap. Listeners were tested individually in a quiet room.

Procedure. The session started with a measurement of their spontaneous motor tempo (spontaneous motor tempo, SMT1). Participants were requested to tap in a regular fashion at a rate that seemed most spontaneous and natural to them. They were then asked to tap as fast as possible (fast motor tempo, FMT) and then as slow as possible (slow motor tempo, SLMT) to obtain an indication of the limits of their motor control and echoic memory. Each recording was made during a 30-second period. The main part of the experiment then began. They heard a training excerpt from their own culture and then each of the 12 selected songs (presented in a counterbalanced order). First, the participants listened to the song and provided a "familiarity" judgment on a scale of 1–7 (1 = very unfamiliar, 7 = very familiar). For both cultures, all the familiar songs received ratings of 6 or 7, and all the unfamiliar songs received ratings of 1, 2 or 3, confirming the selection criteria obtained during the pretests. None of the participants was familiar with the stimuli from the other culture. They were then asked to tap in a regular fashion in time to the music at a rate that seemed to go most naturally with the music (spontaneous synchronization tempo, SST). Half the participants were then asked to tap faster and then faster still until they were unable to synchronize with the music (fast synchronization tempo, FST). The other half of the participants were asked to tap slower and then slower still until they were unable to synchronize with the music (slow synchronization tempo, SST). They then switched tasks. This order was counterbalanced within and between participants. These synchronization rates were recorded during the 60 seconds of each musical excerpt. The session, which lasted about 90 minutes, finished with a second measurement of their spontaneous motor tempo (SMT2).

Data analysis. Three dependent measures were recorded:

Spontaneous motor tempo (SMT1, SMT2), an indication of the referent period. The mean IOI durations (and standard deviations) were recorded over the 30-second tapping period.

Spontaneous fast and slow motor tempi (FMT and SLMT), an indication of motor limits and echoic memory. The mean IOI durations (and standard deviations) were recorded over the 30-second tapping period.

Spontaneous synchronization tempo (SST), an indication of the referent level. The mean IOI duration of synchronized taps was recorded over the 60-second musical excerpt for periods of regular tapping (within a tolerance window of $\pm$ 20%).

Number of hierarchical levels synchronized (NHL), an indication of the range of accessible hierarchical levels. The number of hierarchical levels with which the participants were successfully able to synchronize their taps ($\pm$ 20%) was calculated from the faster and slower synchronizations.

RESULTS

Spontaneous Motor Tempo (SMT). TABLE 2 shows the mean SMT (and standard deviations) for the four groups of participants. An ANOVA of the SMT by nationality (French, Tunisian) and expertise (musicians, nonmusicians) revealed a significant difference between the French and Tunisian participants ($F(1,44) = 5.08$, P <0.03) but not between the musicians and nonmusicians, and no significant interac-

TABLE 2. Spontaneous motor tempo (SMT), fastest motor tempo (FMT), and slowest motor tempo (SLMT) for the four groups of participants

Subjects	SMT (ms IOI)	FMT (ms IOI)	SLMT (ms IOI)
French musicians	702 (187)	187 (36)	3675 (956)
French nonmusicians	712 (243)	192 (47)	2229 (1109)
Tunisian musicians	847 (286)	164 (17)	3417 (1326)
Tunisian nonmusicians	847 (286)	192 (52)	2660 (1152)

tion between these two factors. The French tapped on average every 707 ms and the Tunisians tapped on average every 851 ms. We had neither hypothesis nor explanation about this effect.

Fastest Motor Tempi (FMT) and Slowest Motor Tempi (SLMT). TABLE 2 also shows the fastest and slowest rates that the participants were able to tap. An ANOVA of the FMT by nationality (French, Tunisian) and expertise (musicians, nonmusicians) revealed no significant differences as a function of either nationality or expertise. The FMT results confirm the absence of problems of motor control in our main tapping task, as all the rates used were slower than these fastest possible rates for all four groups. However, an ANOVA of the SLMT revealed a significant difference as a function of expertise $(F(1,44) = 11.13, P < 0.01)$: musicians tapped on average every 3,546 ms and the nonmusicians every 2,445 ms. This result confirms previous observations of a greater echoic memory span in musicians (Drake *et al.*, 2000).

Spontaneous Synchronization Tempo (SST). The experimental design is one of interaction. The validity of the stimulus and participant selection relies on there being no main effects. As predicted, an ANOVA of the SST as a function of nationality (French, Tunisian), expertise (musician, nonmusician), and type of music (French, Tunisian) revealed no significant main effects. The Tunisians synchronized, on average, at the same rate as the French, the musicians at the same rate as the nonmusicians, and all the participants tapped, on average, at the same rate for the French and Tunisian music. This absence of significant effects confirms the validity of our methodological paradigm and the choice of musical stimuli.

As predicted, however, we observed a significant interaction between nationality and type of music $(F(1, 44) = 88.0\ P < 0.0001)$. FIGURE 2 shows that the French subjects tapped slower with the French music than with the Tunisian music and that the Tunisian subjects tapped slower with the Tunisian music than with the French music.

Number of Levels (NHL). FIGURE 3 shows that, on average, Tunisian participants could synchronize with more levels than could the French participants (Tunisian = 5.4 levels, French = 4.9 levels) $(F(1,44) = 4.20, P < 0.05)$. The musicians had access to more levels than did the nonmusicians (musicians = 5.6 levels, nonmusicians = 4.8 levels) $(F(1,44) = 10.5, P < 0.01)$. More importantly, a significant interaction was noted between nationality and type of music $F(1,44) = 46.0, P < 0.0001$: participants used more levels in their own music.

Familiarity. Within each type of music, three of the exemplars had been selected as being highly familiar to the participants from their culture, and three of the exemplars were highly unfamiliar. We had expected that the more familiar participants were with a particular piece of music, the greater would be their awareness of the

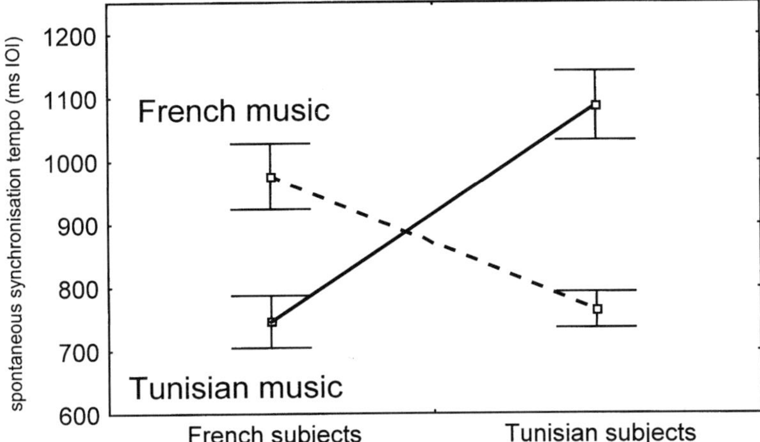

FIGURE 2. Mean spontaneous synchronization tempo (ms IOI) for the French and Tunisians subjects for both French and Tunisian music. Error bars indicate standard errors.

underlying structure and so the slower they would synchronize on average and the greater the number of levels at which they could synchronize. An ANOVA on SST by nationality, expertise, type of music, and familiarity (familiar, nonfamiliar) did not reveal a significant main effect of familiarity or interactions with other factors. However, an ANOVA of NN by nationality, expertise, type of music, and familiarity (familiar, nonfamiliar) revealed a significant main effect of familiarity ($F(1,44) = 11.8$, $P < 0.01$), but no interactions with other factors: participants could synchronize with more levels with familiar music (5.5 levels) than with unfamiliar music (5 levels). These results suggest that familiarity with a particular type of music probably has a greater influence on the way we perceive music than does familiarity with a particular exemplar.

Sequence Tempo. We had also systematically controlled for sequence tempo by selecting one slow, one medium, and one fast exemplar in each subcategory of musical excerpt. Concerning spontaneous synchronization tempo, all groups tapped slowest for the slowest exemplars and fastest for the fastest exemplars ($F (2,88) = 28.06$, $P < 0.0001$); however, a significant interaction between type of music (French, Tunisian) and exemplar ($F(2,88) = 22.4$, $P < 0.0001$) revealed that this effect was more marked for the French than for the Tunisian music. No difference in the number of accessible levels was observed between the different sequence rates.

DISCUSSION

To our knowledge, this is the first cross-cultural investigation of how listeners perceive the hierarchical temporal structure of music. If we accept the principle that the way in which participants synchronize with music provides an indication of how they perceive that music, we have demonstrated that the perception of music from

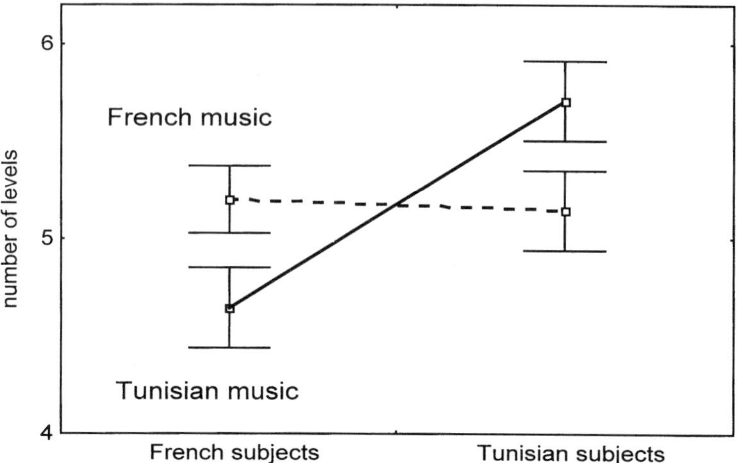

FIGURE 3. Number of levels with which the French and Tunisians were able to synchronize with French and Tunisian music. Error bars indicate standard errors.

our own culture is different from that of music from another culture: our participants tapped slower and with more hierarchical levels with music of their own culture than with music of an unfamiliar musical culture. This interaction was almost completely symmetrical.

Our theoretical framework provides one interpretation. When participants synchronized with music from an unfamiliar culture, they focused spontaneously on the hierarchical level (referent level) closest to their referent period. This was relatively fast. In the present experiment, as previously,[6] the mean spontaneous synchronization tempo (referent level = 750 ms IOI) was extremely close to the mean spontaneous motor tempo (referent period = 779 ms IOI). Basically, they relied on their own spontaneous tempo rather than using their knowledge of musical structure to organize events over longer time spans. Also, the fact that participants had difficulty synchronizing with other hierarchical levels indicates that their knowledge of the musical structure was limited. However, when participants synchronized with music from their own culture, they synchronized spontaneously at slower rates (that is, at higher hierarchical levels), thus providing evidence that they organized events over longer time spans. They were also able to synchronize with more hierarchical levels, indicating superior knowledge of the musical structure.

These main results lead to a clear conclusion: passive acculturation or implicit learning plays an important role in the way in which listeners perceive the music with which they have grown up. Of particular note is the fact that in the present study, this factor has an even stronger influence than the previously described factor of explicit music training. Indeed we were surprised to observe the absence of difference between musicians and nonmusicians in the spontaneous synchronization task, although musicians could synchronize with significantly more hierarchical levels. This reduced effect of musical training may be related to the present choice of musical stimuli. To compare French and Tunisian music, we opted for popular sung music rather than classical excerpts used previously.[6] The previous greater effect of musical training may have been due to the superior exposure of the musicians to classical music. In any case, the choice of using popular music proved successful as it enabled us to demonstrate the influence of passive acculturation.

It was recently demonstrated that one important factor influencing listeners' preference judgments for music is their familiarity with the musical idiom. The present results extend this finding further by suggesting one underlying process: we will tend to prefer a piece of music if we are familiar with its musical structure through years of implicit learning, enabling more efficient (hierarchical) organization of events over time.

REFERENCES

1. JONES, M.R. 1976. Time, our last dimension: toward a new theory of perception, attention, and memory. Psychol. Rev. **83:** 323–355.
2. JONES, M.R. 1987. Dynamic pattern structure in music: recent theory and research. Percept. Psychophys. **41:** 631–634.
3. JONES, M.R. & M. BOLTZ. 1989. Dynamic attending and responses to time. Psychol. Rev. **96:** 459–491.
4. JONES, M.R. 1990. Learning and the development of expectancies: an interactionist approach. Psychomusicology **9:** 193–228.
5. DRAKE, C., M.R. JONES & C. BARUCH. 2000. The development of rhythmic attending in auditory sequences: theory and research. Cognition **77:** 251–288.
6. DRAKE, C., A. PENEL & E. BIGAND. 2000. Tapping in time with mechanically and expressively performed music. Music Percept. **18:** 1–24.
7. DRAKE, C. 1998. Psychological processes involved in the temporal organization of complex auditory sequences: universal and acquired processes. Music Percept. **16:** 11–26.
8. DRAKE, C. & D. BERTRAND. 2001. The quest for universals in temporal processing in music. Ann. N.Y. Acad. Sci. **930:** 17–27.
9. TILLMANN, B., J. BHARUCHA & E. BIGAND. 2000. Implicit learning of tonality: a self-organizing approach. Psychol. Rev. **107:** 885–913.

Music and Learning-Induced Cortical Plasticity

CHRISTO PANTEV,[a] BERNHARD ROSS,[a] TAKAKO FUJIOKA,[a]
LAUREL J. TRAINOR,[b] MICHAEL SCHULTE,[a,c] AND MATTHIAS SCHULZ[a]

[a]*The Rotman Research Institute, Baycrest Centre for Geriatric Care,
Toronto, Ontario, Canada*

[b]*Departmant of Psychology, McMaster University, Hamilton, Ontario, Canada*

[c]*F.C. Donders Centre for Cognitive Neuroimaging, Nijmegen, The Netherlands*

ABSTRACT: Auditory stimuli are encoded by frequency-tuned neurons in the auditory cortex. There are a number of tonotopic maps, indicating that there are multiple representations, as in a mosaic. However, the cortical organization is not fixed due to the brain's capacity to adapt to current requirements of the environment. Several experiments on cerebral cortical organization in musicians demonstrate an astonishing plasticity. We used the MEG technique in a number of studies to investigate the changes that occur in the human auditory cortex when a skill is acquired, such as when learning to play a musical instrument. We found enlarged cortical representation of tones of the musical scale as compared to pure tones in skilled musicians. Enlargement was correlated with the age at which musicians began to practice. We also investigated cortical representations for notes of different timbre (violin and trumpet) and found that they are enhanced in violinists and trumpeters, preferentially for the timbre of the instrument on which the musician was trained. In recent studies we extended these findings in three ways. First, we show that we can use MEG to measure the effects of relatively short-term laboratory training involving learning to perceive virtual instead of spectral pitch and that the switch to perceiving virtual pitch is manifested in the gamma band frequency. Second, we show that there is cross-modal plasticity in that when the lips of trumpet players are stimulated (trumpet players assess their auditory performance by monitoring the position and pressure of their lips touching the mouthpiece of their instrument) at the same time as a trumpet tone, activation in the somatosensory cortex is increased more than it is during the sum of the separate lip and trumpet tone stimulation. Third, we show that musicians' automatic encoding and discrimination of pitch contour and interval information in melodies are specifically enhanced compared to those in nonmusicians in that musicians show larger functional mismatch negativity (MMNm) responses to occasional changes in melodic contour or interval, but that the two groups show similar MMNm responses to changes in the frequency of a pure tone.

KEYWORDS: music; cortical plasticity; MEG; functional mismatch negativity (MMNm)

Address for correspondence: Dr. Christo Pantev, Ph.D., D.Sc., Rotman Research Institute, Baycrest Centre for Geriatric Care, University of Toronto, 3560 Bathurst Street, Toronto, Ontario, Canada M6A 2E1. Fax: 416-785-2862.
pantev@rotman-baycrest.on.ca

Ann. N.Y. Acad. Sci. 999: 438–450 (2003). © 2003 New York Academy of Sciences.
doi: 10.1196/annals.1284.054

INTRODUCTION

Research, particularly in the last two decades, has revealed that even in the adult brain the functional organization in the mammalian cortex is not statically fixed, but adjusts in response to alteration of behaviorally relevant input and processing. Injuries, such as deafferentation through damage of a section of hair cells in the cochlea, also remodel the cortical representation. The knowledge that cortical representations are dynamic and continuously modified by experience is based on a series of classical animal studies, particularly those conducted over the last two decades by Merzenich and colleagues.[1,2] Studies in humans support and complement these findings. The most relevant conditions for a cortical reorganization to occur are: (1) increased use or behaviorally relevant stimulation of a receptor pool that leads to expansion of the respective cortical representation and (2) heavy training schedule and a high motivational drive in a behaviorally relevant context.[3] The first example of cortical reorganization of the auditory cortex following increased and behaviorally relevant stimulation was provided by studying musicians.[4] Musicians are usually motivated to practice for hours a day over many years to achieve their specific skills. We found a significant increase in the cortical representation for piano tones in musicians compared to nonmusician controls. Musicians and nonmusicians do not differ in their representation of pure tones, which we do not learn and experience either in the natural environment or during musical education; there is no difference between musicians and nonmusicians. The specificity of the enhancement in musicians for musical tones but not for pure tones suggests that it is the consequence of intensive musical training. Furthermore, we found that the cortical representation of the tones of the musical scale is more extended in those musicians who started playing their instrument earlier in life. We also investigated cortical representations for notes of different timbre (violin and trumpet) and found that they are enhanced in violinists and trumpeters, preferentially for the timbre of the instrument on which the musician was trained. Evoked responses from the auditory cortex responses were higher in amplitude when the musicians heard tones of their own instrument than tones from other instruments.[5]

Together, these studies strongly suggest that the extensive training of musicians has a specific and profound effect on the functional organization of the auditory cortex. In this chapter we describe three studies that extend these findings. First, we examine the learning process itself. Second, we examine cross-modal aspects of the functional reorganization and cortical plasticity. Finally, we examine effects of musical training on the neural encoding and discrimination of sequences of notes or melodies.

PLASTICITY OF THE HUMAN AUDITORY CORTEX INDUCED BY LEARNING TO DISCRIMINATE AMBIGUOUS "VIRTUAL" MELODIES

Most studies of musically induced plasticity have compared musicians with years of extensive training to nonmusicians. We were interested in whether we could induce plastic changes through specific relatively short-term training. Plasticity of the human auditory cortex was investigated by having adults learn to discriminate ambiguous "virtual" melodies.[6] A short melody of eight harmonic complex tones was composed. The harmonic complex tones were a sum of sinusoidal sound waves with frequencies that were integer multiples of a certain fundamental frequency (f_0). The perceived pitch of such complex tones corresponds to the f_0 even when the f_0 is absent ("pitch of the missing fundamental" or "virtual pitch" auditory illusion[7]). In the case of virtual pitch perception, the information from the different harmonics must be bound together in order to perceive the virtual pitch, as there is no energy in the stimulus at this frequency. This auditory binding becomes more difficult when few and/or higher harmonics are present. We constructed a melody consisting of eight complex tones, which were composed of three harmonics (FIG. 1a). Each complex tone of the melody could be perceived according to either the spectrum frequencies (spectral pitch) or the virtual pitch corresponding to the missing fundamental frequency. The missing fundamental frequencies of the eight complex tones followed the beginning of the tune "Frère Jacques" (virtual melody), whereas the harmonics were chosen so that the spectral melody had an inverse contour to the virtual one. Thus, the melody was used as an indicator of whether subjects perceived the spectral pitch or the virtual pitch of the complex tones. Ten subjects, who were initially only able to perceive the spectral pitch melody, were investigated. They were intensively trained until they gained the ability to perceive the virtual pitch melody, and we were able to investigate the involvement of plastic reorganizational processes in virtual pitch formation and perception in the auditory cortex. The magnetic auditory evoked fields were recorded contralateral to the ear of presentation over both hemispheres before and after training in a passive listening paradigm by means of MEG.

In all 10 subjects, training resulted in a sudden switch from the spectral to the virtual mode of pitch perception. Group results were determined in seven subjects who trained for about 1 week (two subjects perceived the virtual melody during the first measurement and one subject trained over 7 weeks). The clear change in perception was accompanied by a distinct increase of the transient gamma-band evoked field that has often been found to be associated with integrative cognitive functions, such as the binding process during object recognition. During the perception of a "Gestalt," the different characteristics of an object are combined into a coherent percept, and the oscillatory neural discharges in the gamma frequency range are proposed to be a representation of this binding process.[8,9] A strong increase was found for the evoked global field power in the gamma frequency band (root mean square, RMS) and for the corresponding estimated cortical strength (c.f. FIG. 1b and d). RMS values averaged across subjects increased significantly from 6.2 fT to 7.9 fT for the left and from 4.96 fT to 6.38 fT for the right hemisphere (ANOVA: F = 7.3; $P = 0.031$). A similar behavior was also observed for the estimated cortical strength (dipole moment) of the gamma band response. The estimated cortical strength of the corresponding sources significantly increased from 1.03 ± 0.16 nAm to 1.92 ± 0.71 nAm in the left hemisphere and from 1.59 ± 0.35 nAm to 2.38 ± 0.47 nAm in the right hemisphere between the two corresponding sessions (F = 6.46, $P = 0.039$). However, no significant differences were observed between hemispheres (F = 0.84, $P = 0.39$).

After training, the estimated equivalent cortical sources were significantly displaced towards the midline by about 6.6 mm and 5.1 mm in the left and right hemispheres, respectively. An additional analysis was performed by means of independent component analysis (ICA). The number of independent components in

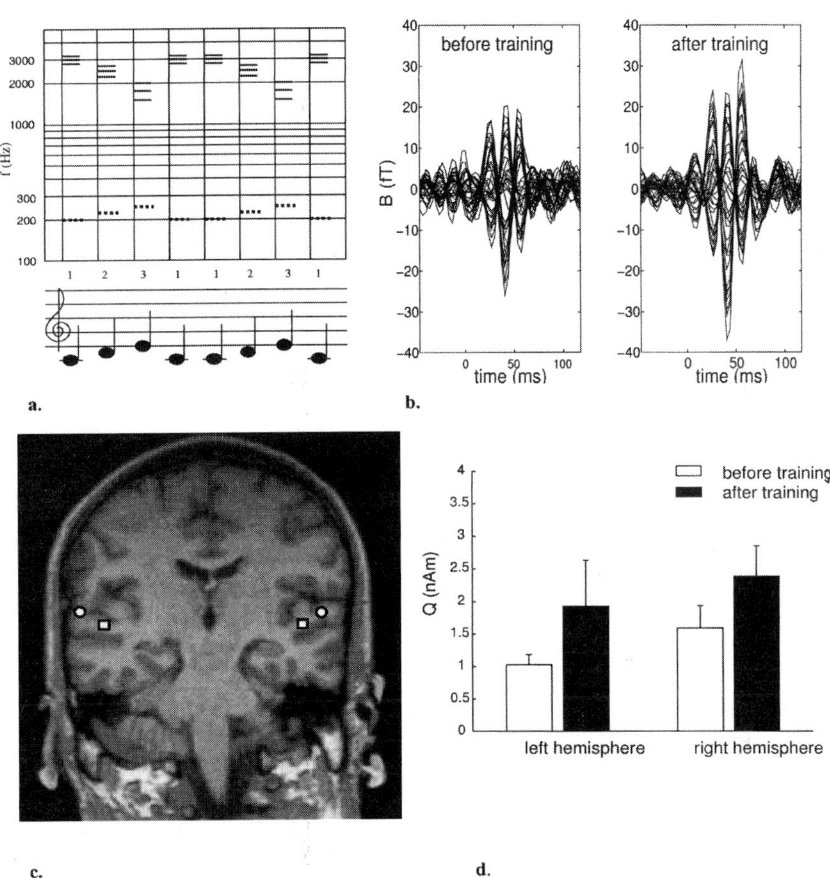

FIGURE 1. (**a**) Stimulus sequence of the virtual melody. The sequence consists of three different complex tones played in the order 12311231. The fundamental frequencies (f_{0i}) were chosen as follow: f_{01}: 200 Hz; f_{02}: 225 Hz; and f_{03}: 250 Hz. The virtual melody follows the tune "Frère Jacques" (shown by the *dotted line*), whereas the spectral melody given by the center frequencies of three harmonics has an inverse pitch contour. (**b**) Averaged gamma band responses of about 600 epochs of one representative subject to tone 1 (band-pass filtered: 24–48 Hz). Data of all 37 measurement channels are plotted together for the session before (*left*) and after training (*right*). (**c**) Location of estimated current dipoles for gamma band responses integrated into an MRI overlay of one individual subject. *Circles* denote the source location before training and *squares* denote the estimated location after training. (**d**) Estimated dipole moments of the gamma band activity for the left and right hemispheres, cross-averaged over all subjects. *Black bars* show the values after training and *white* bars before training. *Error bars* indicate the standard error of the mean.

which the evoked gamma band response could be separated is related to its spatio-temporal variability. After training, fewer independent components were needed to explain the gamma band responses than before. The obtained ICA results can be interpreted as indicating higher synchronization of the cortical networks involved in the generation of the evoked gamma band activity after achieving the ability to perceive the virtual melody.

In this study, the question of learning-induced plasticity in the perception of the virtual pitch of complex tones was directly addressed. The auditory Gestalt recognition was associated with plastic reorganizational processes, expressing themselves as an increase in global field power, cortical strength, shift in source location, and a lower spatiotemporal variability. Together, these results suggest that the integration of different spectral pitches into a single coherent percept is associated with a convergence of harmonically related information. The enhancement of RMS and cortical strength values may be interpreted either by higher synchronization (already suggested by ICA) or by enlargement of the involved cortical networks or, most likely, by both. As the latency of the evoked gamma band activity is about 30–70 ms, this plastic reorganization of neural networks is likely to take place at the level of the primary auditory cortex. We have thus shown that relatively short-term training can lead to plastic changes in the primary auditory cortex.

CROSS-MODAL REORGANIZATION OF CORTICAL FUNCTION IN PROFESSIONAL TRUMPET PLAYERS

Cross-modal plasticity was first described in humans who experienced sensory loss in early life. Areas of cortex, deprived of exposure to the visual or auditory stimuli to which they normally respond, start to process information from the intact senses. For example, tactile processing occurs in the occipital cortex in blind individuals[10] and visual processing in the auditory cortex in those who are deaf.[11] Multisensory integration has been defined by Meredith and Stein[12] as the increase of a neuron's responses to a stimulus combination as compared to its response to an individual stimulus. Multimodal representations are possibly generated through the convergence of information from different sensory systems onto a common group of neurons. These neurons integrate multimodal signals that occur within a certain time frame for a defined receptive field. When multiple sensory cues are available, synthesis of cross-modal information is performed, and the response is modified if stimuli from two or more modalities are presented simultaneously.[13,14] The perception of cross-modal cues, involving cross-modal plasticity mechanisms, must be important for learning and playing a musical instrument, because the perception of the sound produced must interact with the motor programs for playing the instrument. Primary cortices that have classically been thought to respond to one modality might actually be more complex. With regard to this idea we hypothesized that cross-modal plasticity is involved in the process of sensory signal integration over the merged senses to achieve a superior musical performance.

We tested 10 professional trumpeters (3 females, 7 males, 26 ± 2.9 years) and nine nonmusician controls (3 females, 6 males, 25 ± 3.9 years). All subjects were right handed and had normal hearing according to air- and bone-conduction thresholds between 250 and 8000 Hz. The trumpet players had been playing their instrument for an average of 15.3 ± 3 years and reported that they practiced an average of 18.4 ± 8 h/week in the 5 years preceding the study. All stimuli were presented contralaterally to the side of MEG measurement. Unimodal tactile stimuli were applied to either the lateral side of the lower lip or the tip of the index finger. The auditory stimulus was the trumpet tone B2 (American notation). In the bimodal conditions, the same trumpet tone was presented simultaneously with the tactile stimulation of either the lower lip or the index finger. The intensity of the digitally sampled trumpet tone was 60 dB above individual sensation level. The sound was delivered through echoless plastic tubing and a silicon earpiece to the subject's ear canal. Balloon membranes of 1 cm in diameter driven by impulses of compressed air applied the tactile stimuli. Each stimulus occurred 256 times in a pseudo-randomly presented sequence with a stimulus onset asynchrony of 1900 ± 200 ms.[15]

Stimulus-related epochs of the evoked magnetic fields were averaged with respect to the different stimuli. The magnetic field maxima of the somatosensory responses were identified for all subjects near 20 and 40 ms after stimulus onset for the lip and about 50 ms for the index finger. Magnetic source analysis was performed using the data points around these field maxima, and the origins of the cortical sources were estimated based on an equivalent current dipole (ECD) model. The field of the ECD was fitted to the measured magnetic field distribution in the time interval with the maximal field power (measured as RMS across all channels). The obtained source coordinates and orientations were used to form a spatial filter that collapsed the time series of the MEG sensors into a single waveform of a magnetic dipole moment. This method of source space projection allowed the observation of the time-course of the cortical source.

For all subjects, the arithmetical sum of the unimodal auditory and somatosensory responses was subtracted from the bimodal response using the tone and the tactile stimulus to the lip or the finger, respectively. The resulting interaction waveforms were tested statistically and were compared between groups. The bootstrap method[16] was applied to the multimodal interaction waveforms in order to estimate the group averages and their 95% confidence limits. The strengths of all waveforms (uni- and bimodal responses, sum of unimodal responses, and multimodal interaction) were evaluated by analyses of variance (MANOVA) to examine the effects of stimulus conditions and experimental groups.

In both the musician and the control groups, the source waveforms obtained from the somatosensory finger area did not show significant differences between the combined condition and the sum of responses to the tone and the finger stimuli, indicating no cross-modal interaction in both groups and no significant difference between the groups as well. By contrast, the source waveforms obtained from the somatosensory lip area showed distinctive differences between the responses to the combined lip and tone stimulation and the responses obtained by summating responses from the unimodal lip and tone stimulation. These differences of bimodal interaction waveforms (multimodal response minus the sum of unimodal responses) are illustrated in FIGURE 2. The group mean interaction waveforms are displayed in combination with their bootstrapped 95% confidence limits. Centered around 33 ms the musician's response waveforms show a clear amplitude increase in the multimodal lip and tone condition compared to summed waveforms of responses in both single modalities (F(1,34) = 7.2, P = 0.011). The multimodal interaction was more pronounced in the musician group than in the control group especially on the right (P =

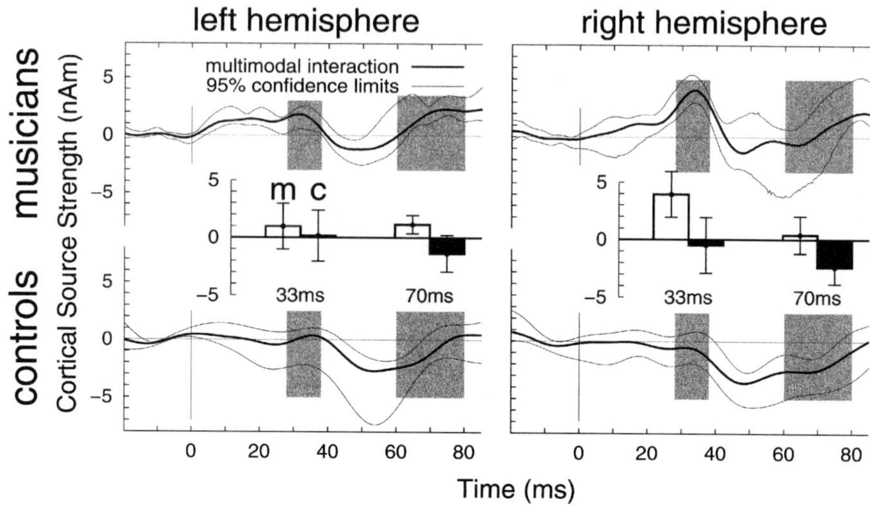

FIGURE 2. Group averaged waveforms of multimodal interaction (*thick lines*), which was calculated as the difference between the multimodal response and the sum of responses to the separately applied auditory and somatosensory lip stimuli. The 95% confidence limits of the mean are shown with *thin solid lines*. The *inserted bar diagram* shows the mean interaction in the *gray* shaded intervals around 33 and 70 ms for the musician group (m, *open bars*) and the control group (c, *filled bars*). The error bars denote the 95% confidence limits as the result of a *t* statistic.

0.012), but it was also significantly larger on the left ($P = 0.037$). However, in the latency interval of 60–80 ms (FIG. 2), the control group showed a noticeable amplitude decrease in the multimodal lip and tone response compared to the summed unimodal responses (F(1,34) = 19.1, $P = 0.00011$). A pair-wise *t* test proved that there was a larger interaction in the control than in the musician group for the right ($P = 0.013$) and left ($P = 0.0026$) hemispheres.

The significant difference found between musicians and nonmusicians is the most important result of this study, illustrating that auditory-somatosensory interaction effects are characteristically different between the two groups. In two distinct time intervals, the multimodal interaction in the musician group was different from that in the nonmusician group. (1) The lip interaction waveform from the musicians showed a positive peak with a maximum around 33 ms, which was missing in the nonmusicians group. (2) In the 60–80-ms interval, the lip interaction waveform in the nonmusician group showed a clear decrease, whereas the corresponding waveform in the musician group did not decrease. Because auditory and somatosensory responses are of opposite polarity in the latency range, the observed enhanced interaction at 33 ms found in musicians can be explained by (i) a decreased auditory response, (ii) an increased somatosensory response, or (iii) an additional multimodal response in the vicinity of the somatosensory cortex, suggesting a qualitatively different way of multimodal information processing.

In contrast to stimulation of the lip, no significant multimodal interactions were noted in either musicians or nonmusicians for stimulation of the index finger. The vocal tract, diaphragm, and particularly the lips are more intensely engaged in playing the trumpet than are the fingers. Therefore, it is most likely that the increased interaction around 33 ms in highly trained professional trumpet players is caused by an increased cortical response of the lip, related to multimodal interaction occurring early in the sensory processing. Although further studies are necessary for a better understanding of multisensory information processing in humans, our study suggests that musical training leads to remarkable modifications in cross-modal processing. The behaviorally relevant stimulation of somatosensory and auditory modalities in trumpet players and the increased use of these modalities during the intense training schedule are the main reasons for the observed multimodal plasticity effects.

NEURAL CODING OF MELODY

In the studies of plasticity discussed thus far, we have focused on the processing of single tones. Meaning in music, however, involves how tones are put together. The question asked here concerns the effect of extensive musical training on the processing of sequences of tones or melodic information. Across different societies, musical structure has two aspects, time structure (rhythm) and pitch structure,[17] although the particular instantiations differ from one musical system to another. From a perceptual point of view, melodic pitch structure has two aspects, a contour and an interval code.[18,19] The contour representation consists of information about the up and down pattern of pitch changes, regardless of their exact size, and is common to both speech prosody and musical melody. The interval representation consists of the more analytic structure of the exact pitch distances between successive tones. Pitch interval is specific to music and is crucial for scales and harmony, where exact intervals define the structure. Behavioral studies have provided evidence that contour is more fundamental than interval in that both infants and musically untrained adults can process contour information, but have difficulties encoding the intervals of unfamiliar melodies.[20,21] However, the pitch interval structure is better perceived by trained than untrained musicians.[22,23]

How is melodic information processed, stored, and retrieved in the brain, and what is the effect of musical training? We investigated the relationship between long-term musical training and automatic melodic processing. Because pitch intervals provide the essence of musical pitch structure, we expected that the underlying neural mechanism would be affected greatly by musical training, whereas because contour structure is common to both speech and music, we expected that musicians and nonmusicians would differ less on this feature. Therefore, we investigated the effect of musical expertise on contour and interval processing and compared it to that in naïve subjects. The stimuli were designed to clearly separate contour and interval encoding. Factors of deviation in musical context involving out-of-key changes, familiarity of melody, and range of pitch leaps that might cause additional cognitive processing were carefully controlled. The magnetic mismatch negativity field (MMNm) was used to investigate the neural mechanisms for the automatic encoding of melodic features. To test the specificity of any enhancements, a control condition examining frequency deviations to single tones without melodic context was also included.

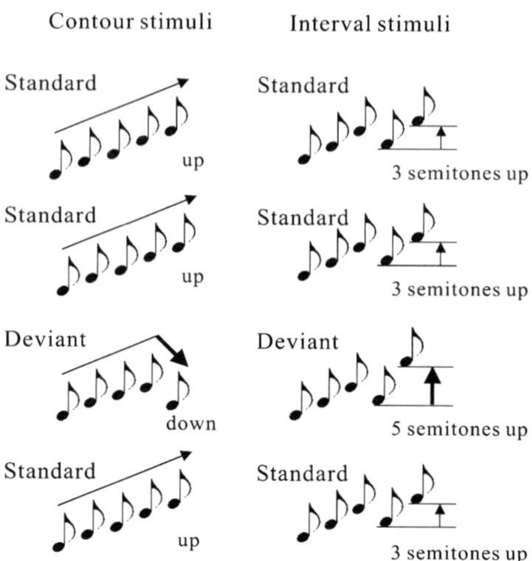

FIGURE 3. Musical stimuli for the contour (*left column*) and interval (*right column*) tasks. In each case, the first four notes of the melodies form a common sequence, which is followed by the *standard* and deviant terminal note. In the contour case, the interval size changes among melodies, but the *standard* terminal notes always rise, whereas the deviant terminal notes always fall. In the interval case, the deviant terminal note is higher than the standard terminal notes by a whole tone, but the contour of the melody does not change.

Twelve musicians (8 females) and twelve nonmusically trained adults (9 females) matched in age (20–40 years) participated in this study. (All participants were right handed and had normal hearing.) None of the subjects in either group had absolute pitch. The stimuli for both the contour and the interval melodic conditions were composed of sequences of 5-note *standard* and *deviant* melodies, with standard melodies occurring 80% of the time. The appropriate five notes for each melody were played in succession with equal note-to-note onsets of 300 ms, for a total melody length of 1,500 ms. The stimuli for the control condition were designed as a sequence of standard and deviant pure tones with different frequencies. The standard contour stimuli consisted of five ascending notes. Eight melody variations were composed with five different starting notes and different intervals between notes, as shown in FIGURE 3. In the corresponding deviant melodies, the first four notes were identical to those of each standard melody. However, the last note was changed to a descending note. Thus, because the intervals changed from trial to trial, the *standard* and *deviant* melodies were only distinguishable by their contour. In the interval condition, the standard stimuli consisted of the same five-note melody that was transposed to eight keys so that the starting notes were in the same range from C5 to G5 (American notation) as those of the contour stimuli. The deviant melodies were derived from the set of standard melodies by raising the final note by a whole tone or major second ($1/6^{\text{th}}$ of an octave), a change that remained within the key of the melody and did not change the contour. The first notes of all contour and interval melodies were between C5 and G5, and the last notes were between G5 and F6. The interval size deviations in the contour melodies ranged from a minor second to a major third ($1/12^{th}$ to 4 of an octave), and its mean value was a major second around the median position of termination (B5), which was the same value of deviation exploited in the interval conditions.

Each melody was 1,500 ms in duration and was separated by a 900-ms silent interval. The experimental session consisted of 900 trials of the contour and 900 trials of the interval condition. In the contour condition, the order of melody variations was pseudo-random. To avoid the appearance of the same note within two successive trials of the interval condition, which might be recognized as "odd" or "primed" despite the interval deviation, the transposition from trial to trial was set to be an upward major third and a downward minor third, until the starting note became G5. Thus, the repeating tonality change resulted in the series "C-E-C#-F-D-F#-D#-G." The control condition consisted of two successive blocks of 500 pure tones of 300-ms duration, presented with an ISI of 450 ms. The frequency of the 80% standard tones was 990.7 Hz (B5) and of the 20% deviant tones 1111.0 Hz (C#6). Thus, the size of the change was a minor second, which equals the mean size of that used in the melodic contour and interval conditions. In all conditions the stimuli were presented at 60 dB above the individual sensation level.

The magnetic field responses were recorded with a 151-channel whole-cortex magnetometer system (OMEGA, CTF Systems Inc, Port Coquitlam, Canada) and the MEG signals were band-pass filtered between 0.1 and 200 Hz. The recordings were performed in a sitting position within the magnetically shielded room. The subjects were instructed not to pay attention to the sound stimuli and to watch soundless movies of their own choice. The recording session began with the control condition for all subjects. For half of the participants, the contour condition was presented first, whereas for the other half the interval condition was presented first. The participants received no explanation about the stimuli. After the MEG recordings, all subjects participated in a behavioral test, consisting of two contour and interval discrimination tasks. The tests were designed as two alternative forced choice tasks (2AFC) with two melodies presented sequentially on each trial. The subjects were instructed to judge whether both melodies were the same or different in terms of contour or interval structure. All melodic stimuli were chosen from the sets of stimuli used in the MEG experiment.

The recorded magnetic field data were averaged selectively for the standard and deviant stimuli and all stimulus types. The analysis technique of signal space projection (SSP)[24] was applied to the MEG data. This results in considerable discrimination against the sensor noise and uncorrelated brain activity from distant brain regions, and it allows the averaging of dipole moment waveforms from repeated measurements in the same subject or between subjects. Grand average dipole moment waveforms across both groups of subjects were obtained selectively for the standard and deviant stimuli. Individual difference waveforms were calculated by subtracting the response to the standard from that to the deviant stimuli. MMNm responses were examined after the onset of the fifth note in the melodic conditions and after the onset of the pure tone stimulus in the control condition. The 95% confidence intervals for the grand averaged response waveforms and the difference waveforms were estimated from nonparametric bootstrap re-sampling analysis.[25] The amplitudes of MMNm were statistically examined by a repeated measures analysis of variance (ANOVA) with one between-subjects factor (group) and two within-subjects factors (condition; hemisphere). The posthoc comparison was calculated with Fisher's PLSD tests using the level of significance as 5%.

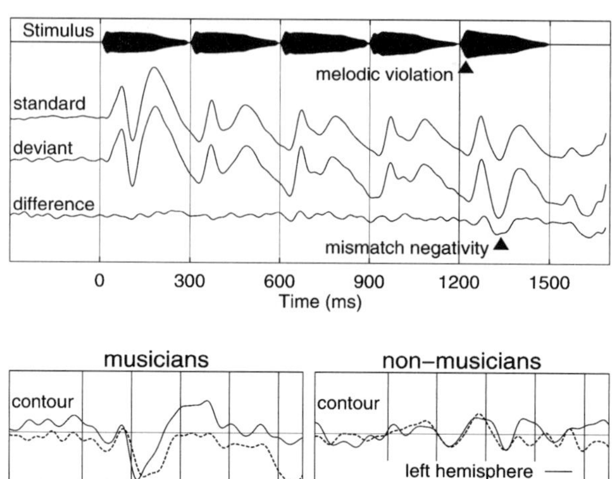

FIGURE 4. (*Top*) Example of a grand average of the source space standard, deviant, and difference (*deviant–standard*, MMNm) waveforms across musicians for the contour-melody stimulus condition, right hemisphere. The y-axis shows the dipole moment (positive upwards), and the x-axis shows the time related to the stimulus onset. The stimuli of both contour and interval melodic conditions do not differ in the range from the 1st note to the 4th note (0.0~1.2 s) but at the last 5th note. The *arrow* indicates the clearly identifiable MMNm responses. (*Bottom*) Grand averages of the source space waveforms of MMNm (difference of the dipole moment time series between deviant and standard response for the contour, interval, and tone frequency stimulus conditions) for musicians (*left column*) and nonmusicians (*right column*). Data of the left hemisphere are indicated by a *solid line*, those of the right hemisphere by a *dotted line*. For both contour and interval condition the timescale on the x-axis refers to the onset of the 5th note.

Clear AEFs were obtained from both musicians and nonmusicians in all stimulus conditions. An example of the grand averaged dipole moment waveforms evoked by the notes of the melodies is displayed in FIGURE 4 (top). Although the signal-to-noise ratio was smaller for the deviants (smaller number of stimuli) than the standards, highly reproducible response patterns were observed. As shown in FIGURE 4 (bottom), after the onset of the fifth note, clear MMNm responses were observed in both hemispheres for both contour and interval conditions in musicians. By contrast, non-musicians showed unclear responses in both contour and interval conditions. However, in the single tone control condition both musicians and nonmusicians have clear MMNm responses to the frequency change of the stimulus. For musicians, the bootstrap statistics revealed highly significant MMNm responses for the deviation of melodies at 100–150 ms after onset of the last note. By contrast, the MMNm responses to the deviation in nonmusicians reached significance only in the right hemisphere for the interval condition. For both groups, the MMN evoked by single tone frequency deviation was highly significant between 95 and 200 ms after stimulus onset. The magnitudes of MMNm for the two melodic conditions in the musician group were compared, and significantly larger MMN was obtained in the interval condition. In addition the MMNm source waveforms show MMNm peaks that are slightly later in the interval condition compared to the contour condition. In general, the MMNm was significantly larger in musicians than nonmusicians according to the ANOVA (F(1, 22) = 12.787, P <0.01) for both contour and interval conditions.

The results of the behavioral tests show that the performance of musicians was significantly better than that of nonmusicians (ANOVA, (F(1,22) = 23.865, P <0.0001). Musicians exhibited good performance in both contour (96.50%) and interval (95.83%) conditions, and they did not show significant differences between tasks. Nonmusicians performed at 63.0% in the interval task, which is above the chance level of 50% (t(11) = 3.564, P <0.01), but below a typical threshold definition of 75% correct. However, their performance of 86.17% in the contour task was significantly better (t(11) = 5.946, P <0.0001) than that in the interval task. Thus, despite good behavioral performance on the contour task in both groups, MMNm responses for automatic contour encoding are much smaller in nonmusicians than in musicians.

In summary, musicians showed significantly larger MMNm responses than did nonmusicians to deviations in melodic contour and interval structure. However, both groups showed similar MMNm responses to pitch deviation with single sine-wave tones. These results strongly support the hypothesis that the musical experience of musicians leads to specific changes in the neural mechanisms for processing abstract melodic information and that long-term musical training particularly enhances the processing of pitch relations between the successive notes of melodies.

CONCLUSIONS

The differences between musicians and nonmusicians observed in our studies contribute to the growing literature on differences in processing between musicians and nonmusicians and suggests that musical training affects a whole network of brain areas, from those involved in stimulus encoding, to those involved in cross-modal integration, to those involved in deviance detection in extended melodies. Furthermore, our training studies indicate that at least some plastic changes in the auditory cortex can occur over the short term with specific laboratory training. We conclude that the brain is constantly changing in response to relevant auditory information in the environment.

ACKNOWLEDGMENT

We thank the students from the Faculty of Music (University of Münster and University of Toronto) who participated in this study. This research was supported by CIHR and the International Foundation for Music Research, California, USA.

REFERENCES

1. MERZENICH, M.M. & W.M. JENKINS. 1993. Reorganization of cortical representations of the hand following alterations of skin inputs induced by nerve injury, skin island transfers, and experience. J. Hand Ther. **6:** 89–104.
2. MERZENICH, M. *et al.* 1996. Cortical plasticity underlying perceptual, motor, and cognitive skill development: implications for neurorehabilitation. Cold Spring Harb. Symp Quant Biol. **61:** 1–8.
3. ELBERT, T. & S. HEIM. 2001. A light and a dark side. Nature **411:** 139.
4. PANTEV, C. *et al.* 1998. Increased auditory cortical representation in musicians. Nature **392:** 811–814.
5. PANTEV, C. *et al.* 2001. Timbre-specific enhancement of auditory cortical representations in musicians. Neuroreport **12:** 169–174.
6. SCHULTE, M. *et al.* 2003. Functional evidence of different modes of pitch perception and learning induced plasticity in the human auditory cortex. Cortical Plasticity. In press.
7. TERHARDT, E. 1974. Pitch, consonance, and harmony. J. Acoust. Soc. Am. **55:** 1061–1069.
8. SINGER, W. & C.M. GRAY. 1995. Visual feature integration and the temporal correlation hypothesis. Annu. Rev. Neurosci. **18:** 555–586.
9. TALLON-BAUDRY, C. & O. BERTRAND. 1999. Oscillatory gamma activity in humans and its role in object representation. Trends Cognit. Sci. **3:** 151–162.
10. SADATO, N. & M. HALLETT. 1999. fMRI occipital activation by tactile stimulation in a blind man. Neurology **52:** 423.
11. FINNEY, E.M., I. FINE & K.R. DOBKINS. 2001. Visual stimuli activate auditory cortex in the deaf. Nat. Neurosci. **4:** 1171–1173.
12. MEREDITH, M.A. & B.E. STEIN. 1983. Interactions among converging sensory inputs in the superior colliculus. Science **221:** 389–391.
13. MEREDITH, M.A., M.T. WALLACE & B.E. STEIN. 1992. Visual, auditory and somatosensory convergence in output neurons of the cat superior colliculus: multisensory properties of the tecto-reticulo-spinal projection. Exp. Brain Res. **88:** 181–186.
14. WALLACE, M.T., M.A. MEREDITH & B.E. STEIN. 1993. Converging influences from visual, auditory, and somatosensory cortices onto output neurons of the superior colliculus. J. Neurophysiol. **69:** 1797–1809.
15. SCHULZ, M., B. ROSS & C. PANTEV. 2003. Evidence for training-induced crossmodal reorganization of cortical functions in trumpet players. Neuroreport **14:** 157–161.
16. EFRON, R. 1993. Scientific misconduct: undue process. Account Res. **3:** 223–238.
17. DEUTSCH, D. 1999. The Psychology of Music, 2nd Ed. Academic Press. San Diego.
18. DOWLING, W.J. 1978. Dichotic recognition of musical canons: effects of leading ear and time lag between ears. Percept. Psychophys. **23:** 321–325.
19. DOWLING, W.J. 1982. Contour in Context: Comments on Edworthy. Psychomusicology **2:** 47.
20. BARTLETT, J.C. & W.J. DOWLING. 1980. Recognition of transposed melodies: a key-distance effect in developmental perspective. J. Exp. Psychol. Hum. Percept. Perform. **6:** 501–515.
21. TREHUB, S.E., L.J. TRAINOR & A.M. UNYK. 1993. Music and speech processing in the first year of life. Adv. Child Dev. Behav. **24:** 1–35.
22. PERETZ, I. & M. BABAI. 1992. The role of contour and intervals in the recognition of melody parts: evidence from cerebral asymmetries in musicians. Neuropsychologia **30:** 277–292.
23. TRAINOR, L.J. 1999. A comparison of contour and interval processing in musicians and nonmusicians using event-related potentials. Australian J. Psychol. **51:** 147–153.
24. TESCHE, C.D. *et al.* 1995. Signal-space projections of MEG data characterize both distributed and well-localized neuronal sources. Electroencephalogr. Clin. Neurophysiol. **95:** 189–200.
25. DAVISON, A.C. & D.V. HINKLEY. 1997. Bootstrap Methods and Their Application. Cambridge University Press. Cambridge, NY.

Biological and Psychological Correlates of Exceptional Performance in Development

ROLF OERTER

Rehkemperstrasse 2, Munich, Germany

ABSTRACT: The issue of exceptional performance is discussed from four perspectives: genetic distribution of high giftedness, gene-environment interaction, neurobiological findings, and the role of deliberate practice as well as environment in general. The genetic perspective is illustrated mainly by the emergenic and epigenetic model of Simonton, followed by the three types of gene-environment interaction (Scarr and McCartney). The neurological perspective focuses on the role of early practice for neurological representation and on hormonal changes during development. The perspective of deliberate practice summarizes the present state of affairs within this field of research. Finally, a distinction between different levels of excellent performance is proposed, arguing against the neglect of genius as a specific phenomenon among excellent performance.

KEYWORDS: deliberate practice; gene-environment interaction; genetic components of musical abilities; expertise

A FIRST GLANCE: MUSICIANS AS A HIGHLY SELECT GROUP

During the last 10–15 years, the former prevailing opinion of music talent as a miraculous gift has been revised by the view that talent is the result of intense learning and training processes, whereby permanent assistance by parents and teachers as well as economic resources play an important role.[1–3] The assessment of the amount of training clearly showed a remarkable correlation between time of practice and level of excellence in music. Therefore, the concept of deliberate practice became the main explanation for excellent performance.[4,5] At present, there is general agreement that the higher the level of musical performance, the higher the amount of deliberate practice. This finding is often reversed by the inference that the only or the main condition for exceptional performance is deliberate practice.

The main reason why this conclusion is premature lies in the statistical nature of the sample of excellent musicians. Research on musicians deals with a highly select group. If, in addition, this select group is again reduced to the small sample of excellent musicians, results cannot be generalized to a large population. Therefore, the great amount of practice is always combined with an extremely select group, for example, the top level with a ratio of about 1:1,000,000. Deliberate practice of mu-

Address for correspondence: Prof. Dr. Rolf Oerter, Emeritus, Rehkemperstrasse 2, D 81247 Munich, Germany. Voice: 0049-89-8112390.
oerter@edupsy.uni-muenchen.de

Ann. N.Y. Acad. Sci. 999: 451–460 (2003). © 2003 New York Academy of Sciences.
doi: 10.1196/annals.1284.055

sicians is probably always combined with higher levels of musical talent. If the dropout rate of learners who are completely neglected in research is included, highly selected learners are with a high probability also musically gifted subjects. We do not even know if the great amount of deliberate practice is due to genetic factors (persistence, energy, and endurance) and therefore a function of innate factors or if deliberate practice is the main precondition for high performance independently from genetic factors of deliberate practice.

AN EMERGENIC AND EPIGENETIC MODEL

Human beings are a species of animal and therefore obey biological laws. A very general biological law is the variation of physical and psychological traits. Every biological characteristic varies among a population of individuals. With regard to music and sports, we observe a skewed distribution of talents that becomes more extreme when excellent performance is under consideration.

Simonton[6] tries to explain the extremely skewed distribution of talents and their development through his model of emergenic and epigenetic mechanisms that are all considered genetically controlled.

The emergenic component of the model starts with the assumption that high capacity consists of multiple components. These components entail all physical, physiological, cognitive, and dispositional traits that facilitate the manifestation of superior expertise in a talent domain; in our case, in music. Note that the term "facilitate" includes a broader understanding of innate conditions. Thus, genetic factors for early interest in music and unusual energy may also account for the development of a high level of performance. Simonton postulates that (a) each of the k components (k stands for specific number) is assessed on a ratio scale, with a true zero point that represents the complete absence of this trait, and (b) the k components are uncorrelated. For the domain of music, we know that both assumptions are too simple. However, Simonton shows that his basic model will not be remarkably affected if both assumptions are modified, because at least some of the k components hold both of these conditions. Contrary to most models in statistics that assume an additive composition of weighted components (variables), Simonton proposes a weighted multiplicative model because it is more consistent with current conceptions of creativity, genius, and outstanding achievement.[7–9] Simonton claims that the two assumptions of independence and multiplicity of traits have six major implications: (1) the domain specificity of talent, (2) the heterogeneity of component profiles within a talent domain, (3) the skewed frequency distribution of talent magnitude, (4) the attenuated predictability of talents, (5) low family heredity, and (6) the variable complexity of talent domains.

The Domain Specificity of Talent. One single component for a specific domain cannot equate the complexity of that domain. In the domain of music, many highly weighted components are necessary for music achievement. On the other hand, other components outside of the domain are not essential. Therefore, a rather low level of achievement in one domain may be combined with a high level of achievement in another domain. Beethoven provides an illustrative example. He showed very low mathematical achievement and only passable skills in the verbal domain. However, the genetic components for high performance are not exclusively domain specific.

Other components such as introversion,[10] high energy, and intensive interest in music also contribute to the outcome of high performance. Note that all components in this model are conceived as genetic variables.

Heterogeneity of Component Profiles within a Talent Domain. Since talent potential consists in the model of multiple components that enter into the potential, the trade-off level of achievement may result from a combination of differently weighted components. "One piano prodigy may have a better rhythmic sense, another a superior feel for melodic line, but their overall talent–as judged by performance in piano competitions–may be indistinguishable.[10, p.439] In most cases, however, we observe variation also in excellent performance, especially with regard to personal style of performance.

Skewed Frequency Distribution of Talent Magnitude. The model predicts that exceptional talent is extremely rare. According to the assumption of multiplicative combination of components, the rarity of musical talent can be predicted. Suppose that there exist only two components for musical talent that can be measured on a 5-point scale, ranging from 0 to 4. If both components are distributed according to the binomial law (a plausible assumption for innate components), we get a percentage distribution of 6.25, 25.00, 37.50, 25.00, and 6.25 for both variables. This distribution is a discrete approximation of the continuous normal distribution. A multiplicative combination of the two variables results in an asymmetric distribution, namely, 49.60, 36.00, 10.90, 3.10, and 0.40. Almost half the subjects would fall into the lowest group, whereas only half a percent of the subject is to be found at the highest possible talent potential. Even if we assume for each subject a weight of more than zero for each variable (e.g., 0.5), the percentage of low talent potential remains high and that of high potential low. Even if it is plausible to assume the presence of the whole range of musical components for each person (see results of early musical development, e.g., Refs. 11 and 12), they will vary across the population of infants. Furthermore, other variables such as temperamental factors of energy and personality factors of introversion and sensitivity[10] are necessary conditions and presumably independently, that is, in a multiplicative way, combined with musical components.

Attenuated Predictability of Talent. Biographical research shows that musical talent is not very predictable.[3,13,14] The emergenic model explains this fact. Since a talent potential assumes many components for performance, a child may show high performance in one measured or observed component (e.g., motor skills in piano playing) and therefore be assessed as a musical talent. However, another component might have a zero weight, which results in a later stage of development when creative interpretation and musical sensitivity become important in disappearance of talent. In general, the multiplicative combination of variables as a nonlinear function makes it difficult to predict the future state of talent in childhood and adolescence. Whereas in an additive model of musical components predictability is much higher, the multiplicative combination of variables opens many more possible outcomes of the talent potential und therefore lowers the predictability.

Low Familial Heredity. Paradoxically, the assumption of innate components of musical talent predicts a low familial heredity. Exceptional musical talent is not expected to exhibit genetically based family pedigrees. Even if all components had heritability coefficients of unity, it cannot be expected that all siblings inherit all necessary traits. Monozygotic twins, however, would allow for assessment of heritability, because they must have the same components with equal weights.

Variable Complexity of Talent Domains. It seems reasonable to assume different numbers of components for different domains. This is also true within the domain of music. Rock and Pop music might demand other and maybe less talent-related components than classical music. Within classical music the components necessary for a conductor and a violinist are different. Composers might, in addition, perform with excellence on an instrument (like Bach on the organ and Mozart on the piano) or might be poor instrumentalists (like Richard Wagner).

Simonton completes his model by the assumption of epigenetic development. The two main assumptions of multiplicative combination of components and their independence are expanded by two crucial features of the hypothesized development of talent:

1. The various innate components often develop more or less independently following an epigenetic program.

2. This epigenetic program is, to a large extent, subject to individual differences. A child can develop one component rather quickly and may show a slow development in another. Therefore, talent development becomes a highly idiosyncratic effect.

These postulates explain in the author's view the following phenomena:

- the occurrence of early and late bloomers. According to the speed of the development of components, talent can appear early or late.

- Potential absence of early talent indicators. If it requires a decade of intensive training and practice to acquire world-class competence, an early start might be highly advantageous. But if some components of musical talent develop later, early indicators of talent remain hidden.[15,16] This can be the case with composers (e.g., Bruckner).

- Possibility of talent loss by declining of a component or by illness that influences the further development of genetic components.

- Possible age dependency. The optimal talent domain may transform with maturation. So an instrumentalist may switch to composition or a would-be composer may become a conductor (see studies about child prodigies studied by Feldman and Goldsmith[17]).

- Increased obstacles to the prediction of talent. The diversity of component profiles within the domain of music lowers the potential utility of various predictor variables. This basic difficulty, already mentioned above, is increased by the epigenetic approach, which predicts early and late bloomers, the coming and going of talent, and the growing complexity of the domain.

GENOTYPE-ENVIRONMENT INTERACTION

Even though Simonton's model represents an extreme position, it explains many observations about the development of musicians. His model is an attempt only to consider genetic factors as a static (emergenic) as well as a dynamic (epigenetic) mechanism of the growing talent. He tries to explain as much as possible through genetic effects. Nonetheless, such a genetic position is not in line with the paradigm of deliberate practice[4] and the overrepresentation of middle-class members and of males.[18]

Scarr and McCartney[19] propose a model for gene-environment interaction that is closer to the reality of musicians' development. They distinguish between passive, evocative, and active genotype-environment effects. Passive effects arise through the genetic kinship between parents and child. Because the child shares 50% of the genes with each of the parents, the child is also sensitive to the environment presented by the parents. They themselves prefer and construct an environment that is in accordance with their own gene potential. This environment also fits at least partially the genotype of the child. Passive effects are regularly observable in the domain of music where parents offer a stimulating environment for their children.

The second kind of effect is called "evocative" by the authors. The child's genotype is putting through in the phenotype a specific musical behavior and thus evokes supporting behavior by the parents who offer the child stimulating situations for their interests and activities. In this way, the child's aptitude is recognized and fostered.

The third kind of genotype-environment effect, the active effect, is the most interesting one. The genotype actively selects and changes the environment in order to develop its potential. Biographies of famous musicians demonstrate this effect. Gershwin learned piano playing by imitating the key sequence of an electric piano. Frank Sinatra left his workplace against the will of his father, an Italian immigrant. Many famous singers and instrumentalists put through their will to become musicians against environmental forces. Even Bach left Arnstadt for a visit of Buxtehude in Luebeck and stayed there several months, even though he was only allowed to be absent for a few weeks.

The active genotype-environment effect is a modern version of Nietzsche's sentence: Werde, der du bist! (Become who you are!).

The approach of Scarr and McCartney has one aspect that makes us uncomfortable. The genotype is conceived of as an active subject who selects and organizes the environment. Even if we should take this as an anthropomorphism, it seems more convincing to transfer the term of an actor to the self that indeed organizes the interaction between organism and environment from a very early age on (see, e.g., Ref. 20).

NEUROBIOLOGICAL FINDINGS ON THE DEVELOPMENT OF MUSICAL TALENT

Thanks to the new methods of brain imaging such as positron emission tomography (PET), transcranial Doppler sonography (TCD), and magnetic resonance imaging (MRI), we also begin to understand the interplay between innate conditions and environment at the neurobiological level. Since this entire volume is concerned with neurological processes connected with music, I focus on some developmental aspects of neurobiological findings. Schlaug et al.,[21] using MRI, showed that the frontal part of the corpus callosum of musicians was larger than that of non-musicians, but this difference was detected only in those musicians who had begun their instrumental practice before the age of 7 years. The same authors also investigated the planum temporale and found a stronger left-hemispheric asymmetry in musicians than in nonmusicians, but this was only true in musicians with absolute pitch. Since the asymmetry of the planum temporale develops between the twenty-ninth and thirty-first week of gestation, the authors assume that the asymmetry in musicians is already innately predetermined and stimulated after birth by a stimulating environment. In this case of interaction the active genotype-environment effect may also be involved. Elbert *et al.*,[22] also using MRI, found differences in the primary somatosensory cortex of musicians compared with nonmusicians. In their often cited study, string players showed enlarged areas for the representation of the small finger and the thumb of the left hand. Again, this increase of area representation occurred only when training began before the age of 7 years. Later learning of a string instrument did not increase the representation for the fingers of the left hand. Pantev *et al.*[23] used the magnetic encephalogram (MEG) to measure cortical representation of tones. Musicians showed an increase of 25% of cortical representation of piano tones but no increase of pure sinus tones compared to nonmusicians. Age was a crucial factor, because children younger than 7 years showed the highest increase. The authors speculate that there might be a genetic component that promotes later learning. Micheyl *et al.*[24] studied the influence of musical training on a deeper brain region, the cochlea, which is responsible for the translation of acoustic waves into neural signals. Musicians showed better ability to process the loudness of sounds, which means that their auditory perception is closer to the real source of sound than is that of nonmusicians. The authors assume that musicians have a predisposition for sound processing, but they cannot exclude that implicit learning in early childhood alone could be responsible for the development of sound perception and its representation in the cochlea.

Other exciting results refer to hormonal differences between musicians and nonmusicians. Hassler and Nieschlag[25] and Hassler[26] found significant lower levels of testosterone in male composers than in control subjects, whereas female composers had a higher level of testosterone than did control subjects. Testosterone is a hormone determining masculinity; estradiol, in turn, is the main hormone responsible for femininity. Therefore, the hormone specificity of musicians points to androgyny, which is also confirmed by personality tests (e.g., Refs. 10 and 27). Geschwind and Galaburda[28] proposed a hypothesis of brain development according to which changes in brain anatomy occur during the prenatal period through the influence of testosterone, whereby musical ability or talent is only a by-product.

The last mechanism that probably controls musical development is the hormone melatonin. Musicians show a significantly higher level of melatonin than do nonmusicians.[29] This hormone has a multifaceted impact on the organism. For example, it suppresses the production of sex hormones, thus contributing to androgyny. Furthermore, melatonin protects from severe infectious diseases, but it weakens the autoimmune system. For example, allergic reactions are an autoimmune disease, and it is actually found more frequently in musicians. Hassler[30] argues that differences in hormonal levels between musicians and nonmusicians are mainly innate and already develop during the prenatal period. Be it as it may, the neurobiological findings of the last decade offer a better understanding of the well-tuned interplay between nature and nurture.

An important contribution of neurological evidence is related to the role of early musical training. From psychological research on deliberate practice[5] as well as from longitudinal measurement of musical abilities,[31] it can be inferred that early training of musical skills is crucial to the final level of performance. As shown above, there is much neurobiological evidence for the importance of early musical stimulation. In essence, the main structural changes of the brain caused by musical training and practice occur before puberty, sometimes before the age of 7–9 years. Singer,[32] like many other neurologists, presents an additional explanation for this phenomenon. During brain development, a huge number of synaptic connections are established, but only those connections that are activated before puberty are maintained, whereas unused connections disappear. Therefore, the learning of specific skills is most effective during childhood. Nonetheless, brain plasticity is maintained throughout our whole life.

EXCELLENT PERFORMANCE AND EXPERTISE

Keeping in mind the three different perspectives presented hitherto, we can turn to our last piece of evidence that will complete this picture of the genesis of excellent performance.

The expertise program in psychological research is not interested in the proof of genetically preformed abilities but rather (a) in the process of becoming an expert in a specific domain and (b) in the nature of achievement in a given domain. Expertise is always domain specific and is independent of general intelligence. As Schneider[33] argues, there exists only a threshold value of intelligence. When this threshold is reached, expertise can develop to a rather high level. The study of musical expertise confirms this position and presents results that can be summarized in the following points:

- Musical expertise is a function of the amount of deliberate practice.[4] The earlier deliberate practice begins, the greater the probability of reaching a high level of expertise.

- Expertise, especially within the domain of high performance in mastering an instrument, usually shows three distinct stages[2,3,14,18,34,35]: (1) A period of play and fun with music. (2) A period of deliberate practice with high assistance of family and teachers. Comparison with peers at the same performance level is also important. (3) Intensification of deliberate practice. Music becomes the core of life planning.

- The route towards musical expertise is accompanied by (a) the parents' intensive assistance and willingness to sacrifice and (b) the monitoring of competent teachers.

- For highly creative tasks, such as composing, contact with the musical culture and construction of musical knowledge are decisive. A composer is always a member within a historical range of creative musicians. But culture also determines expertise in instrumental training, because the kind of instrument and the selected music to be learned are completely dependent on the culture in which the learner lives.

SUMMARY AND CONCLUSIONS

Combining all the components of the puzzle of "excellent performance," we can draw a picture that brings together evidence from different research areas that hitherto took little notice of one other. Starting with the last aspect, that of expertise, deliberate practice and a supportive social network seem indispensable in most cases of musical excellence. Consider also, however, cases such as those of Gershwin, Frank Sinatra, and many jazz musicians.

These facets of musical expertise are necessary but not sufficient conditions for gaining musical excellence. Genetic conditions of musical talent must join the positive environmental niche for musical development. When a combination of many genetic factors is needed for musical excellence, the appearance of a high musical talent becomes a rare event, because a multiplicative rather than an additive combination of factors is at work.

The genotype-environment interaction works as a passive, evocative, and active interplay in which the self coordinates genetic potential with environmental affordances.

Psychological as well as neurobiological research justifies the assumption of a temporal "window" in childhood for some musical domains during which musical development towards expertise is optimal. This seems trivial, but it is not true for each musical domain. Brass players start their training much later than do pianists and violinists and also gain high virtuoso levels.

Finally, I try to answer the question of whether the development towards exceptional performance is merely a quantitative process as the successive increase of basic competencies or whether it is better understandable as a successive qualitative change. Research into the astonishing musical competencies of infants during the first year of life[11] suggests that parallel to the language acquisition device (LAD) may be a music acquisition device (MAD) with the whole potential for the acquisition of music expertise in any musical domain. These basic skills have only to be promoted, which is close to a simple quantitative increase. Where the qualitative change comes in is the establishment of structures that develop successively to higher order complexity. Those structures may include musical schemas (as in sight-reading), motor skills (such as playing an instrument), and auditory skills (such as discerning between major and minor chords within milliseconds). With increasing musical competence, long compositions such as symphonies and operas are saved as structures in memory.[36] On the whole, it seems that exceptional performance is always characterized by a cluster of many components building a complex structure. A high level of performance without complexity might be a contradiction per se. Thus, the synthesis attempted in this presentation also offers an understanding of the interaction of quantitative and qualitative change in the development towards expertise and exceptional performance.

In a discussion of excellent performance in music, no distinction is made between the different levels of excellence. Are there only quantitative differences between high performance, excellent performance, and genius? Is it only a greater amount of some components? If one looks for a most parsimonious explanation, two assumptions remain necessary: (1) The assumption of different levels of structural complexity. Think of Bach's incredible capacity for composing fugues. (2) The assumption of different speed of information processing. For example, Schubert in his short life created so many compositions that it would take more time to copy them by hand than Schubert needed to invent and to write them down.

REFERENCES

1. SLOBODA, J.A. 1990. Musical excellence: how does it develop? *In* Encouraging the Development of Exceptional Skills and Talents. M.J.A. Mowe, Ed. :165–178. British Psychological Society. Leicester.
2. SLOBODA, J.A. 1994. Music performance: expression and the development of excellence. *In* Musical Perceptions. R. Aiello & J.A. Sloboda, Eds. :152–169. Oxford University Press. New York.
3. SOSNIAK, L.A. 1990. From typo to virtuoso: a long-term commitment to learning. *In* Music and Child Development. F.R. Wilson & F.L. Roehmann, Eds. :274–290. MMC Music. San Louis, MO.
4. ERICSSON, K.A., R.T. KRAMPE & C. TESCH-RÖMER. 1993. The role of deliberate practice in the acquisition of expert performance. Psychol. Rev. **100:** 363–406.
5. ERICSSON, K.A., Ed. 1996. The Road of Expert Performance: Empirical Evidence from the Arts and Sciences, Sports and Games. Erlbaum. Mahwah, NJ.
6. SIMONTON, D.K. 1999. Talent and its development. Psychol. Rev. **106:** 435–457.
7. EYSENCK, H.J. 1995. Genius: The Natural History of Creativity. Cambridge University Press. Cambridge, England.
8. LYKKEN, D.T., M. McGUE, A. TELLEGEN & T.J. BOUCHARD. 1992. Emergenesis: genetic traits that may not run in families. Am. Psychologist **47:** 1565–1577.
9. STERNBERG. R.J. & T.I. LUBART. 1995. Defying the Crowd: Cultivating Creativity in a Culture of Conformity. Free Press. New York.
10. KEMP, A.E. 1996. The Musical Temperament. Psychology and Personality of Musicians. Oxford University Press. Oxford.
11. TREHUB, S.E. 2001. Musical predispositions in infancy. Ann. N.Y. Acad. Sci. **930:** 1–16.
12. TREHUB, S.E., E.G. SCHELLENBERG & D. HILL. 1997. The origin of music perceptiom and cognition. *In* Perception and Cognition of Music. I. Deliège & J. Sloboda, Eds. :103–128. Psychological Press. East Sussex, UK.
13. HOWE, M.J.A., J.W. DAVIDSON & J.A. SLOBODA. 1998. Innate talents: reality or myth? Behav. Brain Sci. **21:** 399–442.
14. SLOBODA, J.A. & M. HOWE. 1992. Transitions in the early musical careers of able young musicians: choosing instruments and teachers. J. Res. Music Ed. **40:** 283–294.
15. SLOBODA, J.A. 1996. The acquisition of musical performance expertise: deconstructing the "talent" account of individual differences in musical expressivity. *In* The Road of Expert Performance: Empirical Evidence from the Arts and Sciences, Sports, and Games. K.A. Ericsson, Ed. :107–126. Erlbaum. Mahwah, NJ.
16. WINNER, E. 1996. Gifted Children: Myth and Realities. Basic Books. New York.
17. FELDMAN, D.H. & L.T. GOLDSMITH. 1986. Nature's Gambit: Child Prodigies and the Development of Human Potential. Basic Books. New York.
18. SLOBODA, J.A. & M.J.A. HOWE. 1991. Biographical precursors of musical excellence: an interview study. Psychol. Music **19:** 3–21.
19. SCARR, S. & K. McCARTNEY. 1983. How people make their own environments: a theory of genotype -> environment effects. Child Dev. **54;** 424–435.
20. RAUH, H. 2002. Vorgeburtliche Entwicklung und frühe Kindheit. *In* Entwicklungspsychologie. R. Oerter & L. Montada, Eds. :131–208. Beltz/PVU. Weinheim.
21. SCHLAUG, G. *et al.* 1995. In vivo evidence of structural brain asymmetry in musicians. Science **267:** 699–701.
22. ELBERT, T. *et al.* 1995. Increased cortical representation of the fingers of left hand in string players. Science **270:** 305–307.
23. PANTEV. C. *et al.* 1998. Increased cortical representation in musicians. Nature **392:** 811–814.
24. MICHEYL, C. *et al.* 1997. Difference in cochlear efferent activity between musicians and non-musicians. Neuroreport **8:** 1047–1050.

25. HASSLER, M. & E. NIESCHLAG. 1989. Masculinity, femininity, and musical composition. Psychological and psychoendrocrinological aspects of musical and spatial faculties. Arch. Psychol. **141:** 71–84.

26. HASSLER, M. 1992. Creative musical behavior and sex hormones: musical talent and spatial ability in the two sexes. Psychoneuroendocrinology **17:** 55–70.

27. HASSLER, M., N. BIRBAUMER & A. FEIL. 1985. Musical talent and visual spatial abilities: a longitudinal study. Psychol. Music **13:** 99–113.

28. GESCHWIND, N. & A.M. GALABURDA. 1985. Cerebral lateralization. biological mechanism, associations and pathology, I and II. Arch. Neurol. **42:** 428–521.

29. HASSLER, M. & D. GUPTA. 1993. Functional brain organization, handedness, and immune vulnerability in musicians and non-musicians. Neuropsychologia **31:** 655–660.

30. HASSLER, M. 2000. Die Musikerpersönlichkeit aus neurobiologischer Sicht. Jahrb. Musikpsychol. **15:** 47–59.

31. GORDON, E. 1986. Musikalische Begabung. Schott. Mainz.

32. SINGER, W. 1998. "Früh übt sich." Zur Neurobiologie des Lernens. *In* Ungenutzte Potentiale. Wege zu konstruktivem Üben. G. Mantel, Ed. :43–53. Schott. Mainz.

33. SCHNEIDER, W. 1992. Erwerb von Expertise. Zur Relevanz von kognitiven und nicht-kognitiven Voraussetzungen. *In* Begabung und Hochbegabung. E.A. Hany & H. Nickel, Eds. :105–122. Huber. Bern.

34. BASTIAN, H.G. 1989. Leben für Musik. Schott. Mainz.

35. MANTURZEWSKA, M. 1990. A biographical study of the life-span development of professional musicians. Psychol. Music **18:** 112–139.

36. DOWLING, W.J. 2003. Entwicklung des Melos-, Klang- und Harmonieverständnisses. *In* Enzyklopädie der Musikpsychologie. R. Oerter & T.H. Stoffer, Eds. Vol. 2: pp. im Druck. Hogrefe. Göttingen.

Prosodic and Melodic Processing in Adults and Children

Behavioral and Electrophysiologic Approaches

CYRILLE MAGNE, DANIELE SCHÖN, AND MIREILLE BESSON

Institute for Physiological and Cognitive Neuroscience, CNRS, Marseilles, France

ABSTRACT: The results of a series of experiments aimed at directly comparing the prosodic level of processing in language with the melodic level of processing in music are reported. The first series of experiments was conducted on adults, musicians and nonmusicians, and the second one on 7- to 9-year-old musician and nonmusician children. However, as this last study is still in progress, only preliminary results will be presented. The theoretic framework within which these experiments are taking place is described. The first problem concerns the specificity of the perceptive and cognitive computations necessary to perceive and understand language. We argue that comparing language with music can provide interesting insights into this complex issue. The second problem is linked to the relationship between different types of learning. Does early musical training influence the way in which musicians process some aspects of language as prosody? These two problems are considered and the results of the experiments are described.

KEYWORDS: prosody; melody; behavior; electrophysiology

THE SPECIFICITY PROBLEM

Specificity is a fundamental problem that is not restricted to language. As soon as we are involved in any perceptual, cognitive, or motor tasks and are required to perceive, understand, and interact with the world around us, we may ask about the specificity of different brain structures and functions that are necessary to perform these tasks. For example, let us take two of the most studied perceptual functions: vision and audition (even if the brain structures and neurophysiologic mechanisms that support vision are better understood than those that support audition). We know that some parts of the cortex, the occipital areas, are more specialized for visual processing, whereas the temporal areas are more specialized for auditory processing. Moreover, we also know that two different pathways, the dorsal and the ventral visual streams, seem to be responsible for processing two different aspects of the objects that compose the world around us: where they are and what they are. Interestingly, recent findings have shown that a similar dissociation may be found in

Address for correspondence: Cyrille Magne, CNRS-INPC, 31 Chemin Joseph Aiguier, 13402 Marseilles Cedex 20, France. Voice: (33) 491 164113; fax: (33) 491 774969.
emagne@Lnf.cnrs-mrs.fr

Ann. N.Y. Acad. Sci. 999: 461–476 (2003). © 2003 New York Academy of Sciences.
doi: 10.1196/annals.1284.056

the auditory system, with a "where" system, useful for localizing where the sounds are coming from, and a "what" system, necessary for identifying the nature of the sounds.[1] Does this mean that the functional organization is similar in vision and in audition and that general perceptual principles are at play in both cases even if, from a structural point of view, the two systems are localized in different brain areas, with different fine-grained anatomies? Although it is too early to answer this question precisely, it is a fundamental problem that scientists are currently trying to solve.

Let us take another example, still comparing visual and auditory perceptual functions. Derived from the Gestalt theory, the principle of proximity, following which elements close to each other tend to be grouped in the same category, has also been shown to play an important role in the visual modality. Recently, Bregman,[2] in his fundamental work on auditory scene analysis, demonstrated that such a proximity principle also applies in the auditory modality, so that auditory events that are close in pitch will tend to be grouped together, whereas two auditory streams of different pitch will tend to segregate.

Specificity of Language Processing

With language, tracking the specificity issue becomes even more complicated for many reasons, one being that language is a typically human cognitive ability that cannot easily be studied in animals. Also, language cannot be considered as a single entity; rather, it encompasses different structural and functional levels of organization that need to be fully considered when addressing the specificity issue. Indeed, linguists generally consider that languages[a] are organized in terms of their phonologic, morphologic, syntactic, semantic, and pragmatic structures, and they try to characterize each of these structural levels within and across languages to better understand what is similar and what is different. At the functional level, psycholinguistic research is aimed at understanding the different types of processing involved at each of these structural levels and at specifying the functional organization between them. The psychological validity of these structural levels remains one of the most important questions to be solved (see also Ref. 3). In other words, are the individuals who speak and understand a language doing so according to the structural levels posited by linguists? Although this question is still unresolved, the structural framework proposed by linguists has been extremely useful in conducting psycholinguistic research. Still at the functional level, the ultimate goal of neurolinguistic research is to uncover the neurophysiologic mechanisms that underlie phonologic, morphologic, syntactic, semantic, and pragmatic processing, to determine where (in which networks of brain structures) they are implemented and how they are functionally organized in relation to one another. Note also that insofar as the psychologic validity of linguistic structures remains an open question, their biologic validity also remains even more open. Again, the scientific study of language has been fruitfully organized from linguistic to psycholinguistic to neurolinguistic research.

---

[a]In this chapter, we focus on the comprehension and not on the production aspects of language, and we are mainly concerned with the comprehension of spoken language.

Comparing Language and Music

To summarize, the problem of the specificity of the brain computations involved in perceiving and comprehending language clearly cannot be examined without taking into consideration the different structural and functional levels that have been posited on the basis of linguistic and psycholinguistic research. Moreover, to address the specificity problem, it is necessary to compare language with other cognitive functions that present similarities in their structural and functional organizations. For reasons to be outlined, we chose to compare language with music.[4] Like language, music is a typically human complex cognitive ability[5] that relies on sequential organization of its basic elements (pitch, duration, intensity, timbre, etc.). Both language and music are wonderful systems that generate expectations at both the perceptive and the cognitive levels.[6] Most importantly, for the line of research that we are conducting, music, like language, relies on different levels of structural and functional organization. Musicologists have analyzed the structure of music (and of different types of music) and have isolated several levels of particular importance: harmony, melody, and rhythm. Again, similar to language research, psychomusicologists are trying to understand the different types of processing involved at each of these structural levels and at specifying the functional organization between them. Finally, rapidly developing research in the field of the neuroscience of music[7,52] is trying to uncover the neurophysiologic mechanisms that underlie harmonic, melodic, and rhythmic processing, to determine in which networks of brain structures they are implemented and how they are functionally organized. As for language, psycho- and neuromusicologists are confronted with the same problem of the psychobiologic validity of the musical structures described by musicologists, and this issue still remains unresolved.

PROSODIC PROCESSING

Our current research is focused on a comparison between the different levels of processing in language and music, as we try to isolate the brain structures and processes that are similar or different between them. In this chapter, we focus on the specific aspects of phonologic processing, prosodic processing. Although some investigators consider prosody as independent from other aspects of phonologic processing, we present it here as part of the same level of processing, mainly for simplicity. Prosody, poetically speaking, can be defined as the music of language, but it certainly encompasses many different aspects, such as accents, stress patterns, and prosodic boundaries at the segmental level and intonation and rhythm patterns at the suprasegmental level.

Interestingly, whereas semantic and syntactic information typically plays a central role in linguistic theory and psycholinguistic models of language comprehension, the role of prosody is more controversial. Behavioral results showing that speech can still be interpreted even when prosodic cues are severely degraded led to the conclusion that prosody is peripheral to spoken language comprehension.[9] By contrast, prosody has been shown to play an important role in many instances: it is a central feature in language acquisition,[10–12] and it may be one of the most primary features of human language.[13]

Behavioral research has demonstrated that variations in fundamental frequency (F0) play a central role in prosodic processing and explain, for instance, most of the perceived differences between interrogative and declarative sentences. Indeed, the direction of the pitch variation (i.e., the F0 rise vs the F0 fall) provides sufficient cues for the listener to differentiate between them.[14,15] Parametric manipulation of the F0 of the sentence's final words and of the pitch of the final notes of musical excerpts was implemented in the present experiments to determine if such manipulations would have similar effects in language and music. Results of a neuropsychologic study, conducted by Patel *et al.*[16] with two amusic patients, tend to show that it may indeed be the case. These two patients showed similar levels of performance in prosodic and musical discrimination tasks, thus suggesting shared neural resources across domains. However, other cases have also been reported of patients with impaired melodic but not prosodic pitch-processing abilities.[17–19] These contradictory results may be explained by differences in task difficulty, because a small change in F0 (i.e., a quarter tone) is readily noticeable by musicians, whereas the rise in intonation to produce a question, for instance, is much larger, around one octave. In experiments to be described, the material was pre-tested to equate task difficulty.

INFLUENCE OF EARLY MUSICAL EXPERIENCE ON LANGUAGE PROCESSING

The second problem we are addressing is linked to the relationship between different types of learning. Is it the case, for instance, that learning to read is influenced by the way in which you learned how to write? (See Ref. 53 for some evidence that it may indeed be the case.) Closer to our concerns, does early musical training influence the way in which musicians process some aspects of language, like prosody? In other domains, results of several behavioral studies have indeed shown improvement of spatial and/or mathematical abilities due to musical training in children. However, these findings are not always conclusive, mainly because of methodologic problems (for a review see Ref. 6). For example, although 6-year-old children who were taught music for 7 months by means of the Kodály method showed improvements in mathematical and reading abilities compared to control children,[20] this method also included some nonmusical training aspects that confound the interpretation of the results. To study the influence of musical training on linguistic abilities in adults, Chan *et al.*[21] compared musicians who started playing music at an early age with nonmusicians without musical training. They found that musicians were better than nonmusicians at verbal-memory tasks. However, these findings were again somewhat difficult to interpret because the group with music training had a higher level of education, thus confounding the interpretation of the verbal advantage.[6] Finally, whereas some investigators have also emphasized the short-term effects of musical exposure on spatial reasoning,[22] others have argued that these effects may be epiphenomena of mood or arousal.[23,24]

The recent development of brain imaging methods offers new possibilities to test the hypothesis that intensive musical training influences performance in other cognitive domains. Previous studies using magnetic resonance imaging have shown structural differences in brain organization between musicians and nonmusicians, with a larger corpus callosum[25] and planum temporale in musicians,[26] especially if musical practice started at an early age. At the functional level, using electromagnetic recordings, increased representations for musicians have been described for somatosensory and auditory stimuli in both the motor[27] and the auditory cortex.[28,29] Using the event-related brain potentials (ERPs) method, previous studies have demonstrated differences in the early ERP components due to musical expertise[32,33] and better perception of the spectral and temporal characteristics of musical sounds among musicians than nonmusicians, even when they are not actively listening to the sounds (See Tervaniemi *et al.*[30] for a review of ERP research on the mismatch negativity [MMN] paradigm). Finally, the amplitude of late positive components, elicited by pitch violations in music, is longer and their onset latency shorter in musicians than in nonmusicians.[31] However, although these studies together with many behavioral experiments[34] have clearly demonstrated the effect of musical expertise in auditory and musical tasks, they did not tackle the question of the influence of musical training on other cognitive skills. To determine if musical training influences pitch processing not only in music but also in language, we conducted a series of experiments with adults and children, musicians and nonmusicians, recording both behavioral and ERPs measures while they performed the same task (detection of pitch contour violations) in music and in language.

DESCRIPTION OF THE EXPERIMENTS

Material and Procedure

In the language condition, participants were presented with short sentences from children's books (FIG. 1), and the F0 of the sentence's final word was manipulated to create a weak (35% increase) or a strong (120% increase) pitch violation. Similarly, in the music condition, the final notes of short musical phrases, taken from children's repertoire or composed for the experiment following the same melodic rules, were manipulated to create a weak (1/5 of a tone) or a strong (1/2 of a tone) pitch violation (FIG. 2). Participants were asked to determine, as quickly as possible while making as few errors as possible, if the pitch of the final words or notes was congruous or incongruous and to press one of two response keys accordingly. EEG was recorded from 28 scalp electrodes (FIG. 3), and EEG acquisition was synchronized with final note or word onset (see Schön *et al.*[35] for further details).

Participants

Half the participants were musicians and half were nonmusicians. In the experiment conducted with the adults, musicians had 15 years of musical training on average and regularly practiced their instruments. Most were professors or students at the National School of Music in Marseilles. Nonmusicians had no specific musical training and were not particularly music lovers. In the experiment conducted with 7- to 9-year-old children, musicians started practicing an instrument as early as 3 to 5 years of age, and all continued to practice regularly (half an hour or longer every day). Nonmusicians had no specific musical training. To ensure that musician and nonmusician children had a similar sociocultural background, the socioeconomic level of the parents was carefully controlled and was similar across children. Fur-

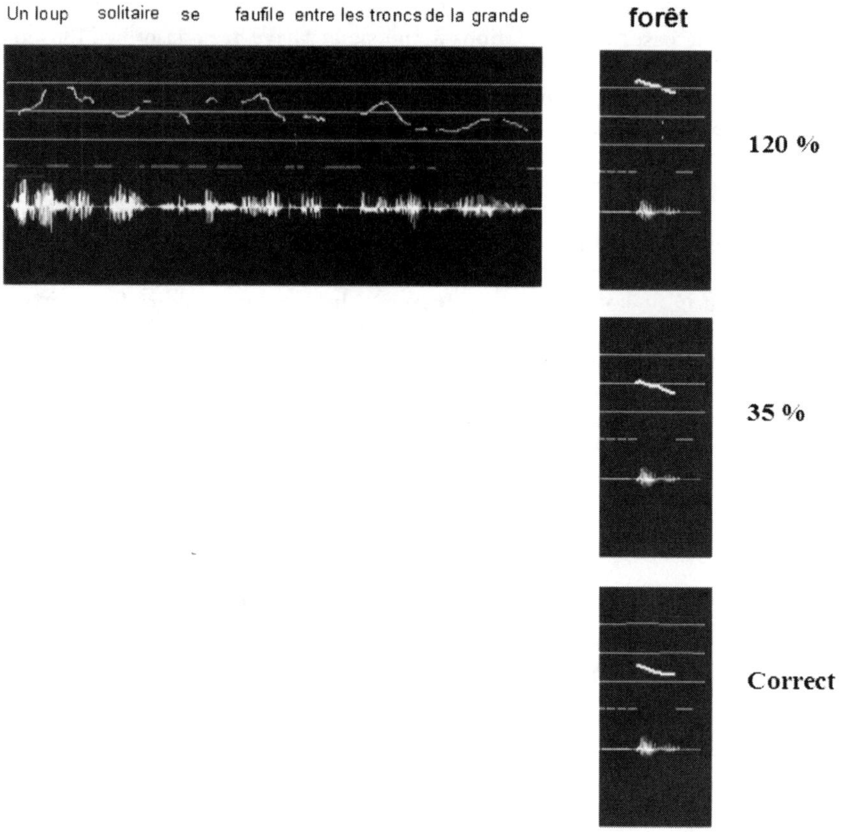

FIGURE 1. Examples of linguistic stimuli used in the experiment. The speech signal is illustrated in the sentence: *"Un loup solitaire se faufile entre les troncs de la grande forêt."* (Literal translation: *"A lonely wolf worked his way through the trees of the large forest."*) The fundamental frequency of the final word of the sentence was increased by 30% (weak pitch violation) or 120% (strong pitch violation).

FIGURE 2. The musical notation is illustrated for the tune "Happy Birthday." The pitch of the final note of the musical excerpt was increased by 1/5 of a tone (weak pitch violation) or 1/2 of a tone (strong pitch violation).

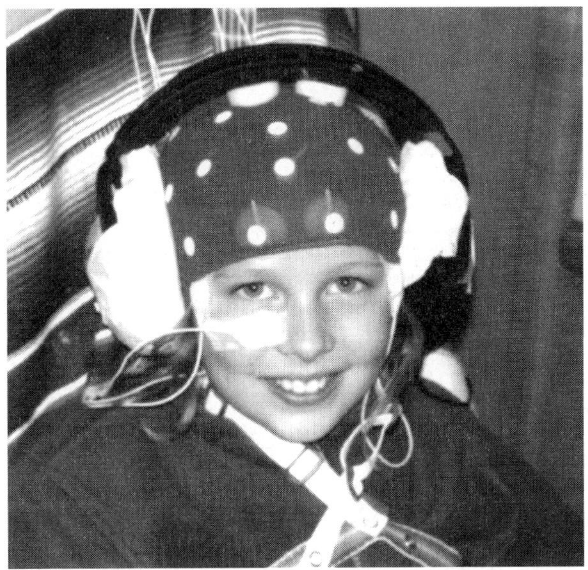

FIGURE 3. Photograph of a child who participated in the experiment (printed with the permission of the parents).

thermore, children were drawn from schools in Marseilles that are known for their high priority in child personal creativity development.

Behavioral Data

As can be seen in FIGURE 4, the main difference in error rates between adult musicians and nonmusicians is mainly found in weak incongruity. Interestingly, this difference was significant not only for music, as expected on the basis of previous results, but also for language with musicians being more accurate than nonmusicians in detecting small pitch violations in both language and music. Overall, reaction times (RTs) were faster for language than for music, but there were no clearcut differences between musicians and nonmusicians, except for the important finding that in both music and language musicians detected weaker incongruities faster than congruous endings, whereas nonmusicians showed no such difference. This last result seems to indicate that nonmusicians were unable to differentiate weak incongruities from congruous endings.

As can be seen in FIGURE 5, the same pattern of results is found with children. However, as we mentioned previously, these results are preliminary and have not yet been tested statistically. Therefore, they should be considered with caution and only as an encouraging indication that musician children may detect weak pitch violations in language better than nonmusician children.

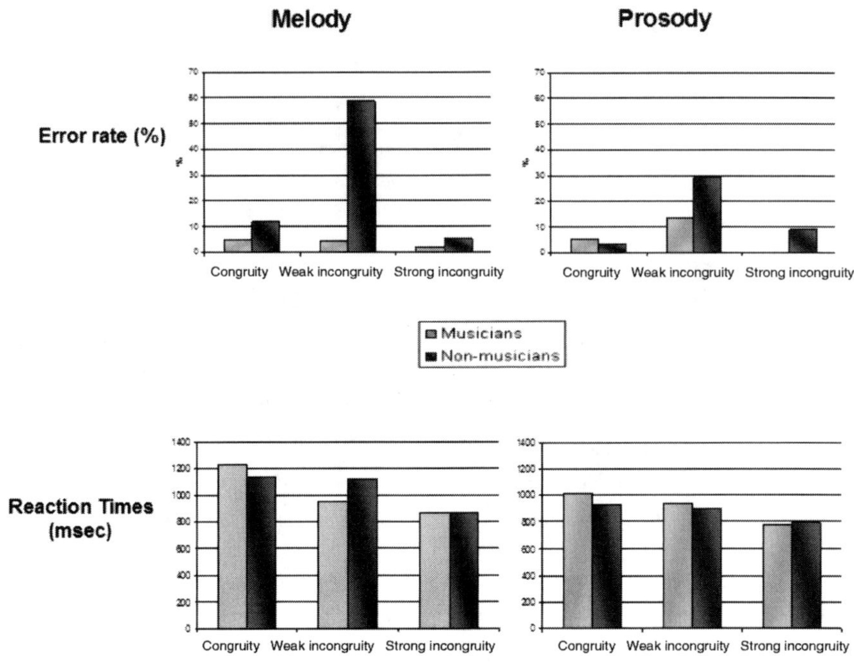

FIGURE 4. Percentage of error rates *(top panel)* and reaction times (RTs) in milliseconds *(bottom panel)* for congruous final notes or words and for weak and strong pitch violations in music and language are presented separately for musician and nonmusician adults. Clearly, the percentage of errors was highest for nonmusicians in response to weak violations in both music and language. Moreover, whereas for musicians the stronger the incongruity, the shorter the RTs, for nonmusicians no differences between congruous items and weak incongruities were found either in language or in music. Only RTs to strong incongruities were faster than those in the other two conditions.

Electrophysiologic Data with Adults

As can be seen in FIGURE 6, large positive components with a parietal scalp distribution and peaking around 400–500 ms postfinal tone or words onset, are elicited for both musician and nonmusician adults in the three experimental conditions. However, both the amplitude and the latency of these late positivities clearly differ among experimental conditions, specifically for musicians. The amplitude is largest for strong incongruity, intermediate for weak incongruity, and smallest for congruous endings for both language and music. By contrast, for nonmusicians, the pattern of results is not so clearcut, mainly because the ERPs to weak incongruities and congruous endings did not differentiate as clearly, and the results of statistical analysis revealed that these differences were not significant. Most interesting is the analysis of the latency of these positivities in the language condition. As can be seen in TABLE 1, the ERPs to strong incongruity start to diverge from the ERPs to congruous endings as early as 150–200 ms for musicians, but only 50 ms later (200–250 ms) for

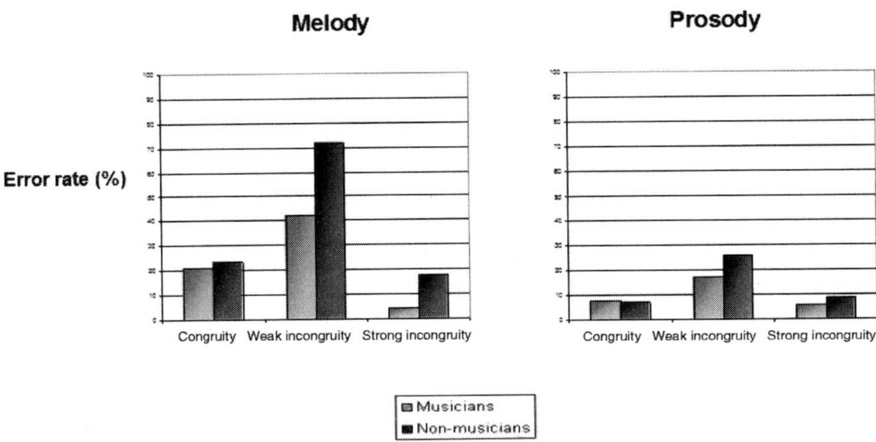

FIGURE 5. Percentage of error rates for congruous final notes or words and for weak and strong pitch violations in music and language are presented separately for musician and for nonmusician children. Clearly, the percentage of errors was highest for nonmusicians in response to weak violations in both music and language.

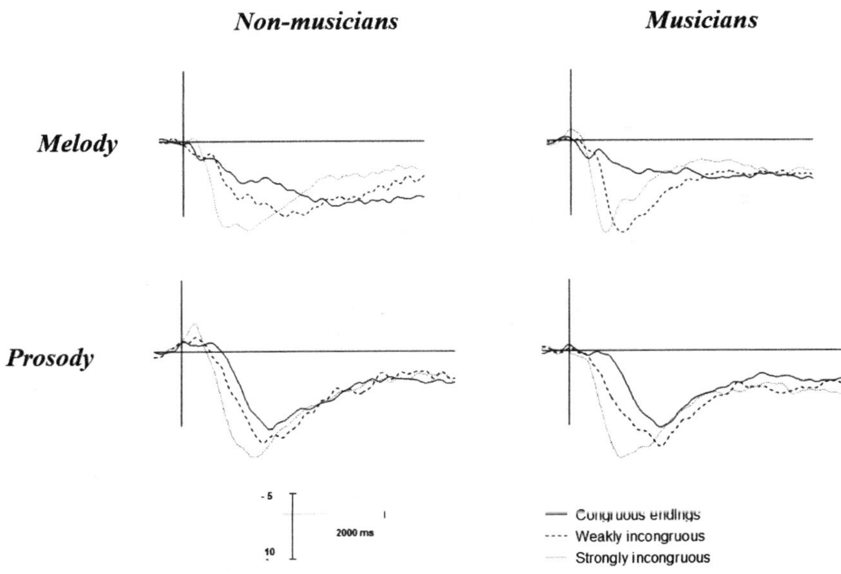

FIGURE 6. Variations in the brain electrical activity of musician and nonmusician adults, time-locked to final notes or words, elicited by congruous, weak or strong violations of pitch. Whereas the EEG was recorded from 28 electrodes, only one electrode is presented for simplicity. On this figure, as on the following ones, the amplitude (in microvolts) is plotted in the ordinate (negative up) and the time (in milliseconds) is in the abscissa.

nonmusicians. Even more striking are the differences as a function of expertise for weak incongruities. In this case, the ERPs to weak incongruity start to diverge from the ERPs to congruous endings as early as 200–250 ms for musicians but only 100 ms later (300–350 ms) for nonmusicians. Thus, musical expertise clearly seems to facilitate the detection of pitch violation in language.

Finally, interesting differences between musicians and nonmusicians were also found in the scalp distribution of early negative effects (FIG. 7). Indeed, these early

TABLE 1. Time course of the congruity effect

	Musicians		Nonmusicians	
	Strong vs Congruous	Weak vs Congruous	Strong vs Congruous	Weak vs Congruous
150–200 ms	**	-	-	-
200–250 ms	**	*	*	-
250–300 ms	**	**	**	-
300–350 ms	**	**	**	*

**P <0.001; P <0.05.

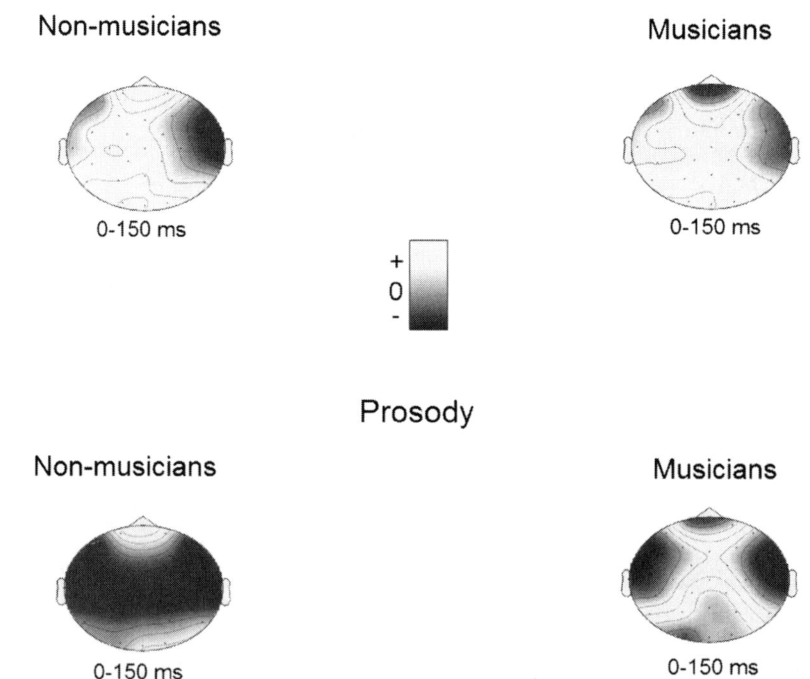

FIGURE 7. Topographic maps computed as an integration of mean amplitude values across time in the strong incongruity condition in specific time windows. These maps illustrate the comparison between musicians and nonmusicians in the language and music conditions.

negative components, which developed in the 0–150 ms latency band in response to the strong pitch violation were larger over right temporal sites (T4) for both musicians and nonmusicians in the music condition. Such a scalp distribution is in line with Zatorre's[36] hypothesis that the right auditory cortical areas are specialized in frequency analysis. However, in the language condition, the distribution of these early negativities was bilateral across temporal electrodes for musicians and was largest over left temporocentral sites for nonmusicians. Although such a left temporal distribution for nonmusicians may fit well with the general idea of left lateralization for language processing, based on the hypothesis just mentioned,[36] a similar distribution of the early negativity for language and music may have been expected because participants were paying attention to pitch changes in both cases. Perhaps analysis of the temporal aspects of language, described by Zatorre and colleagues as mainly involving the left auditory cortex, overrides the frequency analysis required by the task. Most importantly, the finding of differences between musicians and nonmusicians in the scalp distribution of the early negativity is in line with results demonstrating functional and structural experience-dependent plasticity in the auditory system in animals[37] and as just mentioned, in different aspects of music processing in musicians.[25–27,38] Thus, the intriguing possibility that musical expertise influences the anatomic-functional architecture of prosodic processing in language is an issue that needs to be pursued in further experiments.

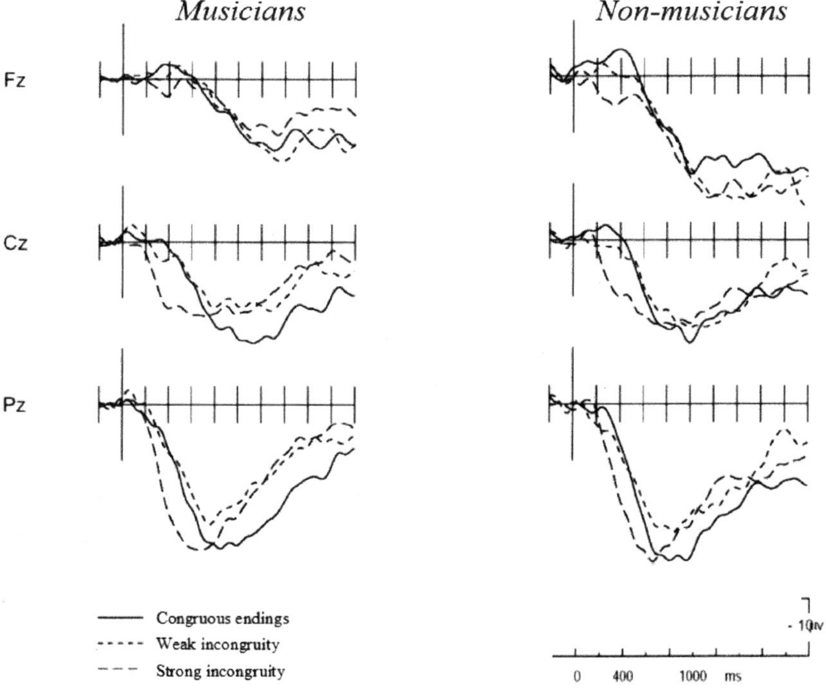

FIGURE 8. Variations in the brain electrical activity of musician and nonmusician children with the linguistic materials. ERPs are time-locked to final word onset in the congruous, weak, or strong violations of pitch conditions (see text for details).

Electrophysiologic Data in Children

The first general remark based on visual inspection of FIGURES 8 and 9 is that children present much larger evoked potential responses than do adults (note that the amplitude scale is 10 μV for children, whereas 5 μV is used for adults). More interesting is the finding that, as for adults, large positive components are elicited in the three experimental conditions in both language and music and for both musician and nonmusician children. In the language condition (FIG. 8), some interesting differences seem to emerge as a function of expertise, mainly regarding the amplitude of an early negative component that is reminiscent of the N400 component found in adults. Indeed, in the congruous condition, the amplitude of this negativity is larger for non-musicians than for musicians. This may reflect the fact that nonmusician children may be focusing their attention on the meaning of the words (thus eliciting a large N400 component) rather than on pitch manipulation. The reverse could be the case for musician children (thus the increased positivities). However, as already mentioned, these data are preliminary and have not yet been tested statistically because of the small number of participants in each group. This can clearly be seen when inspecting, for instance, the ERPs of nonmusicians in the music condition (FIG. 9). The fact that they had difficulties detecting the weak pitch violations in music is reflected by the very small number of trials included in the average in this con-

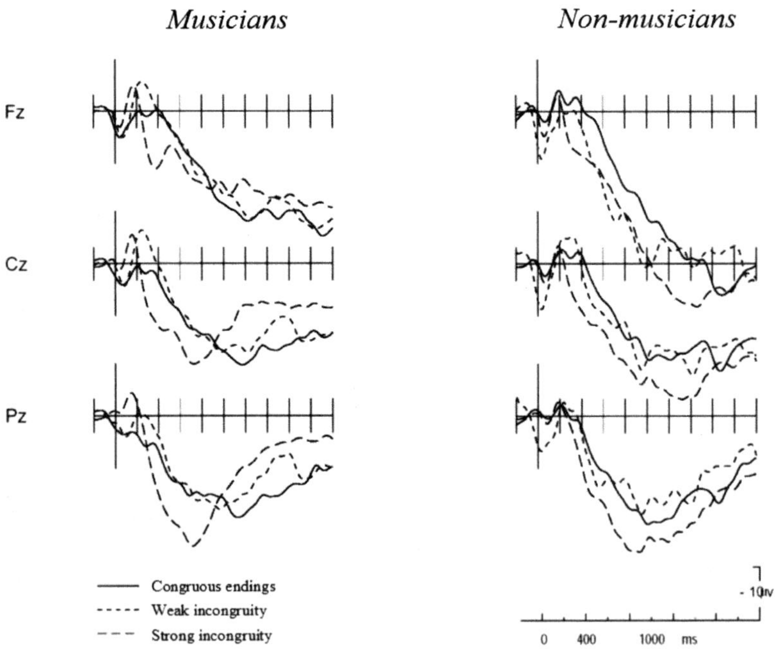

FIGURE 9. Variations in the brain electrical activity of musician and nonmusician children with the musical materials. ERPs are time-locked to final note onset in the congruous, weak, or strong violations of pitch conditions (see text for details).

dition, that consequently presents a much lower signal to noise ratio than do the other conditions. While it is therefore premature to draw any conclusions from these data, it is interesting to note the occurrence of early negative components, peaking around 200 ms after the onset of the weak and strong pitch violations in the musician but not in the nonmusician ERPs. Moreover, the latency of this early negativity seems to vary according to the degree of incongruity of the pitch violation, with shorter latency to strong than weak pitch violations. To summarize, although the electrophysiologic data recorded from musician and nonmusician children in response to pitch manipulations in both language and music are very promising, it is too early to draw firm conclusions. However, this work is in progress and we hope to present complete results in the very near future.

CONCLUSION

Studying the prosodic aspects of speech using brain imaging methods is a new and promising field of research. We see two main reasons why prosody is a particularly important aspect of spoken language. First, prosody is used to convey different types of information that cut across the different levels of language processing that we reviewed herein. Prosody is useful for the structural segmentation of speech and can help disambiguate syntactically ambiguous sentences.[8,39,40] Prosody also entertains strong relations with the semantic and pragmatic levels of language processing.[41] Maybe even more importantly, emotional prosody, defined as the ability to express emotions through variations in the different parameters of human speech (pitch contour, intensity, and duration), is probably one of the most basic features of language. This is illustrated by the fact that infants prefer their mother's voice to a stranger's voice as early as 2–4 days after birth.[42] Emotional prosody also carries important information related to the sex, age, and emotional state of the speaker.[43] Studies on the encoding of emotional states in the speech signal have shown clear correlations with global properties, such as loudness, speech rate, and pitch contour.[44–46] For instance, depressed and schizophrenic patients typically show reduced emotional prosody expression[47] and depressed children (9 to 11 years old) also identify affective prosody less accurately than nondepressed children.[48]

Recent studies using emotional speech (semantically neutral German sentences spoken with happy or sad intonation) in an elegant cross-modal priming experiment in which ERPs were recorded have demonstrated interesting gender differences between women and men.[49] Basically, the results showed that women were more sensitive than men to emotional speech. This is not to say, however, that men were insensitive to emotional prosody; rather, they just needed more time to process it! Whereas the interpretation of such intriguing results is not entirely clear,[50] they open new and exciting lines of research. Moreover, using functional magnetic resonance imaging (fMRI) with the same sentence materials, Kotz *et al.*[51] found that positive and negative intonations in normal speech conditions elicited a bilateral fronto-temporal and subcortical pattern of activation. Although these results certainly need to be pursued in further experiments, they are important in that they call into question an exclusive right hemisphere lateralization of emotional prosody and expand patient data on the functional role of the basal ganglia during the perception of emotional prosody.

The second reason why it seems very interesting to study prosody is linked with the strong similarities that may exist between the prosodic aspects of language processing and the melodic aspects of music processing. As we tried to highlight in the present series of experiments, both prosody and melody rely on similar acoustic parameters, which clearly allows the designing of numerous experiments. Consequently, it will be of great interest in future studies to manipulate specific acoustic features, such as pitch, intensity, or duration, to determine the extent to which parametric manipulations of these different factors produce similar or different effects in language and in music. The present results clearly showed interesting similarities and differences in pitch processing across both domains. They also highlighted the influence of early musical training on the detection of pitch variations not only in music, but also in language. This line of research, which was initiated thanks to the help of the International Foundation for Music Research, will certainly be continued in our laboratory in the years to come.

ACKNOWLEDGMENTS

This research was supported by a grant from the International Foundation for Music Research to Mireille Besson (IFMR: RA #194). Cyrille Magne benefited from a Cognitive Science Fellowship from the French Ministry of Research and Industry, and Daniele Schön was supported by the IFMR to conduct this research (2001–2002). The authors thank Monique Chiambretto and Martien Schrooten for their methodological assistance.

REFERENCES

1. RAUSCHECKER, J.P. & B. TIAN. 2000. Mechanisms and stream processing of 'what' and 'where' in auditory cortex. Proc. Natl. Acad. Sci. USA **97:** 11800–11806.
2. BREGMAN, A.S. 1990. Auditory Scene Analysis. MIT Press. Cambridge, MA.
3. BESSON, M. *et al.* 2003. Le traitement du langage. *In* L'imagerie fonctionnelle électrique (EEG) et magnétique (MEG): ses applications en sciences cognitives. B. Renault, Ed. Hermés. Coll. Sciences Cognitives. In press.
4. BESSON, M. & D. SCHÖN. 2002. Comparison between language and music. *In* The Biological Foundations of Music. R. Zatorre & I. Peretz, Eds. :232–258. Science and Medical Publications. Oxford University Press. Oxford.
5. KIVY, P. 1990. Music Alone. Philosophical Reflections on the Purely Musical Experience. Cornell University Press. Ithaca, NY.
6. SCHELLENBERG, E.G. 2001. Music and nonmusical abilities. Ann. N.Y. Acad. Sci. **930:** 355–371.
7. NAKADA, T. 2000. Integrated Human Brain Science: Theory, Method, Application (Music). Elsevier. Amsterdam.
8. PYNTE, J. 1998. The role of prosody in semantic interpretation. Music Percept. **16:** 79–98.
9. CUTLER, A. 1984. Stress and accent in language production and understanding. *In* Intonation, Accent and Rhythm: Studies in Discourse Phonology. D. Gibbon & H. Richter, Eds. :77–90. Walter de Gruyter. Berlin.
10. JUSCZYK, P. W. *et al.* 1992. Perception of acoustic correlates of major phrasal units by young infants. Cognit. Psychol. **24:** 105–109.
11. MEHLER, J. *et al.* 1988. A precursor of language acquisition in young infants. Cognition **29:** 143–178.

12. CUTLER, A. *et al.* 1997. Prosody in the comprehension of spoken language: a literature review. Lang. & Speech. **40:** 141–201.
13. MOLINO, J. 2000. Toward an evolutionary theory of music and language. *In* The Origins of Music. N.L. Wallin, B. Merker & S. Brown, Eds. :165–176. MIT Press. Cambridge, MA.
14. CUTLER, A. & D.R. Ladd. 1983. Prosody: Models and Measurements. Springer. Heidelberg.
15. LIBERMAN, M. & J. PIERREHUMBERT. 1984. Intonational invariance under changes in pitch range and length. *In* Language Sound Structure. M. Aronoff & R. Oehrle, Eds. :157–233. MIT Press. Cambridge. MA.
16. PATEL, A.D. *et al.* 1998. Processing prosodic and musical patterns: a neuropsychological investigation. Brain Lang. **61:** 123–144.
17. AYOTTE, J. *et al.* 2002. Congenital amusia: a group study of adults afflicted with a music-specific disorder. Brain **125:** 238–251.
18. PERETZ, I. *et al.* 2002. Congenital amusia: disorder of fine-grained pitch discrimination. Neuron **33:** 185–191.
19. SCHÖN, D. *et al.* 2001. Naming of musical notes: a selective deficit in one musical clef. Cortex **37:** 407–421.
20. GARDINER, M.F. *et al.* 1996. Learning improved by arts training. Nature **381:** 284.
21. CHAN, A.S. *et al.* 1998. Music training improves verbal memory. Nature **396:** 128.
22. RAUSCHER, F.H. *et al.* 1993. Music and spatial task performance. Nature **365:** 611.
23. NANTAIS, K.M. & E.G. SCHELLENBERG. 1999. The Mozart effects: an artifact of preference. Psychol. Sci. **10:** 370–373.
24. THOMPSON, W.F. *et al.* 2001. Arousal, mood, and the Mozart effect. Psycholog. Sci. **12:** 248–251.
25. SCHLAUG, G. *et al.* 1995. In vivo evidence of structural brain asymmetry in musicians. Science **267:** 699–701.
26. SCHLAUG, G. *et al.* 1995. Increased corpus callosum size in musicians. Neuropsychologia **33:** 1047–1055.
27. ELBERT, T. *et al.* 1995. Increased cortical representation of the fingers of the left hand in string players. Science **270:** 305–307.
28. PANTEV, C. *et al.* 1998. Increased auditory cortical representation in musicians. Nature **392:** 811–814.
29. PANTEV, C. *et al.* 2001. Representational cortex in musicians: plastic alterations in response to musical practice. Ann. N.Y. Acad. Sci. **930:** 300–314.
30. TERVANIEMI, M. 2001. Musical sound processing: evidence from electric and magnetic recordings. Ann. N.Y. Acad. Sci. **930:** 259–272.
31. BESSON, M & F. FAÏTA. 1995. An event-related potential (ERP) study of musical expectancy: comparison of musicians with non-musicians. J. Exp. Psychol.: Human Percept. and Perf. **21:** 1278–1296.
32. KOELSH, S. *et al.* 1999. Superior pre-attentive auditory processing in musicians. Neuroreport **10:** 1309–1313.
33. REGNAULT, P. *et al.* 2001. Different brain mechanisms mediate sensitivity to sensory consonance and harmonic context: evidence from auditory event-related brain potentials. J. Cognit. Neurosci. **13:** 241–255.
34. KISHON-RABIN, L. *et al.* 2001. Pitch discrimination: are professional musicians better than nonmusicians? J. Basic Clin. Physiol. Pharmacol. **12:** 125–143.
35. SCHÖN, D. *et al.* The music of speech: music facilitates pitch processing in language. Psychophysiology. In press.
36. ZATORRE, R.J. 2002. Structure and function of auditory cortex: music and speech. Trends Cognit. Sci. **6:** 37–46.
37. KILGARD, M.P. & M.M. MERZENICH. 1998. Plasticity of temporal information processing in the primary audtory cortex. Nat. Neurosci. **1:** 727–731.
38. BEVER, T.G. & R.I. CHIARELLO. 1974. Cerebral dominance in musicians and nonmusicians. Science **185:** 537–540.
39. MAGNE, C. *et al.* 2003. On-line analysis of prosody and syntax. Submitted.
40. STEINHAUER, K. *et al.* 1999. Brain potentials indicate immediate use of prosodic cues in natural speech processing. Nat. Neurosci. **2:** 191–196

41. ASTESANO, C. *et al.* 2003. Brain potentials during semantic and prosodic processing: underspecification of prosody. Cognit. Brain Res. Submitted.
42. HEPPER, P.G. *et al.* 1993. Newborn and fetal response to maternal voice. J. Reprod. Infant Psychol. **11:** 147–153.
43. JOHNSTONE, T. & K.R. SCHERER. 2000. Vocal communication of emotion. *In* Handbook of emotions. M. Lewis & J.M. Haviland-Jones, Eds. :220–235. Guilford Press. New York.
44. ALTER, K. *et al.* 2003. Affective encoding in the speech signal and in event-related brain potentials. Speech Commun. **40:** 61–71.
45. LADD, D. *et al.* 1985. Evidence for the independent function of intonation contour type, voice quality, and F0 range in signalling speaker affect. J. Acoust. Soc. Am. **78:** 435–444.
46. SCHERER, K.R. *et al.* 1991. Vocal cues in emotion encoding and decoding. Motiv. Emot. **15:** 123–148.
47. ALPERT, M. *et al.* 2001. Reflections of depression in acoustic measures of the patient's speech. J. Affect. Disord. **66:** 59–69.
48. EMERSON, C.S. *et al.* 1999 Investigation of receptive affective prosodic ability in school-aged boys with and without depression. Neuropsychiatry Neuropsychol. Behav. Neurol. **12:** 102–109.
49. SCHIRMER, A. *et al.* 2002. Sex differentiates the role of emotional prosody during word processing. Brain Res. Cognit. Brain Res. **14:** 228–233.
50. BESSON, M. *et al.* 2002. Emotional prosody: sex differences in sensibility to speech melody. Trends Cognit. Sci. **6:** 405–407.
51. KOTZ, S.A. *et al.* 2003. On the lateralization of emotional prosody: an event-related functional MR investigation. Brain Lang. **86:** 366–376.
52. ZATORRE, R.J. & I. PERETZ, Eds. 2001. The Biological Foundations of Music. Ann. N.Y. Acad. Sci. **930:** 1–462.
53. LONGCAMP, M. *et al.* 2003. Visual presentation of single letters activates a premotor area involved in writing. NeuroImage **19:** 1492–1500.

Young Children's Harmonic Perception

EUGENIA COSTA-GIOMI

The University of Texas, School of Music, Austin, Texas 78712-1435, USA

ABSTRACT: Harmony and tonality are two of the most difficult elements for young children to perceive and manipulate and are seldom taught in the schools until the end of early childhood. Children's gradual harmonic and tonal development has been attributed to their cumulative exposure to Western tonal music and their increasing experiential knowledge of its rules and principles. Two questions that are relevant to this problem are: (1) Can focused and systematic teaching accelerate the learning of the harmonic/tonal principles that seem to occur in an implicit way throughout childhood? (2) Are there cognitive constraints that make it difficult for young children to perceive and/or manipulate certain harmonic and tonal principles? A series of studies specifically addressed the first question and suggested some possible answers to the second one. Results showed that harmonic instruction has limited effects on children's perception of harmony and indicated that the drastic improvement in the perception of implied harmony noted approximately at age 9 is due to development rather than instruction. I propose that young children's difficulty in perceiving implied harmony stems from their attention behaviors. Older children have less memory constraints and more strategies to direct their attention to the relevant cues of the stimulus. Younger children focus their attention on the melody, if present in the stimulus, and specifically on its concrete elements such as rhythm, pitch, and contour rather than its abstract elements such as harmony and key. The inference of the abstract harmonic organization of a melody required in the perception of implied harmony is thus an elusive task for the young child.

KEYWORDS: perception of harmony; tonality; development; children; early childhood; music education

INTRODUCTION

Infants less than 1 year old are sensitive to many characteristics of the music they hear. They can perceive melodic contour, rhythm, tempo, style, timbre, scale structure, and intervalic dissonance.[1–3] These findings are relevant to education, because they suggest that by the time children start formal music instruction, they have learned much about music in an implicit way. The school music curriculum recognizes the music knowledge and abilities that children bring to the classroom and provides them with opportunities to learn formally about diverse aspects of music. Timbre, tempo, contour, pitch, texture, form, style, rhythm, loudness, and meter are

Address for correspondence: Dr. Eugenia Costa-Giomi, The University of Texas, School of Music, 1 University Station E3100, Austin, TX 78712-1435. Voice: 512- 471-2495.
costagiomi@mail.utexas.edu

Ann. N.Y. Acad. Sci. 999: 477–484 (2003). © 2003 New York Academy of Sciences.
doi: 10.1196/annals.1284.058

formally introduced in school music even from kindergarten. However, some musical elements are not included in the school curriculum until much later. Harmony and tonality have traditionally been considered among the most difficult elements for children to perceive and manipulate. In the United States, these are seldom taught until approximately third grade (8–9-year-old children).

The perceptions of harmony and tonality are intrinsically intertwined in Western music and are discussed in general terms in this short article. Whereas the perception of the internal structure and components of chords could be analyzed independently from the parameters that govern tonal music, the perception of the function of the chords is dependent on the tonal context in which they are embedded. The term "harmonic perception" will be used in the broadest sense to include these two types of perception.

DEVELOPMENT OF HARMONIC PERCEPTION

Moog[4] asserted that young children are deaf to harmony until they are 6. This strong statement was based on young children's lack of preference for consonant over dissonant accompaniments to a melody. Despite numerous studies corroborating that children younger than 6 do not prefer consonant accompaniments and are inconsistent in their consonance/dissonance preferences,[5–9] it is now clear that even 2- to 7-month-old infants have a clear sensitivity and predisposition to consonance. This apparent contradiction between infants' preference for consonant intervals and young children's lack of preference for consonant accompaniments may be explained by the focus of these studies on different aspects of musical consonance: *sensory consonance* in the case of infants and *harmonic consonance* in the case of young children. Sensory consonance, a universal rejection of the roughness produced by specific components of sound, has a biological basis, whereas harmonic consonance, the rejection of the roughness produced by the violation of tonal principles, is culture based.[10]

Children's harmonic perception seems to develop slower than the perception of other musical elements. For example, the singing of 5-year-olds, although very accurate in terms of melodic contour and rhythm, is often characterized by an unstable sense of tonality.[11–13] At this age, children seem to be sensitive to key membership but not to implied harmony, which only becomes apparent at age 7.[14] The responses of 5-year-olds in probe tone[15] and cadence perception studies[16] indicate that they have not assimilated the traditional tonal hierarchy of our Western music language. At 6, however, children notice the lack of a conclusive cadence if it is missing from a progression. By 8, they can discriminate between conclusive and inconclusive cadences and have a vague sense of the function of the dominant chord, and by 10, they are capable of a more analytic approach to the perception of cadences and the tonal function of tonic and dominant chords. The development of harmonic and tonal perception is also reflected in the type of errors that children make in performance tasks. For example, when playing the piano, 7–9-year-olds make significantly more chord errors that violate tonality than do 10–12-year olds, even when the length of the musical training of the two groups is comparable.[17] Some studies suggest that a rapid development in harmonic perception occurs between the ages of 7 and 9 with children being able to detect the harmonic changes introduced to a stimulus.[18,19]

IMPLICIT AND EXPLICIT LEARNING OF HARMONY AND TONALITY

Children's gradual harmonic and tonal development has been attributed to their cumulative exposure to Western tonal music and their increasing experiential knowledge of the rules that govern this music.[6] Two questions that are of much relevance to educators are: (1) Can focused and systematic teaching accelerate the learning of the tonality principles that seem to occur in an implicit way throughout childhood? (2) Are there cognitive constraints that make it difficult for young children to perceive and/or manipulate certain harmonic and tonal principles, constraints that, when overcome through normal development, result in rapid improvement in harmonic perception?

The documented attempts to teach harmony and tonality to children are scarce. Rowntree[20] found that melodic and harmonic instruction for a period of 7 weeks improved children's performance in the chord test of Bentley's *Measures of Musical Abilities,* and Taylor[18] concluded that music instruction helps children to complete certain harmonic tasks. Despite these indications that focused instruction may improve children's harmonic perception, American music teaching materials place little emphasis on harmonic concepts until the age of 10.

It is apparent that in addition to lacking formal instruction in harmony and tonality, young children are provided with limited opportunities to manipulate or even direct their attention to tonal and harmonic elements in their daily lives. While children receive frequent feedback from their experimentation with many musical elements such as loudness, timbre, or tempo (e.g., "Softer! The baby is sleeping!"; "We like the sound of the drum, don't we?"; "Slower please."), they seldom receive specific reinforcement for their tonal and harmonic explorations. Parents and caregivers can more easily direct the child's attention to the rhythm, beat, melodic contour, or form of the music through verbal cues and descriptions, movements, visual representations, and modeling than to its implied harmony. In fact, adults themselves do not seem to focus spontaneously on the harmonic development of the music they hear. Their descriptions of music rarely include references to harmony or tonality,[21,22] perhaps because they lack precise terms to express their perception of these elements.[23] Although adults may not purposely direct their attention to harmony when listening to music, they do perceive implied changes in harmony in short melodies when prompted[24] to detect harmonic incongruencies even in a preattentive state.[25,26] Event-related brain potentials (ERPs) indicate that adults perceive a deviant chord in short progressions despite directing their attention to another musical aspect of the stimulus.[25,26] They can also perceive differences in harmonic cadences when listening for irrelevant verbal cues (Bigand). Although this seems to indicate that harmonic discrimination is an automatic and simple task for adults, this is in fact not the case. The type of stimulus and the characteristics of the deviant chord affect adults' perception of the harmonic incongruence. Only 10% of nonmusicians could detect a mistuned chord within a progression when specifically asked to do so.[28]

Does music instruction affect harmonic perception? Differences in neurological responses to harmonic incongruencies of musicians and nonmusicians indicate that it does. Certain cognitive late-latency ERPs (i.e., N2b and P3) elicited by musicians when purposely discriminating chords are not present in nonmusicians when performing the same task.[28] The presence of these ERPs, associated with high order cognitive processes, reflected the musicians' accuracy in identifying deviant chords (80% accuracy). Even in preattentive states such as reading a book while listening to progressions that included deviant chords did ERP components differ between musicians and nonmusicians: musicians elicited mismatch negativity, whereas nonmusicians did not.[28]

Behavioral measures also indicate that musicians are more accurate than nonmusicians in identifying implied harmony by detecting harmonic changes through harmonic cues (whether a pitch belongs to a certain chord) rather than interval cues.[24] It would seem unthinkable that jazz musicians could modulate and follow the implied harmonic cues provided by other players with such immediacy and apparent effortlessness without instruction. In fact, jazz theory knowledge and jazz experience were found to be good predictors of the quality of jazz vocal improvisation.

Sensitivity to harmony seems to develop naturally by mere exposure to music and without specific training. This may explain the fact that most musical aptitude batteries include tests of harmonic perception. However, through formal instruction, performance in harmonic tasks improves and the cognitive processes involved in such tasks are modified. Would young children's performance in harmonic tasks improve if they were provided with specific training?

FORMAL INSTRUCTION ON HARMONIC FUNCTIONS DURING EARLY CHILDHOOD

In a series of studies, 5–10-year-old children completed a series of perceptual and performance tasks after receiving instruction on tonic and dominant chords that ranged from two 15-minute to twenty 30-minute sessions.[29–32] The comparison of children with more or less harmonic instruction at each age level allowed the study of the effects of instruction and development on their performance in the tasks.

In all studies, children who were 5 or older performed above chance in tasks requiring the discrimination between two chords. Children clapped when they heard a chord change in progressions composed of repeated I5 (tonic triad in root position) and V6 (dominant triad in first inversion) that included eight chord changes. They detected the chord changes accurately and quickly. It is not surprising that even the youngest children performed well in this task given the clear pitch differences between tonic and dominant chords: two pitches change when going from a tonic to a dominant triad. Children could have succeeded in this task simply by clapping whenever they heard *any* pitch change. A new task was devised so that children could not rely exclusively on pitch discrimination to differentiate chords. Children listened to progressions in which there were many pitch changes, but only a few indicated a chord change. In the chord progressions used, each chord was played successively in different positions (e.g., I5, I6, I6/4) before changing to another chord. This resulted in pitches changing among different positions of the chord and between chords, Children performed above chance in this task regardless of age, suggesting that by 5, the sense of implied harmony is developed.[31] Older children responded more accurately and faster than younger children, and instruction on tonic and dominant chords improved children's performance in this task.[31] Although these results show that young children can detect implied harmony and get better at it with focused instruction, other results indicate that this ability is indeed quite limited.

When a simple melody was superimposed on the two-chord progressions (I5/V6), the younger children could not detect the chord changes. They could not perceive the implied harmony of a melody even when an accompaniment explicitly marked the changes in harmony. These results suggest that young children's ability to detect chord changes in a progression may be inhibited by the presence of a melody. Children's spontaneous attention to the melody of the music they hear[16] may prevent them from focusing on the accompaniment. Efforts to help them disassociate the melody from the accompaniment by changing the register and/or the timbre of the accompaniment improved young children's detection of chord changes.[30] However, 5-, 6-, and 7-year-olds performed still at chance level even when the melody and the accompaniment were played with distinct timbre and reverse register (melody in the lower register and accompaniment in the higher register).

The performance of children in other tasks supports the idea that their sensitivity to implied harmony improves dramatically after age 7. Children were asked to play the tonic and dominant chords on the omnichord to a familiar tune sung by the researcher, an activity that they had practiced during 10 or 20 weeks of instruction in their public schools.[31] The children had unlimited time to find the proper accompaniment and could record their performances as many times as they wished until satisfied with the results. The older children (9- and 10-year-olds) performed significantly better in this task than did the younger children. The younger children could not play any complete phrases of the accompaniment accurately (a phrase consisted of four chords), whereas most of the older children played both phrases correctly. Instruction did not improve children's performance in this task. The youngest children's dependence on the individual pitches of the melody was evident from the rhythm of their invented accompaniments: they often played a chord for each note of the melody. The older children, instead, played the chords on the beat, as they had learned to do during weeks of instruction.

The task just described, although based on only two chords, had a strong memory component: children needed not only to identify the proper chords of the accompaniments while practicing, but also to remember the correct sequence of chords while the melody was being sung by the teacher. To reduce the memory demands of the task, children were asked to identify the chords and write them down after repeated listenings.[31] The youngest and the oldest children's performances in this task were similar to those in the previous task. Most 6-year-olds could not identify any complete phrases of the accompaniment, whereas most 9- and 10-year-olds identified both. What is interesting is that the performance of the 7- and 8-year-olds was better in this memory-reduced task than in the memory-demanding one. Most of these children identified at least one of the phrases accurately and, on average, six of the eight chords of the accompaniment. This suggests that sensitivity to implied harmony in a meaningful musical context such as a complete melody may be observable in children as young as 7.

Another task that placed low memory demands upon the children required them to sing a familiar song to a modulating accompaniment played in the omnichord.[31] Children sang the nonsense words of the refrain of the song after the teacher sang a whole verse in the usual key. The accompaniment played in the omnichord changed keys abruptly at the end of the verse, transposing the song an ascending whole step. Children's sensitivity to the accompaniment and the key shift was studied by observing whether and how their singing adapted to the unexpected modulation. Of those children who could maintain a stable tonal center when singing the original song (song without the modulation), approximately half were able to modulate to the new key. Children's awareness of the shift in tonal center of the accompaniment improved with age. The percentage of children in kindergarten through fourth grade that could modulate to the new tonality in their singing ranged from 46% in kindergarten to 60% in fourth grade. Instruction improved the performance of the oldest children.

In a simpler task related to the one just described and that did not require mastery at singing, children listened to a singer performing songs that they had learned during 10 weeks of instruction.[32] The tunes were sung with various accompaniments including the one based on tonic and dominant chords familiar to the children and three other tonic/dominant accompaniments that were unfamiliar and inappropriate. For example, one of the accompaniments replaced all tonic chords with dominant chords and vice versa, and another presented the abrupt ascending modulation described earlier. Five- and six-year-olds responded differently from the older children. They were not only less accurate than the older children in identifying whether the accompaniment sounded "right," but they were also more affected by familiarity with the stimulus and the type of accompaniment.

DEVELOPMENT AND INSTRUCTION

The results of these studies suggest that the improvement in harmonic perception is developmental in nature and that instruction has limited effects on young children's perception of implied harmony and the tonal function of chords. Although it is possible that intense formal instruction on an instrument, for example, may affect young children's perception of harmony more dramatically, musical instruction provided in the context of a public school curriculum only improves children's performance in those tasks in which they are already successful. Instruction improved young children's discrimination of chords and perception of chord membership, but it did not improve children's perception of implied harmony. Children did not learn to manipulate tonic and dominant chords functionally better as a result of instruction but as a result of development. I argue that it is the development of general cognitive abilities that allows children to learn implicitly or explicitly about the harmonic organization of music.

A drastic improvement was noted in children's perception of implied harmony and explicit harmonic organization between the ages of 8 and 9. It is apparent that at this age children had less memory constraints and more strategies to direct their attention to the relevant cues of the stimulus. They could disassociate the melody and accompaniment better and could extract harmonic information from a melody. In order to do this, they needed to organize melodic information into differentiated and regular harmonic units (chords), a task that relies on the use of abstraction and attention to multiple aspects of the stimulus (e.g., beat and pitch). Younger children, instead, focused on the melody whenever present and on its individual elements (notes). They seemed unable to direct their attention to the accompaniment that explicitly provided the relevant harmonic information they needed and to group the pitches of a melody according to the underlying harmonic unit to which they belong. The younger children could identify pitch membership to chords in the absence of a melody, but they could not do this for the pitches of a melody.

It seems that young children direct their attention to the more concrete features of the melody such as pitch, rhythm, and contour and that it is the abstract nature of the chords that makes harmonic perception difficult for young children. Chords, and all their components, are not necessarily present in a given musical stimulus, yet its underlying harmonic organization may be inferred. The abstract nature of a chord is similar to that of the beat, which does not need to be present to be inferred, or key, which may be implied by the presence of only a few of its components. The level of abstraction necessary to make these inferences is not developed until the end of early childhood, making the perception and manipulation of these more abstract concepts elusive to the young child.

REFERENCES

1. TREHUB, S. 2001. Musical predispositions in infancy. Ann. N.Y. Acad. Sci. **930:** 1–16.
2. TRAINOR, L. *et al.* 2002. Preference for sensory consonance in 2- and 4-month-old infants. Mus. Percept. **20:** 187–194.
3. ILARI, B., L. POLKA & E. COSTA-GIOMI. 2002. Infants' long-term memory for complex music. Presented at the 143rd Meeting of the Acoustical Society of America. Pittsburgh, PA, USA. June.
4. MOOG, H. 1976. The Musical Experience of the Pre-school Child. Schott & Co., Ltd. London.
5. BRIDGES, V. 1965. An exploratory study of the harmonic discrimination ability of children in kindergarten through grade three in two selected schools. Unpublished Doctoral dissertation, Ohio State University, Columbus.
6. SLOBODA, J.A. 1985. The Musical Mind: The Cognitive Psychology of Music. Clarendon Press. Oxford.
7. TEPLOV, B.M. 1966. Psychologie des aptitudes musicales. Presses Universitaires de France. Paris.
8. REVESZ, G. 1954. The Psychology of Music. University of Oklahoma Press. Norman .
9. YOSHIKAWA, S. 1973. Yoji no waon-Kan no hattatsu [A developmental study of children's sense of tonality]. Ongaku-Gaku **19:** 5–72.
10. TRAMO, M.J. *et al.* 2001. Neurobiological foundations for the theory of harmony in Western tonal music. Ann. N.Y. Acad. Sci. **930:** 92–116.
11. KREUTZER, N.J. 2001. Song acquisition among rural Shona-speaking Zimbabwean children from birth to 7 years. J. Mus. Ed. **49:** 198–211.
12. RAMSEY, J.H. 1983. The effects of age. Singing ability and instrumental experiences on preschool children's melodic perception. J. Mus. Ed. **31:** 133–145.
13. DOWLING, W.J. 1985. Development of musical schemata in children's spontaneous singing. *In* Cognitive Processes in the Perception of Art. W.R. Crozier *et al.*, Eds. :145–163. North-Holland. Amsterdam.
14. TRAINOR, L. & S. TREHUB. 1994. Key membership and implied harmony in Western tonal music: developmental perspectives. Percept. Psychophys. **56:** 125–132.
15. KRUMHANSL, C. & F. KEIL. 1982. Acquisition of the hierarchy of tonal functions in music. Mem. Cognit. **10:** 243–251.
16. IMBERTY, M. 1969. L'acquisition des structures tonales chez l'enfant. Klincksieck. Paris.
17. GUDMONSDOTTIR, H.R. 2003. Music reading errors of young piano students. Unpublished Doctoral Dissertation. McGill University, Canada.
18. TAYLOR, S. 1969. The musical development of children aged seven to eleven. Doctoral Dissertation. University of Southampton, UK.

19. O'HEARN, R.N. 1984. An investigation of the response to change in music events by children in grades one, three, and five. Dissertation Abstracts Int. **46:** 371A.

20. ROWNTREE, J.P. 1970. The Bentley "Measures of Musical Abilities": a critical evaluation. Bull. Council Res. Mus. Ed .**22:** 25 –32.

21. FLOWERS, P. 1983. Aspects of concert attendance behavior of undergraduate nonmusic majors. Contrib. Mus. Ed. **10:** 19–26.

22. FLOWERS. P. 1984. Attention to elements of music and effect of instruction in vocabulary on written descriptions of music by children and undergraduates. Psychol. Mus. **12:** 17–24.

23. HAIR, H.I. 1981. Verbal identification of music concepts. J. Res. Mus. Ed. **29:** 11–21.

24. PLATT, J.R. & R. RACINE. 1994. Detection of implied harmony changes in tradic melodies. Mus. Percept. **11:** 243–264.

25. KOELSCH, S. *et al.* 2000. Brain indices of music processing: "nonmusicians" are musical. J. Cognit. Neurosci. **12:** 520–541.

26. KOELSCH, S. *et al.* 2001. Neapolitan chords activate the area of Broca: a magnetoencephalographic study. Ann. N.Y. Acad. Sci. **930:** 420–421.

27. BIGAND, E. & B. TILLMAN. 2001. The effect of harmonic context on phoneme monitoring in vocal music. Cognition **81:** B11–B20.

28. TERVANIEMI, M. 2001. Musical sound processing in the human brain: evidence from electric and magnetic recordings. Ann. N.Y. Acad. Sci. **930:** 259–272.

29. COSTA-GIOMI, E. 1994. Recognition of chord changes by 4- and 5-year-old American and Argentine children. J. Res. Mus. Ed. **42:** 68–85.

30. COSTA-GIOMI, E. 1994. Effect of timbre and register modifications of musical stimuli on young children's identification of chord changes. Bull. Council Res. Mus. Ed. **121:** 1–15.

31. COSTA-GIOMI, E. & R.A. DOS SANTOS, 2001. The effects of instruction on young children's perception of tonic and dominant chords. Presented at the meeting of the Society for Music Perception and Cognition. August.

32. COSTA-GIOMI, E. 2000. Young children's identification of simple harmonic accompaniments. *In* Proceedings of the 6th International conference for Music Perception and Cognition. C. Woods, G. *et al.,* Eds. Keele University Department of Psychology. Keele.

Do Mental Speed and Musical Abilities Interact?

WILFRIED GRUHN,[a] NIELS GALLEY,[b] AND CHRISTINE KLUTH[a]

[a]Music Education Department, University of Music Freiburg, Freiburg, Germany

[b]Department of Clinical Psychology and Psychotherapy, University of Cologne, Cologne, Germany

ABSTRACT: The relation between mental speed and musical ability was investigated. Seventeen subjects aged 3–7 years were divided into two subgroups: one (G1; $n = 9$) consisted of children who participated in an early childhood music program and who received informal musical guidance, but no special training; the other (G2; $n = 8$) consisted of highly talented young violin players who received intensive parental support and special training by daily deliberate practice. Mental and musical abilities of both groups were controlled by standardized tests (Kaufman's ABC and Gordon's PMMA) and compared with data taken from recordings of saccadic eye movement using online identification from an electrooculogram (EOG). Results of EOG measurement are referred to as "mental speed," which correlates highly with general mental abilities (intelligence). These results were compared with EOG scores taken from a larger sample of children of the same age range ($n = 82$) who received no music instruction. The grand average of their scores served as a reference line for mental speed, which is normally expected to be performed by an equivalent age group. Data in the two experimental groups did not differ statistically; however, all musically experienced children had a highly significant advantage in mental age ($P < 0.01$) compared to the reference line of the normal population who did not exhibit any effect of training and practice. This indicates strong interaction between mental speed and music ability, which can be interpreted in terms of the expertise model and cognitive transfer effects.

KEYWORDS: expertise; electrooculogram; g factor; intelligence; mental speed; music aptitude

INTRODUCTION

In recent years the expertise model has been favored by several researchers in music psychology when investigating music achievement and the growth of musical talent. This research paradigm focuses mainly on the effects of practice time on expert performance.[1–3] Since Ericsson et al.'s[1] first report, several studies[4,5] have demonstrated a strong relation between deliberate practice and attained performance in music. Observational inquiries demonstrated for experts an accumulated amount

Address for correspondence: Wilfried Gruhn, Music Education Department, University of Music Freiburg, Freiburg, Germany. Voice: +49-761-319.1529; fax: +49-761-319.1542.
w.gruhn@mh-freiburg.de

Ann. N.Y. Acad. Sci. 999: 485–496 (2003). © 2003 New York Academy of Sciences.
doi: 10.1196/annals.1284.059

of practice time of above 10,000 hours within 10 years (10-year rule). In view of these findings, we might assume that there is a causal link between practice time and expertise. However, the question of whether practice time is a crucial factor in developing musical expertise is still unresolved. Although children who continue musical studies successfully report that they "had at least one parent or teacher who cared deeply about their talent, and who worked with them daily, sitting with them as they practiced and establishing a structure of discipline,"[6] why these children go on practicing while others drop out has to be investigated. What makes them stay with music and continue their daily efforts to perform better and gain higher executive skills on their instrument? Obviously, other factors of personality engaged in this development such as endurance, ability to concentrate, discipline, commitment to work, intrinsic motivation, and ease in mastering technical skills play an important role when applied to a given talent.[7] However, the expertise model is highly recognized in music psychology, although the biases and weaknesses of this approach have also been addressed.[8] Therefore, the question of whether other individual components can be identified that are responsible for the development of musical abilities remains unresolved. Music achievement is clearly based on music aptitude or talent, but what makes talent grow and succeed? Practice time is an important factor, but is it sufficient to explain musical expertise?

Another paradigm for the description and evaluation of differences in human mental ability is the model of a multilevel hierarchy with a general factor at the apex and many specific abilities at the bottom. Spearman[9] in 1904 first attributed the neutral label g to this general mental ability. This was based on the empirical result that the first unrotated principal component accounts for a large proportion of the variance in mental test batteries and explains the high positive correlation between two variables. Therefore, g describes a mental aptitude rather than achievement in terms of accumulated knowledge.[10,11] Since Spearman's two-factor model with g as a group factor and s as specific factors, the g-factor paradigm reappeared in the 1990s.[12–14] The g factor is a psychometric construct that cannot be described in terms of knowledge or skills or even theoretical cognitive processes. In essence, it is not a psychological variable, but a biological one that which represents a property of the brain.[13]

The empirical finding of a g factor refers to basic cognitive abilities that are not specific to any particular domain of mental skills and are independent of cultural context and content. If intelligence refers to the ability to process complexity, one criterion is speed of information processing, which is needed to recognize, integrate, and evaluate information. In this respect, individual differences in g may therefore result from differences in mental speed and the efficiency of neural information processing. An empirical measure of mental speed, then, may indicate the potency of g.

In view of these theories, the appearance of music achievement can be seen primarily as either an outcome of a general biological potential (Jensen), a consequence of the amount of time spent on musical practice (Ericsson), or an environmentally supported development of a domain-specific musical aptitude.[15] The goal of this study, therefore, was to investigate whether the interaction between general mental abilities (g) and musical practice or training can be found. As an approach to this question, saccadic eye movements were recorded on an electrooculogram (EOG). In several studies it was demonstrated that some parameters of saccadic eye movements function as a valid indicator of mental speed.[16–18] For example, fixation duration (i.e., the interval between two saccades) and saccadic reaction time positively correlate with a person's IQ. Additional indicators derived from a saccadic tracking paradigm, such as the percentage of omissions and anticipations, significantly correlate with IQ. In general, mental speed as measured by basic cognitive tasks is widely accepted as an underlying mechanism of the *g* factor,[14,19–22] and it seems appropriate to evaluate the cognitive potential of young children in addition to IQ measurements, because it is widely independent from formal instruction and regular school education. These indicators of mental speed were collected from two groups of musically, differently trained children and compared with their scores of music aptitude and general intelligence as well. Finally, all measures can be related to the mental scores of a larger population of peers with no musical training that provide a reference line of mental speed that is expected to be normally found in an equivalent age group.

THE EXPERIMENT

Subjects

Two groups of young children were recruited from the local Freiburg area. Group 1 ($n = 9$; M = 5.33 years; SD = 0.82) comes from an early childhood program in music. Children participate in this program from the age of 2.5 years. The program offers informal guidance in music with emphasis on neither musical practice nor instrumental instruction.[23,24] Rather, the children were exposed to a natural musical setting such as songs and tonal and rhythmic patterns. Group 2 ($n = 8$; M = 6.25 years; SD = 0.78) consists of musically gifted young violin students who participate in a program to promote talented string players.[a] These students have taken at least 1 year of private lessons (twice a week) and practice their instrument every day (M = 0.5 hours) monitored by one parent. Both groups are related to a reference group (G3) consisting of "normal" middle-class children from the Cologne area ($n = 82$; M = 5.66 years; SD = 0.85) without any musical background who provide the norms of electrooculographic data that can be expected in an equivalent age group.

Measures

Both experimental groups (G1, G2) as well as the reference group (G3) underwent EOG measurements for mental speed and a finger-tapping task. Additionally, both experimental groups (G1, G2) were tested by Gordon's Primary Measures of Music Audiation (PMMA 1979), which is standardized for children from kindergarten to primary grade as an indicator of their music aptitude and by Kaufman's Assessment Battery for Children (K-ABC) as a measure of intelligence (IQ).[25,28]

To determine mental speed, saccades were identified online from an electrooculogram (EOG). An amplifier transmits corneoretinal potentials of the subjects with a bandwidth of 0–50 Hz to a 12-bit analog digital converter. After online iden-

---

[a]Children of this group are enrolled in the *Pflüger Foundation Freiburg,* which is exclusively devoted to teaching highly talented young violin players. Children take group and individual lessons and play in mixed ensembles supervised by selected musicians.

tification of all saccades and blinks, data were stored on the hard disk of a 486 PC with a time resolution of 1 ms. To record EOG data, four Medi-Trace-Mini electrodes (Ag/AgCI) were fixed above and below one eye for vertical movement and on both temples close to the eyes for horizontal eye movement. Visual stimuli were presented on a computer screen (17"). The distance from eye to screen, the angle between eyes and stimulus, and the width of the line on which the stimulus moved were adjusted for every child. More details of the methods applied for saccadic parameters are described in Refs. 16 to 18.1.

All subjects performed the following tasks:

- *Finger-tapping task* on a Morse key with indication and middle finger of both hands for 20 seconds to determine the mean tapping interval.

- *Looking task for adaptation*: the average reaction time was counted for 100 saccades while children observed a picture.

- *Running spot*: children were asked to track a horizontal running spot on a TV screen as closely as possible (sinus form). The stimulus was accelerated from 0.2 to 1.0 Hz. Here, mental load increases with velocity.

- *Jumping spot*: children were asked to track a vertical jumping spot (square wave form), which was accelerated in the same frequency range. In this task, subjects initially (i.e., a frequency of 0.2 Hz) tend to respond with reactive responses saccades, whereas later at the middle (0.6 Hz) and higher range (0.8–1.0 Hz), anticipatory responses dominate[25,26] until an increasing percentage of omissions occurs. The mental load increases correspondingly to the successively shortened intervals of the sinus task.[18] The jumping spot task has shown the closest relation to mental performance; therefore, the following parameters are used as the most solid measures of mental speed: saccadic reaction time, saccadic anticipatory latency, omissions of response saccades, and percentage of anticipatory response saccades.

To determine intelligence, Kaufman's Assessment Battery for Children (K-ABC), which is standardized for children aged 2.6–12.6 years, was taken individually by children in groups 1 and 2, separately from sessions of oculomotor measurements. According to their age, children performed at most 13 of 16 subtests in a quiet room given by the experimenter. Measures were taken using three scales: the scale of holistic thinking, the scale of particular thinking, and general intellectual abilities, which results from both scales.

For measuring music aptitude, Gordon's Primary Measures of Music Audiation (PMMA) consisting of two subtests, tonal and rhythmic tests, were given individually after EOG measurements. The musical items were presented by a CD player. The child was seated at a table in front of loudspeakers. The task was to judge whether the second part of a pair of items was the same or different. Children who needed help in filling out the test sheets (circling same or different faces) were assisted by the researcher.

Statistics

All data of saccadic parameters of each individual subject were summarized as medians, because time measures are often distributed in a nonlinear mode. Group means were calculated as arithmetic means or medians, and differences were proofed for significance between groups with parametric or nonparametric statistics. Because both experimental groups could not be fully matched for age, raw data were age adjusted for statistical analysis: for the entire sample of all subjects (total $n = 99$), a linear regression was calculated. Individual values were computed as the difference between raw data and expected values for the particular age. The saccadic parameters of groups 1 and 2 were correlated with those collected from PMMA and K-ABC. EOG data of both experimental groups were then related to the reference line of G3. Because of the small sample, original data of the experimental subjects were transformed into best-fitting curves following the Lowes algorithm, which makes no assumption about the underlying relationship.

RESULTS

First and probably most surprisingly, oculomotor variables of the jumping spot task did not show any significant differences between both experimental groups (G1, G2), who were differentiated by their musical training, but similar with respect to their general musical treatment. However, significant differences appeared for the reference group (G3), who are not musically experienced. Secondly, age-adjusted values revealed the strongest differences for the youngest children of G1 who did not receive specific musical training, but who participated the longest in an informal early music program. The differences between G1 and the reference group (G3) were highly significant: saccadic reaction time ($P = 0.000$), percentage of omissions ($P = 0.008$), and saccadic latency ($P = 0.004$). Furthermore, a single factor analysis of variance showed significant effects between groups 2 and 3 for saccadic reaction time (df = 2; F = 6.901; $P = 0.002$), omission (df = 2; F = 3.874; $P = 0.024$), and saccadic latency (df = 2; F = 4.188; $P = 0.018$).

Tapping

For the mean tapping interval, the musically trained children (G2) revealed a significant effect: the mean interval was significantly smaller ($P = 0.049$; age adjusted) than that of the other groups. Musically trained violin players performed at a higher tapping speed as a consequence of their daily practice. The development of controlled finger movement as indicated by regression lines (FIG. 1) reflects a clear preponderance of the violin players (G2) over the other groups.

Oculomotor Variables

Saccadic Reaction Time

Reaction time serves as an indicator of mental speed.[19,21] The global mean saccadic reaction time revealed a strong advantage of both experimental groups (G1 + G2) compared to the reference group (G3). The time being ahead (mental age advantage) for the musically trained children (G2) was about 1.5 years (FIG. 2). Age-adjusted data exhibited a significant difference between G1 and G3 ($P = 0.000$) and between G2 and G3 ($P = 0.007$). A comparison of age-adjusted data demonstrated that subjects in the reference group (G3) were 13 ms slower and children from the

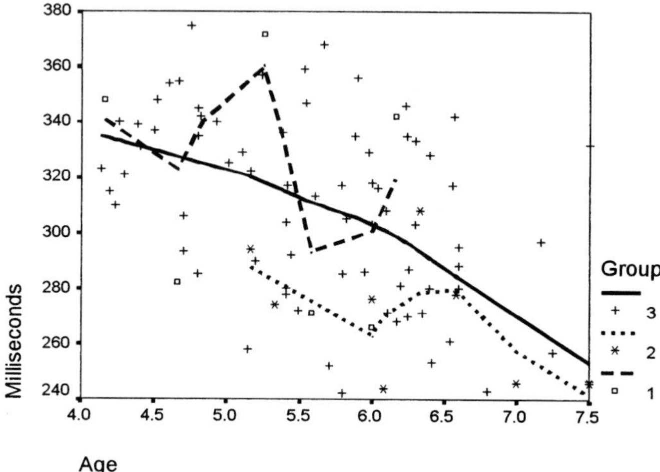

FIGURE 1. Mean tapping intervals with regression lines that show the shortest intervals for G2 with a decreasing tendency by age in all groups. Age-adjusted data reveal a significant difference only between G2 and G3 ($P = 0.049$).

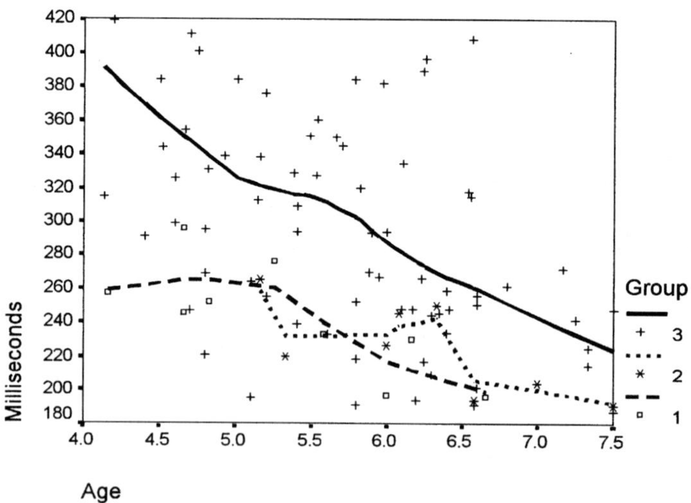

FIGURE 2. Saccadic reaction time over age for the jumping spot task for experimental groups (G1, G2) and control group (G3).

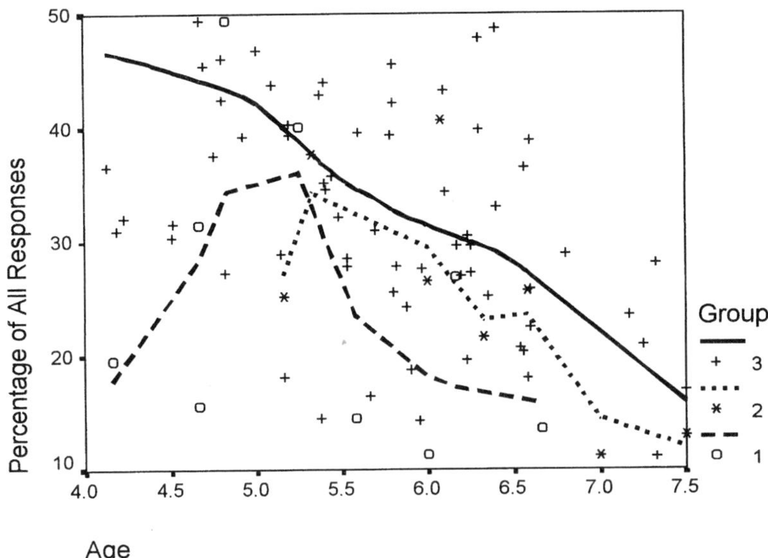

FIGURE 3. Percentage of omissions over age for the jumping spot task for musically talented (G1, G2) and control groups (G3). The advantage of mental age exceeds 1 year. Difference between pooled experimental groups (G1, G2) and control group is highly significant ($P = 0.003$).

experimental groups (G1, G2) were 62 ms faster than expected, a difference that is highly significant (t test with independent samples $P < 0.01$).

Omissions of Saccadic Responses

Omission of saccadic responses can be seen as the best indicator of mental abilities.[18] Omissions along with cognitive development decrease with age. Here, both experimental groups (G1, G2) exceed the reference group (G3) significantly (FIG. 3). From this, we might conclude that advantage in mental age is immediately reflected by a smaller amount of omissions.

Anticipatory Saccadic Responses

The percentage of anticipations in the jumping spot task is another valid indicator of mental growth, especially with respect to the midst frequency of accelerated movement. The percentage of anticipatory responses (in short anticipations) is a function of stimulus frequency: at the lowest range, most responses are reactive, because one cannot surely anticipate when the next stimulus will occur. However, when the intervals between jumps become smaller, we start to predict the next jump, and most responses turn into anticipations. At the highest frequency range of jumps, the task becomes too difficult for most children; therefore, anticipations decrease and omissions increase. The increase at the 0.6-Hz frequency range is relevant for 30% in the youngest children and 60% in the older ones. Early anticipations at the

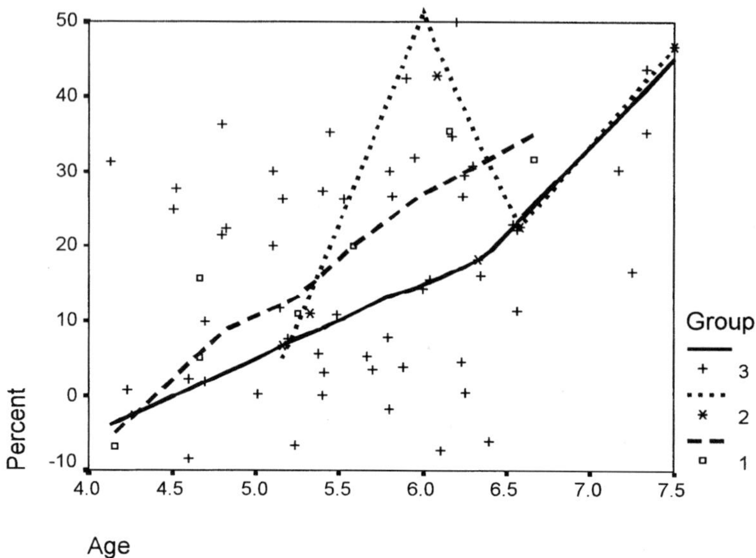

FIGURE 4. Improvement in anticipation from 0.2 to 0.6 Hz over age demonstrates an advantage for both experimental groups (G1, G2) but notable differences in the violin players (G2), which create big fluctuations in the Lowes best-fitting curve for this group.

0.2-Hz frequency range refer to a simple behavior of mere guessing, which is inhibited by age and intellectual growth, whereas later anticipations at the 0.6-Hz frequency range indicate the degree of achievement. Therefore, we calculated the improvement in anticipation from 0.2 to 0.6 Hz (difference in percentage at 0.6 Hz minus the percentage at 0.2 Hz), which is shown in FIGURE 4. The difference of age-adjusted group means showed only marginal significance between experimental groups ($P = 0.051$).

Anticipatory Saccadic Latency

Anticipatory latency is the counterpart of reaction time. An increase over age was recognized for all groups (FIG. 5). Musically experienced groups (G1, G2) showed an advantage in anticipatory latency. Based on age-adjusted means, the difference between G1 and G3 revealed probability at the 0.01 level.

Saccadic Eye Movement and Intelligence

Only one significant Spearman rank correlation was found between saccadic indicators and the results of Kaufman's Assessment Battery for Children (K-ABC), namely, the scale of intellectual abilities (SIA) and the age-adjusted measures for omissions ($r = -0.529$; $P = 0.029$), which therefore can be seen as the strongest indicator of mental speed. The next highest correlation was found between SIA and age-adjusted anticipations ($r = 0.447$), which, however, is statistically not significant

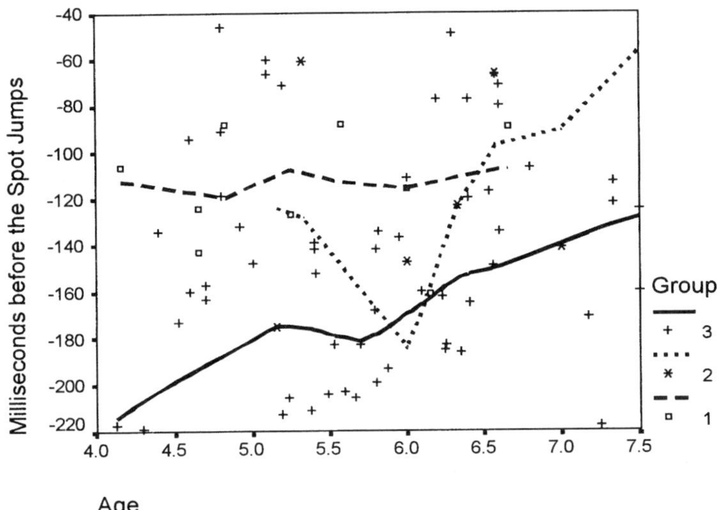

FIGURE 5. Anticipatory latency over age for the jumping spot task for musically talented (G1, G2) and control groups (G3). The advantage of mental age exceeds 1 year for group 1.

($P = 0.072$) mainly because of the small number of subjects ($n = 17$) participating in the study. With respect to the data for omissions, both experimental groups were significantly superior to the reference group. This was confirmed by the K-ABC. Both musically talented groups ranked slightly above average (SIA = 100): children with no special training (G1) achieved a median of 105; highly gifted violin students (G2) achieved a median of 110.5.

Saccadic Eye Movement and Musical Aptitude (PMMA)

The same tendency was reflected by the data of the Primary Measures of Music Audiation (PMMA). Here, the percentile rank of the total of both subtests (tonal and rhythm test) significantly correlated with the age-adjusted data from omissions ($r = 0.539$; $P = 0.026$). Children of the early music program group (G1) rank above 65% of their peers. Not surprisingly, children violinists (G2) ranked much higher than G1 children because of their music training; their percentile rank was above 99% of their peers. An interaction between mental speed, cognitive potential, and music ability was evident.

Intelligence Test and Musical Aptitude (K-ABC and PMMA)

Musical and intellectual development in both experimental groups (G1, G2) demonstrated a nearly perfect linear progression, which indicates an obvious interaction between both domains with age. A gradual progression in intellectual abilities was replicated by the same progression of music audiation.

DISCUSSION

The collected data reveal a strong linear effect of age, that is, all indicators of mental speed increase with age. Therefore, age-adjusted data were used for analysis, so that measures of different ages could be equally related to one another. In general, findings indicate that musical potential enables people to achieve musically, but specific musical training and intensive practice have no manifest effect on general mental abilities and do not enhance mental speed. On the contrary, this may indicate that general mental abilities function as a breeding ground for the development of domain-specific abilities. Music achievement, then, would not depend only on the intensity and extension of deliberate practice; conversely, these efforts can perhaps only succeed when based on mental abilities. Otherwise, practice causes frustration or students drop out.

However, at the moment, our data do not allow us to clearly determine whether mental speed or musical practice prompts music achievement and which role music aptitude plays in this context. The data only show a strong correlation–which is no causal link–between some parameters of mental speed (= general mental abilities) and musical experience. The longer practical experience continues, the more correlation increases. Therefore, we can assume that mental speed and music potential collaborate to cause achievement. Because mental speed is closely related to the speed of stimulus conduction and information processing, which is part of intelligence, it seems clear that mental speed supports mastering complex tasks. One of the most complex activities to perform, however, is music. Therefore, mental speed can be seen as a prerequisite to developing proficiency.

On the other hand, children of the early music program with no specific musical training (G1) who also perform better in oculomotor tasks, but lack high music aptitude as reflected by PMMA scores, exceed average peers by their mental speed, more likely because they come from a socially highly selected intellectual population with parents who are highly engaged in serving their children with the best possible education. These children also participate in many other extracurricular activities such as horseback riding, ballet, foreign languages, and second instruments. Therefore, we assume that general cognitive development and strong parental stimulation mutually interact. Although we cannot decide if mental speed or music aptitude is prominent in the development of music abilities, we recognize that those children who participate in musical activities are excelled by a higher degree of mental speed as their peers. Daily practice, monitored by at least one parent, contributes to music achievement within the limits of an individual child's music potential, but it does not prompt higher mental speed. Nevertheless, the interaction between general cognitive abilities (mental speed) and special musical abilities is evident.

Although the part that intelligence and training play in promoting musical achievement cannot yet be decided, data confirm the significant interaction between general aspects of intelligence (reflected by mental speed) and the development of special skills (indicated by music aptitude test and music performance). At the moment, it seems reasonable to assume that without basic cognitive potential, special aptitudes cannot develop to their full degree. We know the importance of particularly sensitive phases for learning. Mental speed probably determines the degree to which special abilities may develop during the sensitive phase. Only a larger population with a broader variety of intellectually and musically differentiated children who are observed in a longitudinal study can help determine the musical and mental factors that possibly interact in a causal relation.

Another question that remains open refers to musical aptitude: does it represent a separable type of "multiple intelligences" that exists on its own or must it be seen as an integrated part of a person's general cognitive potential. With respect to our data, it seems reasonable to assume that individual intelligence factors and music aptitude factors can only develop within a given cognitive potential for which mental speed may function as an appropriate indicator. Additionally, personality factors contribute to musical achievement. Mental speed represents only one cognitive measure among others that define the limits within which special achievement may develop. This mental condition may lead to musical expertise if deliberate practice and parental support fit into an ensemble of conditions that direct and reinforce individual efforts.

REFERENCES

1. ERICSSON, K.A., R.T. KRAMPE & C. TESCH-RÖMER. 1993. The role of deliberate practice in the acquisition of expert performance. Psychol. Rev. **100:** 363–406.
2. ERICSSON, K.A. & A. LEHMANN. 1996. Expert and exceptional performance: evidence on maximal adaptations on task constraints. Annu. Rev. Psychol. **47:** 273–305.
3. JØRGENSEN, H. & A.C. LEHMANN, Eds. 1997. Does Practice Make Perfect? Current Theory and Research on Instrumental Music Practice. The Norwegian State Academy of Music. Oslo.
4. SLOBODA, J.A., J.W. DAVIDSON, M.J.A. HOWE & D.G. MOORE. 1996. The role of practice in the development of performing musicians. Br. J. Psychol. **87:** 287–309.
5. O'NEILL, S. 1997. The role of practice in children's early musical performance achievement. *In* Does Practice Make Perfect? H. Jørgensen & A. Lehmann, Eds. :53–70. The Norwegian State Academy of Music. Oslo
6. WINNER, E. & G. MARTINO. 2000. Giftedness in non-academic domains: the case of the visual arts and music. *In* International Handbook of Giftedness and Talent. 2nd Ed. K.A. Heller, F.J. Mönks, R.J. Sternberg, *et al.*, Eds. :95–109. Elsevier Science. Amsterdam.
7. SCHNEIDER, W. 2000. Giftedness, expertise, and (exceptional) performance: a developmental perspective. *In* International Handbook of Giftedness and Talent. :165–177. Elsevier Science. Amsterdam.
8. STERNBERG, R.J. 1996. Costs of expertise. *In* The Road to Excellence: The Acquisition of Expert Performance in the Arts and Sciences, Sports and Games. K.A. Ericsson, Ed. Erlbaum Associates. Mahwah, NJ.
9. SPEARMAN, C. 1904. 'General intelligence' objectively determined and measured. Am. J. Psychol. **15:** 201–293.
10. BRAND, C. 1979. General intelligence and mental speed: their relationship and development. *In* Intelligence and Learning. J.P. Das & N.O'Connor Eds. :589–593. Plenum. New York.
11. GOTTFREDSON, L.S. 1998. The general intelligence factor. Scientif. Am. (German Ed.) **3.1999:** 24–31.
12. BRAND, C. 1996. The g Factor. General Intelligence and its Implications. Wiley. Chichester UK.
13. JENSEN, A. 1998. The g Factor: The Science of Mental Ability. Praeger. Westport, CT.
14. DEARY, I.J. 2000. Looking Down on Human Intelligence. Oxford University Press. Oxford.
15. GORDON, E.E. 1979. Primary Measures of Music Audiation (PMMA). GIA Publ. Inc. Chicago.
16. GALLEY, N. 1993. Augenbewegungen, Antizipation und Leistung: Auf dem Wege zu einem neuropsychologischen Konzentrationsmodell. *In* Aufmerksamkeit und Energetisierung. J. Beckmann, H. Strang & E. Hahn, Eds. :229–246. Hogrefe. Göttingen.

17. GALLEY, N. 1998. An enquiry into the relationship between activation and performance using saccadic eye movement parameters. Ergonomics **40:** 698–712.
18. GALLEY, N. & L. GALLEY. 1999. Fixation durations and saccadic latencies as indicators of mental speed. *In* Personality Psychology in Europe. I. Mervielde, I.J. Deary, F. De Fruyt & F.Ostendorf, Eds. **7:** 221–234. University Press. Tilburg.
19. KAIL, R. 1993. Processing time decreases globally at an exponential rate during childhood and adolescence. J. Exp. Child Psychol. **56:** 254–265.
20. CARROLL, J.B. 1993. Human Cognitive Abilities. Cambridge University Press. Cambridge, MA.
21. NEUBAUER, A. 1995. Intelligenz und Geschwindigkeit der Informationsverarbeitung. Springer. Wien, New York.
22. SALTHOUSE, T.A. 1996. The processing-speed theory of adult age differences in cognition. Psychol. Rev. **103:** 403–428.
23. GRUHN, W. 1999. The development of mental representations in early childhood: a longitudinal study on music learning. *In* Music, Mind, and Science. Suk Won Yi , Ed. :434–453. Nat. Univ. Press. Seoul.
24. GRUHN, W. 2002. Phases and stages in early music learning. A longitudinal study on the development of young children's musical potential. Music Ed. Res. **4:** 51–71.
25. KAUFMAN, A.S. & N.C. KAUFMAN. 1994. Kaufman-Assessment Battery for Children (K-ABC). Deutschsprachige Fassung von P. Melchers & U. Preuß. Swets & Zeitlinger. Frankfurt.
26. GAYMARD, B., C.J. PLONER, S. RIVAUD, *et al.* 1998. Cortical control of saccades. Exp. Brain Res. **123:** 159–163.
27. GALLEY, N. 2001. Physiologische Grundlagen, Meßmethoden und Indikatorfunktion der okulomotorischen Aktivität. *In* Enzyklopädie der Psychologie. Biologische Psychologie. F. Rösler, Ed. Band **4:** Grundlagen und Methoden der Psychophysiologie. :237–316. Hogrefe. Göttingen.
28. RAVEN, J.C. 1979. Colored Progressive Matrices. German Version by H. Kratzmeier. Beltz. Weinheim.

Dyslexia and Music

From Timing Deficits to Musical Intervention

KATIE OVERY

Music and Neuroimaging Laboratory, Department of Neurology, Beth Israel Deaconess Medical Center and Harvard Medical School, Boston, Massachusetts 02215, USA

ABSTRACT: The underlying causes of the language and literacy difficulties experienced by dyslexic children are not yet fully understood, but current theories suggest that timing deficits may be a key factor. Dyslexic children have been found to exhibit timing difficulties in the domains of language, music, perception and cognition, as well as motor control. The author has previously suggested that group music lessons, based on singing and rhythm games, might provide a valuable multisensory support tool for dyslexic children by encouraging the development of important auditory and motor timing skills and subsequently language skills. In order to examine this hypothesis, a research program was designed that involved the development of group music lessons and musical tests for dyslexic children in addition to three experimental studies. It was found that classroom music lessons had a positive effect on both phonologic and spelling skills, but not reading skills. Results also indicated that dyslexic children showed difficulties with musical timing skills while showing no difficulties with pitch skills. These apparent disassociations between spelling and reading ability and musical timing and pitch ability are discussed. The results of the research program are placed in the context of a more general model of the potential relationship between musical training and improved language and literacy skills.

KEYWORDS: developmental dyslexia; music; language; timing; rhythm; temporal processing

INTRODUCTION

Developmental dyslexia is characterized by specific learning difficulties in the domain of literacy despite otherwise normal intelligence and schooling.[1] Research into the cause of this learning disability has established a number of associated problem areas, including phonological awareness,[2] aspects of auditory and visual perception,[3] rapid naming,[4] and fine and gross motor skills.[5] Such a variety of problem areas is not easily explained, but current causal theories propose that deficient timing skills, particularly rapid timing skills, may be a key underlying factor.

The "rapid temporal processing"[6] hypothesis proposes that dyslexic children experience difficulties in perceiving rapidly presented sounds, thus impairing as-

Address for correspondence: Dr. Katie Overy, Music and Neuroimaging Laboratory, Department of Neurology, Beth Israel Deaconess Medical Center and Harvard Medical School, Boston, MA 02215, USA. Voice: 1-617-632-8929; fax: 1-617-632-8920.
kovery@bidmc.harvard.edu

Ann. N.Y. Acad. Sci. 999: 497–505 (2003). © 2003 New York Academy of Sciences.
doi: 10.1196/annals.1284.060

pects of phonological perception and consequently literacy development. The "magnocellular theory of dyslexia"[7] suggests that such rapid perception problems are also experienced in the visual domain, causing difficulties with visual aspects of reading. The "double deficit hypothesis"[8] shifts emphasis away from perception to cognition to suggest that dyslexic children have problems with rapid naming skills, which compound their phonological difficulties to cause problems with word recognition and fluent reading. Motor timing difficulties have also been noted by several researchers in areas such as finger tapping[9] and rapid articulation.[10] The "cerebellar deficit hypothesis"[11] proposes that these and other motor deficits are coupled with automatization problems to create a wide range of difficulty areas, from poor balance and untidy handwriting to weak phonological representations. Recently, it has been hypothesized that dyslexic children's difficulties with speech perception are caused by a "deficit in the perceptual experience of rhythmic timing,"[12] resulting in an impaired ability to segment continuous speech into syllables.

Temporal cues are certainly important in speech perception,[13] and reading proficiency is notably affected by temporal fluency.[14] Timing also plays a key role in motor control, particularly in the development of fluent, automatized motor skills.[15] Thus it is reasonable to propose that a remediation method focusing on timing skills may be of particular benefit to dyslexic children. Music offers an ideal, multisensory learning environment for such remediation. Music-making is an enjoyable activity, requiring highly accurate timing skills, particularly in the auditory and motor domains, while the act of singing naturally emphasizes the sound patterns of speech. A growing amount of evidence from a range of research disciplines also suggests that musical experience can have a positive effect on language and literacy abilities.[16–18] In addition, there is increasing elucidation of the numerous shared features of music and language, from developmental characteristics to perceptual processes and common neural substrates.[19–21] With all of these factors taken into consideration, a research program was designed to examine the potential effects of music lessons on the language and literacy skills of dyslexic children.

RESEARCH PROGRAM

The primary aim of this research program was to evaluate the potential of music lessons as a support tool for dyslexic children, while a secondary aim was to explore the specific nature of dyslexic children's difficulties with musical timing. It is often reported that dyslexic children can experience difficulties with rhythm skills,[22] but it is not clear whether these difficulties arise from motor deficits, perception deficits, difficulty extracting the underlying meter, or other factors. Such information could be valuable when designing a musical program for dyslexic children and might also provide further insights into the nature of dyslexic children's language difficulties. To address these research aims, two specific working hypotheses were proposed:

H1: Dyslexic children experience difficulties with musical timing skills, but not with musical pitch skills.

H2: Classroom music lessons can have a positive effect on dyslexic children's phonological skills, reading skills, and spelling skills.

These hypotheses were tested in three different studies: an exploratory study conducted with a class of children screened for dyslexia, a music intervention study conducted with dyslexic children, and a final study comparing the musical skills of dyslexic children with those of control children. To conduct this research, a collection of musical tests was designed specifically for one-to-one use with dyslexic children, and a program of classroom-based musical activities was developed for small groups of dyslexic children.

Musical Tests for Dyslexic Children

Although musical aptitude tests are commercially available,[23–25] they did not meet the purposes of this research, as they are generally limited to the domain of auditory skills. Furthermore, such tests often require long periods of concentration and pencil and paper responses, both of which can cause particular difficulties for dyslexic children. Consequently, an original collection of musical aptitude tests was developed, which included motor skill tasks and placed a strong emphasis on timing skills. Pitch tasks were also included in the collection as a contrasting measure of musical skill. The tests and test procedures were developed and refined over the course of the three studies. The final version measured a wide range of musical skills, was enjoyed by the children, and was quick and easy to conduct in a school setting. The tests were conducted using a portable electronic keyboard, a laptop computer, and a CD player.

Study One

The first study[26] was essentially exploratory, designed to evaluate the strength of the two hypotheses and to suggest directions for subsequent studies. A class of 28 children (mean age 6.7 years) was selected from a school that was participating in a music project organized by The Voices Foundation. As part of this music project, the classroom teacher was trained to conduct singing-based music lessons with her pupils for approximately 1 hour per week, usually divided into three 20-minute sessions. At the beginning of the school year, the children were screened for "risk of literacy difficulties" using the age-appropriate Dyslexia Screening Test (DST)[27,28] and were also tested on the musical tests and pre-tested on the WORD tests of single word reading and spelling.[29] At the end of the school year, the children were post-tested on the WORD literacy tests and the phonologic segmentation task from the DST.

To address H1, the children were divided into three groups according to their score on the dyslexia screening test: strong risk ($n = 6$, mean age 6.8 years), mild risk ($n = 6$, mean age 6.8 years), or no risk ($n = 16$, mean age 6.7 years) of literacy difficulties. The musical test scores of the "strong risk" children were then compared with those of the "no risk" children. Results showed that the strong risk children performed significantly worse on all of the timing tasks, but not the pitch task, thus lending support to H1. The most significant difficulties were evident on the rhythm copying task ($P <0.001$), suggesting that rhythmic motor skills might be a key area of difficulty.

To address H2, the pre-test and post-test scores of the language and literacy tests were compared across all three groups of children. In the absence of a control group the raw scores were converted into standardized scores, which take into account development due to time (or age). This introduced an "implied" control group of the children on which the tests were "normed." Results showed that no significant improvements were made in reading skills, but the strong risk and mild risk groups made significant improvements in spelling skills ($P < 0.05$), and all three groups made significant improvements in phonologic segmentation skills ($P < 0.01$). H2 was thus partially supported in that phonologic skills were dramatically improved during a period of music training, apparently leading to spelling improvements but not reading improvements. It was concluded that a study conducted by a musically trained teacher, working with small groups of dyslexic children and using dyslexia-appropriate, rhythmic musical material, might have a positive effect on reading skills as well as phonological and spelling skills.

Musical Training Program for Dyslexic Children

Before designing a special program of musical activities for dyslexic children, three different approaches to classroom music education were consulted: Education Through Music,[30] Growing with Music,[31] and Earwiggo.[32] The first two approaches focus on group singing games, with strong emphasis on rhythmic and melodic understanding as well as enjoyment and inventiveness. Both methods are adaptations of the Kodaly approach to music education, which has been reported to lead to improvements in other areas of the school curriculum, including language skills.[33] Earwiggo was designed specifically for children with special needs and places similar importance on musical enjoyment, along with the development of listening skills, memory skills, sequencing skills, and motor coordination. Drawing on elements from all three of these approaches, a series of musical games was developed for dyslexic children, focusing particularly on rhythm and timing skills, while taking into account potential difficulties in these areas. The activities were designed to progress gradually from a very basic level to a more advanced level of skill, with opportunities for variation and creativity within each particular game. For the purposes of this research program, the musical curriculum was designed to cover a period of 15 weeks.

Study Two: Music Intervention

The aim of the second study[34] was to further test H2 by measuring the effects of specially designed music lessons on the language and literacy development of dyslexic children. Nine dyslexic boys (mean age 8.8 years) from two different schools were included in the study. Both schools offered high levels of literacy support and regular school music lessons, thus providing a stringent test environment in which to measure the effects of additional group music lessons. Considering the particular difficulties of finding a suitable dyslexic control group for the study (matched for age, IQ, level of reading difficulties, level of literacy support, amount of musical experience, and so forth), it was decided to monitor a "control period" during which the children received no extra music lessons. Development was compared between this 15-week control period and a subsequent 15-week music intervention period. During the intervention period, music lessons were conducted three times per week in sessions of 20 minutes, resulting in a total of 1 hour of training per week. During the control period, the children were visited in their classroom for 1 hour per week, when possible, to limit the possibility of Hawthorne effects. Development was monitored using the WORD[29] tests of single word reading and spelling, selected tests from both the Phonological Abilities Test (PAT)[35] and the Dyslexia Early Screening Test (DEST),[28] and the musical tests outlined above.

The test results indicated that the music program had a significant positive effect on four areas of skill: rhythm copying (P <0.05), rapid auditory processing (P <0.05), phonological ability (P <0.01), and spelling ability (P <0.05). These results lend further support to H2 and also strengthen the proposal that timing skills may play a key role in the transfer of musical abilities to language abilities.[36] Interestingly though, no significant improvements were identified in reading skills, thus partly challenging H2. This pattern of results is similar to that of study 1, in which phonological and spelling skills were significantly improved while reading skills were not. It is possible that reading improvements would have followed the phonological and spelling improvements if the music intervention had lasted longer or involved more intense training. Spelling improvements have been found to precede reading improvements in other intervention studies with dyslexic children.[37] It could also be argued that phonological improvements are initially more likely to lead to spelling improvements than reading improvements, as spelling is a phoneme-to-grapheme skill, whereas reading is a grapheme-to-phoneme skill that can also rely on visual recognition and context as decoding strategies. It has been demonstrated that young children can often spell words that they cannot read[38] and suggested that the relationship between spelling and phonological awareness is particularly strong in the early stages of literacy learning.[39] Such an interpretation would require further investigation, preferably with larger numbers of dyslexic children and longer periods of music training. Future studies of this nature would also benefit from more precise knowledge of the nature of dyslexic children's musical timing difficulties, allowing for more specific focus on these areas during music lessons.

Study Three: Testing Musical Skills

The goal of the third study[40] was to examine dyslexic children's musical timing difficulties in more detail, thus further testing H1. Fifteen dyslexic boys (mean age 9.0 years) and 11 controls (mean age 8.9) were compared on a range of musical tests, including rhythm skills, metre skills, rapid auditory processing skills, and pitch skills. The children also undertook the WORD reading and spelling tests and were scored according to their musical experience, based on a short interview.

Results showed a clear tendency for the dyslexic group to score lower than the control group on the tests of timing skills, while scoring higher than the control group on the tests of pitch skills. Two tests showed significant differences between the groups, with dyslexic children performing higher on the task of pitch discrimination (P <0.05) and lower on the tasks of rapid auditory processing (P <0.05). Interestingly, a subgroup of five dyslexic children was found to account for all the significant difference on the latter task, lending support to suggestions that rapid temporal processing difficulties affect only a subset of the dyslexic population.[41] Performance on the tempo copying task was difficult to assess, as the dyslexic children tended not to complete the required number of taps (P <0.05). Out of five tested tempi (48, 60, 80, 120, and 240 bpm), the dyslexic group only showed significant difficulty at 80 bpm (P <0.05).

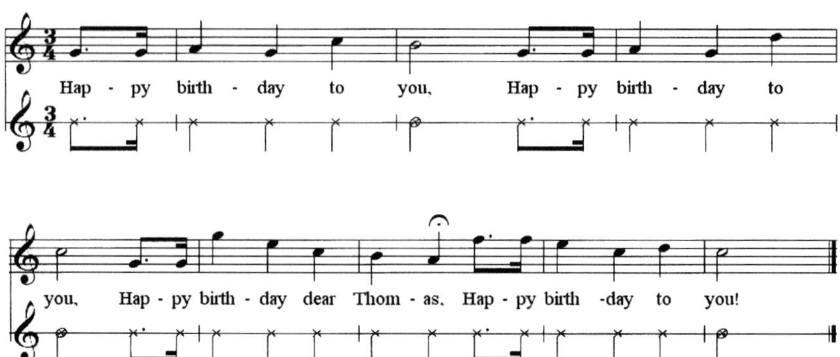

FIGURE 1. Diagram showing how the rhythm of a song corresponds to the syllables of the text.

A correlation analysis was also conducted across all children and all tests, and a number of strong correlations was found between tests involving similar skills, such as rhythm perception and rhythm copying ($r = .64$, $P < 0.001$). A particularly interesting correlation was found between scores on the "song-rhythm tapping" test and the WORD spelling test ($r = .54$, $P < 0.005$). Although these two skills may seem unrelated, both in fact depend on syllable segmentation skills (FIG. 1), thus providing an intriguing link between performance on the two tasks. On further analysis, it was found that this correlation was almost twice as strong in the control group ($r = .66$, $P = 0.02$) as in the dyslexic group ($r = .37$, $P = ns$), implying that the dyslexic group may not have been using the strategy of syllable segmentation as effectively as controls. This result provides additional indirect support for H2, as it suggests that learning to tap to the rhythm of a song could be an effective learning tool in the development of syllable segmentation skills and subsequent spelling skills.

Overall, H1 was strengthened rather than challenged by the results of this study, although the difference in performance between the two groups was not as great as anticipated, and the superior pitch skills exhibited by the dyslexic group were unpredicted. One potential explanation for this pattern of results is the fact that the dyslexic group scored higher than the control group on the measure of musical experience. It is possible that such experience gave the dyslexic children an advantage on the musical tests, thus increasing their scores on the pitch tasks and reducing the differences between the two groups on the timing tasks. Alternatively, it is also possible that the dyslexic groups' high performance on the pitch tests reflects a particular strength in this area. Backhouse,[42] for example, has noted a heightened sense of pitch in some dyslexic musicians. It has also been proposed that dyslexia can be associated with "right hemisphere" strengths[43] as well as "left hemisphere" weaknesses,[44] while neurological research has suggested that pitch skills are processed predominantly in the right hemisphere[45] and musical timing processes are processed predominantly in the left hemisphere.[46] It would be interesting to examine this relationship further in future research.

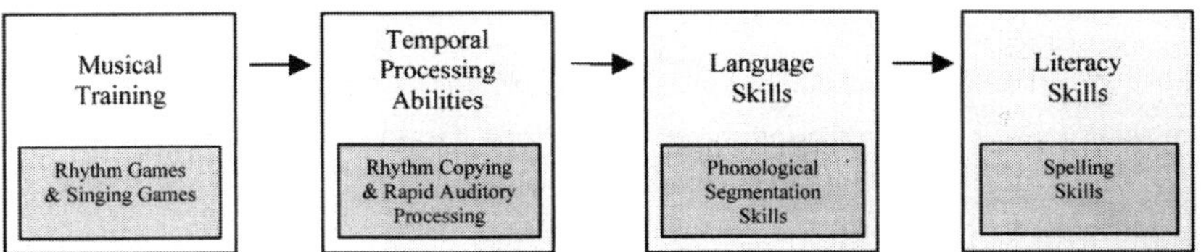

FIGURE 2. Diagram showing a general model of the possible relationship between musical training and the development of literacy skills. The results of the research program reported here are inserted into the model.

CONCLUSIONS

The results of this research program support suggestions that dyslexic children can experience difficulties with temporal aspects of auditory and motor skills. The most apparent areas of musical difficulty involved rhythmic motor skills and rapid auditory processing skills, reflecting the similar weaknesses experienced by dyslexic children in rhythmic and rapid aspects of language processing.[6,10,12] No difficulties were found with musical pitch skills, further indicating that timing may be a particular problem area for dyslexic children.

The proposal that classroom music lessons might help to improve dyslexic children's language and literacy skills was partially supported. In two different studies, phonological skills and spelling skills were significantly improved, while reading skills were not significantly affected. In the second of these studies, rhythm copying and rapid auditory processing skills were improved alongside the phonological and spelling improvements. Although this result does not demonstrate a causal relationship, it does lend support to the hypothesis that temporal processing may be a key mechanism in the potential transfer of musical abilities to language abilities.[25,36,47] FIGURE 2 shows how the results of this research program fit into a more general model of the possible relationship between musical training and the development of literacy skills. Of course it is also possible that other processes are involved in such a transfer effect; for example, singing might lead directly to phonologic development, while learning to read music might help with learning to read text. Further research will be required to expand and refine the general model outlined here, preferably with different kinds of musical training and different tests of musical skills, perceptual and cognitive skills, and language and literacy skills.

In summary, this research strengthened the argument that music lessons have the potential to provide a valuable multisensory learning environment for dyslexic children. A particular advantage of music lessons as a language support tool is that they can be used at any stage of literacy development and at any age, from pre-school to high school. There is also a wealth of musical material available for such a remediation approach, from nursery rhymes and folk music to popular music and art music, along with an established tradition of classroom music education that can be modified and adapted as necessary. Furthermore, the potential enjoyment and skill development gained from music lessons offer valuable opportunities for lifelong learning.

ACKNOWLEDGMENTS

This research was supported by a scholarship from the Economic and Social Research Council and was conducted at Sheffield University, UK, under the guidance of my Ph.D. supervisors Prof. Rod Nicolson, Dr. Angela Fawcett, and Prof. Eric Clarke.

REFERENCES

1. WORLD FEDERATION OF NEUROLOGY. 1968. Report of Research Group on Dyslexia and World Illiteracy. Dallas. W.F.N.
2. BRYANT, P.E. & L. BRADLEY. 1985. Children's Reading Problems. Basil Blackwell. Oxford.
3. WITTON, C. *et al.* 1998. Sensitivity to dynamic auditory and visual stimuli predicts nonword reading ability in both dyslexics and controls. Curr. Biol. **8:** 791–797.
4. WOLF, M. 1991. Naming speed and reading: the contribution of the cognitive neurosciences. Reading Res. Q. **26:** 123–141.
5. NICOLSON, R.I. & A.J. FAWCETT. 1994. Comparison of deficits in cognitive and motor skills in children with dyslexia. Ann. Dyslexia **44:** 147–164.
6. TALLAL, P., S. MILLER & R.H. FITCH. 1993. Neurological basis of speech: A case of the preeminence of temporal processing. Ann. N.Y. Acad Sci. **682:** 27–47.
7. STEIN, J. & V. WALSH. 1997. To see but not to read; the magnocellular theory of dyslexia. Trends Neurosci. **20:** 147–152.
8. WOLF, M. & P. BOWERS. 2000. The question of naming-speed deficits in developmental reading disability: an introduction to the double-deficit hypothesis. J. Learn. Disabil. **33:** 322–324.
9. WOLFF, P.H. 2002. Timing precision and rhythm in developmental dyslexia. Reading & Writing **15:** 179–206.
10. FAWCETT, A.J. & R.I. NICOLSON. 2002. Children with dyslexia are slow to articulate a single speech gesture. Dyslexia **8:** 189–203.
11. NICOLSON, R.I., A.J. Fawcett & P. Dean. 1995. Time estimation deficits in developmental dyslexia: evidence for cerebellar involvement. Proc. Roy. Soc. **259:** 43–47.
12. GOSWAMI, U. *et al.* 2002. Amplitude envelope onsets and developmental dyslexia: a new hypothesis. Proc. Natl. Acad. Sci USA **99:** 10911–10916.
13. MARTIN, J.G. 1986. Aspects of rhythmic structure in speech perception. *In* Rhythm in Psychological, Linguistic & Musical Processes. J.R. Evans & M. Clynes, Eds. :79–95. Charles C Thomas. Springfield, IL.
14. HANES, M.L. 1986. Rhythm as a factor of mediated and nonmediated processing in reading. *In* Rhythm in Psychological, Linguistic and Musical Processes. J.R. Evans & M. Clynes, Eds. :99–130. Charles C Thomas. Springfield, IL.
15. THAUT, M.H. *et al.* 2001. Auditory rhythmicity enhances movement and speech motor control in patients with Parkinson's disease. Funct. Neurol. **16:** 163–172.
16. KILGOUR, A.R. *et al.* 2000. Music training and rate of presentation as mediators of text and song recall. Mem. Cognit. **28:** 700–710.
17. SUTTON, J. 1995. The sound-world of speech- and language-impaired children: the story of a current music therapy research project. *In* Art and Music, Therapy and Research. A.L. Gilroy, Ed. :153–163. Routledge. London.
18. DOUGLAS, S. & P. WILLATTS. 1994. The relationship between musical ability and literacy skill. J. Res. Reading **17:** 99–107.
19. SLOBODA, J. 1985. The Musical Mind: The Cognitive Psychology of Music. Clarendon Press. Oxford.
20. BESSON, M. & D. SCHON. 2001. Comparisons between language and music. *In* The Biological Foundations of Music. R.J. Zatorre & I. Peretz, Eds. Ann. N.Y. Acad. Sci. **930:** 232–258.
21. PATEL, A. *et al.* 1998. Processing syntactic relations in language and music: an event-related potential study. J. Cognit. Neurosci. **10:** 717–733.

22. MILES, T.R. & J. WESTCOMBE, Eds. 2001. Music and Dyslexia: Opening New Doors. Whurr Publishers. London & Philadelphia.
23. GORDON, E. 1979. Primary Measures of Music Audiation. GIA Publications. USA.
24. WING, H.D. 1968. Tests of musical ability and appreciation. Cambridge University Press. UK.
25. OVERY, K. 2000. Dyslexia, temporal processing and music: the potential of music as an early learning aid for dyslexic children. Psychol. Music **28:** 218–229.
26. SEASHORE, C.E. 1960. Measure of Musical Talents. The Psychology Corporation. New York.
27. FAWCETT, A.J. & R.I. NICOLSON. 1996. The Dyslexia Screening Test. The Psychological Corporation. London.
28. NICOLSON, R.I. & A.J. FAWCETT. 1996. The Dyslexia Early Screening Test. The Psychological Corporation. London.
29. WECHSLER, D. 1993. Wechsler Objective Reading Dimensions. The Psychological Corporation. UK.
30. RICHARDS, M.H. 1977. Aesthetic Foundations for Thinking. Richards Institute of Music Education and Research. California.
31. STOCKS, M. & A. MADDOCKS. 1992. Growing with Music. Longman Group UK Ltd. Harlow, Essex.
32. WEST, & HOLDSTOCK. 1985. Earwiggo Again: Rhythm Games. Yorkshire and Humberside Association for Music in Special Education. UK.
33. HURWITZ, I. et al. 1975. Non-musical effects of the Kodaly music curriculum in primary grade children. J. Learn. Disabilities **8:** 45–52.
34. OVERY, K. et al. 2003. Music Intervention with Dyslexic Children. Submitted.
35. MUTER, V. et al. 1997. Phonological Abilities Test. The Psychological Corporation. UK.
36. OVERY, K. 2000. How can music affect cognitive ability? The case of improved language skills. Presented at the SRPMME Annual Conference, Leicester, UK, April 9. http://www. sempre.org.uk/conferences/apr2000/#Overy
37. NICOLSON, R.I. & A.J. FAWCETT. 1994. Spelling remediation for dyslexic children: a skills approach. In Handbook of Spelling: Theory, Process and Intervention. G.D.A. Brown & N.C. Ellis, Eds. :505–528. Wiley. Chichester.
38. Frith, U. 1985. Beneath the surface of developmental dyslexia. In Surface Dyslexia. K. Patterson et al., Eds. :505–528. LEA. London.
39. ELLIS, N. & S. CATALDO. 1992. Spelling is integral to learning to read. In Psychology, Spelling and Education. C. Sterling & C. Robson, Eds. :122–142. Multilingual Matters. Clevedon.
40. OVERY, K. et al. 2003. Dyslexia and Music: Measuring Musical Timing Skills. Dyslexia **9:** 18–36.
41. MARSHALL, C.M. et al. 2001. Rapid auditory processing and phonological ability in normal readers and readers with dyslexia. J. Speech Lang. & Hear. Res. **44:** 925–940.
42. BACKHOUSE, G. 2001. The piano tuner. In Music and Dyslexia. T.R. Miles & J. Westcombe, Eds. Whurr Publishers. Gateshead, UK.
43. WEST, T.G. 1991. In the Mind's Eye: Visual Thinkers, Gifted People with Learning Difficulties, Computer Images, and the Ironies of Creativity. Prometheus Books. Buffalo, NY.
44. TEMPLE, E. et al. 2000. Disruption of the neural response to rapid acoustic stimuli in dyslexia: evidence from functional MRI. Proc. Natl. Acad. Sci. USA **97:** 13907–13912.
45. ZATORRE, R. 2001. Neural specializations for tonal processing. In The Biological Foundations of Music. R.J. Zatorre & I. Peretz, Eds. Ann. N.Y. Acad. Sci. **930:** 193–210.
46. SAMSON, S. et al. 2001. Cerebral substrates for musical temporal processes. In The Biological Foundations of Music. R.J. Zatorre & I. Peretz, Eds. Ann. N.Y. Acad. Sci. **930:** 166–178.
47. OVERY, K. 2002. Dyslexia and music: from timing deficits to music intervention. PhD Thesis. University of Sheffield, UK.

Effects of Musical Training on the Auditory Cortex in Children

LAUREL J. TRAINOR, ANTOINE SHAHIN, AND LARRY E. ROBERTS

Department of Psychology, McMaster University, Hamilton, Ontario, Canada L8S 4K1

ABSTRACT: Several studies of the effects of musical experience on sound representations in the auditory cortex are reviewed. Auditory evoked potentials are compared in response to pure tones, violin tones, and piano tones in adult musicians versus nonmusicians as well as in 4- to 5-year-old children who have either had or not had extensive musical experience. In addition, the effects of auditory frequency discrimination training in adult nonmusicians on auditory evoked potentials are examined. It was found that the P2-evoked response is larger in both adult and child musicians than in nonmusicians and that auditory training enhances this component in nonmusician adults. The results suggest that the P2 is particularly neuroplastic and that the effects of musical experience can be seen early in development. They also suggest that although the effects of musical training on cortical representations may be greater if training begins in childhood, the adult brain is also open to change. These results are discussed with respect to potential benefits of early musical training as well as potential benefits of musical experience in aging.

KEYWORDS: auditory cortex; plasticity; learning; musical training; children; development; musicians; event-related potential (ERP)

INTRODUCTION

Recent studies have shown that the auditory cortex responds differently to sound in musicians than in nonmusicians. For example, neuromagnetic (MEG) studies have shown that the N1m response, which occurs about 100 ms after the sounding of a musical tone, is enhanced in musicians,[1] and the effect is larger for tones in the timbre of the musician's primary instrument.[2] Event-related potentials (ERPs) derived from EEG recordings have shown enhanced P2 responses at about 180 ms after sound onset and enhanced N1c responses at about 140 ms.[3] Later ERP potentials associated with the conscious interpretation of sound are also enhanced in musicians compared to nonmusicians.[4,5] These studies of functional differences between musicians and nonmusicians are paralleled by anatomic studies showing enlargement of areas important in music perception. Specifically, musicians have enlargement of the anteromedial region of Heschl's gyrus,[6] the right-sided planum,[7] and the anterior corpus callosum.[8] At the same time, nonmusicians also show remarkable

Address for correspondence: Laurel J. Trainor or Larry E. Roberts, Department of Psychology, McMaster University, Hamilton, Ontario, Canada L8S 4K1. Voice: 905-525-9140, ext. 23007; fax: 905-529-6225.

LJT@mcmaster.ca

Ann. N.Y. Acad. Sci. 999: 506–513 (2003). © 2003 New York Academy of Sciences.
doi: 10.1196/annals.1284.061

sensitivity to musical input. Even when they are not paying attention, ERP responses from the auditory cortex (specifically the mismatch negativity response) indicate that the brains of nonmusicians notice changes in the melodic contour (up/down pattern of pitch changes) and melodic interval (pitch distances between tones) of melodies.[9]

In this chapter, we question how these differences between musicians and non-musicians arise. Because genetic and environmental factors interact in intricate ways from the earliest stages of development,[10] it is difficult to quantify the separate contributions of these factors. However, the question we can ask is whether the ERP components seen to be enhanced in musicians can also be enhanced through auditory training. A positive answer to this question would suggest that musical experience and practice contribute substantially to the musician/nonmusician differences seen. A second question that can be asked is how early in development differences between future musicians and nonmusicians can be seen. Recent studies indicate that auditory cortical evoked responses do not mature until late adolescence.[11,12] At the same time, there is suggestive evidence of a critical period for musical development, in that musicians who began musical training early show the largest N1m enhancement and those starting after the age of about 10 years tend not to show any N1m enlargement.[1] In the first part of this chapter, we compare musician/nonmusician cortical differences with cortical changes induced by auditory training in non-musician adults. In the second part of this chapter, we present data comparing the cortical responses of young children with extensive musical experience to those of young children without extensive musical experience. In the final part of the chapter, we consider the implications of these findings for musical education.

PLASTICITY OF EVOKED RESPONSES: LABORATORY TRAINING IN COMPARISON WITH NATURALISTIC MUSICAL TRAINING

In Western society, people differ greatly in the amount of musical training and practice they have. This provides the grounds for a naturalistic experiment comparing nonmusicians with minimal musical experience to musicians who have had years of experience practicing for hours virtually every day. Shahin *et al.*[3] compared 11 professional violinists who were members of Canada's National Academy Orchestra, 9 pianists who had at least Grade 10 certification from Canada's Royal Conservatory of Music, and 14 university students who did not play an instrument and had no formal musical training. EEG was recorded in each participant as they listened to violin, piano, and sine wave tones (A3 and C3, American notation) matched in loudness (500-ms tones; SOA = 3,000 ms). EEG traces for all 28 channels of recorded activity distributed across the scalp are shown in FIGURE 1A for the musical tones. The dipolar patterns, where the electrical activity recorded at some sites is positive and at some sites is negative, indicate significant brain events. Clear N1b, N1c, and P2 events can be seen. Furthermore, source localization analyses using BESA software locate the generators of the activation to secondary auditory cortical areas in the temporal lobe (FIG. 1), with the P2 event located consistently more medial than the other components and the N1b more medial than the N1c.

The most interesting aspect of Shahin *et al.*[3] in the present context is that both the N1c and the P2 components were significantly larger in the musician groups than in

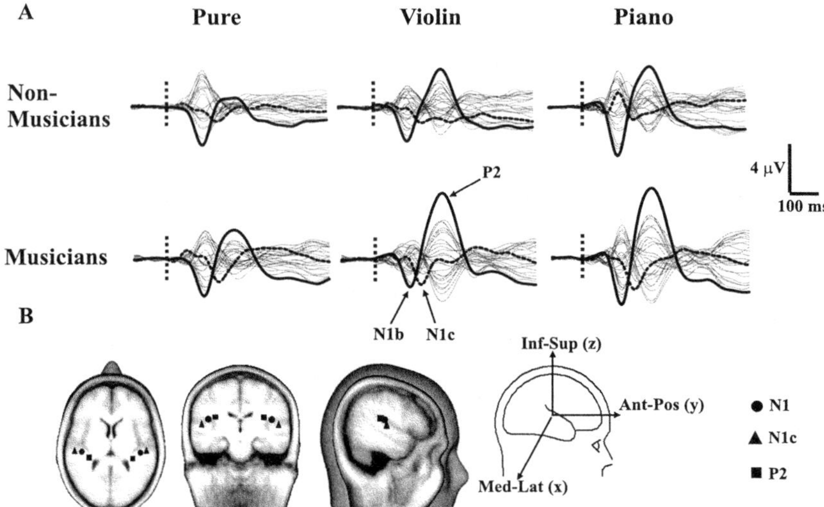

FIGURE 1. (***Upper Panel***) EEG traces for all 28 channels evoked by the sine wave, violin, and piano tones for musicians and nonmusicians. The *dashed vertical line* indicates onset of the tones. The Cz electrode is shown in *bold*, and the T8 electrode is a *dashed line*. (***Lower Panel***) The locations of the regional sources for grand average data with a hemispheric symmetry constraint for the N1b, N1c, and P2 components determined by BESA and superimposed on the average brain of BESA.

the nonmusicians. This suggests that these brain events are critical for musical processing. However, it does not tell us if musicians were simply born with brains that give good N1c and P2 responses or if the years of musical training contributed to the observed enhancement. To examine the role of experience, the degree to which these components are neuroplastic needs to be addressed. The most stringent test of such neuroplasticity would be to train nonmusician adults, who are well beyond any critical period for musical training if one exists, in a task such as pitch discrimination and examine whether this auditory experience enhances their N1c and P2 responses. This is exactly what was done by Bosnyak *et al.*[13]

Bosnyak *et al.*[13] presented nonmusicians with sine wave tones around 2.0 kHz, amplitude modulated at 40 Hz. Initially, ERPs were measured in response to 1.8-, 2.0-, and 2.2-kHz tones. Participants were then given half an hour of training per day for 15 days in discriminating small frequency changes from the 2.0-kHz tone only. Their ERPs were then measured again at the end of training for tones at all three frequencies, and behavioral discrimination without EEG again 7 weeks later. Amplitude modulation allowed observation of the steady-state response which, because of its 40-Hz rate of modulation, is related to the middle latency responses (occurring 25 ms after stimulus onset) known to be generated in the primary auditory cortex. Interestingly, although training appeared to have no effect on the amplitude of the steady-state response (but did affect temporal properties of the steady-state response), it did have a substantial effect on the N1c and P2 components, suggesting that, in adults at least, the secondary auditory cortex is more plastic than the primary auditory cortex.

Behaviorally, the participants' ability to discriminate fine frequency differences improved for all three frequencies, but improved more for the trained 2.0-kHz tones than for the other two. Electrophysiologically, both the N1c and the P2 were larger after training, with the largest effects seen for the trained frequency of 2.0 kHz.

This training study corroborates the naturalistic musical experience study just described[3] in that the components found to be enhanced in musicians than non-musicians were the same components found to be enhanced after frequency discrimination training. There are also two other reports that P2 amplitude increases with training. Tremblay *et al.*[14] found increased P2 amplitude after auditory temporal training, and Atienza *et al.*[15] found increased P2 amplitude after training in pitch deviance detection.

Little is yet known about what particular process the P2 brain event represents. However, a positive surface potential such as that found in the P2 is likely associated with the depolarization of pyramidal cell bodies (a sink) in deeper cortical layers (IV–VI), whereas a negative potential such as the N1b is likely to be associated with current sinks occurring on apical dendrites in the superficial layers (II–III). The observed neuroplastic changes, then, are likely to be associated with synaptic changes affecting depolarization of neurons. One possibility for the learning mechanism involves projections from the basal forebrain, which appear to modulate synaptic activity in the auditory cortex (see Shahin *et al.*[3] and Bosnyak *et al.*[13] for a discussion). In any case, the training study in conjunction with the study comparing musicians and nonmusicians suggests that learning and experience play a large role in sculpting the musical brain. In the next section, we consider how early such effects can be seen.

EARLY DEVELOPMENT OF ENHANCED RESPONSES: EVOKED POTENTIALS IN YOUNG SUZUKI MUSIC STUDENTS

Recent studies have made it clear that the auditory cortex undergoes a very protracted maturational process in normal development.[11,12] Data from our laboratory, presented in FIGURE 2, agree with these findings. In response to a violin tone, young children show a prominent P1 component, but an adult-like N1b is not seen until well into adolescence. These results suggest that the maturational level of the auditory cortex can be assessed by examining the N1b component.

Shahin *et al.*[16] asked how early in development the effects of musical experience could be seen. Would children exposed to music from an early age show more rapid development in the auditory cortex? Could this be seen in children younger than 6 years of age? To this end, Shahin *et al.*[16] compared seven 4-year-old children taking Suzuki music lessons (6 pianists and 1 violinist) with six age-matched control children who were not studying music. Parents of all the Suzuki music students reported that their children listened to a substantial amount of music in the home, and six of the seven children were in a family where at least one parent or sibling played a musical instrument and practiced at home. For the control children, only one child had a parent who played a musical instrument, and this person never practiced at home.

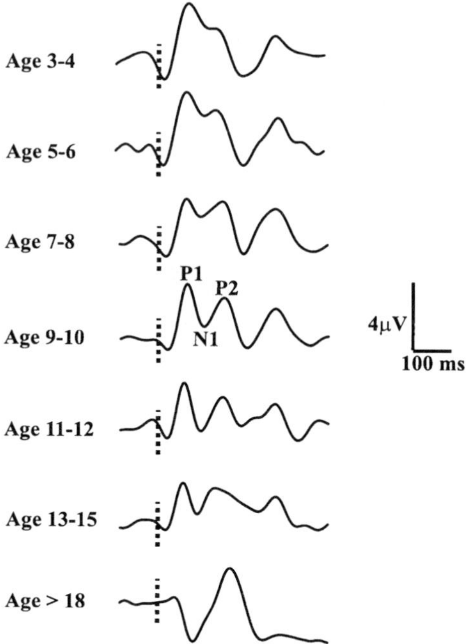

FIGURE 2. Evoked responses of children between 4 years and adulthood to a violin tone measured at a frontocentral site (Fz). The P1 component diminishes with increasing age, whereas the N1b component increases with increasing age. The number of children in each group is as follows: 10, 3–4 years; 11, 5–6 years; 14, 7–8 years; 7, 9–10 years; 10, 11–12 years; 6, 13–15 years; 14, over 18 years.

The children were presented with the same violin, piano, and sine wave tones and the same SOA (3000 ms) used in the Shahin *et al.*[3] study just discussed. Their ERPs were measured twice, once at 4 years of age, when the Suzuki music students were just starting music lessons, and once a year later when they were 5 years of age. There were no significant differences across the two measurements, so the waveforms are shown collapsed across this factor for the pure tones, violin tones, and piano tones in FIGURE 3. However, significant differences were noted between the two groups of children and between the type of tone presented.

Interestingly, ERP responses to piano tones were the most mature, in that clear P1, N1b, and P2 components could be identified, and those responses to the pure tones were least mature, in that only clear P1 components were present. The differences between the sine tones and the musical tones might reflect the increased cortical response to broad-band signals, and the differences between the violin and piano tones might reflect the more abrupt onset of the piano tones, which could be expected to generate more synchronous activation in the auditory cortex and thus lead to larger peaks in the components of the evoked potentials. Interestingly, studies of the maturation of auditory evoked potentials[11,12] showing no N1b response in

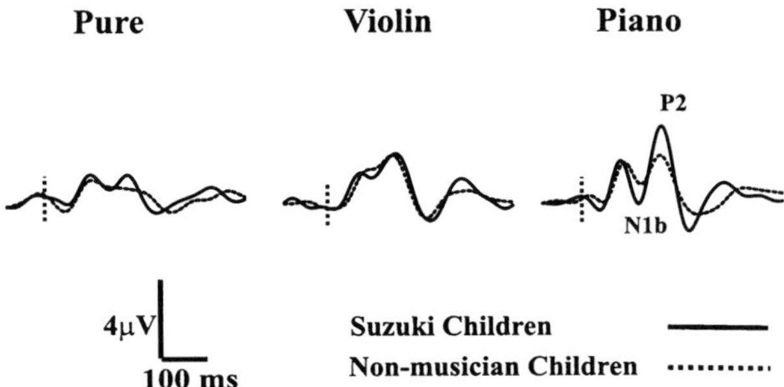

FIGURE 3. Responses of Suzuki-trained and control children to sine tones, violin tones, and piano tones from a central site on the scalp (Cz). *Dashed vertical line* indicates the onset of tones.

children of this age used much shorter SOAs (600 ms or less compared to our 3,000 ms).

Of most interest in the present context are the differences between the two groups of children. The P2 response is most clearly seen at the central vertex electrode. As can be seen in FIGURE 3, this response is also larger in the Suzuki-trained than in the control children for the piano tones. The beginnings of an N1b response can also be seen in response to the piano tones, and again FIGURE 3 shows that this response is larger in the Suzuki-trained than in the control children. It is interesting that the Suzuki-trained and the control children only differ in their responses to piano tones, because six of the seven Suzuki pupils were studying piano.

Thus, by age 4–5 years, we can differentiate two groups of children who have had different musical experiences on the basis of their auditory evoked responses. First, the N1b component appears to emerge earlier in the Suzuki group than in the control group, suggesting that this component might serve as an index of auditory cortical maturation. Second, the P2 component, which we found to differentiate musicians and nonmusicians[3] and to be enhanced with auditory training in nonmusician adults,[13] was also larger in the Suzuki-trained children than in the control children for tones of their instrument of practice. This adds to the hypothesis that this component is particularly neuroplastic.

We have shown that auditory ERPs can differentiate young children with more from those with less musical experience. However, they do not tell us if these differences are largely controlled by genetic factors or if they are largely the result of musical training. However, the fact that the same component, P2, is implicated in adult musician/nonmusician differences, is affected by auditory training in nonmusician adults, and also differentiates young children with or without extensive musical experience, suggests that experience plays a significant role in the differences seen between Suzuki-trained and control children. However, a definitive answer to this question must await studies with younger children and studies in which musical experience is explicitly controlled in the experimental design.

CONCLUSIONS AND IMPLICATIONS FOR MUSICAL TRAINING

We conclude that cortical sound representation is affected by auditory experience. Auditory training in adult nonmusicians leads to enhancement of the P2 auditory evoked potential. This same potential is enhanced in adult musicians compared with nonmusicians and is enhanced in children as young as 4 years of age who have had extensive musical experience in comparison with children who have not. Although more work is needed to clarify the role of specific experience in this enhancement, the data suggest that early musical training has a substantial effect on auditory cortical representations.

The tentative implications of this work for musical education, then, are that musical training will have a substantial effect on auditory cortical development. The optimal age for commencement of musical training remains unclear, however. Pantev *et al.*[1] found a significant correlation between the age of onset of music lessons and the size of the N1b response in musicians, with those starting lessons after age 10 not showing the effect. Trainor *et al.*[4] showed a similar correlation between the age of onset of music lessons and the size of a later component, the P3. Thus, earlier training appears to lead to larger changes in the auditory cortex. However, the relation between cortical responses and behavior is complex. The musicians in Pantev *et al.*'s study and in Trainor *et al.*'s study who commenced musical training later were not inferior musicians by behavioral standards. Thus, it is possible that early training is not essential for becoming a highly skilled musician, but that it is helpful and that those who start training later may solve the problem in a different way because they may have less plasticity in the auditory cortex.

The effects of musical training on the auditory cortex are also of interest with respect to the relation between musical training and domains other than music. For example, learning to read a language such as English involves auditory processing because in order to learn to read, children must be able to break a word into its component phonemes (e.g., "cat" is composed of three phonemes, /c/, /a/, /t/) and associate each with a written symbol. Indeed, there are reports that preschool children's phonemic awareness and early reading skill are correlated with their musical training.[17,18] In conjunction with our findings of earlier maturation of evoked responses in children with extensive musical training, these studies suggest that early musical training may enhance language and reading skills as well as musical skills. Again, however, cause-and-effect studies have not yet been done.

We conclude that auditory evoked responses can differentiate children with extensive musical training from those without it at ages as young as 4 to 5 years. In particular, the N1b and P2 components appear to emerge earlier in children with musical training. The nature of the sound appears to have a great effect on children's evoked potentials, with children of this age only showing clear components to meaningful sounds rich in harmonics and with abrupt onsets (i.e., piano tones). Interestingly, the effects of experience are not limited to children. Even adult nonmusicians given auditory training showed enhancement of the same cortical components seen to be enlarged in adult musicians and children with extensive musical experience. These results imply that musical training in adulthood and even in old age may confer benefit. Indeed, it is possible that continued musical practice may help ameliorate the effects of aging.

ACKNOWLEDGMENTS

This research was supported by the Canadian Institutes of Health Research and the Natural Sciences and Engineering Research Council of Canada.

REFERENCES

1. PANTEV C., R. OOSTENVELD, A. ENGELIEN, *et al.* 1998. Increased auditory cortical representation in musicians. Nature **392**: 811–814.
2. PANTEV C., L.E. ROBERTS, M. SCHULZ, *et al.* 2001. Timbre-specific enhancement of auditory cortical representations in musicians. Neuroreport **12**: 1–6.
3. SHAHIN, A., D. BOSNYAK, L.J. TRAINOR, *et al.* 2003. Enhancement of neuroplastic P2 and N1c auditory evoked potentials in musicians. J. Neurosci. **23**: 5545–5552.
4. TRAINOR L.J., R.N. DESJARDINS & C. ROCKEL. 1999. A comparison of contour and interval processing in musicians and nonmusicians using event-related potentials. Aust. J. Psychol. **51**: 147–153.
5. BESSON, M. & F. FAÏTA. 1995. An event related potential (ERP) study of musical expectancy: comparison of musicians with nonmusicians. J. Exp. Psychol. Hum. Percept. Perform. **21**: 1278–1296.
6. SCHNEIDER, P., M. SCHERG, M. DOSCH, *et al.* 2002. Morphology of Heschl's gyrus reflects enhanced activation in the auditory cortex of musicians. Nat. Neurosci. **5**: 688–694.
7. SCHLAUG, G., L. JANCKE, Y. HUANG, *et al.* 1995. Increased corpus callosum size in musicians. Neuropsychologia **33**: 1047–1055.
8. SCHLAUG, G., L. JÄNCKE, Y. HUANG, *et al.* 1995. In vivo evidence of structural brain asymmetry in musicians. Science **267**: 699–701.
9. TRAINOR, L.J., K.L. MCDONALD & C. ALAIN. 2002. Automatic and controlled processing of melodic contour and interval information measured by electrical brain activity. J. Cognit. Neurosci. **14**: 430–442.
10. ELMAN, J.L., E.A. BATES, M.H. JOHNSON, *et al.* 1996. Rethinking Innateness: A Connectionist Perspective on Development. MIT Press. Cambridge, MA.
11. PONTON C.W., J.J. EGGERMONT, B. KWONG, *et al.* 2000. Maturation of human central auditory system activity: evidence from multi-channel evoked potentials. Clin. Neurophysiol. **111**: 220–236.
12. PANG W.E. & M.J. TAYLOR. 2000. Tracking the development of the N1 from age 3 to adulthood: an examination of speech and non-speech stimuli. Clin. Neurophysiol. **111**: 388–397.
13. BOSNYAK, D.J., R.A. EATON & L.E. ROBERTS. 2002. Modification of distributed auditory cortical representations in nonmusicians by training for pitch discrimination with 40-Hz amplitude modulated tones. Submitted for publication.
14. TREMBLAY, K., N. KRAUS, T. MCGEE, *et al.* 2001. Central auditory plasticity: changes in the N1-P2 complex after speech-sound training. Ear Hear. **22**: 79–90.
15. ATIENZA, M., J.L. CANTERO & E. DOMINGUEZ-MARIN. 2002. The time course of neural changes underlying auditory perceptual learning. Learn. Mem. **9**: 138–150.
16. SHAHIN, A., L.J. TRAINOR & L.E. ROBERTS. 2003. Enhanced auditory evoked potentials in children receiving Suzuki musical training. Submitted for publication.
17. ANVARI, S., L.J. TRAINOR, J. WOODSIDE, *et al.* 2002. Relations among musical skills, phonological processing, and early reading ability in preschool children. J. Exp. Child Psychol. **83**: 111–130
18. LAMB, S.J. & A.H. GREGORY. 1993. The relationship between music and reading in beginning readers. J. Educ. Psychol. **13**: 13–27.

Gray Matter Differences between Musicians and Nonmusicians

CHRISTIAN GASER[a,b] AND GOTTFRIED SCHLAUG[b]

[a]Department of Psychiatry, University of Jena, Jena, Germany

[b]Department of Neurology, Music and Neuroimaging Laboratory,
Beth Israel Deaconess Medical Center and Harvard Medical School,
Boston, Massachusetts 02215, USA

ABSTRACT: Musicians learn complex motor and auditory skills at an early age and practice these specialized skills extensively from childhood through their entire careers. Using a voxel-by-voxel morphometric technique, we found gray matter volume differences in motor as well as auditory and visuospatial brain regions comparing professional musicians (keyboard players) with matched amateur musicians and nonmusicians. These multiregional differences might represent structural adaptations in response to long-term skill learning and repetitive rehearsal of these skills. This is supported by finding a strong association between structural differences, musician status, and practice intensity as well as by a wealth of supporting animal data showing structural changes in response to long-term motor training.

KEYWORDS: morphometry; plasticity; skill acquisition; keyboard players

INTRODUCTION

Musicians are skilled in performing complex physical and mental operations such as translating visually presented musical symbols into complicated movements of fingers and hands and memorizing long musical phrases. Playing a musical instrument typically requires the integration of multimodal sensory and motor information and multimodal sensory feedback mechanisms to monitor the performance. The neural correlates of most of these musical operations are not fully understood, and no firm associations between specialized musical skills and particular brain regions or a characteristic brain anatomy are established. Although several functional and a few structural imaging studies have reported regional brain differences between musicians and nonmusicians,[1-6] no study has yet searched across the whole brain space for structural brain differences between these two groups that could be linked to the exceptional and specialized skills of musicians as well as to the long-term and extensive rehearsal of these skills.

Address for correspondence: Gottfried Schlaug, M.D., Ph.D., Department of Neurology, Music and Neuroimaging Laboratory, Beth Israel Deaconess Medical Center and Harvard Medical School, 330 Brookline Avenue, Boston, MA 02215. Voice: 617-632-8912; fax: 617-632-8920.

gschlaug@bidmc.harvard.edu

Ann. N.Y. Acad. Sci. 999: 514–517 (2003). © 2003 New York Academy of Sciences.
doi: 10.1196/annals.1284.062

MATERIAL AND METHODS

Material. We compared 20 male professional musicians and 20 male amateur musicians (all keyboard players) to a group of 40 male nonmusicians (all matched with regard to age, handedness, and verbal skills). A professional musician was defined as someone whose main profession was to be a musician (either a performing artist, music teacher, or music student at a conservatory) with an average practice time of at least 1 hour per day. An amateur musician was defined as someone who plays a musical instrument regularly but has a profession other than that as a professional musician. The amateur musicians had an average daily practice time that was half that of professional musicians (1.15 vs 2.23 h/d). Nonmusicians were defined as those who never played a musical instrument.

Magnetic Resonance (MR) Data Acquisition and Image Analysis. Isotropic (1 mm^3) T$_1$-weighted MR data sets were acquired from each subject. Voxel-by-voxel t tests using the general linear model[7] were used to search for gray matter differences between professional musicians, amateur musicians, and nonmusicians. We assessed the correlation between musician status and gray matter differences by modeling the musician status as a 3-level gradation, in which professional musicians were highest and were assigned a value of 1, amateur musicians were intermediate and were assigned a value of 0.5, and nonmusicians were the lowest and were assigned a value of 0.

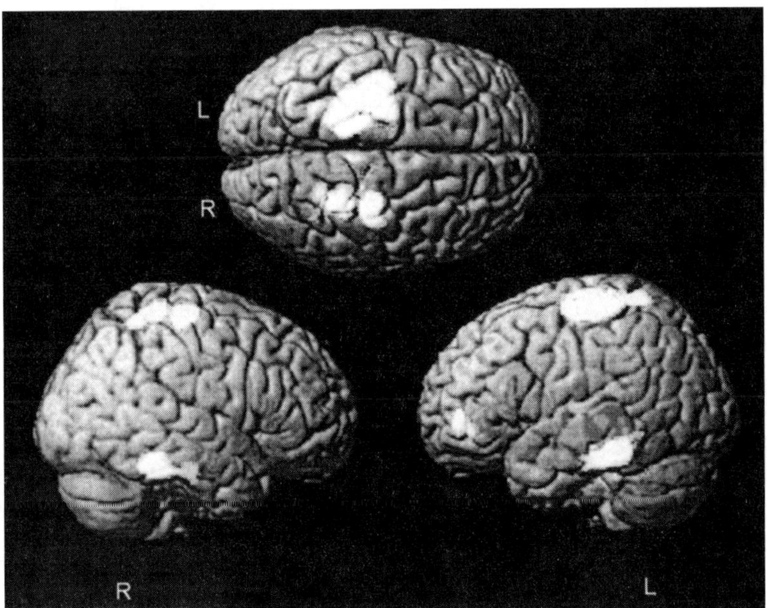

FIGURE 1. Brain regions with gray matter differences between professional musicians, amateur musicians, and nonmusicians (P <0.05, corrected for multiple comparisons; only clusters of voxels consisting of at least 225 voxels are displayed corresponding to a spatial extent threshold of P <0.1).

RESULTS

Areas of significantly increased gray matter volume comparing professional musicians (keyboard players) with amateur musicians (keyboard players) and nonmusicians were found in peri-rolandic regions including primary motor and somatosensory areas, premotor regions, superior parietal regions, and the inferior temporal gyrus bilaterally (FIG. 1). Additional significant differences were seen in the left cerebellum, left Heschl's gyrus, and left inferior frontal gyrus. No areas showed a significant decrease in gray matter volume comparing professional with amateur musicians and amateur musicians with nonmusicians, respectively.

CONCLUSIONS

Differences in the distribution of gray matter were mainly seen in brain regions that are of critical importance for a performing musician. These brain regions (e.g, primary sensorimotor region, premotor region, and cerebellum) are either part of a motor learning and skill acquisition network or they are part of a network that may play a crucial role in translating musical notation into complicated motor commands (e.g., superior parietal, inferior temporal lobe). There are two possible explanations for our findings: either individuals with a particular brain anatomy are drawn to becoming musicians or structural differences are due to use-dependent brain growth during a critical period of brain maturation. This second explanation is supported by a wealth of animal experiments[8] showing increases in glial, capillary, and synapse densities after long-term motor learning and by additional results of our study showing more gray matter differences depending on practice intensity. Overall, the results of our study and other studies[5–6,9–11] provide strong links between specialized skills and certain brain structures.

REFERENCES

1. AMUNTS, K., G. SCHLAUG, L. JAENCKE, *et al.* 1997. Motor cortex and hand motor skills: structural compliance in the human brain. Human Brain Mapping **5:** 206–215.
2. SCHLAUG, G., L. JAENCKE, Y. HUANG, *et al.* 1995. In vivo evidence of structural brain asymmetry in musicians. Science **267:** 699–701.
3. SCHLAUG, G., L. JAENCKE, Y. HUANG, *et al.* 1995. Increased corpus callosum size in musicians. Neuropsychologia **33:** 1047–1055.
4. SCHLAUG, G. 2001. The brain of musicians. A model for functional and structural adaptation. Ann. N.Y. Acad. Sci. **930:** 281–-299.
5. ZATORRE, R.J., D.W. PERRY, C.A. BECKETT, *et al.* 1998. Functional anatomy of musical processing in listeners with absolute pitch and relative pitch. Proc. Natl. Acad. Sci. USA **95:** 3172–3177.
6. SCHNEIDER, P., M. SCHERG, H.G. DOSCH, *et al.* 2002. Morphology of Heschl's gyrus reflects enhanced activation in the auditory cortex of musicians. Nat. Neurosci. **5:** 688–694.
7. ASHBURNER, J. & K.J. FRISTON. 2000. Voxel-based mophometry–the methods. Neuroimage **11:** 805–821.
8. ANDERSON, B.J., P.B. ECKBURG & K.I. RELUCIO. 2002. Alterations in the thickness of motor cortical subregions after motor-skill learning and exercise. Learn. Mem. **9:** 1–9.
9. ELBERT, T., C. PANTEV, C. WIENBRUCH, *et al.* 1995. Increased cortical representation of the fingers of the left hand in string players. Science **270:** 305–306.

10. Pantev, C., R. Oostenveld, A. Engelien, *et al.* 1998. Increased auditory cortical representation in musicians. Nature **392:** 811–814.
11. Keenan, J.P., V. Thangaraj, A. Halpern & G. Schlaug. 2001. Absolute pitch and planum temporale. Neuroimage **14:** 1402–1408.

Toddlers' Musical Preferences

Musical Preference and Musical Memory in the Early Years

ALEXANDRA LAMONT

Department of Psychology, Keele University, United Kingdom

ABSTRACT: The current paper reports a pilot study of the preferences of children aged 2–3.5 years for different kinds of music. With the use of a novel interactive procedure to measure active preferences, preliminary results indicate general preferences for fast and loud music irrespective of style.

KEYWORDS: musical preference; infancy; toddlers

Previous research has shown that infants prefer consonance over dissonance, coherent over incoherent phrases, and beginnings over endings. My own earlier work has also shown preferences among 12–18-month-old children for fast loud music compared to slow quiet music, irrespective of style. It is known that children's musical preferences can be affected by a range of nonmusical features such as artist, band, fashion, friends, and so on, and adults prefer a broad range of music depending on factors such as personality and experience.

Methods for studying musical preference vary, depending on the age of the participants. In infancy, discrimination studies are based on either familiarity or preference. The conditioned head-turn procedure used by Trehub and colleagues[1] depends on the recognition of a difference. The contingent head-turn preference procedure, as used by Saffran and colleagues,[2] depends on a demonstrated preference for one stimulus over another. With older children and adults, preferences are typically studied through questionnaires or interviews that ask participants to nominate artists or styles of music, without having the chance to hear them. The current study adapts the infant contingent head-turn preference procedure to allow musical preferences to be measured by the participant pressing a key in order to hear a given extract of music (from a choice of four possible keys/musics). The data thus show discrimination among a number of musical styles, and preferences can also be measured.

A toy keyboard with flashing lights was connected to a portable computer with specially written software, enabling the presentation of real musical extracts to be linked to the toy keys and to be heard only when the key is depressed. Participants

Address for correspondence: Dr. Alexandra Lamont, Department of Psychology, Keele University, UK. Voice: +44 1782 583323.
a.m.lamont@keele.ac.uk

Ann. N.Y. Acad. Sci. 999: 518–519 (2003). © 2003 New York Academy of Sciences.
doi: 10.1196/annals.1284.063

heard a range of music including two extracts similar in style, key, and tempo as well as contrasting extracts; the music was not the same for all participants. The extracts were randomly assigned to keys to counterbalance color, order, and position, and the experimenter demonstrated each musical extract first. The participant then played with the toy for up to 10 minutes, exploring the musical extracts. Key presses were recorded in milliseconds, and preferences were calculated as a proportion of the total time allocated to all four extracts.

The results indicated a general preference for fast and loud music (61%) over slow and quiet music (39%). Furthermore, when two extracts were similar in key, tempo, and style, participants also preferred well-known music (e.g., Vivaldi's *Four Seasons* compared to a less well-known Vivaldi string concerto). Participants adopted a variety of strategies, including the methodical (pressing each key one at a time until the extract ends), the intuitive (pressing keys for shorter periods of time for less preferred music), and the scientific (exploring the keyboard and pressing more than one key at once).

The study indicates that preferences for loud and fast music are found in toddlerhood and that preferences for well-formed pieces of music can begin as early as toddlerhood. It also demonstrates an effective procedure for measuring musical preferences in an ecologically valid way without needing language and provides reliable data amenable to systematic analysis.

REFERENCES

1. TREHUB, S., E.G. SCHELLENBERG & S.B. KAMENETSKY. 1999. Infants' and Adults' Perception of Scale Structure. J. Exp. Psychol.: Human Percept. Perf. **25:** 965–975.
2. SAFFRAN, J.R., M.M. LOMAN & R.R.W. ROBERTSON. 2000. Infant memory for musical experiences. Cognition **77:** 15–23.

Long-Term Memory for Pitch in Six-Month-Old Infants

JUDY PLANTINGA AND LAUREL J. TRAINOR

Psychology Department, McMaster University, Hamilton, Ontario L8S 4K1, Canada

ABSTRACT: We examined 6-month-old infants' long-term memory representations for the pitch of familiar melodies. Infants remembered the relative pitch of the melodies, but the absolute pitch was either not remembered or not a particularly salient attribute.

KEYWORDS: infant; memory; melody; absolute pitch; relative pitch

INTRODUCTION

Understanding pitch perception in infancy is basic to understanding the development of music perception. It has been proposed that absolute pitch is a primitive ability and that infant perception is dominated by absolute pitch.[1,2] According to this view, the ability to encode the relative pitch of a melody develops with experience.[2–4] To evaluate this proposal, we carried out studies to determine the nature of infants' long-term memory representations. Previous research has shown that infants remember musical pieces over long periods of time.[5,6] Furthermore, infants remember the timbre and tempo of these compositions.[6] We expected, then, that infants would remember the absolute pitch.

MATERIAL AND METHODS

Experiment 1. Sixteen infants heard six repetitions each day of one of two old English folk songs, "The Country Lass" or "The Painful Plough, " at home for seven days. On the eighth day they were given a visually based preference test in which trials of the novel and familiar pieces alternated. Each trial began when the infant looked at a flashing toy on one side, causing the music for that trial to begin playing. Each trial ended when the infant looked away, causing the music to stop. The measure of preference was the average looking time per trial in order to hear the novel versus the familiar piece. Infants preferred the novel over the familiar melody, $F(1,12) = 7.20$, $P < 0.02$, indicating that they remembered the familiarized melody.

Address for correspondence: Dr. Laurel J. Trainor, Psychology Department, McMaster University, Hamilton, Ontario L8S 4K1, Canada. Voice: 905-525-9140, ext. 23007; fax: 905-529-6225.

ljt@mcmaster.ca

Ann. N.Y. Acad. Sci. 999: 520–521 (2003). © 2003 New York Academy of Sciences.
doi: 10.1196/annals.1284.064

Experiment 2. We tested whether infants would recognize a familiar song in transposition. The familiarization and testing were the same as in experiment 1, except that 32 infants were tested in one of four conditions where, compared to familiarization, both the novel and the familiar melodies were: (1) transposed up a perfect fifth, (2) transposed down a perfect fifth, (3) transposed up a tritone, or (4) transposed down a tritone. Infants preferred the novel over the familiar melody, $F(1,28) = 5.54$, $P = 0.02$, regardless of whether the transposition was up or down by a perfect fifth or a tritone, indicating that they encoded the relative pitch of the melodies in long-term memory.

Experiment 3. We tested whether infants remembered the absolute pitch of the melodies. After familiarization, 16 infants were tested in one of two conditions where the familiar melody was tested against (1) the familiar melody transposed up a perfect fifth or (2) the familiar melody transposed down a perfect fifth. Infants showed no preference for the familiar song at a novel pitch (transposed version of the song) over the familiar song at the familiar pitch, $F(1,14) = 0.003$, $P = 0.95$.

CONCLUSIONS

Contrary to our original hypothesis, we conclude that 6-month-old infants remember melodies in terms of relative pitch, that they do not remember absolute pitch or it is not salient to them, and that if a developmental shift in perception from absolute to relative pitch occurs, it takes place before 6 months of age.

ACKNOWLEDGMENTS

The research was supported by a grant from the National Sciences and Engineering Research Council of Canada (to L.J.T.) and a National Sciences and Engineering Research Council of Canada graduate scholarship (to J.P.).

REFERENCES

1. MIYAZAKI, K. 1995. Perception of relative pitch with different references: some absolute-pitch listeners can't tell musical interval names. Percept. & Psychophys. **57:** 962–970.
2. SAFFRAN, J.R. & G.J. GRIEPENTROG. 2001. Absolute pitch in infant auditory learning: evidence for developmental reorganization. Dev. Psychol. **37:** 74–85.
3. SERGEANT, D.C. & S. ROCHE. 1973. Perceptual shifts in auditory information processing of young children. Percept. Mus. **1:** 39–48.
4. TAKEUCHI, A.H. & S.H. HULSE. 1993. Absolute pitch. Psychol. Bull. **113:** 345–361.
5. SAFFRAN, J.R., M.M. LOMAN & R.R.W. ROBERTSON. 2000. Infant memory for musical experiences. Cognition **77:** B15–B23.
6. TRAINOR, L.J., L. WU & C.D. TSANG. 2003. Long-term memory for music: infants remember tempo and timbre. Dev. Sci. In press.

Absolute Pitch Does Not Depend on Early Musical Training

DAVID A. ROSS,[a] INGRID R. OLSON,[b] AND JOHN C. GORE[c]

[a]*Department of Diagnostic Radiology, Yale School of Medicine, New Haven, Connecticut 06520, USA*

[b]*Department of Psychology, University of Pennsylvania, Philadelphia, Pennsylvania, 19104-6196, USA*

[c]*Department of Radiology and Radiological Sciences, Vanderbilt University Medical Center, R-1032 MCN, Nashville, Tennessee 37232, USA*

ABSTRACT: The etiology and defining characteristics of absolute pitch (AP) have been controversial. To test the importance of musical training in the development of this skill, we developed a new paradigm for identifying AP that is independent of a subject's musical experience. We confirm the efficacy of the paradigm using classically defined AP and non-AP musicians. We then present data from a nonmusician who nevertheless appears to possess AP. We conclude that musical training is not necessary for the development of AP.

KEYWORDS: absolute pitch (AP); learning; tonal memory

INTRODUCTION

For most individuals the perception of melodic sequences is based on understanding the relationships between different tones. This skill is referred to as "relative pitch" (RP). A small subset of the population is also capable of identifying quickly and accurately the absolute frequency of tonal stimuli. This skill is referred to as "absolute pitch" (AP).

A central point of controversy in the study of AP has been the relative roles of heredity and early learning. Early models of AP suggested that "true" AP was a genetically determined trait.[1,2] Recent evidence supports this model.[3,4] A second model advocates that AP is the result of early musical experiences[5,6] in which some individuals memorize the exact frequency of each musical note. Thus, according to this model, AP is largely a categorical phenomenon. Indirect evidence in support of this theory indicates that all individuals known to have AP have had musical training[2,6] and that the degree of AP tends to correlate with the years of musical experience.[5]

Historically, most paradigms that test for AP ask subjects to name a series of tones according to the Western musical scale. Because performance on such a test is

Address for correspondence: Dr. David A. Ross, Department of Diagnostic Radiology, Yale School of Medicine, Box 208043, New Haven, CT 06520. Voice: 203-785-5296; fax: 203-785-6534.

david.a.ross@yale.edu

Ann. N.Y. Acad. Sci. 999: 522–526 (2003). © 2003 New York Academy of Sciences.
doi: 10.1196/annals.1284.065

predicated on a basic level of musical knowledge, these methods are insensitive to the possibility of discovering AP in either a nonmusician or a musician of non-Western training. As such, there is no convincing evidence that only musicians possess AP.[7]

To address these issues, we designed a new paradigm that qualifies subjects as AP possessors or nonpossessors independent of their musical experience. Using this method, we tested the following predictions of the early learning model: (1) AP is a musical, and therefore categorical, phenomenon; and (2) it is impossible to develop AP without musical training.

PART 1

Method

Participants. Twenty-seven experienced musicians were recruited (6 with AP, 21 with non-AP [NAP], as categorized with a classic note-naming test). Additionally, we tested subject R.M., a 25-year-old man with minimal musical experience. R.M. was incapable of accurately naming musical notes, but nevertheless claimed to have AP.

Stimuli and Procedure. Subjects were played a pure sine tone followed by a silent interval (2, 8, or 16 seconds), after which they were required to reproduce the original target using a digital sine function generator. Individuals were not given feedback on their performance until the completion of all tasks, and at no time was a reference tone provided.

Data Analysis. Responses were octave corrected and distance was calculated in half-steps away. Constant error (CE) was used as a measure of bias and standard deviation (sd) as a measure of variability. Thus, accurate performance would yield a normal distribution centered at zero (low CE and low sd) and random performance would produce a uniform distribution of responses (low CE but high sd).

Results

AP vs NAP comparisons (FIG. 1). The AP and NAP groups differed significantly with respect to sd ($F(1,75) = 45.60$; $P < 0.0001$) but not with respect to CE ($F(1,75) = 0.024$; $P > 0.80$). Neither group was affected by the time delay, and both groups differed significantly from the prediction of the null hypothesis ($P < 0.01$).

Subject R.M. vs AP and NAP Groups. With respect to CE, subject R.M. did not differ from either group ($P > 0.40$). With respect to sd, subject R.M. was mildly different from the NAP group ($t(20) = 2.00$; two-tailed $P < 0.10$), but not from the AP controls ($t(5) = 1.89$; $P > 0.10$).

These data demonstrate that both AP and NAP subjects were able to reproduce the target stimulus accurately and that performance did not decay for intervals of up to 16 seconds.

PART 2

Motivation

Passive exposure to interfering tones is known to differentially affect performance of AP possessors and nonpossessors, presumably by destroying individuals'

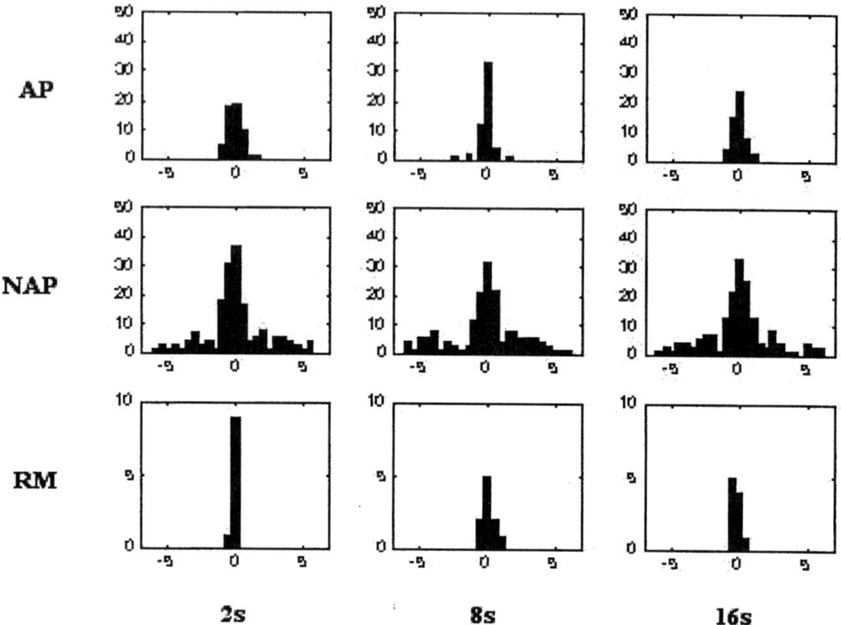

FIGURE 1. Histogram of responses by the absolute pitch (AP) and non-absolute pitch (NAP) groups and subject R.M. for silent delays of 2, 8, and 16 seconds.

short-term memory traces of stimuli.[8,9] We sought to test whether a pre-categorical paradigm of this sort could be used to distinguish explicitly between AP and NAP individuals.

Method

Experiment 2b differed from Experiment 2a by filling the interstimulus interval (ISI) with a variable number of distracting tones (either a 2-second ISI with 1 distracting tone, an 8-second ISI with 31 distractors, or a 16-second ISI with 71 distractors). All other parameters were the same as those in Experiment 2a.

Results (Fig. 2)

AP vs NAP Comparisons. The AP and NAP groups differed significantly with respect to sd ($F(1,82) = 133.37$; $P < 0.0001$) but not with respect to CE ($F(1,82) = 1.05$; $P > 0.30$). The AP distributions differed significantly from the null hypothesis for all time intervals ($P < 0.01$), whereas the NAP group differed from chance for the first interval only.

Subject R.M. vs AP and NAP Groups. With respect to CE, subject R.M. did not differ significantly from either group ($P > 0.40$). With respect to sd, subject RM differed significantly from the NAP group ($t(20) = 5.34$, $P < 0.001$) but not the AP group ($t(5) = 1.20$, ns).

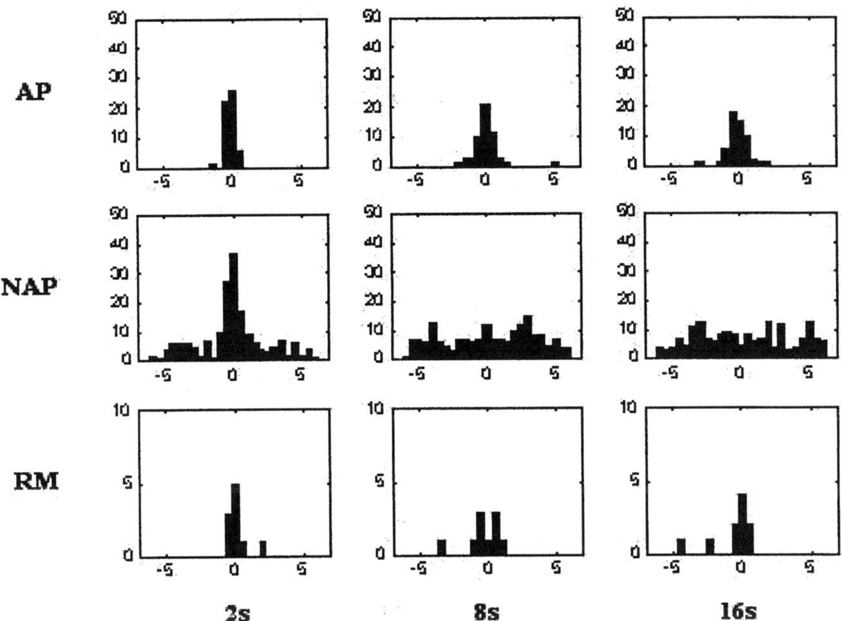

FIGURE 2. Histogram of responses by the absolute pitch (AP) and non-absolute pitch (NAP) groups and subject R.M. for conditions with 1, 31, and 71 interfering tones.

These data show that AP and NAP subjects were differentially affected by interfering tones: both groups were accurate without interference, but only AP subjects were accurate with interference.

CONCLUSIONS

The early learning theory states that absolute pitch (AP) is a categorical phenomenon and that early musical training is required for its development. However, this theory has been largely untestable in that paradigms to identify AP possessors have been predicated on a basic level of musical expertise. Based on extensive literature on pitch memory,[8–11] we constructed a new paradigm to test for AP that is pre-categorical and does not require any musical expertise. Whereas both AP and NAP subjects were able to reproduce target stimuli after a silent delay, we predicted that only AP possessors would be accurate in the presence of tonal interference. Our data strongly support this hypothesis and thereby demonstrate the efficacy of the method.

We also present the case report of a nonmusician, subject R.M., who is incapable of naming notes. Nevertheless, R.M.'s performance on our paradigm was significantly better than that of the NAP musicians and indistinguishable from that of the AP controls. Based on these data, we conclude that subject R.M. possesses AP. To the best of our knowledge, this is the first report of a nonmusician with AP. These data disprove the hypothesis that early musical training is required for the development of AP. Instead, they suggest that true AP may be pre-categorical and, at least in some cases, dependent on hereditary factors.

REFERENCES

1. SEASHORE, C.E. 1938. Psychology of Music. McGraw-Hill. New York, NY.
2. BACHEM, A. 1955. Absolute pitch. J. Acoust. Soc. Am. **27:** 1180–1185.
3. BAHARLOO, S. *et al.* 2000. Familial aggregation of absolute pitch. Am. J. Hum. Genet. **67:** 755–758.
4. GREGERSEN, P.K. *et al.* 2001. Early childhood music education and predisposition to absolute pitch: teasing apart genes and environment. Am. J. Med. Genet. **98:** 280–282.
5. TAKEUCHI, A. & S.H. HULSE. 1993. Absolute pitch. Psychol. Bull. **113:** 345–361.
6. MIYAZAKI, K. 1988. Musical pitch identification by absolute pitch possessors. Percept. Psychophysiol. **44:** 501–512.
7. GREGERSEN, P.K. *et al.* 1999. Absolute pitch: prevalence, ethnic variation, and estimation of the genetic component. Am. J. Hum. Genet. **65:** 911–913.
8. PECHMANN, T. & G. MOHR. 1992. Interference in memory for tonal pitch: implications for a working-memory model. Mem. Cognit. **20:** 314–320.
9. DEUTSCH, D. 1970. Tones and numbers: specificity of interference in immediate memory. Science **168:** 1604–1605.
10. KELLER, T., N. COWAN & J.S. SAULTS. 1995. Can auditory memory for tone pitch be rehearsed? J. Exp. Psychol. Learn Mem. Cognit. **21:** 635–645.
11. BACHEM, A. 1954. Time factors and absolute pitch determination. J. Acoust. Soc. Am. **26:** 751–753.
12. SIEGEL, J.A. 1974. Sensory and verbal coding strategies in subjects with absolute pitch. J. Exp. Psychol. **103:** 37–44.
13. PROFITA, J. & T.G. BIDDER. 1988. Perfect pitch. Am. J. Med. Genet. **29:** 763–771.

Investigating the Musical Qualities of Early Infant Sounds

BARBARA RUZZA,[a,b] FRANCESCA ROCCA,[b,c] DANIELA LENTI BOERO,[a,d] AND CARLO LENTI[a,b,e]

[a]San Paolo Hospital, University of Milan, Milan, Italy

[b]Laboratory of Bioacoustics and Comparative Analysis of Behaviour and Cognitive Functions, University of Milan, Milan, Italy

[c]Doctorate of Metabolic Diseases, Department of Paediatrics, University of Milan, Milan, Italy

[d] Faculty of Pedagogical Sciences, University of Urbino, Urbino, Italy

[e]Chair of Child Neuropsychiatry, University of Milan, Milan, Italy

ABSTRACT: Early vocalizations in Italian and Moroccan infants are examined and the results presented.

KEYWORDS: early vocalization; speech sounds; acoustical analyses

INTRODUCTION

In the second month of life, human infants begin to produce sounds and protophones that are different from crying. In the past, most protophones were generically categorized as "babbling."[1] Other researchers[2] believe that this term is too ambiguous and prefer to use special words to designate the entire class of speech precursors. We agree with them. The aim of this study was to explore the difference between early vocalizations in two different ethnic groups.

MATERIAL AND METHODS

The subjects of this study were five 2-month-old infants, three of whom were Moroccan (1 female, 2 male) and two Italian (1 male, 1 female). All of the infants were born full term (>37 weeks' gestional age) and had normal parameters at birth (weight, length, c.c., Apgar). The infants were audiorecorded in their home while interacting with the mother, the investigator, and, only for the Moroccan babies, a cultural mediator. The duration of each session was variable and was subordinated by the infant's will to vocalize. Vocalizations were recorded with a DAT sound recorder and a unidirectional microphone (positioned between 5 and 10 cm from the

Address for correspondence: Dr. Barbara Ruzza, San Paolo Hospital, University of Milan, Milan, Italy.
uonpiamilano@tin.it

Ann. N.Y. Acad. Sci. 999: 527–529 (2003). © 2003 New York Academy of Sciences.
doi: 10.1196/annals.1284.066

mouth of the crying babies) and were acquired by means of the Software Pro Tools. Recordings were subdivided in frames lasting from 6–12 seconds each. Each frame was analyzed with Software Canary 1.2. We obtained a total of 109 and 54 vocalizations, respectively, for Moroccan and Italian subjects. We used software SYSTAT 10 for statistical analysis of acoustic parameters.

Investigators have disagreed on the distinction between speech-like and nonspeech sounds (2, 3, and 4). In the present study we separated vegetative sounds (such us coughing, burping, and sneezing) and fixed vocal signals (crying, laughter, and moaning) from protophones (1).

The speech sounds (protophones), as formulated in this study, refer to cooing, gooing, vowel-like, consonant-like, marginal babbling, and canonical babbling, according to Oller.[1] We considered two parameters of time: duration of emission and inspiration pause between two vocalizations. We also analyzed separately the values of the start, end, and maximum and minimum frequencies measured on the fundamental frequency (F_0) of nonspeech and speech sounds. We considered the melodic contours with different morphologies: rise or rise and fall (⌒), flat (——), falling or falling-rising (⌣), and vibrato (〰). Finally, the vocalizations may show a shift contour with interruption of the fundamental frequency that then starts again with a higher frequency (⌒⌒). Furthermore, the speech-like sounds mostly show a mixed contour made up of two different melodies.

RESULTS AND CONCLUSIONS

In both samples (Italian and Moroccan), speech utterances were present in a higher percentage of infants than were nonspeech sounds: 74% speech versus 26% nonspeech for Italian infants and 66% speech versus 34% nonspeech for Moroccan infants. Both Italian and Moroccan infants used the same typologies of vocalizations, but the relative percentages were different: in the Italian sample, quasi-vowel (cooing, 45%) and vowel-like (27.7%) sounds predominated, whereas in the Moroccan infants, vowel-like (46.5%) and cooing sounds (36.6%) were prevalent. A comparison of speech-sound typologies showed that the two ethnic groups had different vocal expressions perhaps due to different kinds of stimulation from the mother.

We observed that the value of the means of time and frequency parameters of the speech sounds was not different between the two samples (TABLE 1). This should be

TABLE 1. Means of time (ms) and frequency (kHz) parameters of speech sounds in two groups of infants

	Italians	Moroccans
Duration	740.1	727.4
Pause	2,066.8	1,974.5
Frequency start	365.1	362.2
Frequency end	305.2	343.2
Frequency maximum	432.4	496.3
Frequency minimum	270.9	257.8

ascribed to the comparable age of the infants, specifically to the length of the larynx and to pulmonary compliance.

Discriminant function analysis of the same frequency parameters and of speech and nonspeech sounds was performed in each subject in both ethnic groups. In all cases, the percentage of vocalizations successfully assigned was high; this could be ascribed to the individual features of the each subject and/or to their ethnic group. The nonspeech utterances successfully assigned were 60% between Italian and Moroccan subjects and 67% between Moroccan and Italian subjects, whereas the speech utterances successfully assigned were 80% between Italian and Moroccan infants and 57% between Moroccan and Italian infants.

This study underscores the existence of speech sounds in the first 2 months of life.

ACKNOWLEDGMENTS

The present study was supported by the Pierfranco and Luisa Mariani Foundation.

REFERENCES

1. OLLER, D. & R. EILERS. 1999. Precursors to Speech in Infancy: The Prediction of Speech and Language Disorders. :223–245. Elsevier Science Inc.
2. ROBB, M., H. BAUER & A. TYLER. 1994. A quantitative analysis of the single-word stages. First Language **14:** 37–48.
3. OLLER, D. 1980. The emergence of sounds of speech in infancy. Child Phonol. **1:** 93–112.
4. STARK, R. 1980. Stage of speech development in the first few years of life. Child Phonol. **1:** 73–90.

Perceiving Prosody in Speech

Effects of Music Lessons

WILLIAM FORDE THOMPSON, E. GLENN SCHELLENBERG, AND
GABRIELA HUSAIN

*Department of Psychology, University of Toronto at Mississauga, Mississauga,
Ontario L5L 1C6, Canada*

ABSTRACT: In two experiments, musically trained and untrained adults were
tested on their ability to match spoken utterances with their tonal analogues
(tone sequences that retained the pitch and temporal patterns of the utterances). In both cases, musical training was associated with superior
performance, indicating an enhanced ability to extract prosodic information
from spoken phrases.

KEYWORDS: speech; prosody; music lessons

INTRODUCTION

We conducted two experiments that tested whether formal training in music is
associated with an enhanced ability to perceive prosody in speech. In other words,
we examined the possibility that music lessons have positive transfer effects that
influence speech perception. Relatively few studies have examined the effects of
music lessons on nonmusical auditory skills. Nonetheless, training in music undoubtedly refines and develops mechanisms involved in processing music. Such
mechanisms could extend to other complex auditory signals that are characterized
by variation in dimensions relevant to music. In particular, prosody in speech
involves the use of paralinguistic cues (e.g., pitch variation, speed, and amplitude
changes) that serve to highlight the syntax and the semantics of an utterance as well
as the speaker's emotional state. Paralinguistic cues, such as pitch height, speed of
talk, and loudness, are essential for accuracy in decoding emotion in speech.[1,2]
These cues are also known to be important in conveying emotion through music.[3]

Decoding speech prosody involves an initial stage in which prosodic information
is extracted and represented and a subsequent stage in which meaning is attributed
to that information. The experiments reported here examined whether music lessons
yield an enhanced ability to *extract* prosodic cues in speech.

Address for correspondence: Dr. William Forde Thompson, Department of Psychology, University of Toronto at Mississauga, Mississauga, Ontario L5L 1C6, Canada. Voice: 905-569-4733;
fax: 905-569-4734.

b.thompson@utoronto.ca

Ann. N.Y. Acad. Sci. 999: 530–532 (2003). © 2003 New York Academy of Sciences.
doi: 10.1196/annals.1284.067

METHODS AND RESULTS

In Experiment 1, musically trained ($n = 22$) and untrained ($n = 16$) listeners were tested on their ability to extract pitch information from short phrases spoken in English (e.g., "the boy went to the store"). On average, trained listeners had more than 15 years of music lessons. Stimuli included four semantically neutral but happy-sounding utterances taken from the Florida Affect Battery (FAB) as well as intonation melodies (tone sequences) derived from these utterances. Each intonation melody consisted of tones that matched the modal pitch and duration of the spoken syllables. Tone syllables had equal amplitude and their pitch varied discretely (unlike continuous pitch variations in speech). In short, the intonation melodies were an abstraction of the pitch and temporal variations in the utterances. "Incorrect" melodies that did not match the utterances were formed by switching the location of two tone syllables. On each trial, listeners heard a spoken phrase followed by an intonation melody. They judged whether or not the intonation melody matched the prosody of the phrase. Musically trained participants outperformed their untrained counterparts, $t(36) = 2.47$, $P <0.02$, with both groups exceeding chance levels of performance.

Experiment 2 was similar to the first except that: (1) the speech stimuli were taken from a foreign language (Spanish), (2) the paralinguistic cues of the speech stimuli conveyed a happy or a sad emotion, and (3) the intonation melodies were created by means of high-pass filtering, such that the pitch of the intonation melodies varied continuously rather than discretely. We tested 12 musically trained ($M = 16$ years of lessons) and 8 musically untrained listeners, none of whom spoke Spanish. Mismatching intonation melodies were derived from novel sentences spoken in the same emotional tone as the target sentences. There was no main effect of emotion (happy and sad) and no interaction between emotion and training group. As in Experiment 1, the musically trained group performed better than the untrained group, $F(1,18) = 4.76$, $P <0.05$, with both groups exceeding chance levels of responding.

DISCUSSION

In both experiments, musically trained participants were better than untrained participants at extracting prosodic information from speech. This enhanced ability to extract prosodic information extended to phrases spoken in an unfamiliar language, and the effect was evident regardless of whether the pitch information was represented discretely or continuously. Such evidence of cognitive transfer between music and speech raises the possibility that speech prosody and music are processed with shared neural resources.[4]

Other evidence suggests that music lessons not only improve the ability to *extract* prosodic cues, but also improve the ability to *interpret* speech prosody. In one study,[5] music and law students listened to auditory signals comprised of the fundamental frequencies of voice samples recorded from depressed and nondepressed individuals (i.e., the words were inaudible). Music students were better than law students at identifying the emotional state of the speakers. More recently,[6] we confirmed that children and adults with music training have enhanced skill at interpreting the emotional significance of prosodic information. Musical expertise requires the ability to express emotional information through variations in loudness, timing, and pitch, and this ability appears to be reflected in an enhanced skill at *extracting* and *interpreting* emotional meaning in speech.

ACKNOWLEDGMENTS

This research was supported by a grant (to W.F.T.) from the International Foundation for Music Research. We thank Cory Sand and Don Smith for research assistance.

REFERENCES

1. FRICK, R. 1985. Communicating emotion: the role of prosodic features. Psychol. Bull. **97:** 412–429.
2. PLANALP, S. 1998. Communicating emotion in everyday life: cues, channels, and processes. *In* Handbook of Communication and Emotion: Research, Theory, Applications, and Contexts. P.A. Andersen & L.K. Guerrero, Eds. :29–48. Academic Press, Inc. San Diego, CA.
3. JUSLIN, P.N. & J.A. SLOBODA. 2000. The influence of musical structure on emotion expression. *In* Music and Emotion. Theory and Research. P.N. Juslin & J.A. Sloboda, Eds. :223–248. Oxford University Press. New York, NY.
4. PATEL, A.D., I. PERETZ, M. TRAMO & R. LABRECQUE. 1998. Processing prosodic and music patterns: a neuropsychological investigation. Brain & Lang. **61:** 123–144.
5. NILSONNE, A. & J. SUNDBERG. 1985. Differences in ability of musicians and nonmusicians to judge emotional state from the fundamental frequency of voice samples. Music Percept. **2:** 507–516.
6. THOMPSON, W.F., E.G. SCHELLENBERG & G. HUSAIN. 2004. Decoding speech prosody: do music lessons help? Emotion **4**(1). In press.

List of Additional Posters Presented

Day 1: Cerebral Organization of Music-Related Functions

Stability and accuracy of tempo perspicacity among everyday ordinary listeners: metronomic adjustment versus motor responses.
Warren Brodsky

Music Science Research, Department of the Arts, Ben-Gurion University of the Negev, Beer-Sheva, Israel

Brain activation due to unresolved chords differs from that due to dissonant chords.
Norman D. Cook

Department of Informatics, Kansai University, Osaka, Japan

Detection of perceived and imagined rhythm from ERP.
Peter Desain[a] and Henkjan Honing[a,b]

[a]Music, Mind, Machine Group, NICI, University of Nijmegen

[b]Music Department, University of Amsterdam

Relevance of neutral melodies in the diagnosis of musical emotional impairment.
Nathalie Ehrlé

Maison-Blanche Hospital, Reims

Mapping musical invention in the brain.
Donald A. Hodges,[a] Lawrence M. Parsons,[b,c] Steven Brown,[b] Michael J. Martinez,[b] Carol L. Krumhansl,[d] Jinhu Xiong,[b] and Peter T. Fox[b]

[a]Institute for Music Research, University of Texas at San Antonio

[b]Research Imaging Center, University of Texas Health Science Center, San Antonio, Texas

[c]Cognitive Neuroscience, National Science Foundation

[d]Department of Psychology, Cornell University, Ithaca, NY

Discrimination of weak and strong metric rhythms in children with spina bifida and cerebellar dysmorphology.
T. Hopyan-Misakyan, E.G. Schellenberg, and M. Dennis

University of Toronto, and The Hospital for Sick Children, Department of Psychology and Brain and Behaviour

The effects of a music listening session on emotion-related mental and physiological responses.
Kari Kallinen

Helsinki School of Economics, Knowledge Media Lab, Helsinki

Ann. N.Y. Acad. Sci. 999: 533–537 (2003). © 2003 New York Academy of Sciences.
doi: 10.1196/annals.1284.099

An electrophysiological marker for phrasing in music.
Thomas R. Knösche,[a] Christiane Neuhaus,[b] Jens Haueisen,[b] Burkhard Maess,[a] Kai Alter,[a] Angela D. Friederici,[a] and Otto Witte[b]

[a]*Max-Planck-Institute of Cognitive Neuroscience, Leipzig*
[b]*Friedrich-Schiller-University, Jena*

Electrophysiological measurement of pitch discrimination in tonal language speakers.
Nikolai Novitski,[a] Rika Takegata,[a] Mari Tervaniemi,[a,b] Risto Näätänen[c]

[a]*Cognitive Brain Research Unit, Department of Psychology, University of Helsinki, and Helsinki Brain Research Centre*
[b]*Institute of General Psychology, University of Leipzig*
[c]*BioMag Laboratory, Medical Engineering Centre Helsinki University, Central Hospital, Helsinki*

Mapping the invention of music and language in the brain.
Lawrence Parsons,[a,b] Steven Brown,[a] Michael Martinez,[a] Donald Hodges,[c] Carol Krumhansl,[d] Jinhu Xiong,[a] and Peter Fox[a]

[a]*University of Texas Health Science Center, San Antonio*
[b]*National Science Foundation, Arlington*
[c]*University of Texas, San Antonio*
[d]*Cornell University, Ithaca*

An fMRI study of music sight-reading.
Daniele Schön,[a,b] Jean Luc Anton,[c] Muriel Roth,[c] and Mireille Besson[a]

[a]*INPC – CNRS, Marseille*
[b]*Department of Psychology, University of Trieste*
[c]*IRMf Center, CHU Timone, Marseille*

A study of action observation in male symphony orchestra string players using efMRI.
Vanessa Sluming

Department of Medical Imaging, The University of Liverpool

fMRI study of language processing and 3D mental rotation in male symphony orchestra musicians.
Vanessa Sluming

Department of Medical Imaging, The University of Liverpool

Increased grey matter density in Broca's area in brains of male symphony orchestra musicians.
Vanessa Sluming

Department of Medical Imaging, The University of Liverpool

Singing vs. language in persons with non-fluent aphasia.
Concetta M. Tomaino

Institute for Music and Neurologic Function, Music Therapy Services, Beth Abraham Family of Health Services

Pre-attentive grouping of sequential sounds: an ERP study comparing musicians and non-musicians.
Titia L. van Zuijen,[a] E. Sussman,[b] I. Winkler,[c] R. Näätänen,[a] and M. Tervaniemi[a]

[a]Cognitive Brain Research Unit, Department of Psychology, University of Helsinki
[b]Albert Einstein College of Medicine, New York
[c]Institute for Psychology, Hungarian Academy of Sciences, Budapest

Day 2: Brain Sciences versus Music

Subjective tempo.
Gabriella Giordanella Perilli

[In private practice and public institutions for mental health, Rome]

The influence of harmonic relations and temporal regularities on chord processing.
Géraldine Lebrun and Barbara Tillmann

UMR-CNRS 5020, Laboratory of Neurosciences and Sensory Systems, Lyon

DNA and composition.
Alexander Mihalic

IMEB, Paris

Day 3: Music and Development

Learning statistical relations between chords by SRNs.
Bedin Nart Atalay

Cognitive Science/Informatics Institute, Ankara, Turkey

Neuropsychology and music therapy in the treatment of post-comatose patients in the acute rehabilitative phase.
Dario Benatti

ARICO (Associazione riabilitazione comatosi), Milan

Grey parrot as a musician: a contribution to human studies.
Luciana Bottoni, Paolo Pioli, Simone Masin, Daniela Lenti Boero, and Renato Massa

Department of Environmental and Landscape Sciences, University of Milan Bicocca

Music therapy as a possible way of intervention for children with severe disability: theoretical approach, setting and model of intervention in five multi-handicapped children.
Angelo Colletti and Marina Rodocanachi

Istituto Don Calabria-Centro Peppino Vismara-Milan

E.M.M.S.A.—Educational, Musical and Multimedial Software Archives in Florence
Stefania Di Blasio

Music Technology—E.M.M.S.A., Centro Studi Musica & Arte, Firenze

Music therapy as non-traditional therapy, what prospects/links with neurosciences?
Carmen Ferrara and Maria Emerenziana D'Ulisse

Association Anni Verdi—Service and School of Music Therapy

The perception of pitch direction in congenital deficits of musical perception.
J.M. Foxton,[a] J.L.Dean,[a] R. Gee,[a] A. Patel,[c] I. Peretz,[d] and T.D. Griffiths[a]

[a]Institute of Neurosciences, Newcastle University

[b]Department of Medical Physics, Newcastle Freeman Hospital

[c]Neurosciences Institute, California

[d]University of Montreal

A relationship between reading and the perception of pitch contour.
J. M. Foxton,[a] H. Brace,[a] F. McIntyre,[a] J.B. Talcott,[b] C. Witton,[b] and T.D. Griffiths[a]

[a]Institute of Neurosciences, Newcastle University

[b]Neurosciences Research Institute, Aston University

The role of attention and metaphor in guided imagery and music guided imagery and music.
Gabriella Giordanella Perilli

[In private practice and public institutions for mental health, Rome]

Behavioural patterns of two-year-olds in music learning.
Catherine Hapke

Department of Music Education, Seminar for School Pedagogy, Tübingen

How and why a young, South African musician / music educator uses music to improve and develop the following physical skills in teaching children 18 months to 5 years of age.
Alet Koch-Lochner

Music School, George, South Cape, South Africa

Rhythm, melody, and the infant cry.
Daniela Lenti Boero,[a,b] and Carlo Lenti[b,c]

[a]Laboratory of Bioacoustics and Comparative Analysis of Behaviour and Cognitive Functions, University of Milan

[b]Faculty of Pedagogical Sciences, University of Urbino

[c]Chair of Child Neuropsychiatry, University of Milan

The development of the rhythmic synchrony: a study at the ages 3–12.
Silvia Malbrán

Faculty of Fine Arts, University of La Plata, Argentina

The impact of musical aptitude in foreign language acquisition.
Riia Milovanov,[a] Mari Tervaniemi,[b] and Satu Nordlund[c]

[a]*Department of English, University of Turku, Finland & Centre for Cognitive Neuroscience, University of Turku*

[b]*Cognitive Brain Research Unit, Department of Psychology, University of Helsinki and Institute of General Psychology, University of Leipzig*

[c]*Centre for Cognitive Neuroscience, University of Turku*

Singing a song: melodic accuracy in children aged 2–3.
Tafuri Johannella

Conservatorio di Musica "G. B. Martini," Bologna

Subject Index